Y0-BYB-363

Triazine Herbicides: Risk Assessment

ACS SYMPOSIUM SERIES **683**

Triazine Herbicides: Risk Assessment

Larry G. Ballantine, EDITOR
Covance Laboratories, Inc.

Janis E. McFarland, EDITOR
Novartis Ciba Crop Protection

Dennis S. Hackett, EDITOR
Novartis Ciba Crop Protection

American Chemical Society, Washington, DC

Library of Congress Cataloging-in-Publication Data

Triazine herbicides : risk assessment / [edited by] Larry G. Ballantine, Janis E. McFarland, Dennis S. Hackett.

p. cm.—(ACS symposium series, ISSN 0097–6156; 683)

Outgrowth of a symposium held in March 1996, New Orleans, La.

Includes bibliographical references and indexes.

ISBN 0–8412–3542–2

1. Triazines—Toxicology—Congresses. 2. Triazines—Environmental aspects—Congresses.

I. Ballantine, Larry Gene, 1944– . II. McFarland, Janis E., 1956– . III. Hackett, Dennis S., 1950– . IV. American Chemical Society. V. Series.

RA1242. T642T75 1998
571.9′56—dc21

98–10369
CIP

This book is printed on acid-free, recycled paper.

Distributed by Oxford University Press

PRINTED IN THE UNITED STATES OF AMERICA

Advisory Board

ACS Symposium Series

Foreword

THE ACS SYMPOSIUM SERIES was first published in 1974 to provide a mechanism for publishing symposia quickly in book form. The purpose of the series is to publish timely, comprehensive books developed from ACS sponsored symposia based on current scientific research. Occasionally, books are developed from symposia sponsored by other organizations when the topic is of keen interest to the chemistry audience.

Before agreeing to publish a book, the proposed table of contents is reviewed for appropriate and comprehensive coverage and for interest to the audience. Some papers may be excluded in order to better focus the book; others may be added to provide comprehensiveness. When appropriate, overview or introductory chapters are added. Drafts of chapters are peer-reviewed prior to final acceptance or rejection, and manuscripts are prepared in camera-ready format.

As a rule, only original research papers and original review papers are included in the volumes. Verbatim reproductions of previously published papers are not accepted.

ACS BOOKS DEPARTMENT

Contents

Dietary and Worker Exposure to Triazines

Triazine Exposure via Water

Triazine Water Issues: Regulatory, Risk, and Ecotoxicology

Mammalian Toxicology and Human Risk Assessment

INDEXES

Preface

AGRICULTURAL PRODUCTIVITY in the United States has dramatically increased over the past 40 years. While one farmer fed only 27 people in 1950, in 1992 one farmer supplied the food for 150 people. The following changes have also taken place:

- corn yields are up 150%
- wheat and soybean yields are up approximately 40%
- cotton yields have increased more than 50%

Although there have been many technological breakthroughs that have contributed to gains in agricultural production, pesticides have played a significant role. As a specific example, farmers have widely used triazine herbicides for over 35 years for weed control in more than 30 crops as an integral component of their production strategy. The extent of triazine use is evident from 1992 USDA figures that indicate that herbicides were applied to 94% of the U.S. corn acreage and atrazine was applied to 66% of this acreage. Of the no-till corn acres planted, approximately 83% are treated with atrazine. Additionally, USDA surveys conducted between 1992–1994 estimated an average of 1.2 applications of a triazine herbicide per acre of no-till corn. Published data, computer model simulations, and field studies confirm the agricultural importance of the triazines, especially in conservation tillage systems. Corn and sorghum yield increases range up to greater than 10 bu/A with atrazine versus alternative herbicides. Triazine herbicides have provided efficient, low-cost, effective weed control on a high percentage of our nation's corn, sorghum, sugarcane, citrus, grapes and fruit, and nut tree acres.

Water stewardship programs initiated recently throughout the United States are showing that site-specific best management practices (BMPs) are effective in improving water quality by reducing runoff of triazines. BMPs have additional advantages in that they reduce runoff of other pesticides, nutrients, and soil into water.

Almost since the discovery of triazine herbicides, scientists in industry, academia, and government have studied the toxicology, environmental fate and behavior of triazines and their potential risk. Even with this abundance of information, the research directed toward better understanding the triazine herbicides has greatly intensified in the 1990s. The Environmental Protection Agency (EPA) and Novartis Crop Protection (formally Ciba-Geigy Corporation) have been drivers in this research effort; the EPA through their initiation of the Special Review process to evaluate the risks and benefits of the triazine herbicides, and Novartis through internal research and funding of external research to generate state-of-the-art information to best assess potential risks and benefits.

The Special Review is initiated with the issuance of Position Document 1 (PD-1) and is a regulatory process providing a mechanism for public participation in EPA's deliberations on a pesticide. In this process, studies and other information on risks and benefits are evaluated and reviewed, prior to the determination of whether a chemical poses any unreasonable adverse effects to humans and the environment. The triazine special review was initiated by EPA in November 1994, and since then Novartis and university scientists, as well as agricultural specialists, have conducted more than 100 new studies on the safety and benefits of atrazine and simazine. These new studies provided a wealth of information to add to the thousands of documents published to date on the triazine herbicides.

In addition to animal and environmental studies required by EPA for reregistration of pesticides, new and advanced studies were conducted for the Triazine Special Review. This new work was conducted by experts in these fields and utilized state-of-the-art toxicology mode-of-action studies, comprehensive dietary, water and worker exposure analyses and Monte Carlo probabilistic modeling on aggregate and cumulative risks. These data were also collectively reviewed by independent expert panels. The new toxicology and exposure data continue to show wide safety margins for the triazines, even when analyzed using the guidelines of the 1996 Food Quality Protection Act and the new EPA health standards.

The ACS symposium from which this book was written, was held to provide a forum to present and discuss the most recent research on benefits, exposure, toxicity, and corresponding risk–benefit analysis of the triazine herbicides. Another important objective was to use the extensive triazine database as a benchmark to help continually improve the risk assessment procedures for pesticides.

Specific topics included in this publication:

- role of triazines herbicides in agriculture
- benefits of triazine herbicides to the American farmer
- triazine herbicides' role in resistance management
- metabolism of triazines
- dietary and worker exposure to triazines
- triazine exposure via water
- triazine water issues: regulatory, risk, ecotoxicity, best management practices
- mammalian toxicology and probabilistic risk assessment

A glossary of nomenclature and chemical structures is included at the end of the book chapters.

This ACS Symposium Series book is designed to communicate to the scientific community, the latest and most advanced information on the triazine herbicides. Additionally, numerous chapters describe the new tools developed to assess the benefits and potential risk associated with crop protectants.

This volume is an outgrowth of a four-day symposium held under the direction of the Agrochemicals Division of the ACS in the March 1996, annual meeting held in New Orleans, Louisiana. We have updated the symposium presentations as appropriate to assure the data and information are current with this publication. The opinions and data interpretations of the contributing authors do not necessarily reflect the position of the editors, the American Chemical Society, Covance Laboratories, Inc. or Novartis Crop Protection, Inc

Acknowledgments

We thank the contributors to this volume for their outstanding efforts in presenting at the symposium and in preparation of the respective manuscripts. We also acknowledge the time and expertise of the many reviewers who helped prepare the manuscripts for publication. We thank Ms. Sara Farlow for the creative book cover design and Ms. Carolyn Barham for her expertise in formatting many of the manuscripts. We give special thanks to Ms. Andrea Elder, Associate Editor, for maintaining her good humor through the review and edits of each chapter and her tireless assistance as our focal point to gather, track, and finalize the book chapters.

LARRY BALLANTINE
Covance Laboratories, Inc.
P.O. Box 7545
Madison, WI 53707–7545

JANIS MCFARLAND
Novartis Crop Protection, Inc.
P.O. Box 18300
Greensboro, NC 27419

DENNIS HACKETT
Novartis Crop Protection, Inc.
P.O. Box 18300
Greensboro, NC 27419

Chapter 1

Benefits of Triazine Herbicides

Leonard P. Gianessi

National Center for Food and Agricultural Policy, 1616 P Street, NW, Washington, DC 20036

Widespread use of the triazine herbicides for weed control on corn, sorghum, sugarcane and fruit crop acreage has resulted in efficient, low-cost effective weed control on tens of millions of acres for the past thirty years. Many growers make a single application of a triazine herbicide at planting. The triazines stay active in the soil for five to six weeks and control most important broadleaf weeds and several important grass weeds. By mixing the triazine herbicides with an herbicide active on grass weeds, the single application is often all that is required for season-long weed control.

The triazines cost approximately $8-10 less per acre of corn than available alternatives. For the minor acreage crops, such as sugarcane and fruit crops, for which there are fewer alternatives, the savings of using triazines amount to $50-70 per acre.

One reason that growers readily adopt the triazines is their excellent crop safety record. U. S. farmers have used the triazine herbicides on a majority of U. S. crop acres for thirty years because of the weed control benefits they provide at an economical cost.

Triazine herbicides are widely-used to control weeds in U. S. crop production. Since 1976, atrazine has been used consistently on approximately two-thirds of the nation's corn acreage (*1,2*). Atrazine also is used on approximately 90% of the acreage of sugarcane and 67% of the acreage of sorghum on a yearly basis. In addition to being used on about 25% of the nation's field corn acreage, cyanazine is used on 25% of the nation's cotton acreage. Although registered for use on field corn, simazine is most widely-used to control weeds in tree fruit and nut orchards and in vineyards. From 25% to 50% of the acreage of the nation's almonds, avocados, blueberries, citrus, hazelnuts, grapes, nectarines, peaches, pears, apples, raspberries and walnuts receive simazine applications. U. S. farmers use the triazine herbicides because of the weed control benefits that they provide. The benefits of the triazines result from four basic characteristics:

- A lengthy period of residual control of germinating weed seeds;
- A broad spectrum of weed species that are controlled;
- A record of excellent crop safety; and
- Economical pricing.

These benefits can be discerned clearly by examining the choices that farmers have in selecting weed control alternatives. Basically, available alternatives offer a shorter period of residual control, control a narrower spectrum of weed species, are more likely to damage the crop and cost more to use. The benefits of the triazines are examined further in the following sections that focus on particular crops and uses: field corn, sugarcane, grapes, citrus, cotton, sweet corn, fallowland and sorghum.

Sugarcane

In Louisiana, sugarcane is planted in August/September, goes dormant in the winter and starts to grow again in the spring. The first harvest for sugar is in the fall of the following year. A wide range of cool-season weeds often becomes established in sugarcane fields after planting. At one time, winter weeds were not considered competitive with sugarcane since they grow when sugarcane is dormant and are removed early in the spring (*3*). Experiments with triazine herbicides applied in October provided residual control of germinating weeds as they emerged throughout the fall and winter. Research demonstrated that sugarcane cannot recover completely from intense winter weed competition and that the residual weed control provided by the triazine herbicides increased sugarcane yield by 12% (*4*).

Sugarcane is related closely to many of the weeds that infest it, and it has been problematic to find herbicides that control the weeds without damaging the sugarcane. Extensive tests in Texas in the 1970's showed that under the worst case conditions, atrazine would lower sugarcane yields by about 5% while alternatives would lower sugarcane yields by 20 - 40% (*5*). Alternatives to atrazine in Louisiana have been estimated to cost from $4 to $5 more per acre. In Texas, Hawaii and Florida, alternatives to atrazine are estimated to be $30 to $70 more per acre. Alternatives generally have less residual control and/or are less efficacious on particular weed species. As a result, combinations of active ingredients or multiple applications would be required as effective atrazine replacements (*6*).

Grapes

Simazine use in grape vineyards is limited to the berm - the raised area on either side of the vine trunks. The remainder of the soil area, the middles, is either disked or mowed. Prior to the development of simazine and other residual herbicides in the 1960's, the chief methods of weed control in the vine row were hand pulling and the use of the French plow, that plows the soil out of the vine row. It has a trip wire that hydraulically moves the plow around the vine. Residual herbicides, including simazine, largely replaced hand hoeing and the French plow in California vineyards because of the following factors: 1) herbicide use requires less labor; 2) mechanical injury to the vine is eliminated; and 3) herbicides provide longer lasting residual weed control (*7*).

Simazine provides four to six months of residual control of most germinating weed species in California vineyards. Simazine controls 36 weed species that are important in grape vineyards. The other available residual herbicides, oryzalin and oxyfluorfen, control 26 to 27 weed species (*8*). Simazine needs to be applied only once at a cost of $10 per acre. To match the broad spectrum control, the other two residual herbicides would have to be combined in an application. The cost would double to $21 per acre. Another possibility is to substitute a non residual herbicide, glyphosate, that would require three applications at an increased cost of $15 per acre (*9*).

Grapes are produced by organic methods on several thousand vineyard acres in California (*10*). Weeds are controlled without herbicides, and there is little or no yield loss (*11*). Organic grape growers use the French plow with the risk that the implement can knock down a vine or two as it progresses. In addition, organic grape growers use hand laborers with hoes to remove the weeds in the row. Since there is no residual control with these methods, they must be repeated to remove new flushes of weeds. The University of California has prepared budgets of the costs of producing organic grapes. The budgets assume two hand hoeings at a cost of $50 to $100 per acre plus an in-row cultivation at $18 per acre (*12*). The use of simazine and other chemical herbicides is considerably less costly.

Citrus

Florida's subtropical temperatures and 54 inches average annual rainfall are conducive to rapid germination of weed seeds and vigorous weed growth. Until residual herbicides, including simazine, became available in the 1960's, citrus groves were tilled or hand weeded. The widespread use of simazine and other herbicides instead of tillage and hand weeding is mainly the result of: 1) lower machinery and labor costs; 2) the need to minimize damage to tree trunks and surface feeders in root systems; 3) the unsatisfactory length of acceptable control of weed species by tillage; and 4) research during the early 1960's that showed significantly better tree growth, earlier production and less physical damage to trees under herbicide programs than those under tillage programs (*13*). Experiments in a young grove showed that trees treated with simazine, starting one year after planting, made significantly more growth than trees cultivated with tillage equipment (*14*). When they began bearing, simazine-treated trees yielded significantly more than cultivated trees.

Simazine typically is applied in citrus orchards in the fall and provides several months of residual weed control. In addition to providing a wide spectrum of weed species control, simazine also prevents the establishment of many troublesome vines that, once established, cannot be controlled by available post-emergence herbicides.

In Texas, research has demonstrated that simazine has no equal in controlling broadleaf weeds. Substituting other products for simazine would likely lead to less effective weed control with resulting citrus yield reductions of 20% (*15*). In Texas citrus orchards available herbicide treatments are estimated at $30 to $50 per acre greater than current simazine usage (*15*).

Sorghum

Early weed control is essential for efficient sorghum production. Sorghum grows slowly in the seedling stage and is especially susceptible to weed interference in the first three to four weeks after emergence (*16*). Weeds must be controlled during this period to achieve maximum yields. Atrazine is the major component in tank mixes for sorghum because it is more effective than all the other herbicides in controlling broadleaf weeds. A recent survey of sorghum farmers indicated that the major alternatives that would be used if atrazine were no longer available are cultivation and 2,4-D with an associated sorghum yield loss of 33% (*17*). 2,4-D and cultivation are used post-emergence to weed growth and do not prevent weed seed germination. As a result, early weed interference would not be prevented and yield losses would result. Research has indicated the 2,4-D injures sorghum in most years. 2,4-D makes sorghum plants brittle for several weeks after application and plants are extremely subject to breaking by winds (*18*).

Field Corn

The introduction of 2,4-D in the mid 1940's ushered in the modern weed control era. It provided good post-emergence control of broadleaf weeds and reached its peak in the 1950's with nearly half of the corn acreage treated. However, 2,4-D had some inherent problems with occasional injury to corn and movement outside the target area. The introduction of atrazine in the late 1950's provided corn growers with a very economical and effective herbicide with excellent corn tolerance. By the 1970's, it had the major market share and has continued its dominance, being used on over two-thirds of the U. S. corn acreage in 1993, more than all the other herbicides for broadleaf weed control combined (*19*).

One of the key advantages for atrazine is its flexibility. It can be soil applied prior to planting, either on the surface or incorporated, and provides season-long control of germinating weed species. It can be applied pre-emergence, at planting or soon after, or it can be applied post-emergence.

Not only does atrazine control most annual broadleaf weeds usually found in corn, it also has some activity for control of grass weeds to complement other herbicides that are used primarily for control of grass weeds (*19*). Atrazine is used quite advantageously with essentially every other herbicide for field corn. If some of these other herbicides were to be used without atrazine, there is a likelihood of a decrease in spectrum of control, increased risk of corn injury, increased risk of movement outside the target area, a decrease in length of control and increased cost (*19*).

Cyanazine was used on 20% of the field corn acreage in 1993, with the majority used in combination with atrazine. This combination has been relatively popular for broad spectrum control. For no-till corn acreage, the combination of cyanazine and atrazine provides good burndown of existing weeds, as well as giving appropriate length of residual control. Atrazine helps to improve pigweed control, while the addition of cyanazine helps to increase control of velvetleaf and some annual grasses. Atrazine costs about $3 per acre. Other alternatives average approximately $10 more in cost per acre. In many areas, since the length of residual control is shorter with atrazine alternatives, two applications would have to be made to substitute for a single atrazine application.

Numerous research studies have been conducted to examine the potential of weed control options as substitutes for cyanazine and atrazine. In Wisconsin, growers have faced restrictions on the use of atrazine that are more stringent than any other state. A three-year study in Wisconsin was conducted to identify the best alternatives and to quantify the cost of the atrazine restrictions to Wisconsin growers. Net returns for all 13 alternative treatments were lower than for the atrazine standard. All alternatives were more expensive and resulted in measurably lower corn yields *(20)*. The least loss in net return ($10/A) resulted from the use of pendimethalin and dicamba. However, growers may be hesitant to use this treatment because of the potential for crop injury.

Sweet Corn

Weeds are more competitive with sweet corn than with field corn since sweet corn does not grow so rapidly nor so tall as field corn *(21)*. Considerably fewer herbicide alternatives exist for sweet corn in comparison to field corn because sweet corn cultivars are only partially tolerant of many herbicides. Many of the "super-sweet" corn hybrids are particularly sensitive to newer herbicides registered for field corn acreage *(22)*. 2,4-D and linuron are registered for sweet corn; however, both can be phytotoxic to the sweet corn plants. Other herbicides registered for sweet corn, such as butylate, do not have an equivalent broad spectrum of control of broadleaf and grass weeds. In addition, in Florida, where much sweet corn acreage is on soils high in organic matter, alternative herbicides often are rendered ineffective. Atrazine has proven to be an effective herbicide for use in Florida's high organic (muck) soils.

Atrazine has been the basis of almost all weed control programs in sweet corn for many years. It controls a wide range of common weeds, can be used in many different combinations and can be applied in many different ways *(23)*. Sweet corn has a high tolerance to atrazine.

Cultivation of sweet corn does not control weeds within the crop row. Also, cultivation is only possible when the soil is dry. During wet periods of the year, it is impossible to cultivate. Also, in Florida, weed generation and growth is extremely rapid. In most cases it is potentially impossible to cultivate the number of times necessary to control weeds during the year.

In Massachusetts, experiments with cultivation of sweet corn indicated that four cultivations were required at a cost that was double the cost of using herbicides *(24)*.

Fallowland

Almost all of the wheat in states such as Colorado, Montana and Nebraska is grown in an annual rotation with fallow. A wheat/fallow rotation is also practiced on 50% of the acreage in Kansas. By keeping the acreage out of crop production for an entire year, the soil stores moisture that is then available for the following year's wheat crop. Uncontrolled weed growth can deplete soil moisture during the fallow year. Growers used to cultivate fallow ground to kill emerged weeds. However, the practice of cultivation dried out the soil, depleted moisture and promoted soil erosion. Research indicated that by applying herbicides to fallow ground, weed growth would be controlled, soil moisture would increase and the following year's wheat yield would increase. The combination of atrazine and cyanazine provided weed control (more than 85% of the plot weed free) for 386 days followed by atrazine alone with 369 days of control *(25)*. Atrazine use produced wheat yields 39% higher than the conventional tillage treatment.

Cotton

In the late 1980's, the use of cyanazine on U. S. cotton increased significantly: from 7% acreage treated to 20% acreage treated *(1,2)*. Cyanazine is one of the most cost effective herbicides cotton farmers can use, particularly for late and layby post-directed herbicide treatments. Cyanazine provides burndown of weeds that are present at the time of application, plus residual control of weeds that germinate after the application. At the same time, the residual control of cyanazine is short enough not to restrict farmers' rotational crop options. While other herbicides may be able to out perform cyanazine in a single feature, such as providing longer residual control of a specific weed, no other herbicide available to cotton farmers offers all of these benefits in one product at a reasonable cost per acre.

Most cyanazine used on cotton is applied as a layby treatment. This application is made when the cotton plants are over twelve inches tall, but have not yet "canopied," or bushed out to the point where they cover the middle of each row and their leaves deprive any weeds under them of the sunlight needed to grow. At the layby stage, cultivation and application equipment is still able to get through the field without doing damage to the cotton plants.

The reason cotton farmers apply a layby treatment to cover the entire row width is that this application is the last opportunity they have to control grasses and broadleaf weeds before the cotton gets too big to spray or cultivate. After the layby application, farmers do not have any good options to come back and clean up any weed problems that arise.

Summary

Uncontrolled weeds would lower significantly U. S. crop yields through competition for sunlight, space, nutrients and moisture. Prior to the development of residual herbicides, U. S. farmers used mechanical tillage and hand hoeing methods for weed control. Residual herbicides, such as the triazines, were adopted because they produce a long period of weed control at a significantly reduced cost in comparison to earlier methods of weed control. For most U. S. crops, more than one herbicide active ingredient is registered to control a similar spectrum of weeds. For field corn, numerous registered alternatives exist (greater than 20 active ingredients). The U. S. herbicide market is extremely competitive, and numerous products have been introduced for corn during the past 30 years. The dominant share of treated acres that triazine herbicides have maintained is indicative of their benefits. In comparison to alternatives (other chemical or non-chemical methods), the triazine herbicides provide superior weed control benefits in terms of spectrum of control, length of control, crop safety and cost.

Literature Cited

1. USDA, *Agricultural Chemical Usage: Field Crops Summary* (**1990-1994**), National Agricultural Statistics Service.
2. USDA, *Inputs; Outlook and Situation*, Economic Research Service, IOS-2, October 1983.
3. Matherne, R. J., *et al., Culture of Sugarcane for Sugar Production in the Mississippi Delta*, USDA Agriculture Handbook No. 417, **1977**.
4. Millhollon, R. W., "Influence of Winter Weeds on Growth and Yield of Sugarcane," *Proceedings of International Society of Sugarcane Technologists*, **1971**.
5. Reeves, Sim A., Jr., "Evaluation of Selected Pre- and Postemergence Chemicals on Weed Control and Phytotoxicity to Eight Sugarcane Varieties," *Proceedings Southern Weed Science Society*, **1977**.
6. Dusky, J. A., and D. L. Colvin, *Weed Management in Sugarcane - 1996*, University of Florida, Cooperative Extension Service, SS-AGR-09, December 1995, revised.
7. Lange, A. H., *et al., Chemical Weed Control in Vineyards*, California Agricultural Experiment Station, leaflet 216, June 1974.
8. *UCIPM Pest Management Guidelines*, Division of Agriculture and Natural Resources, University of California, Publication 3339.
9. Kempen, Harold M., letter to USEPA, Public Docket OPP-30000-60, March 18, 1995.
10. "Bugs, Weeds, and Fine Wine," *Business Week*, August 10, 1992.
11. "Organics on the Rise," *Fruit Grower*, October 1995.
12. Klonsky, Karen, *et al., Sample Costs to Produce Organic Wine Grapes in the North Coast*, UC Cooperative Extension Service, December 21, 1993.
13. Tucker, D. P. H., et al., "Two Weed Control Systems for Florida Citrus," *Proceedings Florida State Horticultural Society*, **1980**.

14. Mersie, W., and M. Singh, "Benefits and Problems of Chemical Weed Control in Citrus," *Reviews of Weed Science*, **1989**.
15. Swietlik, D., letter to USEPA, Public Docket OPP-30000-60, February 14, 1995.
16. Stahlman, Phillip W., letter to USEPA, Public Docket OPP-30000-60, February 16, 1995.
17. Morrison, W. M., *et al., The Biologic and Economic Assessment of Pesticides in Grain Sorghum*, USDA, National Agricultural Pesticide Impact Assessment Program, **1994**.
18. Regehr, David, letter to USEPA, Public Docket OPP-30000-60, February 26, 1995.
19. Knake, Ellery L., letter to USEPA, Public Docket OPP-30000-60, February 13, 1995.
20. Harvey, R. Gordon, "Weed Control Options without Atrazine or Bladex," *Proceedings 1996 Wisconsin Fertilizer, Ag Lime and Pest Management Conference*, Volume *35*.
21. Doersch, R. E., *Weed Control in Commercial Sweet Corn Production*, University of Wisconsin Cooperative Extension Service, Fact sheet A2345, July 1974.
22. Beste, C. Edward, letter to USEPA, Public Docket OPP-30000-60, February 6, 1995.
23. Cooperative Extension Service, *Western Washington Weed Control Guide: Sweet Corn*, Washington State University, EM 3981, July 1975.
24. Hazzard, Ruth, ed., *Proceedings of the Northeast Farmer to Farmer Information Exchange: Sweet Corn Meeting*, Northeast Organic Farming Association, **1994**.
25. Anderson, Randy L., and Darrryl E. Smika, *Chemical Fallow in the Central Great Plains*, Colorado State University Experiment Station, Bulletin 5883, January 1984.

Chapter 2

The Role of Triazines in Managing Weeds Resistant to Other Herbicides

H. M. LeBaron

Independent Consultant in Agricultural Science, P.O. Box 285, Heber, UT 84032

Triazine-resistant weed biotypes first appeared 28 years ago, about 10 years after the commercial introduction of this class of herbicides. Resistant biotypes of 59 weed species have now been reported. However, since 1984, the number of occurrences per year, the areas infested, and the severity of infestations have generally decreased. Triazine resistant biotypes are seldom serious problems for farmers, even though most of them continue to use triazines on the same acres. Several reasons for this phenomenon will be discussed. Unlike triazine-resistant weed, biotypes which have evolved resistance to other herbicide classes, especially to the ALS/AHAS and ACCase inhibitors, have occurred much more rapidly, are very difficult to avoid or manage, and any control methods other than triazines will be much more costly. These newer herbicides are now used frequently in corn, soybean, cereal grains and in other crop rotations. Without triazine herbicides, the repeated use of these new single target site herbicides on the same weeds will quickly render the new classes of herbicides useless, which may jeopardize our current agricultural systems and our farm economy.

The triazine herbicides have been used for effective weed control for over thirty five years, but recently a number of concerns regarding this class of herbicides have been raised by the EPA and others. Among the issues raised is the concern that many of the weeds once completely controlled by triazine herbicides are now less affected by them, having evolved a resistance to these chemicals.

In fact, few of these triazine resistant weed biotypes have been serious problems for farmers, and are now being easily managed. These triazine-resistant biotypes are actually decreasing in importance and acreage in the United States. However, this is not the case with the resistance that is evolving in weeds that were formerly controlled with some of the newer herbicides, including the ALS- and ACCase-inhibitors. These resistant weed biotypes pose a serious problem to farmers, and without the use of triazine herbicides as an alternative method of control, modern agriculture in the United States would be greatly impaired. The triazine herbicides are the most effective method of weed control on the market as well as the most economical. The recent move toward no-till farming, resulting in a great reduction in soil erosion, depends on the use of triazine herbicides. The fact that farmers can now use a variety of chemically different products for weed control is the best hope that the problem of herbicide resistant weed biotypes can be successfully combated. The continued use of triazine herbicides is a cornerstone of this approach. Without the option of using these herbicides, it is difficult to calculate the long-term consequences to the farmer and consuming public, but the cost will surely exceed our current estimates.

Brief History of Weed Management

It may be difficult for some to understand the extent of progress in the management of weeds that has occurred in the last half century. Until the 1950s "weed science" was not even a meaningful term at universities where agriculture was taught. Prior to development of chemical herbicides, weeds were largely controlled the old fashion way. Mechanical improvements in the process brought us the disk, harrow, cultivator and other methods of tillage. As someone who worked on a farm in the era before modern herbicides were used, perhaps I have a better perspective on the subject of weed control than most. My fifty years of experience in the area of weed control, with both hoe and herbicide, reinforce my appreciation for the triazine herbicides. I would like to share the insights provided by my experience with those who may be making decisions on the future of triazines, in the hope that I can provide some guidance to them.

I grew up on a farm in Canada, in the province of Southern Alberta. Some of my earliest recollections as a child are of me with a hoe in my hand; for cutting down the weeds in the fields was an almost constant concern, and a constant chore. For most of my first twenty years, whenever I was free from other farm duties, I worked alongside my family in the fields hoeing weeds. We worked at this from sunrise to sunset, in spring, summer and fall. Despite this nearly constant labor, and the fact that my family was large and the farm small, we were never able to get ahead of the weeds and other pests. We would have welcomed any effective alternative method that would have helped us, but there was none. Perhaps it is interesting to note that by today's standards we would have been considered organic farmers, with produce that was one hundred percent natural. Our type of farming might now be called "sustainable" or "low

input", although at that time no one in their right mind would have used those terms. And in fact, they would not have been accurate; the input was enormous in petroleum energy, management, human labor and sweat, and the farming methods sustained neither the family nor, in many cases, the land. It is also worth noting that the produce we grew with such effort on our small farm in Canada was inferior in quantity, quality, and appearance to what is produced today. It was a search for an alternative to this way of farming that motivated me to begin my study of agricultural science. I hoped in my studies to find a better way to control weeds and other pests in order to better provide for my family, and to help other farmers.

At the time I began my study of agricultural science, not all viewed the future with optimism. Virtually no serious or knowledgeable expert in the 1940s dared to predict that there was any hope of adequately feeding the almost two billion people then on earth, and nothing but mass famine, wars and other catastrophes could be expected for our increasing population. I encountered the thoughts of one such expert in William Vogt's "Road To Survival" (*1*). Vogt argued that the stark limits on productivity imposed by current agricultural practices, with the resulting soil erosion and other abuses of the land, would lead to a dire future of pestilence and want. He, and others, even argued that corn should be prohibited as a crop because it required so much cultivation, which in turn resulted in the most serious soil erosion of any agricultural land use. These fears were not realized, but at the time they strongly reinforced my commitment and conviction that there had to be some solution to this hopelessness.

Seeking an alternative to the costly hand labor and damaging tillage then employed in agriculture, I began testing chemicals and other methods. This was about forty-five years ago, when I was an undergraduate student. The results looked promising. I can remember well my great excitement following results of tests I conducted on controlling quackgrass with atrazine. Few today remember what a plague this weed was to the farmers in Canada and the northern United States, especially in corn. Atrazine and the other triazine herbicides provided immediate and effective control of this pestilence, thus sparking a revolution in agriculture. In other crops where triazine herbicides could be used selectively, such as sorghum, sugarcane, perennial fruits and nuts, and others, the results were also impressive. This early success led to further research efforts and progress in discovering and developing many other types of herbicides for weed control in crop and non-crop acreage.

Occurrence and Risk of Herbicide Resistant Weeds

The introduction of herbicides to modern agriculture, and their subsequent widespread use, has resulted in many benefits for our society. Increased food production and farm profitability are perhaps the two greatest advantages. But widespread herbicide use has also brought problems; chief among them is the occurrence of weeds that have become resistant to the very herbicides once used to control them. At the current time, at least 107 weed species have been identified that have evolved biotypes with resistance to at least one or more of 21 classes of herbicides. This rise of herbicide resistant biotypes is of great concern. It is a greater problem than the evolution of resistance to pesticides among insects, and actually poses a greater problem to agriculture than plant disease.

Modern agriculture, with its large yields and efficient production, depends on herbicides for control of weeds. It is not economically possible to return to the old ways of controlling weeds that were used before the introduction of chemical control. The labor, machinery, crop varieties and agricultural systems that were used then are no longer available, and were too inefficient, costly and unproductive. Farms are also larger now; many farms are 400 acres or more in size. This is a direct result of the farmer's ability to control weeds so efficiently and dependably with herbicides. Just as there is no return to the past, there are no alternatives currently available or on the horizon for most of our crop and weed management needs. Also, many of the older herbicides that would be useful for the management of resistant biotypes have been discontinued, mainly for economic reasons.

Triazine Resistance

In order to justify the continuing use of atrazine and other triazine herbicides, we must objectively and fairly review and consider the problems of triazine resistant biotypes. Although it was not the first weed biotype to show resistance to modern herbicides, there was confirmation in 1969 that a biotype of *Senecio vulgaris* (common groundsel) had evolved resistance to simazine, after several years of repeated use in a nursery in western Washington (*2*). The further discovery that isolated chloroplasts of this resistant biotype were insensitive to triazine herbicides (*3*) led the way for much important research and greater understanding of the mechanism triazine resistance and herbicide mode of action (*4*).

By 1981, about 30 weed species had developed resistance to triazine herbicides (*5*). Eleven of these had been observed within the United States. By 1989, a total of 57 weed species had been reported to have triazine resistant biotypes, 23 of which were found in the United States (*6*). In the last seven or eight years, only eight more species have been reported. Of these 65 triazine-resistant biotypes, 20 broadleaf and eight grass species have been reported in the United States, with one or more species occurring in 34 of the 50 states.

However, the infestations and seriousness of other resistant weeds have been increasing much more rapidly than triazine-resistance (*7*). At the present time at least one or more weed species have evolved resistance to 21 classes of herbicides, with a total of 198 biotypes and 107 species. Table I gives a summary of this trend for several of the most important herbicide classes over the past 25 years. It is readily seen that biotypes resistant to the ACCase- and ALS-inhibitors, which have come into greatest use within the past five to eight years, have increased most rapidly. Resistance of these biotypes to ACCase- and ALS-inhibitors will likely be very problematic for growers in the years ahead.

Table I. Number of Weed Species Which Have Evolved Resistance To Major Classes of Herbicides

Herbicide Class (Example of Class)	*To 1981 (5)*	*To Sept. 1989 (6)*	*To March 1996*
Triazines (atrazine)	29	57	65
Ureas (diuron)	2	6	19
Phenoxys (2, 4-D)	4	6	10
Bipyridiliums (paraquat)	5	13	19
Dinitroanilines (trifluralin)	2	3	6
Carbamates (triallate)	1	2	8
*ACCase-inhibitors (diclofop)	1	4	12
*ALS-inhibitors (chlorsulfuron)	0	6	30

*First introduction in the early 1980s.

To put the issue of weed resistance to triazine herbicides in perspective, we need to understand how farmers and crop producers have been able to deal with triazine-resistant biotypes, and what they are facing from herbicide resistant biotypes and other herbicide classes in the future.

From a recent survey I have made by phone, in person, or from recent literature or scientific reports, every weed scientist and expert contacted within the United States agrees that triazine-resistant weeds are localized, are being effectively or adequately managed, that they are not increasing in number, acreage or intensity, and that they should not present a serious problem in the future. These experts usually responded that triazine herbicides are and will continue to be used by growers who have resistant biotypes in order to control the spectrum of weeds which have not developed triazine resistance.

In addition to managing resistant weeds, farmers and herbicide users have generally stopped growing continuous corn or using only triazine herbicides for several other reasons. Rotations of corn with soybeans and other crops have long been encouraged and often followed, which in the past required that the farmer use other classes of herbicides or methods of weed control. To improve control of grasses, atrazine has usually been combined with other classes of herbicides for more than 20 years.

Triazine resistant biotypes have seldom occurred except after at least 10 years of repeated use where crop and herbicide rotations have not been used. Only three to six species of weeds have been of significant importance, and they have been limited in acreage, mostly in non-crop situations. After 40 years of use, triazine resistant biotypes occur on limited acreage, and triazine herbicides are still used even where resistance occurs in order to control susceptible weeds. The percent of atrazine-treated acres have remained constant since 1984, with a very high level of grower confidence.

It is also well-documented that most triazine-resistant biotypes are less fit relative to their triazine susceptible biotypes (*4,8*). This lack of photosynthetic fitness renders triazine-resistant biotypes competitively disadvantaged. Thus, based on selection pressure, these resistant biotypes often do not become dominant in a population.

ALS- and ACCase-Inhibitor Resistance

The above conclusions and predictions are not true of resistant biotypes to non-triazine types of herbicides. There is no evidence that biotypes resistant to herbicides other than the triazines, have reduced fitness. All data on the ALS-inhibitor resistant biotypes show that they are as equally fit and vigorous as the susceptible native populations (*7,9-14*). Several inheritance mechanisms appear to be involved in ALS-inhibitor resistance, but all cases involve insensitive acetolate synthase enzyme systems (*7,15*). This resistance is dominant, so both the heterozygous (RS) as well as the homozygous (RR) individuals are resistant (*15*). Therefore, once the initial incident of resistance occurs, the resistant subpopulation will dominate the population in a very short period of time (*9,15*). Although the initial frequency of resistance-conferring alleles for ALS-inhibitors is not known, it is considerably higher than that for the triazine herbicides.

The first case of ALS-resistant weeds was reported in 1987 when kochia control failures occurred in Kansas after five consecutive years of chlorsulfuron use (*7*). This kochia biotype proved to be cross-resistant to six other ALS-inhibitor herbicides, including both sulfonylureas and imidazolinones. Within five years, sulfonylurea-resistant kochia had been identified at 832 sites in 11 states of the United States and three Canadian provinces. Numerous cases of broadleaf weed resistance to the ALS-inhibitors have been reported, including cases of cocklebur resistance to imidazolinones herbicides, after only two to four consecutive years of use in soybeans. Since 1989, the number of species evolving resistance to ALS-inhibitors has increased over two-fold in soybeans, rice and roadsides (*7,12,16*).

In a recent classic example, DeFelice conducted a study in Missouri on corn-soybean rotation in which he used only ALS-inhibitor herbicides, i.e., Pursuit (imazethapyr) in soybeans, and Beacon (primisulfuron-methyl) in corn. Within five years, a common waterhemp biotype resistant to 5X rates of ALS-inhibitors was flourishing. Within the same state, Kendig reported one case of atrazine resistant common waterhemp, which developed where a farmer grew continuous corn and used only atrazine for at least 10 years (*17*).

The ALS-inhibitors are at highest risk for the evolution of resistance in weeds because they have a single target site, are effective against a wide spectrum of weeds, and are relatively persistent, often providing season-long control of germinating weed seeds. Also, the various sites of mutations for resistance are not near the active site of the enzyme and thus there is no fitness loss due to a lower affinity for the normal substrates (*15*).

The first confirmed incident of an ACCase-inhibitor-resistant grass weed species attributed to herbicide selectivity was rigid (or annual) ryegrass (*Loluim rigidum*, Gaudin) in Australia in 1982 (*18*). Since that time, the number of occurrences and spread of this resistant weed has exploded (>3,000 sites throughout Australia), and includes a wide variety of biotypes with different mechanisms of resistance, some of which are capable of cross-resistance to almost all herbicides via metabolic detoxification. ACCase-inhibitor-resistant wild oat (*Avena fatua* L.) and green foxtail (*Setaria viridis* L.) which first occurred in Canada (*11*), is now scattered over more than 100 sites in Canada, Australia, United States and Europe (*10,11,16,18*).

The Role of Triazines in Managing Herbicide Resistant Weed Biotypes

Atrazine and other triazine herbicides are essential as an integral part of the strategy for future herbicide-resistant weed management. The rotation of crops and herbicides is a must. Atrazine is a corn and sorghum herbicide that can be utilized in a rotation with crops such as soybeans, small grains, canola, or forages. Atrazine inhibits photosynthetic electron transport. It should be rotated or combined with families of herbicides which exhibit other modes of action. The use of ALS-inhibitor herbicide products in soybeans could be rotated with the use of atrazine in corn. To understand the urgent biological risk our present agricultural systems are facing, one needs to simply review the partial list of ALS-inhibitor herbicides, as presented in Table II and their probable use year after year in corn and soybean rotations.

During my recent survey, I also asked each weed scientist which herbicides represent the most serious threat to agriculture from development of resistant biotypes. Without exception or hesitation, they always responded that resistant biotypes to ALS- and ACCase-inhibitor herbicides will soon become, if not already, the most serious of such threats. Not only have triazine-resistant biotypes been relatively easy to avoid or control, but triazines will play an essential role in the future to manage resistant biotypes which have developed from other herbicide classes. In support of the continued use of atrazine and simazine, weed scientists throughout the United States have reported and confirmed in many research reports and letters to EPA that weeds evolving resistance to the newer and alternative herbicides with target sites at the ALS/AHAS and ACCase enzymes are rapidly increasing (see Table III).

Another question I asked is: "Would the triazine herbicides be a valuable help in controlling the weeds that escape or evolve resistance to these other herbicides? Again, the immediate and enthusiastic responses were always, "Yes!" Few weed scientists or biologists who understand the potential time bomb which is even now ticking away would question the conclusion that where ALS- and ACCase-inhibitors are now used repeatedly, these new and low-rate herbicides will be of no or limited use within a few years if atrazine and other triazine herbicides are not available as alternate methods of weed control. In a risk-benefit analysis, it must be recognized that some minor risk from triazine herbicides would be much better than losing these essential low-rate herbicides completely.

TABLE II. ALS-INHIBITOR HERBICIDES: THREE SEPARATE CLASSES OF CHEMISTRY INHIBIT ACETOLACTATE SYNTHASE (ALS), AN ENZYME INVOLVED IN THE SYNTHESIS OF CERTAIN AMINO ACIDS[1]

Imidazolinone Herbicides	
Contour	imazethapyr + atrazine
Detail	imazaquin + dimethenamid
Pursuit	imazethapyr
Pursuit Plus	imazethapyr + pendimethalin
Passport	imazethapyr + trifluralin
Resolve	imazethapyr + dicamba
Scepter	imazaquin
Squadron	imazaquin + pendimethalin
Tri-Scept	imazaquin + trifluralin
Sulfonamide Herbicides	
Broadstrike + Dual	flumetsulam + metolachlor
Broadstrike + Treflan	flumetsulam + trifluralin
Broadstrike Plus	flumetsulam + clopyralid
Broadstrike Plus Post	flumetsulam + clopyralid + 2,4-D
Sulfonylurea Herbicides	
Accent	nicosulfuron
Basis	nicosulfuron + thifensulfuron methyl
Beacon	primisulfuron methyl
Canopy	chlorimuron ethyl + metribuzin
Classic	chlorimuron ethyl
Concert	chlorimuron ethyl + thifensulfuron methyl
Exceed	prosulfuron + primisulfuron methyl
Gemini	chlorimuron ethyl + linuron
Lorox Plus	chlorimuron ethyl + linuron
Peak	prosulfuron
Permit	halosulfuron
Pinnacle	thifensulfuron methyl
Preview	chlorimuron ethyl + metribuzin
Synchrony STS	chlorimuron ethyl + thifensulfuron methyl

[1]The ALS-inhibitor in prepackaged mixtures is underlined.

TABLE III: SUMMARY QUOTES ON MANAGEMENT OF HERBICIDE RESISTANT WEEDS FROM U.S. WEED SCIENTISTS

Arkansas

Dr. Ford L. Baldwin, University of Arkansas. Atrazine replacements in corn are primarily ALS inhibitors. Weed resistance to these compounds has developed rapidly. Both modes of action are needed for resistance management.

Colorado

Dr. Philip Westra, Colorado State University. Triazine resistant pigweed and kochia have been documented in Colorado irrigated cropland. However, these resistant weeds have had little impact on atrazine use because so many other susceptible weeds are still economically controlled by atrazine. Of much greater concern, at least 50% of all kochia samples in Colorado are now resistant to ALS herbicides. Kochia from very limited number of locations is also showing increased level of tolerance to dicamba.

Connecticut

Dr. John F. Ahrens, The Connecticut Agricultural Experiment Station. Triazine resistance has not been a serious problem because of combination treatments, follow-up treatments with post products, and alternating the use of preemergence herbicides in the production of nursery, ornamental and forest crops.

Illinois

Dr. Ellery L. Knake, University of Illinois and Dr. Loyd M. Wax, ARS, USDA. Although a little weed resistance has been reported for atrazine in some of the major corn producing states, after about 35 years of use, it is not considered very prevalent or very serious. Atrazine resistant biotypes have not developed as major problems in Illinois, because of crop and mode of action rotations. Most of the new alternatives to atrazine have ALS enzyme inhibition as their mode of action, and development of resistance to these herbicides occurred after as few as three to four applications.

Dr. George Kapusta, Southern Illinois University. Triazines are photosynthetic inhibitors. In contrast, most all alternatives to the triazines are ALS inhibitors with a single site of action. Within a few years of the introduction of the ALS inhibiting herbicides, resistant weeds developed. If the triazine herbicides were not available, the ALS herbicides would become the dominant family in use. Almost certainly, within a few years, there would be widespread resistance to the ALS herbicides, resulting in greatly diminished value of these very valuable products.

Dr. Ronald F. Krausz, Southern Illinois University. Atrazine revolutionized no-till and reduced till in corn production. Weed resistance management also would be adversely affected if triazines were banned. New herbicides inhibit the ALS enzyme system. Since these types of herbicides are

Continued on next page.

Table III. *Continued*

also used for soybeans, the expansion of resistant weeds would result. Multiple applications would be required to control resistant weeds thus increasing herbicide costs and total herbicide volume.

Dr. Marshal D. McGlamery, University of Illinois. If atrazine is banned, the result will be greater potential for ALS resistant weeds! Some triazine resistant weeds have occurred, but they are not a major problem in the central corn belt where farmers have used a multiple mode of action concept. Triazine resistant weeds are a problem only where farmers have used simplistic solutions. However, ALS resistant weeds are a great concern since ALS herbicides dominate the soybean "broadleaf" market and are now trying to dominate the corn "broadleaf" market.

Dr. Stephan E. Hart, University of Illinois. New alternatives (ALS inhibitors) are prone to generate herbicide resistant weeds with only a few years of continuous use. If triazines are taken out of the crop rotation, it will mean disastrous consequences for corn and soybean production because of ALS resistant weeds.

Indiana

Dr. Thomas T. Bauman, Purdue University. Atrazine, cyanazine, and simazine are critical to corn weed management programs. Even if the technology exists to replace these with other newer herbicides, this will only be successful for a very limited time. ALS-resistant weeds are rapidly spreading and will severely limit these newer herbicides in the central corn belt. I believe that it would be wise to continue the use of the triazine herbicides because there are no other herbicides that work as well as they do, while providing an alternate mode of action that will allow farmers to manage weed resistance.

Kansas

Dr. David L. Regehr, Kansas State University. Alternatives to atrazine are more expensive, have less crop safety, and are susceptible to selection for ALS resistance among weed biotypes.

Dr. Phillip W. Stahlman, Kansas State University. Triazine resistant weed populations occur throughout the state, but are effectively controlled by one of more of the postemergence products.

Kentucky

Dr. Michael Barrett, University of Kentucky. Many triazine resistant weeds can be managed by the use of chloroacetamide herbicides in mixture with the triazine. Because of the use of ALS inhibitors for both soybeans and corn, the potential is great for the rapid development of weeds resistant to these herbicides.

Nebraska

Dr. Alex R. Martin, University of Nebraska. Herbicide resistant weeds are an increasing risk today. Weed populations have a much greater probability

Table III. ***Continued***

of developing resistance to the new herbicides (e.g., ALS-inhibitors) than to atrazine. Diversity in herbicide use is a key in combating the evolution of herbicide resistant weeds. For this reason, retaining atrazine will be important, as it will play a substantial role in herbicide resistance management in the future.

Dr. Fred Roeth, University of Nebraska. Many of the newer herbicides which can be atrazine replacements are ALS inhibitors. This group of herbicides has encountered weed resistance in as few as five years. Herbicide rotation is an accepted practice to avoid or manage herbicide resistant weeds. Atrazine, a photosynthesis inhibitor, is an important alternative herbicide to the ALS inhibitors. Without atrazine, I believe we will see a rapid increase in weed resistance to ALS inhibitor herbicides.

Dr. David L. Holshouser, University of Nebraska. Triazines will continue to play a valuable role in resistant weed management. Many of the newer herbicide families are resistant-prone due to their very site-specific mechanism of action and their long, residual activity. We are beginning to see these resistant weeds in Nebraska. Crop and/or herbicide rotations are our best resistance management strategies, but this requires herbicides with different mechanisms of action. Triazine herbicides should play a major role in delaying potential resistance.

New York

Dr. Russel R. Hahn, Cornell University. The evidence is very strong that resistance to the ALS inhibitors may develop very rapidly. Growers who use a resistance management program, including crop rotation, cultivation, and/or the use of herbicides with different modes of action either in combinations or as sequential treatments, have not had problems with triazine resistance.

North Dakota

Drs. C. G. Messersmith, J. D. Nalewaja, and W. H. Ahrens, North Dakota State University. We are not aware of any triazine resistant weed populations in ND. However, weed populations resistant to the ALS, DNA, and ACCase herbicides have appeared in recent years. The triazines have a significant role to play in resistant weed management, because they increase of our herbicide arsenal. Atrazine provides excellent control of kochia, a species that has developed ALS resistance at numerous sites in ND. Atrazine and cyanazine will certainly be needed where ALS resistance becomes a problem,

Ohio

Dr. John Cardina, Ohio State University. Currently available atrazine alternatives include mostly ALS herbicides with specific activity on a single plant enzyme. Such products are used not only on corn, but on soybeans and wheat. The impact of such widespread use of single site-of-action herbicides could be herbicide resistance on, a scale beyond the current level, which is already difficult for growers to manage. Rotation to other crops and herbicides

Continued on next page.

Table III. ***Continued***

with different modes of action were effective ways to manage resistant populations. ALS resistant weeds will be difficult to manage without triazines.

The Greene County Agronomy Committee Extension. We have increasing concern for the possibility of weeds becoming resistant to the newer classes of herbicides.

Oklahoma

Dr. Mark Hodges, Oklahoma State University. The selection of herbicide resistant species is less likely to occur in systems that combine as many effective weed control strategies as possible. Diversified and integrated management practices that provide maximum possible soil surface protection should be adopted. This will become much harder and certainly much more costly if atrazine is not available to the farmers of the high plains.

Oregon

Arnold P. Appleby, Oregon State University. We are strongly encouraging our growers to avoid using herbicides in the same chemical family more than a year or two in a row in order to prevent or delay the development of herbicide-resistant weed populations. Herbicide resistance is becoming the major problem in commercial weed control. Restricting the herbicides available can only exacerbate the problem.

Tennessee

Dr. Robert M. Hayes, University of Tennessee. Triazine resistance has not been confirmed. We are very concerned with cocklebur resistance to ALS inhibitors. This is even more alarming when we consider that most of the alternatives to the triazines are ALS inhibitors. The triazines offer an alternative mode of action to manage ALS resistance.

Dr. Thomas Mueller, University of Tennessee. Many of the newer herbicides inhibit the ALS enzyme in plants. These herbicides are safe, economical, and valuable production tools to farmers. The problem is that they have identical sites of action and resistance will develop. Atrazine has a different mode of action and its availability helps manage ALS resistant weeds.

Dr. Neil Rhodes, Jr., University of Tennessee. We continue to document more locations of biotypes resistant to the imidazolinones.

South Carolina

Drs. Billy J. Gossett and Edward C. Murdock, Clemson University. No triazine resistant weed biotypes have appeared in South Carolina. No resistance problems, and very few rotation restrictions, makes the triazine herbicides even more popular. We recommend rotations of crops and herbicides to manage or prevent herbicide resistant weeds. Loss of the triazines would represent a major reduction in our options for herbicide rotation.

Table III. *Continued*

Virginia

Dr. Henry P. Wilson, V.P.I. and State University. Over 20 herbicide products classified as ALS inhibiting herbicides are currently marketed for weed control in agronomic crops. To date, 14 weed species have been identified as resistant to these herbicides. Discontinued use of triazines will result in increased use of ALS herbicides and widespread selection for resistant biotypes. The triazines are the best herbicides available for management of ALS resistance.

Washington

Dr. C. Patrick Fuerst, Washington State University. Although there are triazine resistant weeds, they can be easily recognized and managed. The inevitable future development of numerous species with ALS resistance is extremely frightening. The triazines are a crucial alternative mode of action to delay or prevent the development of weed resistance to ALS herbicides. A ban on the triazines would be a serious blunder and would lead to a devastating proliferation of ALS resistant weeds.

Wisconsin

Drs. Chris Boerboom, Jerry Doll, David Stoltenberg and Nelson Balke, University of Wisconsin. Wisconsin has triazine resistant lambsquarters and pigweed species on roughly 500,000 acres (4% of agricultural lands) and atrazine resistant velvetleaf in three known locations (about 300 acres). Triazine resistant kochia also exists in the state, but is of minimal agricultural significance. The occurrence of these triazine resistant weeds has necessitated the use of tank mixtures or alternative mode of action herbicides for their control. Dicamba is effective on all three weed species. The velvetleaf is resistant only to atrazine so other triazines are still equally effective. We have greater concerns about the potential for ALS inhibitor resistant biotypes than triazine resistant biotypes. Most new corn and soybean herbicides have this mode of action, which has high potential for selecting resistant biotypes. They should be used in rotation with herbicides of alternate modes of action to prolong their efficacy. Triazines provide an alternate mode of action when corn is grown.

Mixtures of herbicides have been proposed as strategies to prevent or delay the evolution of resistance to the resistance-prone sulfonylurea and imidazolinone herbicides. For a mixture to be efficacious in preventing resistance, both herbicides should have all or most of the following traits: a) control the same spectra of weeds; b) have the same persistence; c) have a different target site; d) be degraded in a different manner; and e) preferably exert negative cross-resistance. Wrubel and Gressel (*19*) compared the proposed mixing partners for use with several widely used ALS-inhibiting herbicides to these criteria and found that: a) all had somewhat different weed spectra; e.g., none controlled common cocklebur as well as imazaquin or imazethapyr in soybean, or kochia as well as chlorsulfuron in winter wheat; b) all were far less persistent than these vulnerable herbicides; c) most had different target sites; d) in soybean most mixing partners were degraded differently than vulnerable herbicides; and e) none of the mixing partners exerted negative cross-resistance. They concluded that not meeting the key criteria of identical control spectra and equal persistence will aggravate future resistance problems, as has happened with insecticides.

To further aggravate this problem, commercial sale of imidazolinone-resistant corn was initiated in 1993. Extending the use of imidazolinones to corn presents several potential problems in terms of weeds evolving resistance. As much as 60% of the soybean area in the northern corn belt is rotated with corn. A farmer who formerly rotated herbicides along with soybean-corn crop rotations can now use imidazolinones continuously. The same populations of weeds will, therefore, be exposed to the same herbicide chemistry year after year, increasing the probability of the evolution of herbicide-resistant weed biotypes. The situation is further exacerbated by the availability of sulfonylurea herbicides in both soybean and corn.

Research is also underway to develop varieties of wheat, oilseed rape, tomato and other crops resistant to ALS-inhibitors. With the prospect of increasing use of ALS-inhibiting herbicides we should expect the more extensive evolution of resistant weed biotypes to follow quickly.

ALS-inhibitor herbicides are widely used in corn, soybeans, and cereals. In 1995 they were used on more than 70% of the United States soybean acreage. Atrazine is used on about 67% of United States corn acreage. Crop and herbicide rotation practices have been, and will be, very important if the development of resistance to these herbicides is to be managed effectively. Based on research and experience, it can be confidently predicted that resistant biotypes to the ALS-inhibitor herbicides will continue to increase. In some areas of the United States they already are a much more serious problem than triazine resistant weeds. In fact, the increase of new ALS-inhibitor resistance is occurring at a rate equal to or greater than was seen with insecticides during the period from 1955 through 1980. It can be conservatively predicted that weed resistance to herbicides has or will become more important economically than resistance to insecticides and fungicides, especially if we do not properly manage the newer herbicides by retaining the triazines as essential combination or sequential herbicides.

Although they have evolved later than pest resistance to insecticides and fungicides, herbicide-resistant weed biotypes will have an even greater impact on agricultural technology and economics in the future if we do not properly manage the newer highly specific mode of action herbicides. Weeds require longer reproductive cycles, usually with lower numbers, and do not travel as far or as readily as insects and pathogens. Therefore, with the potentially larger number of herbicides having different modes of action, we should be more successful in avoiding or managing resistant weed biotypes. But much will depend on keeping atrazine and other of the older herbicides in our arsenal for many years to come.

Literature Cited

1. Vogt, W.; *Road to Survival;* William Sloane Associates, Inc.: New York, 1948.
2. Ryan, G.F. *Weed Sci.* **1970**, *18*, pp. 614.
3. Radosevich, S.R.; DeVilliers, O.T. *Weed Sci.* **1976**, *24*, pp. 229,.
4. Arntzen, C.J.; Pfister, K.; Steinbeck, K.E. In *Herbicide Resistance in Plants;* LeBaron, H.M., Gressel, J.; Eds.; John Wiley and Sons: New York, 1982.
5. *Herbicide Resistance in Plants. Chapters 2 and 3,* LeBaron, H.M., Gressel, J.; Eds.; John Wiley and Sons: New York, 1982.
6. LeBaron, H.M., *The History and Current Status of Atrazine-Resistant Weeds in the U.S.*, Laboratory Study No. CG-5, submitted to EPA, June 16, 1989.
7. Saari, L.L.; Cotterman, J.C.; Smith, W.F.; Primiani, M.M. *Pestic. Biochem. and Physiol.* **1992**, *42*, pp. 110-118.
8. Cole, D.J. *Pestic. Sci.* **1994**, *42*, pp. 209-222.
9. Thompson, C.R.; Thill, D.C.; Shaffi, B. *Weed Sci.* **1994**, *42*, pp. 172-179.
10. Devine, M.D.; Shimabukuro, R.H., In *Herbicide Resistance in Plants: Biology and Biochemistry*, Powles, S.B., Holtum, J.A.M., Eds.; Lewis Publishers: Boca Raton, FL, 1994, pp. 141-169
11. Morrison, I.N. and M.D. Devine, Phytoprotection 75 (suppl.):5-16, 1994.
12. *Herbicide Resistance in Weeds and* ; Thill, D.C.; Mallory-Smith, C.A.; Saari, L.L.; Cotterman, J.C.; Primiani, M.M.; Caseley, J.C.; Cussans, G.W.; Atkin, R.K., Eds.; Butterworth-Heinemann: Oxford, England, 1990.
13. Holt, J.S.; Thill, D.C. In *Herbicide Resistance in Plants: Biology and Biochemistry;* Powles, S.B., Holtum, J.A.M., Eds.; Lewis Publishers: Boca Raton, FL, 1994, pp. 289-316.
14. Warwick, S.I. and L.D. Black, Phytoprotection 75 (Suppl.):37-49, 1994.
15. Gressel, J., International Symposium on Weed and Crop Resistance to Herbicides. Cordoba, Spain, 1995, p.29.
16. Shaner, D.L. *Weed Tech.* **1995**, *9*, pp. 850-856.
17. Kendig, A., Personal communication, 1996.
18. *Herbicide Resistance in Plants*; Powles, S.B.; Holsum, J.A.M., Eds.; Lewis Publishers: Boca Raton, FL, 1994.
19. Wrubel, R. and J. Gressel, 1994. Weed Technology 8:6352648.

Chapter 3

A Simulation Analysis of the Use and Benefits of Triazine Herbicides

David C. Bridges

Crop and Soil Sciences Department, University of Georgia, Georgia Station, Griffin, GA 30223–1797

A systematic analysis of the use and benefits of triazine herbicides in U.S.-grown corn and sorghum was conducted to determine the potential cost and yield changes that might result if triazine herbicides were not available. Using regional data on weed infestation/incidences, potential yield loss, and yield potential, net return changes were ascertained using comparative field weed efficacy data. Model parameters included efficacy data, treatments (active ingredients), target weeds, percent market share, treatment costs, and regionally specific yield, production, and value data. Models performed a substitution analysis to calculate a protection value, using a sequential (Monte Carlo) numerical procedure, for all products. Cost and yield changes associated with alternative products are relative to the potentially regulated herbicide (Triazine(s)). Cost and yield changes are the principle input value for the aggregate economic impact estimate. This model included 28 weed species and 36 weed control alternatives, including recently registered herbicide classes.

The potential triazine alternatives can be classified generally into 3 groups:

1. Dicamba, bromoxynil and 2,4-D
2. ALS inhibitor herbicides
3. Flumetsulam + metolachlor

Each alternative has one or multiple deficiencies, including limited weed spectrum, higher cost, limited application timing, drift potential, no soil residual activity, ALS-inhibitor leading to resistant biotypes, reduced corn selectivity, weather sensitive, and other economic impact to sorghum and numerous minor crops is also significant without triazines. The yield and cost net return of atrazine was higher than any of the other alternatives. These analyses demonstrated quantitatively that no single atrazine (triazine) replacement exists.

The triazine herbicides are among the most widely used herbicides in the U.S. and the world. Atrazine treatment acres exceed those for any other single herbicide active ingredient and it is by far the most commonly used herbicide in U.S. corn production. Simazine is a mainstay of weed control in more than 30 crops, ranging from citrus to strawberries to turfgrass. In fact, more than 400,000 U.S. farmers rely on atrazine for weed control. The importance of these herbicides to American agriculture is evident, illustrated by the following usage facts:

In 1994 more than

- 67% of the U.S. corn acreage was treated with atrazine,
- 65% of the U.S. grain sorghum acreage was treated with atrazine, and
- 90% of the U.S. sugarcane acreage was treated with atrazine.

While over the past few years the average atrazine application rate (pounds per acre treated) has declined, the percent of corn acres treated continues to increase. This is remarkable given the fact that the unspoken goal for the last 30 years of every major agricultural chemical manufacturer in the U.S. (probably the world) has been to find an 'atrazine replacement'. In fact, most never expected to replace atrazine, they simply desired a small share of the 70 million-acre U.S. corn herbicide business. Every conceivable candidate chemistry has been evaluated in comparison to the triazines, especially atrazine, and the elusive replacement has not been found. The fact that atrazine has remained the most relied-upon herbicide for weed control in corn since its introduction during the 1950s is the strongest possible testimonial to its effectiveness and reliability.

The efficacy, reliability, cost effectiveness, and safety of the triazine herbicides can easily be demonstrated. A benefits assessment attempts to quantify and enumerate benefits associated with use of a specific pesticide, or group of pesticides. In other words, the assessment attempts to enumerate, often in economic terms, why a pesticide is used. Benefits are accrued relative to efficacy, spectrum of activity, reliability, cost, return on investment, crop safety, non-target safety, and any other variable that is a basis for selecting one pesticide over another.

With the Triazine Special Review the requirements for a comprehensive benefits assessment reached unprecedented proportions. In late 1994, the need for developing an assessment approach capable of capturing and quantitatively describing the biologic and economic benefits derived from the use of triazine herbicides in U.S. agriculture became apparent. The objective was not limited to establishing the benefits of triazine use, but to also ascertain the comparative benefits of possible alternative treatments. That result was achieved producing an assessment and response to the Environmental Protection Agency's (EPA) Position Document-1 (PD1) that fairly and representatively illustrates the biologic (agronomic) and economic benefits of a substantial portion of the current triazine uses and conservatively forecasts the potential economic impact should the current uses be eliminated.

The objectives of this paper are to outline the benefits assessment process, focusing primarily on the biologic (agronomic) portions of the assessment and to highlight a few of the conclusions of the study. More details on the aggregate economic assessment are presented in the chapter by Dr. Gerald Carlson of North Carolina State University, included in this book.

BENEFITS ASSESSMENT PROCESS

Overview. When products are as widely used as the triazine herbicides and are so well established as a major component of conventional production practice, it is difficult to know precisely what will happen if those products are removed. However, predicting the economic impact of such an event is precisely what is attempted in a benefits assessment. Several important determinants of use were considered in evaluating the current uses of the triazine herbicides and potential alternative products.

Comparative biological performance. Comparing efficacy, or the ability to control weeds, was a major focus in the assessment. Biological performance was compared by developing 18 models to simulate the effect of weed control in corn and grain sorghum yield with the triazine and other commonly-used herbicides. This became the basis for all cost and yield change estimates for the study.

Performance profiles. Efficacy, spectrum of activity, and crop tolerance were compared using a vast set of data from Ciba Crop Protection's (Ciba) corn weed control database, a database that is comprised of field studies conducted by university scientists and Ciba scientists around the U.S.

Product comparisons. Labels and other technical data were evaluated to determine label requirements, restrictions, and other parameters that dictate how a product must be used.

Hazard profiles. Safety information relating to worker and applicator exposure, protective clothing requirements, reentry, toxicity classification, etc. were compared for all alternatives.

In addition to these, costs and application logistics were considered, but the most important consideration was comparative biological performance, as estimated by the models. Therefore, the remainder of this paper will focus on the determination of comparative biological performance which was accomplished using a series of models to perform a complex substitution analysis. The mechanistic, transparent models were used to determine cost and yield changes associated with the use of each biologically-relevant alternative practice. The study was designed to be inclusive, balanced, systematic, objective, and based on substantial data, and relying on expert opinion to a minimum. The models compared each commonly used corn and sorghum herbicide, cultivation, and unregistered products in advanced stages of development. The models provided two critical types of data: a) relative benefits of active ingredients and treatments, and b) identification of potential alternative treatments based on the premise that the targeted use of a pesticide is what must be satisfied with the loss of a pesticide.

Models - areas of coverage. The overall benefits process is outlined in Figure 1. Several models were developed. Three 'national' models were developed, one for corn using efficacy data obtained from the Ciba database, referred to as the Ciba National Corn Model; one for corn using efficacy data obtained from university extension recommendations, referred to as the University National Corn Model; and one for sorghum using efficacy data obtained from the Ciba database, referred to as the Ciba National Sorghum model. Fifteen (15) regionally-specific models were developed to reflect the site specific nature of corn and grain sorghum production, weed infestations, and weed control practices. Ten (10) regional corn models were developed, one for each of the ten USDA production regions in which field corn is grown. Five (5) regional grain sorghum models were developed, one for each of the five production regions in which grain sorghum is grown.

The core module of each of the models was fashioned after a model that was developed to analyze the use and benefits of pesticides in U.S.-grown peanuts, a study that was published in 1994 (*1*).

MODEL PARAMETERS AND SOURCES OF DATA

Overview: The primary types of data used within the model included: efficacy, treatments (single active ingredients, plus combinations), treatment targets, pest incidence and potential losses, percent market share, treatment costs, and regionally-specific yield, production, and value data.

Efficacy data. Efficacy data for all of the regional models (10 corn and 5 grain sorghum), the Ciba National Corn Model, and the Ciba National Grain Sorghum Model came from the Ciba weed control database, a database that includes field trials conducted by university research/extension scientists and Ciba scientists across the U.S. The database includes all of Ciba's trials conducted from 1980 through 1994 and all university trials that have been conducted from 1984 through 1994. It is probably the most comprehensive corn weed control database in the world. Weed control data from 4,926 trials, approximately 80% of which were conducted by university cooperators and 20% of which were conducted by Ciba scientists, were summarized to obtain weed control data for the 28 weed species and the 36 treatments included in the model. Every major corn-producing state is represented in the database. All data were converted to a common format (rate definitions, days after application, species designations, etc.) before being averaged by species, active ingredient, or treatment.

Efficacy data for the University National Corn Model were obtained from weed control recommendations published in 22 university extension publications, representing 33 states which comprise more than 95% of the U.S. corn acreage. An average efficacy value by species and treatment was calculated by converting

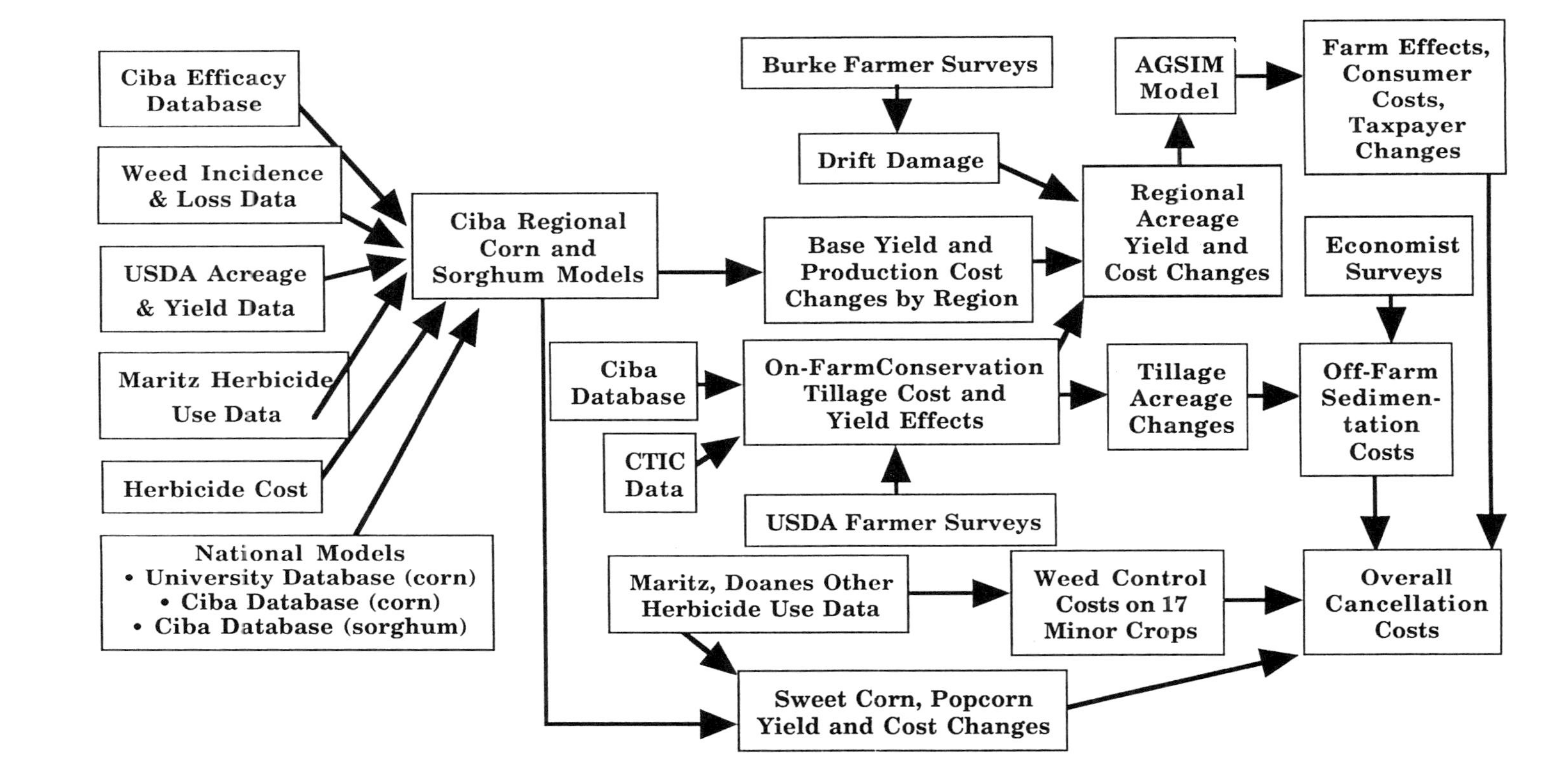

Figure 1. Overall Use and Benefits Analysis Process for the Triazine Herbicides.

the categorical (none, poor, fair, good, excellent) data to percentages using the median range value reported for each individual table.

Herbicide active ingredients and treatments. The Ciba National Corn Model and each of the 10 regional corn models included 36 individual treatments plus all possible permutations of two-way and three-way combinations. Common premix treatments were included, such as atrazine + metolachlor, atrazine + alachlor, atrazine + dicamba, atrazine + cyanazine, etc., were included as individual treatments. The University National Corn Model included 28 individual treatments.

Treatment targets. The basis of the substitution analysis was that it is the targeted use of herbicides that must be replaced when a product is no longer available. This arises from the assumption that the choice to use a particular herbicide is primarily based on the fact that the herbicide will control a particular target weed, or combination of weeds. Therefore, at any site of use, there is a set of weed species that are the primary target for a particular herbicide application. Within each region, target weed species were determined for each herbicide treatment. Several sources of data were considered for determining target species. Correlations between weeds species and treatments within a region, surveys, and the Weed Science Society of America's Crop Losses reports (*2*) were considered. Since no source of data provided a consistent and logical set of targets for all treatments, a panel of university scientists and agricultural specialists assigned five target species for each of the treatments included in each of the regional and national models. Herbicide labels, weed incidence data, and efficacy data were considered in determination of target species.

Pest incidence and potential losses due to weeds. Several sources of data were used to determine pest incidence and potential loss by species. A 1992 Weed Science Society of America survey on crop losses due to weeds stimulated a 1994 survey by the National Center for Food and Agricultural Policy (Gianessi, L., NCFAP, Washington, DC, personal comm.). This state-level survey determined the percentage of corn acres in each state that was infested at a potentially damaging level of each weed species. At the same time respondents provided data on the average percent loss that would be expected if the weed species was not controlled. Similar data were collected by Pike in a 1994 survey of the U.S. North Central Corn Belt with extension specialists as respondents. These data were combined to provide state-level data on weed incidence and damage. Weed incidence data were converted to either a national or regional acreage-weighted basis for use in the models. Incidence data was available representing an area of more than 95% of the U.S. corn acreage.

Loss estimates were available from 15 states. The data were averaged across all sources by species to provide estimates of potential losses from each of the 28 weed species included in the model. The resulting values were compared to values published in refereed literature and to those found within HERB for corn, a computer weed control decision aid from North Carolina State University.

Percent market share data. Market share data from three sources were initially considered: Maritz Market Survey, the National Center for Food & Ag Policy, and USDA National Agricultural Statistics Service (NASS). Maritz market data proved to be the most comprehensive with respect to coverage and sample size. Therefore, 1994 Maritz market survey data were used to calculate national and regional acreage-weighted market share data for each of the herbicide treatments used in the model. Market shares were calculated by dividing the number of acres treated with a given herbicide treatment by the 1994 USDA NASS planted acres. Percent treated acres was rounded to the nearest whole percent.

Treatment costs. Treatment costs were based on herbicide application rate, cost per pound of active ingredient, and application costs. Herbicide costs were based on 1995 anticipated cost per pound of active ingredient which were calculated based on 1994 prices adjusted for inflation to 1995. Several pricing sources were considered in developing the prices used in the model: 1993 AgChem Price Survey (*3*); National Center for Food and Ag Policy; various extension price surveys, and industry price sources. Prices for premix products were determined to be 90% of component prices. Application cost was assumed to be $6.70/acre treated and cultivation cost was assumed to be $6.70/acre.

Yield, production, and value data. Corn and grain sorghum yield, production, pricing, and value data were obtained from USDA NASS. Five-year averages of planted acres, yield, production, and price were used for all model simulations.

Overview of model. The models were developed to perform a substitution analysis. The approach is to calculate to the yield savings, or protection values, for the potentially regulated herbicide, based on the target species identified for that herbicide, the incidence of the target species within the region, the potential loss associated with each of the target species, and the efficacy of the herbicide on each of the target species. The protection value associated with the potentially regulated herbicide is calculated using a sequential (Monte Carlo) numerical procedure that considers the most potentially damaging target species first, i.e., the target weed species with the highest value for the percent acres infested (incidence) multiplied by the potential loss (loss). Each of the remaining target weeds is considered sequentially in decreasing order of potential damage (incidence x loss). Losses are computed with and without treatment with the potentially regulated herbicide. Protection value is calculated as the difference between losses with and without herbicide treatment. Net protection value is calculated as the protection value minus the cost of treatment. Protection and net protection values are then calculated for each of the potential alternative treatments using the same target species, infestation values, and loss values as for the potentially regulated herbicide. Because the same targets, infested acres, and potential damage values are used for the potentially regulated herbicide as for the alternatives a rigorous comparison of benefits can be made. Differential protection values are always used thus negating the importance of absolute values. Since cost and yield changes associated with alternative treatments are relative to those of the potentially regulated herbicide, they are calculated with a high degree of confidence in the final values. It is these cost and yield change values that are used to make estimates of weighted cost and yield changes that are likely to occur within a given region once an assumption of apportionment of treatment acres is made. These cost and yield changes are the principle input values for aggregate economic impact estimates.

RESULTS AND DISCUSSION

Many simulations were run, national versus regional, corn versus grain sorghum, and Ciba efficacy data versus university efficacy data. The remainder of this paper will focus on conclusions drawn from regional simulations for field corn using the Ciba efficacy data.

Comparison of Ciba versus University efficacy data. To establish the objectivity and validity of the efficacy values contained in the Ciba corn weed control database a direct comparison was made of the results obtained from runs of the Ciba National Corn Model and the University National Corn Model. Comparing net return to treatment for all products used on 5% or more of the U.S. corn acreage showed a very small difference between the two sources of data. There was no evidence of bias and mean absolute difference was only 5.89% (Table I).

Table I. Net returns to treatment (NRT), according to the University National Corn Model and the Ciba National Corn Model, for all treatments used on 5% or more of the U.S. corn acreage

	Net Return to Treatment ($ Millions)		
Treatment	*University Model*	*Ciba Model*	*% Diff*[1]
Atrazine (Pre)	2189.189	2246.294	-2.61
Atrazine (Post)	1141.844	1200.926	-5.17
Metolachlor + Atz	1606.918	1697.621	-5.64
Alachlor + Atz	858.552	909.272	-5.91
Cyanazine + Atz	1094.182	1164.630	-6.44
Dicamba + Atz	880.490	952.465	-8.17
Metolachlor	1676.490	1603.639	+4.34
Nicosulfuron	1235.161	1154.183	+6.56
Alachlor	1043.154	1006.429	+3.52
Acetochlor	639.073	646.806	-1.21
Dicamba	1500.878	1407.116	+6.25
Bromoxynil	719.980	753.611	-4.67
2,4-D	1262.058	1058.974	+16.09

[1]Computed as [NRT (University) - NRT (Ciba)]/NRT (University) *100%

This comparison is very important in that it provides the strongest evidence that there is no bias in the Ciba efficacy data, as was expected because the approximately 80% of the records contained in the database are from university-conducted research trials. It also indicates that university extension recommendations, with respect to efficacy, are consistent with and well correlated with aggregate efficacy data. This establishes the validity of the judgements of university weed scientists, the same scientists who provided information on weed incidence and damage.

Identification of possible alternative treatments. Yield and cost changes were determined for each possible alternative herbicide treatment. However, atrazine, simazine, and cyanazine are used alone primarily for the control of broadleaf weeds, or they are used in combination with other herbicides, especially the chloracetamide herbicides metolachlor and alachlor for broad spectrum control of both broadleaf and grass species. For the purpose of discussion alternatives are combined into three groups.

Dicamba, bromoxynil, and 2,4-D. The three products each provide partial replacement value for atrazine, simazine, or cyanazine when they are used for broadleaf weed control in corn. Neither of these three herbicides will control grasses. Therefore, for most corn acres they would have to be used in combination with another herbicide for broad spectrum control, including broadleaves and grasses. The cost for these three herbicides ranges from slightly less to $3.00/acre more than for atrazine applied alone preemergence. The average cost difference is about $1.50/acre more than atrazine alone applied preemergence. Because these herbicides are only used postemergence, they have restricted, narrow windows for application. Dicamba and 2,4-D are both volatile herbicides with significant potential for non-target species sensitivity. Therefore, the potential for drift damage to non-target plant species is significant. In fact, in some regions of the country their application is significantly restricted due to potential drift to sensitive crops like cotton, tobacco, and vegetables. Their utility as triazine replacements is limited due to the fact that they do not provide residual control. Therefore, repeat applications may be required if reinfestation occurs.

Reduced efficacy due to reinfestation and the lack of residual control is borne out in yield reductions compared to yields when triazine herbicides are used. These negative yield effects were demonstrated in the model comparisons. So, even though the average cost of these herbicides is slightly less than for atrazine, considering the yield changes that will occur, the net impact is a loss to triazine users. Consideration of these treatments as alternatives to triazine herbicides must take into account the lack of residual control, application timing restrictions, and the logistics of relying solely on postemergence treatments.

ALS-inhibitor herbicides. In the past ten years several herbicides which share a common mode of action, acetolactase inhibition, have been registered for weed control in corn. Several active ingredients are included in this group: primisulfuronmethyl, prosulfuron, nicosulfuron, thifensulfuronmethyl, rimsulfuron, halosulfuron, and imazethapyr. Collectively, these herbicides have often been touted as the most likely atrazine replacement. As a group they provide relatively wide spectrum weed control, but they also share a significant number of limitations. First, those which are used for corn weed control are primarily postemergence herbicides with very narrow windows of application. They are primarily used for the control of small weeds. They are weather sensitive. They do not work well on drought-stressed weeds, and with excessive rainfall there is the risk of not being able to apply in a timely fashion. Overall crop tolerance of ALS-inhibitors is not as great as with the triazines. Considering the group as a whole, the spectrum of activity is quite good, but the spectrum of activity of the individual products is generally narrow. Some have significant rotational crop limitations relative to carry-over potential.

A final, and significant, consideration with the ALS-inhibitor herbicides is that of resistance management. Resistance developed more quickly with the ALS-inhibitor herbicides than with other groups of herbicides. The rate at which new ALS-inhibitor resistant species has been reported currently exceeds the rate with any previously-registered group of herbicides. They are widely used for soybean weed control throughout the U.S. and due to the almost direct rotational acreage of corn and soybeans within the Corn Belt, difficulties in managing ALS-inhibitor resistant weed species can be anticipated should corn weed control become heavily reliant on these herbicides. To achieve acceptable control of both broadleaf and grass species will require that combinations of these herbicides be used or that they be applied in combination with other products, which will be more costly than using the triazines alone or triazines in combination with grass herbicides.

Flumetsulam plus metolachlor. The final group includes one premix product, a mixture of flumetsulam plus metolachlor. This new product includes flumetsulam, an ALS-inhibitor herbicide, for broadleaf weed control and metolachlor for control of grasses and small-seeded broadleaf weeds. Its use over the past couple of years has been very limited and could be best characterized as a partial atrazine replacement. Assuming that the triazines are commonly used in combination with either metolachlor or alachlor, flumetsulam+metolachlor costs approximately $5.50/acre more than atrazine+metolachlor. However, there is significant potential for crop injury. The treatment is applied preemergence, so it does provide residual weed control, but it is a highly weather-sensitive treatment. Good activating rainfall is required for activity, but excessive rainfall greatly increases the potential for corn injury. Compared to atrazine, its spectrum is incomplete. It has significant soil type limitations with respect to where it can be used safely. There is also the potential for interactions with organophosphate insecticides, which further increases the potential for corn injury. Comparing the yield protection potential of this treatment with that of atrazine-based treatments in the major corn-producing regions indicates that significant yield losses are likely to occur as a result of incomplete weed control. Since flumetsulam is an ALS-inhibitor, the resistance management concerns that are associated with ALS-inhibitor herbicides apply to this potential replacement.

CONCLUSIONS

Among the numerous questions that must be answered relative to any potential regulation or elimination of the use of triazine herbicides within the U.S. is "What will be the impact to growers, consumers, and the agricultural community as a whole?" This analysis of the impact of such regulation included the impacts to the grain sorghum market and that for numerous minor crop uses. The impact on grain sorghum and minor crops is large and has been documented in Ciba's PD1 response to the EPA. However, the largest impact will be derived on the more than 50 million acres of U.S. corn that is treated with atrazine. Considering all triazine uses, approximately 60 million acres of corn are treated annually in the U.S. The question is, are there viable, economical, minimally impacting alternatives to the use of triazine herbicides in the U.S. corn industry. To date the answer is no. There are alternatives for some segments and for portions of the acreage, but for the 60 million acres of corn as a whole, there are not good alternatives.

This is apparent when considering the current level of dependence on triazine herbicides. Atrazine is the most reliable, economical, and flexible herbicide available for weed control in corn. With respect to crop safety, atrazine is among the safest herbicides ever used in corn. It provides residual weed control with a tremendous margin of crop safety under a variety of environmental conditions. It can be premixed or tank-mixed with every herbicide currently registered for use in corn. It provides control of both broadleaf weed and grasses and it can be used in any tillage system.

The economics of triazines, particularly atrazine, use in corn is evident in this and other analyses. It is economical. This analysis showed highly positive net returns for the use of atrazine. The initial cost is low, reliability is high, and there is often no need for a follow-up treatment. Most consider it to be one of the safest herbicides that can be applied to corn, safe in terms of worker/applicators and non-target crops. The triazines are non-volatile, so non-target safety is high. These characteristics, along with low avian, mammalian, and aquatic effects contribute to the widespread use. As previously stated the percent corn acres treated with atrazine continues to increase. Grower satisfaction and dependence is high. It is not surprising that these herbicides, particularly atrazine, remain the most widely used herbicide for corn weed control in the U.S. This analysis demonstrated quantitatively that at the present no single atrazine replacement exists and that the value of the triazines is remarkable.

Literature Cited

1. Bridges, D. C., C. K. Kvien, J. E. Hook, and C. R. Stark. 1994. An Analysis of the Use and Benefits of Pesticides in U.S.-Grown Peanuts: II. Southeastern Production Region. NESPAL Report 1994-002, 220 pgs., University of Georgia.
2. Crop Losses Due to Weeds in the United States - 1992. Weed Science Society of America, 403 pgs., Champaign, IL.
3. 1993. AGCHEMPRICE - March 1993, Summary Edition. DPRA Inc., Manhatten, KS.

Chapter 4

Costs Impacts if Atrazine or Triazines Were Not Available to Growers

Gerald A. Carlson

North Carolina State University, Box 8109, Raleigh, NC 27695

Economic analyses have shown that if atrazine is not an available herbicide choice, there will be a significant, detrimental economic impact on corn and sorghum producers. Corn and sorghum producers in all 10 USDA regions will experience losses when no corn or other price changes are assumed. The losses for corn growers ranges from $5.18 to $37.67 per acre of corn treated, with losses in the Corn Belt of $23.90 per treated acre. For sorghum, losses range from $3.55 to $15.65 per treated acre. In the primary growing regions of the Northern Plains and Southern Plains, the losses are $6.75 and $10.72 per treated acre, respectively.

Analysis from the livestock sector demonstrates a severe impact on the income of livestock producers if atrazine is not available. A minimum annual income loss to growers of $777 million for the eight livestock sectors results from increased corn costs and thus increased feed costs for hogs, turkeys, broilers and fed beef. The impact loss to the hog industry would be $192 million annually, because of increased feed cost and reduced production. Other sectors suffering major income losses are fed cattle at $168 million and dairy at $161 million.

The use and importance of atrazine in conservation tillage is demonstrated by an increasing yield advantage as tillage intensity decreases on an increasing number of conservation tillage acres.

These and other specific analyses show a minimum annual economic impact of $1.66 billion should the triazines not be available for production agriculture.

Atrazine and simazine availability is necessary to prevent this conservative estimate of economic loss to U.S. agriculture and crop production.

Assessing Economic Costs of a Triazine Cancellation

The Environmental Protection Agency's Special Review of the triazine herbicides must include an assessment of "lost benefits" or costs of the prospective cancellation of atrazine or all triazines--not only costs to farmers but to the public as well. Costs of an atrazine or triazine quantified in this study include:

- direct costs to triazine herbicide users
- decreased yields in corn and sorghum
- higher prices to consumers of agricultural products
- increases in sedimentation costs *related to no-till corn production*
- taxpayer costs associated with changes in farm program payments.

Every effort was made to complete a comprehensive study, considering many of the crops on which the triazine herbicides are used and the available herbicides that could serve as alternatives if the triazine herbicides were not available. In the effort to ensure a comprehensive study, particular emphasis was placed on corn and sorghum, which account for the largest acreage on which the triazine herbicides are used.

However, due to the difficulty in accurately quantifying certain costs, some of the potential costs of canceling this important group of herbicides were not included in this study. These costs include:

- costs of increased incidence of weed resistance to alternative herbicides used to replace the triazines
- drift damage to crops other than soybeans (such as cotton and tobacco) by alternative herbicides
- direct labor and management costs to farmers and weed control specialists of adjusting to new methods of weed control
- potential unit price increases of alternative herbicides, resulting from competitive marketplace changes in the absence of the triazines
- potentially lower land values as topsoil erosion increases with less no-till corn or less conservation tillage

Also, for 17 of the crops on which triazine herbicides are used, no yield effect or crop quality decreases resulting from use of alternative herbicides were measured or included except in sweet corn and popcorn. Yield effects were only included for corn (field corn, sweet corn, and popcorn) and sorghum. Thus, the overall costs of cancellation presented should be considered conservative or "under estimates."

A concerted effort was made to include any potential economic gains resulting from an atrazine or triazine cancellation, such as savings to taxpayers in farm program payments. Under the 1990 farm legislation (in effect at the time of this study), corn and sorghum deficiency payments decline as crop production declines and market prices rise. Estimates for this potential gain were made by applying a leading national economic model for the U.S. agricultural sector--AGSIM (*1*). This model was used to simulate national production and prices for all the major commodities with and without the cancellation of the triazines. An average cost to producers, consumers and taxpayers with and without the 1990 Farm Bill in effect was used to approximate conditions for the next five years as the current farm program is phased out.

The entire process to evaluate costs of a triazine or atrazine cancellation is summarized in Figure 1. The details are provided on regional costs to corn and sorghum production resulting from a triazine or atrazine cancellation. In these sections, explanations are provided for assessing direct weed control costs (yield and herbicide cost changes), cost changes related to drift damage from alternative herbicides used in corn, and on-farm costs from reductions in no-till corn production. The sum of the costs related to weed control, drift damage, and reduced no-till production were entered into the AGSIM model. Effects on farmers, consumers, and taxpayers over the next five years were simulated and added to the above corn and sorghum costs. These effects, include off-farm sedimentation costs related to reduced conservation tillage and the changes in weed control costs on 17 other crops. The final section summarizes the findings and provides separate estimates of the overall costs of a triazine or atrazine cancellation at the national level.

Corn and Sorghum Yield and Weed Control Cost Changes

A comprehensive biological and economic study was conducted for the field corn and sorghum sectors to provide cost estimates for a cancellation of atrazine or all triazine herbicides. The study included large amounts of input data, such as university efficacy ratings on major herbicides control of weed species in corn and sorghum, costs of herbicide and cultivation treatments, acres planted, weed densities, and current herbicide use patterns for each of the 10 U.S. Department of Agriculture (USDA) production regions (*2*). These 10 regions include Northern Plains, Corn Belt, Lake States, Delta, Northeast, Southeast, Appalachia, Mountain, Southern Plains, and Pacific.

Results from almost 5,000 field herbicide trials conducted by universities and Ciba Crop Protection (Ciba) provided an alternative source of herbicide efficacy data (*3*). This information, combined with yield loss relationships provided yield and treatment cost changes that would potentially result from use of alternative herbicides following an atrazine or triazine cancellation. Thirty-six different weed control alternatives were included in the analysis. The regional models included in the study provide rankings of the per acre profitability of using each possible alternative herbicide. To obtain regional estimates of yield and cost changes from a cancellation of atrazine or the triazines, current utilization data for these herbicides in *eight* distinct market niches were assembled from Ciba and Maritz Marketing Research surveys (based on pre-emergence (pre) or post emergence (post) treatments (*4*). The market niches included atrazine-pre, atrazine-post, *atrazine combinations-post*, atrazine-broad spectrum, cyanazine and atrazine, cyanazine-pre, cyanazine-post, and simazine-pre.

The adjustments a farmer would make to his or her weed control program, if atrazine or the triazines were not available, will vary according to

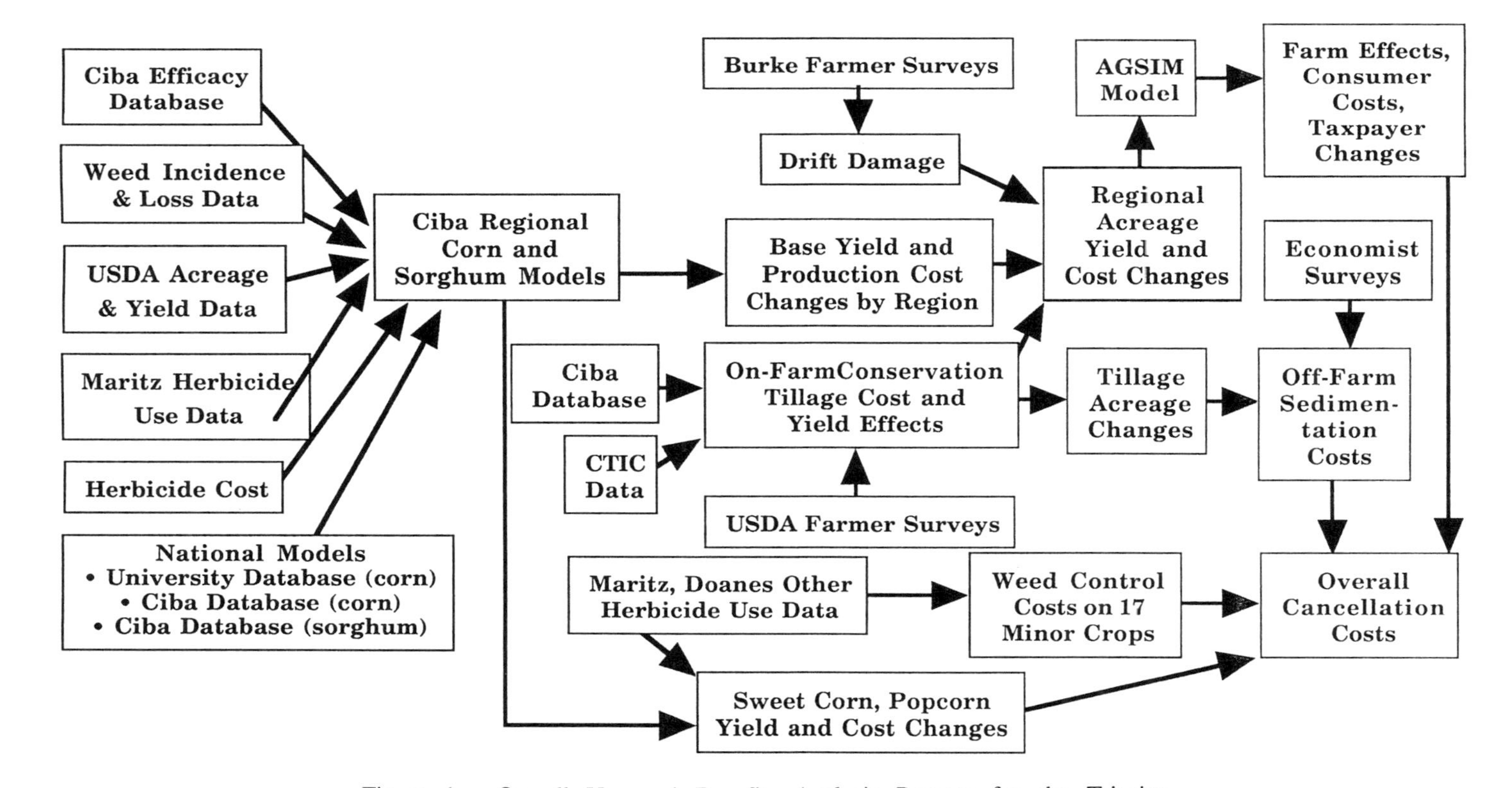

Figure 1. Overall Use and Benefits Analysis Process for the Triazine Herbicides.

weed species, crop yields, and herbicide use patterns because the 10 USDA regions differ in many aspects. Thus, the acres on which atrazine or triazines are currently used were apportioned by market niche and region among a suitable set of potential alternative herbicides considering the following: comparative control of 28 target weed species, return on investment, current market share, timing and method of application, and product label or other restrictions. Weighted average yield and cost changes were then computed for each region for both the atrazine and triazine cancellations by using acreage shares as weights.

Tables I and II give the corn and sorghum yield and cost changes, respectively, resulting from an atrazine cancellation for each region. For corn, an atrazine cancellation would result in higher herbicide costs and lower yields for all regions except the Southern Plains, which would have a small positive yield effect. The main effect of an atrazine cancellation in sorghum is a decrease in yield. Of course, the national effects are heavily influenced by changes in the Northern Plains, Corn Belt, and Lake States since these regions account for the great majority of corn production in the United States.

Table I: Conservative Estimates of Regional Yield and Cost Changes Per Acre of Corn Grown with Atrazine Cancellation

		Atrazine Ban	
Region	*Cost Change ($/A)*	*Yield Change (Bu/A)*	*Net Return Change ($/A Treated)*
Northern Plains	+3.04	-3.30	-17.72
Corn Belt	+3.52	-5.64	-23.90
Lake States	+1.50	-2.45	-18.79
Delta	+7.52	-3.22	-19.64
Northeast	+3.39	-2.80	-15.12
Southeast	+0.87	-1.13	-5.18
Appalachia	+6.99	-8.54	-30.61
Mountain	+2.30	-4.05	-30.58
Southern Plains	+8.40	+1.58	-6.44
Pacific	+0.49	-0.28	-37.67

Major changes center around corn and sorghum declines, but production of soybeans and some other crops would increase slightly. Major increases in corn and sorghum prices are estimated to occur from the reduced yield and higher cost of producing these crops. Including corn acres not treated with atrazine, the national production for all corn acres grown would decrease 4.31% and the

Table II: Conservative Estimates of Regional Yield and Cost Changes Per Acre of Sorghum Grown With an Atrazine Cancellation

		Atrazine Ban	
Region	*Cost Change ($/A)*	*Yield Change (Bu/A)*	*Net Return Change ($/A Treated)*
Northern Plains	-2.97	-4.44	-6.75
Corn Belt	-4.49	-8.54	-15.65
Lake States	-4.49	-8.54	-15.65
Delta	-1.13	-2.09	-3.55
Northeast	-4.49	-8.54	-15.65
Southeast	-1.13	-2.09	-4.73
Appalachia	-4.49	-8.54	-15.65
Mountain	+0.32	-0.90	-7.25
Southern Plains	-1.02	-3.81	-10.72
Pacific	+0.32	-0.90	-7.25

price would increase 10.4%. For sorghum, production for all sorghum acres grown would decrease 7.06% and the price would increase 12.23%.

Costs Associated with Drift Damage

Many of the herbicides that could potentially replace atrazine or the triazines are associated with potential drift damage to adjacent, nontarget crops. To assess the potential increase in crop damage to adjacent crops resulting from use of alternative herbicides if atrazine or the triazines were cancelled, Ciba commissioned Burke Marketing Research, Inc. (*4*) to conduct a survey of corn farmers. Researchers asked 1,316 farmers in 27 states to recall all incidents of drift damage to adjacent crops associated with corn herbicide applications for a 10-year period. If farmers reported damage to adjacent crops from a particular herbicide, they were asked about frequency, acres with drift damage, and bushels lost per acre from the drift.

The results indicated that 26.6% of farmers surveyed had experienced drift damage over a 10-year period. Of the damage reported, most (80.1%) occurred on soybeans, so this was the only crop evaluated in detail. Drift damage also occurred on tobacco and cotton, but the incidences were too few to include in this evaluation. A large share of the damaged soybean acres was associated with drift of dicamba (61%) and 2,4-D (14.7%). *Most incidences resulted in small yield losses (about 4 bu./acre or 10% of potential yield) so losses will be greater where soybeans are grown close to corn. (4)*

Increased Conservation Tillage Costs

The initial corn and sorghum evaluation developed by Ciba in March 1995 (*4*) was based on conventionally-tilled corn and sorghum acreage. However, 41% of the corn production land receives some form of conservation tillage, which helps to prevent soil erosion. There is strong evidence that the loss of the triazines would impact conservation tillage practices, such as no-till in corn, slowing the growth of conservation tillage. The two main effects that impact growers and can be quantified are: (1) an increase in direct tillage costs, and (2) an increase in yield losses on no-till corn.

More tillage trips across the field are required for conventionally-tilled corn than for corn produced using conservation tillage practices. The USDA has surveyed farmers in the 10 major corn-growing states and found that the average trips per acre are 3.47 for conventional tillage, 1.5 for corn produced with ridge-till, and 1.10 for no-till corn (*5*). According to farmer surveys conducted by Doane Marketing Reseach, each trip across the field (for any type of tillage) costs $6.80 per acre (*6*). Therefore, conventional tillage acres would have direct tillage costs $15.64 (acreage weighted average) higher than no-till acres.

The amount of no-till corn acres that would remain in production following an atrazine or triazine cancellation is difficult to estimate. However, the percent of conservation tillage corn acres has increased from about 5% in 1989 to 18% in 1994 (*7*). With the triazines available, this percentage is expected to increase over the next five years at the same average rate to about 30 percent of all corn acres produced under conservation tillage by the year 2000. However, if atrazine or the triazines were cancelled, no-till corn production would be more difficult. It can be assumed that no-till corn production would remain constant at 1994 levels if atrazine were cancelled and would decline slightly to 15% if all triazines were cancelled.

Two factors help explain why there will be a decline in no-till corn production if the triazines are not available. First, no-till corn production is more dependent on triazine use than is conventionally-tilled corn. Table III shows atrazine and triazine use per acre of corn grown under five tillage systems (*7*). It shows that atrazine product acres are about 49% higher with no-till corn than corn produced using conventional tillage with moldboard plow. *The dependency of no-till on all of the triazines rises to 1.204 treatments, or 57% higher than that land which receives moldboard plowing. Ridge-till also has more dependency on the triazines, but mulch till is more similar to conventional tillage without moldboard.*

Some analysts have claimed that conservation tillage is suited primarily to highly erodible land and that conservation tillage practices would decline if the conservation compliance provisions of the Farm Bill were eliminated. However, the USDA survey data do not support this claim. Table IV shows that the percent increase in conservation tillage adoption from 1989 to 1994 is higher on non-highly erodible land than on highly erodible land (*7*). These data lead one to believe that conservation tillage systems are adopted because they are profitable to farmers using currently available herbicides and not because they are legally mandated.

Finally, the yield gains from weed control using the triazine herbicides compared to alternative herbicides are higher on no-till corn than on conventionally tilled corn, as is clearly shown in Table V. Ciba's university database was queried for all trials that contained corn yields and tests conducted under "no-till" or "minimum-tillage." Treatments containing atrazine were defined as those in which atrazine was used at least once, regardless of time of application. Simple average yields were computed across all tests with atrazine and without atrazine treatments. Next, percent changes in yield were computed between those treatments using and those not using atrazine by tillage system. The results demonstrate a yield advantage with atrazine use; that yield advantage increases in less intensively tilled systems. For example, when comparing no-till corn production with and without atrazine use, production with atrazine results in an 11.4% higher yield than production without atrazine. The same comparison in conventionally tilled corn shows that atrazine use in conventional tillage results in only a 4.3% yield increase when compared to production without atrazine.

The yield losses if atrazine were not available for no-till corn were assumed to occur on all no-till corn acreage including that area not planted to no-till. The regional yield changes shown in Table I were expanded to include the no-till corn effects. Also, the extra costs from additional trips across the field on no-till and mulch-till corn were added to the regional weed control costs. The estimates of cost and yield changes with an atrazine or triazine cancellation were used for the aggregate analysis using the AGSIM model.

The AGSIM Model

The AGSIM model has equations projecting the yield and area planted of 12 major field crops in the 10 USDA production regions. With a policy change such as a herbicide cancellation, the economic effects were traced *throughout the economy*. Lower yields per acre and reduced aggregate corn production mean higher unit prices in the feed, industrial, and export markets. Higher production costs of corn or sorghum relative to alternative crops will lead farmers to allocate more land to alternative crops such as soybeans or wheat. Farmers growing crops and feeding livestock adjust to the changes in feed prices with production in the following years. These changes were evaluated year by year for 10 to 12 years beyond the start of the proposed cancellation.

Table III. Triazine Herbicide Treatments by Tillage System in Corn

USDA Average for 1992-1994

	Conventional *Tillage*		*Conservation Tillage*		
	Conv W*	Conv W/O**	*Mulch*	*Ridge*	*No-Till*
% Planted Acres	10%	49%	24%	3%	14%
Atrazine Product Acres/Corn Acres	0.558	0.678	0.664	0.777	0.829
Triazine Product Acres/Corn Acres	0.764	0.872	0.861	0.896	1.204

*Conv W is conventional tillage with use of moldboard plow
**Conv W/O is conventional tillage without use of moldboard plow

Table IV. Corn Acres Planted by Tillage System and Land Erodibility Classification

(Millions/Acres)

	Highly Erodible Land			*Non-Highly Erodible Land*		
	1989	*1994*	*Change*	*1989*	*1994*	*Change*
Conventional Tillage	7.06	5.24	-1.82	32.80	29.63	-3.17
Conservation Tillage	3.48	6.67	3.19	8.20	18.95	10.75
Percent Change in Conservation Tillage			91.7%			131.0%

Table V. Yield Comparisons by Tillage System

	Number Treatments	*Corn Yield (Bu/A)*	*% Change W/O Atrazine*
Conventional Tillage (1, 630 Tests)			
With Atrazine	11,042	117.96	
Without Atrazine	22,467	113.05	-4.3
Minimum-Till (100 Tests)			
With Atrazine	802	142.80	
Without Atrazine	995	135.33	-5.5
No-Till (425 Tests)			
With Atrazine	3,988	106.34	
Without Atrazine	3,249	95.48	-11.4

The difference in average farm income with and without a triazine herbicide cancellation gives an estimate of the costs of the cancellation to all farm commodities for both crop farmers and livestock producers. One of the largest costs of a triazine cancellation is an increase of approximately 12% in corn feed costs and a 14% to 18% increase in sorghum feed costs. These increases result in loss of net income of about $700 to $800 million per year to livestock producers. Consumers of meat, milk products, and other products made from corn and sorghum (ethanol, cereals, corn sweeteners, etc.) will also experience the economic effects of an atrazine or triazine cancellation as they will spend approximately $1.4 billion more for these products subsequent to a cancellation. However, taxpayer costs for maintaining the Farm Program and disaster payments will be reduced by approximately $600 million. Details of these simulation results are described fully in Ciba's initial response to EPA's Special Review (*4*).

Minor Crop and Off-Farm Effects

There are yield effect and weed control costs in other commodities besides corn and sorghum. Sweet corn and popcorn have similar weed pests but fewer herbicide alternatives available for use than field corn. Also, these crops are more valuable crops per acre, so losses would be considered large relative to field corn. Both yield effect and cost of production changes were computed for these two crops using the same agronomic and analytical approach as for field corn. Sweet corn losses are estimated to be $80.5 million for a triazine cancellation and $62.4 million for an atrazine cancellation (Table VI).

Herbicide cost changes were also estimated for 17 other crops. These costs are greatest for sugarcane, grapes and citrus growers. The large number of commodities in which simazine and atrazine are used over relatively large portions of the acreage grown makes the losses in these so called "minor use crops" substantial. The sum of the extra herbicide costs for these 17 commodities and that from sweet corn and popcorn result in total losses estimated to be $160.8 million for a triazine cancellation and $96.1 million for an atrazine cancellation (Table VI).

The final category of costs associated with an atrazine or triazine cancellation considered in this study are the off-farm costs associated with the reduction in no-till and ridge-till corn acreage planted. Economists have made various studies of the damage to freshwater recreation, water storage, navigation, flood control, and water treatment from soil sedimentation associated with agricultural practices. A recent summary of these studies by Smith (*8*) indicates that off-farm costs average 4.57% of gross crop value per acre. This results in an off-farm cost of $15.27 for each corn acre in conventional tillage (compared to no-till), and $5.34 for each corn acre moved

Table VI. Minimum Annual Differences in Treatment Costs if Atrazine or Simazine Was Not Available To Minor Acreage Crops in the United States

Crop	*Triazine Ban (\$000)*	*Atrazine Ban (\$000)*
Apples	1,692.6	0.0
Almonds	2,092.4	0.0
Cherries	1,245.1	0.0
Grapes	12,201.5	0.0
Pears	370.4	0.0
Peaches	1,008.9	0.0
Pecans	2,671.6	0.0
Walnuts	1,061.8	0.0
Macadamia/Guava	147.4	124.7
Sugarcane	21,038.3	21,038.3
Conifers	6,502.0	3,426.3
Nurseries	1,100. 4	0.0
Golf Courses	1,223.3	0.0
Prof. Lawn Care	8,272.0	1,510.6
Sod Farms	2,161.5	2,087.7
Citrus	11,763.2	0.0
Crop Fallow	1,575.2	1,575.2
Popcorn[1]	4,127.9	3,957.7
Sweet Corn[1]	80,519.9	62,414.2
Totals	160,775.4	96,134.7

[1]Includes Yield and Cost Effect

from ridge-till to conventional tillage. The total off-site costs for U.S. corn production for no-till and ridge-till systems are $155.4 million for an atrazine cancellation and $188.3 million for a cancellation of all triazines.

Overall Costs of Atrazine and Triazine Cancellations

The overall costs to farmers, consumers, and taxpayers of either an atrazine or triazine cancellation are extensive (see Table VII). The sum of corn and sorghum sector costs, off-farm costs, and costs to the minor use crops is a minimum of $1.66 billion for a triazine ban, and $1.47 billion for an atrazine ban each year. These figures include calculations of the savings to taxpayers from lower farm program payments based on the best estimate of the farm program provisions over the next five years. Note that the corn and sorghum sector costs are almost the same for an atrazine as for a triazine ban since cyanazine is not an effective alternative for atrazine in controlling many weeds in many locations.

Table VII. Minimum Annual Cancellation Costs (Average of with and Without 1990 Farm Programs)*

	Triazine Ban	*Atrazine Ban*
Corn and Sorghum Sector Costs	$1.3 Billion	$1.2 Billion
Off-Farm Sedimentation Costs	$188 Million	$155 Million
Costs for Minor Crops	$161 Million	$96 Million
Total Costs	$1.66 Billion	$1.47 Billion

*These costs will increase as the Farm Programs phase out.

The $1.47 billion annual loss for an atrazine cancellation is made up of losses to livestock producers, consumers, sedimentation costs and minor crop effects. The approximately $1.2 billion loss from yield and cost changes in the corn and sorghum sectors will fall heavily on livestock producers. The distribution of losses among the various livestock sectors is shown in Table VIII. The largest losses will be felt by the hog, dairy and fed cattle producers. Because of the yield losses in corn and sorghum from the proposed triazine herbicide cancellations, prices of these commodities will increase (approximately 12% in corn feed costs and 14% to 18% in sorghum feed costs) and this will reduce production of most livestock. For example, hog production is estimated to fall by about 1.8%. Although some livestock prices will increase slightly, this will not be sufficient to prevent the higher feed costs from reducing livestock farmer incomes as shown (*9*).

Table VIII. Effects of Atrazine or Triazine Cancellation on Income for Various Livestock Sectors (Free Market, 1996 to 2000, average) estimated from AGSIM in $Million per Year)

Livestock Sector	*Atrazine Cancellation*	*Triazine Cancellation*
	($ million per year)	
Cow/Calf	-92	-99
Fed Cattle	-168	-180
Hogs	-192	-208
Broilers	-103	-111
Turkeys	-26	-28
Eggs	-33	-35
Sheep	-2	-2
Dairy	-161	-173
Total	-777	-836

The costs of an atrazine or triazine cancellation are high primarily because of the resulting substantial yield loss in corn and sorghum combined with the higher costs for alternative herbicides. In addition, there are substantial costs to minor use crops, increased costs to local citizens faced with more soil sedimentation, and increased on-farm costs from added tillage in corn and from efforts to avoid herbicide drift associated with alternative herbicides.

These estimates depend upon comprehensive weed density, yield damage models, and extensive university trial data on relative herbicide efficacy. The AGSIM model allows the inclusion of costs to consumers, livestock producers, and taxpayers. This study shows the extensive data collection and analytical effort needed to carry out a credible assessment. The cancellation of one or a group of herbicides used on many crops and affecting a large part of the agricultural economy requires a detailed assessment of on-farm costs as well as effects beyond the farm gate. The impact to growers if atrazine and simazine were not available for use in production is a minimum of $1.66 billion annually.

Literature Cited

1. Taylor, C. R. 1993. "A Description of AGSIM, and Econometric-Simulation Model of Regional Crop and National Livestock Production in the United States" in C. R. Taylor, S. R. Johnson, and K. H. Reichelderfer, eds., Agricultural Sector Models for the United States: Description and Selected Policy Applications. Ames, Iowa, Iowa State University Press.
2. Bridges, D. (This volume). "A Simulation Analysis of the Use and Benefits of Triazine Herbicides."

3. Ciba Crop Protection. 1996. "Supplement to Ciba's Benefits Analysis of Atrazine and Simazine," Ciba Crop Protection, Greensboro, North Carolina.
4. Ciba Crop Protection. 1995. "Atrazine/Simazine Response to the United States Environmental Protection Agency's Position Document 1: Initiation of Special Review November 23, 1994, " Ciba Crop Protection, Greensboro, North Carolina.
5. Bull, Len, Herman Delvo, Carmen Sandretto, and Bill Linamood. 1993. "Analysis of Pesticide Use by Tillage System in 1990, 1991 and 1992 Corn and Soybeans." Agricultural Resources: Inputs Situation and Outlook Report, AR-32, U.S. Dept. Agr. Econom. Res. Serv.
6. Doane Ag Services. 1994. 1994 Farm Machinery Custom Rates Guide. Vol. 57, No. 19-5. Doane's Agricultural Report. St. Louis, MO.
7. Conservation Technology Information Center. 1989-1994. Executive Summary, National Crop Residue Management Survey. West Lafayette, IN.
8. Smith, V. K. 1992. Environmental Costing for Agriculture: Will it be Standard Fare in the Farm Bill of 2000? "American Journal of Agriculture Economics." Vol. 74: 1076-88.
9. Novartis Crop Protection. 1997. "Supplement to Novartis' Benefit Analysis of Atrazine and Simazine: Additional Information on Weed Control, Yield and Impact to Growers and Livestock Producers." p. 71-74. Greensboro, North Carolina.

Chapter 5

The Role of Best Management Practices in Reducing Triazine Runoff

R. S. Fawcett

Fawcett Consulting, 30500 Doe Circle, Huxley, IA 50124

Best Management Practices (BMPs) effective in reducing runoff of triazines into surface water include conservation tillage, buffers and vegetated filter strips, terraces, contour planting, postemergence application, and mechanical incorporation. Efficiency of BMPs has often depended on site conditions. Reductions in runoff of triazines with conservation tillage and filter strips occur primarily due to increases in water infiltration, rather than reductions in erosion. Soil type and structure, topography, and antecedent soil moisture have all influenced the efficiency of these BMPs. For example, no-till systems have completely eliminated triazine runoff in some studies, but have been ineffective in others. While individual BMPs are not universally appropriate, enough different practices are available to allow the design of effective runoff minimizing systems for any site.

Use of triazine herbicides (atrazine, cyanazine, and simazine) produces economic and environmental benefits. Triazines are especially important in conservation tillage systems which reduce soil erosion, improve water quality, and provide wildlife habitat. While atrazine was used on 68% of all U.S. corn in 1994, it was used on 84% of no-till corn (*1*). Over 70% of conservation plans required by the 1985 Food and Security Act mandate some form of conservation tillage. Adoption of practices which minimize runoff of triazines can allow continued benefits of the use of these products while avoiding standard-exceeding detections in surface water. Studies conducted over the past 30 years have documented the effectiveness of BMPs and factors influencing their efficacy. Adoption of effective BMPs through both voluntary and mandatory programs could reduce levels of triazines detected in surface water and remove a major concern about adverse impacts of these herbicides.

Point Source BMPs

Although nonpoint source runoff accounts for most pesticides detected in surface water, point sources can be important. Frank et al. (*2*) intensively monitored pesticides in 11 agricultural watersheds in Ontario, Canada. They determined that 22% of pesticide detects were due to carelessness associated with operating equipment adjacent to streams. In these cases sudden elevations of pesticide concentrations occurred, followed by sudden declines, independent of surface runoff events. Evidence in watersheds was found for the deposition of pesticides close to or directly into stream water during the process of drawing water, mixing pesticides, spraying, or cleaning equipment, and seepage from discarded containers in or around the spray site.

Point source losses of triazines into surface water can be reduced by avoiding handling and mixing of products near surface water unless containment systems are in place. If surface water sources are used for pesticide spraying, water should be pumped into a nurse tank and hauled to a safe mixing site away from surface water. Water-tight dikes and pads at pesticide storage and handling sites can contain spills or storm water containing pesticides. Most states now require such containment at commercial pesticide handling sites. Sprayer rinsing should either be confined to a water-tight pad so that rinse water can be disposed of properly, or conducted in the field so that rinsate is applied to labeled crop fields. Pesticide containers should be triple-rinsed or pressure-rinsed and recycled or disposed in approved landfills. The increased use of returnable bulk containers has greatly reduced disposal of pesticide containers. In Illinois in 1995, 70% of triazine-containing herbicides were sold in bulk. A similar trend to bulk handling has occurred in the rest of the Corn Belt.

Nonpoint Runoff BMPs

Pesticide loss in surface runoff with sediment and water carriers is determined by the volume of carriers and the concentrations in carriers. Triazines are moderately adsorbed to soil colloids (atrazine k_{oc} = 100 ; cyanazine k_{oc} = 190 ; simazine k_{oc} = 130). Although triazine concentrations are much higher in sediment, because so much more water leaves fields than sediment, water carries the majority of chemical leaving fields. Edge-of-field studies have shown that up to 90% or more of triazines lost in runoff are carried in the water phase (*3,4*). Once triazines are transported into streams, and equilibrium occurs, an even smaller fraction is found in suspended sediment (*5*). Practices which reduce erosion without affecting water runoff thus cannot be expected to produce large reductions in triazine runoff.

Application Timing. Herbicide runoff potential is greatest when heavy rains closely follow application. Most triazine runoff usually occurs with the first one or two runoff events (*2,6*). After initial rain events, the herbicide is moved into the soil, reducing interaction with overland flow and runoff with later events. If applications can be timed to avoid periods when heavy rains are likely or to coincide with periods when gentle rains are likely, runoff risk is reduced. For example, the Kansas Atrazine Management Plan allows higher atrazine rates when application is made before April 15. Long-term weather records and computer simulations indicate that runoff is less likely prior to that date due to the predominance of low intensity rainstorms in early spring, compared to more high intensity thunderstorms after April 15.

Postemergence application has reduced runoff of atrazine compared to soil application, due to the impact of crop and weed growth on water behavior. When planting dates were staggered to allow for identical dates for soil and postemergence atrazine applications, runoff was 70% less with postemergence application (*7*). Techniques which reduce application rates or reduce the area treated, reduce runoff risk. Applying atrazine to a band 50% of the row width resulted in a 69% reduction in runoff, compared to broadcast application (*8*).

Conservation Tillage. Conservation tillage systems leave all or part of the residue from the previous crop on the soil surface, protecting soil from the erosive impacts of rainfall. In some systems (referred to as reduced or mulch tillage) tools such as the chisel plow, disk, and field cultivator are used to perform tillage while leaving part of the crop residue on the surface. No-till systems leave the soil undisturbed prior to planting into a narrow seedbed. Reductions in soil erosion are correlated to percent of the soil surface covered by crop residue.

Conservation tillage affects pesticide runoff by reducing sediment loss and changing water behavior. Surface residue acts as small dams, slowing water runoff and giving more time for surface-applied herbicides to be carried into the soil. Often, total water infiltration is increased. However, studies comparing herbicide runoff with conservation tillage to runoff with conventional tillage (usually employing the moldboard plow and burying most crop residue) have produced variable results. Many studies comparing herbicide runoff under various tillage systems have utilized rainfall simulation techniques on small plots. Almost always very heavy rainstorm events (such as once-in-50-year or once-in-100-year events) are simulated within a day of herbicide application. Under these conditions herbicides are washed from surface crop residue present with conservation tillage, and may become a part of overland flow before infiltrating into the soil. Higher concentrations of herbicide in runoff may offset lower runoff volumes so that total herbicide runoff is sometimes similar or greater with conservation tillage than with conventional tillage. When published rainfall simulation study data through 1991 were summarized (*9*), all conservation tillage systems reduced runoff of pesticides by an average of 23%, compared to conventional tillage (99 treatment-site-years of data). Considering only no-till studies, pesticide runoff was reduced by an average 34% (29 treatment-site-years of data). Recent studies have produced similar results. In a Texas rainfall simulation study (10) no-till reduced atrazine runoff by 42%, compared to chisel plowing.

Conservation tillage systems have usually been shown to have greater benefit in reducing herbicide runoff in natural rainfall studies. Under natural rainfall conditions, usually small rains occur first after herbicide application, washing herbicides off crop residue and into the soil, before heavier, runoff-producing rains occur. Natural rainfall studies are also more likely to be conducted on watersheds with more than a one year history of tillage treatment, unlike simulation studies which are often short-term. Much of the benefit of conservation tillage in reducing herbicide runoff is due to greater water infiltration, and water infiltration benefits of conservation tillage are more likely to occur over several years. For example, in a Maryland study (*11*), atrazine runoff was reduced by 29% by no-till in the first year of the study, but by the third and fourth year runoff was reduced by 100% due to elimination of any water runoff. Because small plot rainfall simulation studies will likely be used to verify effectiveness of BMPs, such as conservation tillage, experimental conditions will need to be carefully selected to avoid misleading results (*12*).

No-till has sometimes dramatically increased water infiltration in fields, reducing surface runoff and the potential to carry contaminants into surface water. Edwards et al. (*13*) compared total water runoff from a 0.5 ha watershed with 9% slope that had been farmed for 20 years in continuous no-till corn to a similar, conventionally tilled watershed. Over four years, runoff was 99% less under the long-term no-till. This decrease in runoff was attributed to increases in infiltration with no-till due to the development of soil macropores in the absence of tillage. Cracks, root channels, and worm holes allow water to bypass upper soil layers when rainfall exceeds the capillary flow infiltration capacity of the soil. Earthworm burrows are especially important in this phenomenon. Nightcrawlers (*Lumbricus terrestris* L.) construct permanent, vertical burrows. Edwards et al. (*14*) found that although earthworm holes greater than 5 mm accounted for only 0.3% of the horizontal area of a no-tilled soil, flow into the holes during 12 rainfall events accounted for from 1.2 to 10.3% of the rainfall from each storm. The tops of these burrows and other macropores are destroyed by tillage.

When all published natural rainfall studies through 1991 were summarized, no-till reduced soil erosion, water runoff, and herbicide runoff by an average 93%, 69%, and 70%, respectively, compared to conventional tillage (*9*). No-till reduced herbicide runoff in 29 out of 32 cases.

The effectiveness of no-till in reducing triazine runoff will depend on local conditions and management, and the length of time fields have been in no-till. In Missouri (*15*), atrazine and cyanazine runoff were compared in watersheds with either a long-term history of disking or no-till. No water or herbicide runoff occurred from the no-till watersheds with the first two rain events producing runoff from tilled treatments. A small amount of runoff occurred from all treatments with a third rainfall event. Totaled over 3 events, no-till reduced total herbicide runoff by 94% and 91% for cyanazine and atrazine, respectively. Total water runoff volume was reduced by 72%.

Infiltration benefits achieved through use of no-till may not be entirely eliminated with shallow tillage. In Iowa (*15*) some watersheds in a long-term no-till field were tilled once with a shallow tillage tool, while others were planted no-till. Total water runoff and atrazine and cyanazine runoff were similar for both tillage systems. The one tillage pass apparently did not destroy the infiltration benefit produced by long-term no-till.

Reduced tillage systems have also reduced herbicide runoff, although often to a lesser extent than long-term no-till. Roughness created by tillage can cause greater ponding of runoff with the first rain events after tillage, reducing runoff until roughness is lost and the soil surface seals. Witt and Sander (*16*) compared moldboard plowing, chisel plowing, and no-till in a 2-year Kentucky study. In a year when 2.5 cm rainfall occurred within 24 hr. of herbicide application, chisel plowing was more effective than no-till in reducing herbicide runoff, because water runoff was reduced most by this treatment. In the following year, when the first rainfall was 1.3 cm, 7 days following treatment, both no-till and chisel plowing reduced atrazine, cyanazine, and simazine losses by more than 90%, compared to moldboard plowing.

Conservation tillage may not increase water infiltration or reduce herbicide runoff if factors such as a high water table, restricting soil layer, or compaction prevent infiltration. Runoff under no-till and moldboard plowing were compared on a soil with a clay layer 1.5 m deep which restricted water movement (*17*). Runoff of water and herbicides was greater from no-till than moldboard plowing when time between rainfall events was less than 7 days, but runoff from moldboard plowing was greater than no-till when 7 or more days passed between rains. When conventional, no-till, and ridge-till were compared on a clay loam soil with < 1% slope in Canada (*18*), tillage had no significant effect on runoff volume, distribution between surface and subsurface runoff, herbicide concentration, or herbicide loss. The appropriateness of conservation tillage as a surface water BMP will depend on local soil conditions. These systems may not be effective or appropriate where water infiltration is greatly limited due to conditions such as claypans.

Drainage Improvement and Compaction Reduction. Improvement of internal soil drainage by installation of drainage tile or other practices can increase water infiltration and reduce surface runoff and pesticide loss. Herbicide losses from high water table fields in Louisiana were compared in fields which were either subsurfaced drained or surface drained only (*19*). Subsurface drainage reduced atrazine losses by 55%. Most of the loss occurred with surface runoff. Compaction can reduce infiltration and increase herbicide runoff. Baker and Laflen (*20*) compared runoff of atrazine, propachlor, and alachlor from plots with and without tractor wheel tracks in a rainfall simulation study. Herbicide runoff was 3.7 times greater from compacted plots. If wheel-track compaction can be avoided through controlled traffic patterns in fields or compaction alleviated through appropriate tillage, herbicide runoff should be reduced.

Mechanical Incorporation. Pesticides are most subject to runoff when they are near the soil surface and can interact with overland flow. After the first few rains following soil surface application, runoff losses of most pesticides decline dramatically as the pesticide is moved below the soil surface by infiltrating water. Mechanically mixing the soil-applied pesticide into the soil or otherwise placing it below the soil surface can reduce runoff. Using rainfall simulation techniques (*20*) incorporation of herbicides with a disk was shown to reduce runoff of atrazine, alachlor, and propachlor by 64, 76, and 76%, respectively, compared to surface application. In a Pennsylvania natural rainfall study of small watersheds with 14% slope, incorporation reduced runoff of atrazine by an average of 49% (*21*).

In two years in an Iowa study conducted in a tile outlet terrace (Baker, J. L., Iowa State University, unpublished data), incorporation reduced runoff of atrazine, metolachlor, and cyanazine, by an average 37% in one year and 36% in another year. In both of these years, over 10 cm of rainfall occurred (in several small events) between herbicide application and the first runoff event. This rainfall resulted in natural incorporation of surface-applied herbicide, reducing potential differences in runoff between incorporation and surface application.

A potential disadvantage of herbicide incorporation is that tillage performed to incorporate the herbicide buries crop residue and increases erosion risk. Conservation plans for many farms with highly erodible land as required by the 1985 Food and Security Act describe the use of various conservation tillage systems and often require from 30 to 70% surface crop residue coverage. Incorporation of herbicides will not be compatible with surface crop residue requirements in some fields. Incorporation tools which leave more crop residue on the soil surface are being developed.

Contour Farming and Terraces. Planting crop rows on the contour (parallel to slopes) rather than up and down hills reduces soil erosion, as rows act as small dams. Because this technique slows and reduces water runoff, it has also reduced pesticide runoff. In an Illinois rainfall simulation study (*4*), contouring plots with 7 to 11% slope reduced water runoff, sediment loss, and total loss of alachlor by 45, 89, and 61%, respectively. When plots were no-till planted, contouring produced herbicide reduction benefits in addition to reductions in runoff due to no-till.

Terraces are constructed to shorten slope lengths, stopping runoff water. Level terraces are constructed where soils are very permeable, and all water infiltrates. Graded terraces direct runoff either into a grassed waterway or into a riser pipe which carries water into an underground drainage tile. These tiles eventually exit into a stream or a grassed waterway.

Because terraces are effective in reducing erosion and cause at least some additional water to infiltrate, they probably reduce herbicide runoff to some extent, although studies to document this benefit are lacking. There is concern that tile outlet terraces could increase surface water contamination, as runoff water is carried from the terrace channel by tiles, often directly to streams. This reduces the chance for interaction of runoff with soil and vegetation which might reduce pesticide concentrations. Alternative BMPs for tile-outlet terraces have been studied.

Untreated setbacks (20 m radius) around tile risers have not reduced herbicide concentrations in water entering risers (*15*) beyond what would be expected due to less treated area (typically 10-15% of the area draining into a riser would be covered by the untreated setback). Because terraces are designed to pond runoff for up to 48 hr, much of the area in setbacks is underwater, preventing the area from behaving like a filter strip. Use of mechanical incorporation or no-till in areas draining into tile risers have both reduced atrazine and cyanazine concentrations in runoff to a greater extent than use of setbacks (*15*). Based on this research, atrazine and cyanazine labeling has been amended to allow either incorporation or no-till as an alternative to setbacks around risers in tile-outlet terraces.

Vegetated Filter Strips and Buffers. The terms filter strip and buffer strip are often used interchangeably to denote an area or strip of land (usually planted to perennial grasses or other vegetation) along the perimeter of land or water used to reduce movement of sediment or other pollutants from fields in runoff. While the ability of filter strips to trap sediment has been well documented (*22*), until recently few studies have investigated the impact of filter strips on pesticide runoff.

Attempts to use computer models to predict the effectiveness of filter strips in removal of pesticides have incorrectly assumed that filter strips only remove sediment, without affecting water infiltration (*23*). Thus, these models predict that removal of moderately adsorbed pesticides such as triazines would be minimal. However, recent studies have shown that filter strips have significant impacts in increasing water infiltration, trapping dissolved pesticides within the strips. This phenomenon explains why controlled field studies have shown reductions in runoff of herbicides such as atrazine by filter strips despite the fact that sediment-bound herbicide accounts for only a small percentage of herbicide contained in runoff.

A Pennsylvania study (*21*) used a 6 m-long area seeded to oats at the base of 22 m-long plots planted to corn and treated with atrazine. Season-long runoff of atrazine with natural rainfall was reduced by 91 and 65% by the oats strip at application rates of 2.1 and 4.5 kg/ha, respectively. Runoff losses of metolachlor and metribuzin were reduced by 50 to 75% by a grass filter strip in a Mississippi study (*24*). The filter strip was 2 m wide, and the plots were 23 m long. Much of the reduction in herbicide runoff was attributed to greater water infiltration into the grass strip.

The effectiveness of bermudagrass and wheat filter strips were studied in Texas (Hoffman, D. R., Texas A & M University, unpublished data). Three 9 m-wide strips of either bermudagrass on winter wheat were established 0, 43, and 88 m uphill from the base of the slope within a 133 m long watershed planted to corn and compared to similar watersheds with no filter strips. Total water runoff was reduced by 57% by bermudagrass and by 50% by wheat strips. Total atrazine runoff in 3 events was reduced by 30% by bermudagrass and 57% by wheat strips.

Field runoff was simulated by adding known concentrations of atrazine to water, based on previous measurements of actual field runoff (*25*). Runoff calculated to simulate runoff from an area 45 m long was applied to the top of 4.5 m- and 9.0 m-long grass filter strips. Thus ratios of treated area to filter strip were 10:1 and 5:1, respectively. A rainfall simulator was used to apply rainfall to the filter strip as simulated runoff were added. The 10:1 filter strip reduced atrazine runoff by 35%, while the 5:1 filter strip reduced runoff by 59.5%. Using similar techniques, runoff with concentrations of either 0.1 or 1.0 mg/L atrazine was applied to filter strips in amounts calculated to represent relative drainage area to filter strip areas of 15:1 and 30:1. Atrazine removal by the 15:1 ratio strip was 31.2% for 0.1 mg/L inflow and 49.8% for 1.0 mg/L inflow. Removal was due both to infiltration of water and to herbicide adsorption. An average 38% runoff water infiltrated into filter strips with the 15:1 ratio, while 32% of water infiltrated into strips with 30:1 ratio.

Using similar techniques (Baker, J. L., Iowa State University, unpublished data), vegetated filter strips were compared to bare ground strips. Herbicides were applied in simulated runoff at the concentration of 1.0 mg/L. Soil was added to some simulated runoff at a concentration of 10,000 mg/L to represent eroded sediment. Simulated runoff was added to filter strips in amounts calculated to represent a 15:1 drainage area to filter strip ratio. Inclusion of sediment reduced both water infiltration and herbicide retention. In absence of sediment, vegetated strips removed 85.2% of atrazine, 82.6% of metolachlor, and 84.1% of cyanazine. With sediment, vegetated strips removed 53.6% of atrazine, 53.3% of metolachlor, and 57.5% of cyanazine. In absence of sediment, bare strips removed 50.7% of atrazine, 45.2% of metolachlor, and 50.1% of cyanazine. With sediment, bare strips removed 33.5% of atrazine, 27.5% of metolachlor, and 35.6% of cyanazine.

Efficiency of grass filter strips under natural rainfall conditions in Iowa (*26*) was studied by collecting runoff from a herbicide-treated cornfield and distributing runoff to 20 m grass strips in amounts equaling drainage area to filter area ratios of 15:1 and 30:1. Filter strip efficiency depended highly on antecedent soil moisture conditions. In one year, results for the first runoff event after herbicide application showed that atrazine removal was 14.1% for the 15:1 area ratio. Wet antecedent soil conditions prevented significant water infiltration into strips, reducing their effectiveness (13% of water runoff infiltrated into the strip). In contrast, later runoff events over 2 years produced atrazine removal rates of from 37.5% to 100%. Reductions in atrazine runoff were highly correlated to water infiltration into filter strips. Analysis of soil within the filter strips confirmed that herbicides were being trapped and held within the strips. Herbicide concentrations declined with time, presumably due to degradation. Perennial grasses (primarily smooth bromegrass) were not adversely affected by herbicides.

Summary.

Numerous BMPs have been shown to be effective in reducing runoff of triazine herbicides. The efficacy of many BMPs is dependent on local conditions and weather. For example, much of the benefit of conservation tillage and filter strips accrues due to increases in water infiltration. If soils are impermeable due to problems such as claypans, or if previous rains have filled the soil profile with water, these BMPs will not be highly effective. BMPs will need to be matched to local conditions. For example mechanical incorporation may be more appropriate than conservation tillage for claypan or high-water-table soils. As many of these soils are not highly erodible, tillage required for incorporation would not violate conservation plans. Conservation tillage may be more appropriate for sloping, erodible fields. Postemergence application or banding could be appropriate for any field.

Because many BMPs also reduce soil erosion and surface water contamination from sediment and nutrients, reducing triazine runoff is compatible with soil conservation and nutrient management goals. Watersheds will need to be evaluated for vulnerability to herbicide runoff and appropriateness of BMPs. Adoption of BMPs in watersheds has reduced atrazine concentrations in surface water to levels below drinking water standards (*27*). Voluntary and incentive programs to adopt BMPs should be able to produce similar benefits in other watersheds. If triazine concentrations remain at levels of concern despite application of BMPs in certain highly vulnerable watersheds, further restrictions may be necessary.

Literature Cited

(1) United States Department of Agriculture. 1995, Publication Ag Ch 1 (95).

(2) Frank, R.; Brown, H. E.; Van Hove Holdrinet, M.; Sirens, G. J.; Ripley, B. D. *J. Environ. Qual.* **1982**, *11*, 497-505.

(3) Baker, J. L.; Johnson, H. P. *Trans. ASAE.* **1979**, *22*, 554-559.

(4) Felsot, A. S.; Mithcell, J. K.; Kenimer, A. L. *J. Environ. Qual.* **1990**, *19*, 539-545.

(5) Pereira, W. E.; Rostad, C. E. *Environ. Sci. Technol.* **1990**, *24*, 1400-1406.

(6) Seta, A. K.; Blevins, R. L.; Frye, W. W.; Barfield, B. J. *J. Environ. Qual.* **1993**, *22*, 661-665.

(7) Pantone, D. J.; Young, R. A.; Buhler, D. D.; Eberlein, C. V.; Koskinen, W. C.; Forcella, F. *J. Environ. Qual.* **1992**, *21*, 567-573.

(8) Gaynor, J. D.; Van Wesenbeeck, I. J. *Weed Tech.* **1995**, *9*, 107-112.

(9) Fawcett, R. S.; Christensen, B. R.; Tierney, D. P. *J. Soil and Water Cons.* **1994**, *49*, 126-135.

(10) Pantone, D. J.; Potter, K. N.; Torbert, H. A.; Morrison, J. E. *J. Environ. Qual.* **1996**, *25*, 572-577.

(11) Glenn, S.; Angle, J. S. *Agric. Ecosytems and Environ.* **1987**, *18*, 273-280.

(12) Basta, N. T.; Huhnke, R. L.; Stiegler, J. N. *J. Soil and Water Cons.* **1996**, *52*, 44-48.

(13) Edwards, W. M.; Norton, L. D.; Redmond, C. E. *Soil Sci. Soc. Am. J.* **1988**, *52*, 483-487.

(14) Edwards, W. M.; Shipitalo, M. J.; Owens, L. B.; Norton, L. D. *J. Soil and Water Cons.* **1989**, *44*, 240-243.

(15) Franti, T. G.; Peter, C. J.; Tierney, D. P.; Fawcett, R. S.; Myers, S. A. *Agric. Eng. Soc. Am.* **1995**, Paper 952713.

(16) Witt, W. W.; Sander, K. W. *Univ. of Ky. Soil Sci. News and Views.* **1990**, *11*, 1-5.

(17) Isensee, A. R.; Sadeghi, A. M. *J. Soil and Water Cons.* **1993**, *48*, 523-527.

(18) Gaynor, J. D.; MacTavish, D. C.; Findlay, W. I. *Arch. Environ. Contam. Toxicol.* **1992**, *23*, 240-245.

(19) Bengtson, R. L.; Southwick, L. M.; Willis, G. H.; Carter, C. E. *Am. Soc. Agr. Eng.* **1989**, Paper 89-2130.

(20) Baker, J. L.; Laflen, J. M. *J. Environ. Qual.* **1979**, *8*, 602-607.

(21) Hall, J. K.; Hartwig, N. L.; Hoffman, L. D. *J. Environ. Qual.* **1983**, 336-340.

(22) Magette, W. L.; Brinsfield, R. B.; Palmer, R. E.; Wood, J. D. *Trans. ASAE.* **1989**, *32*, 663-667.

(23) Flanagan, D. C.; Neibling, W. H.; Foster, G. R.; Burt, J. P. *Am. Soc. Agri. Eng.* **1986**, Paper 86-2034.

(24) Webster, E. P.; Shaw, D. R.; Holloway, J. C., Jr. *Weed Sci. Soc. Am. Abs.* **1993**, *33*, 79.
(25) Mickelson, S. K.; Baker, J. L. Amer. Soc. Agri. Eng. 1993, Paper 932084.
(26) Misra, A. K.; Baker, J. L.; Mickelson, S. K.; Shang, H. *Amer. Soc. Agri. Eng.* **1994**, Paper 942146.
(27) Fawcett, R. S. *Proc. Brighton Crop Prot. Conf.* **1993**, *3*, 1105-1113.

Metabolism of Triazines

Chapter 6

The Metabolism of Atrazine and Related 2-Chloro-4,6-bis(alkylamino)-*s*-triazines in Plants

Gerald L. Lamoureux[1], Bruce Simoneaux[2], and John Larson[3,4]

[1]**Biosciences Research Laboratory, ARS, USDA, P.O. Box 5674, State University Station, Fargo, ND 58105–5674**
[2]**Ciba-Geigy Corporation, Greensboro, NC 27419**
[3]**Corning-Hazelton, Madison, WI 53704**

The metabolism of atrazine and related 2-chloro-4,6-bis(alkylamino)-*s*-triazine herbicides in plants is reviewed and the structures of 13 metabolites recently identified from mature plants grown in the field are reported. The 2-chloro-4,6-bis(alkylamino)-*s*-triazines are initially metabolized in plants by three competing reactions: hydrolytic dehalogenation, *N*-dealkylation, and glutathione (GSH) conjugation. Metabolites produced by *N*-dealkylation can be further metabolized by hydrolytic dehalogenation or GSH conjugation, those produced by hydrolytic dehalogenation can be further metabolized by *N*-dealkylation and it is proposed that those from the GSH conjugation pathway may slowly become hydroxylated at the 2-position of the triazine ring. Ten metabolites of atrazine have been identified from the *N*-dealkylation and hydroxylation pathways and 14 have been identified from the GSH conjugation pathway. Three additional metabolites that have an amino function on the 2-position of the triazine ring have been identified, but their route of formation is uncertain.

2-Chloro-4,6-bis(alkylamino)-*s*-triazine (2-chloro-*s*-triazine) metabolism in plants has been extensively studied and comprehensive reviews have been written on the chemical and physical properties, metabolism and mode of action of these compounds (*1*) and on their metabolism and selectivity in plants (*2*). Based upon numerous studies conducted since the first patent was filed for these herbicides in 1954, the 2-chloro-*s*-triazines such as atrazine (2-chloro-4-ethylamino-6-isopropylamino-*s*-triazine), simazine (2-chloro-4,6-bis[ethylamino]-*s*-triazine), and propazine (2-chloro-4,6-bis[isopropylamino]-*s*-triazine) have been shown to be metabolized in plants by three competing initial reactions: *N*-dealkylation of the side-

[4]**Current Address: ABC Laboratories California, Madera, CA 93638**

chains, hydrolytic dehalogenation, and nucleophilic displacement of the 2-chloro group with glutathione (GSH) (*1,2,3*) (Figure 1). These pathways are interactive, i.e., products from the *N*-dealkylation pathway can enter the GSH or the hydrolytic dehalogenation pathways, products from the hydrolytic dehalogenation pathway can enter the *N*-dealkylation pathway, and it is thought that products from the GSH pathway can become hydroxylated at the 2-position of the *s*-triazine ring (*1,2,3*).

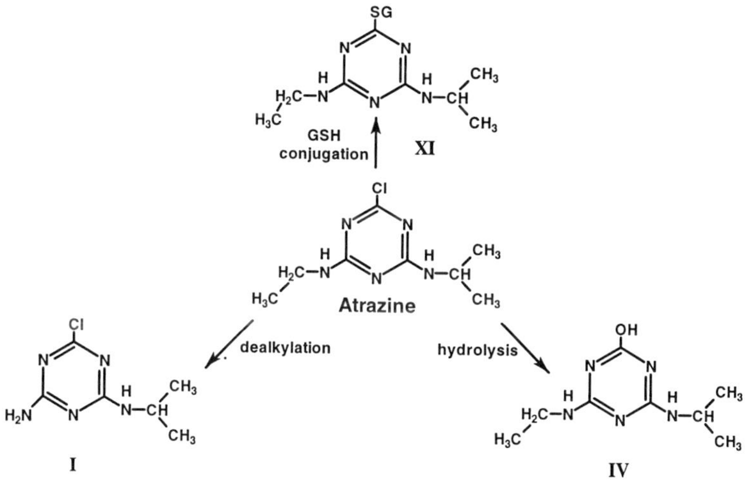

SG
N N
H H CH3
H2C—N N N—CH
H3C CH3
GSH
conjugation XI
Cl
N N
H H CH3
H2C—N N N—CH
H3C Atrazine CH3
Cl OH
N N dealkylation hydrolysis N N
H CH3 H H CH3
H2N N N—CH H2C—N N N—CH
CH3 H3C CH3
I IV

Figure 1. Competing initial reactions in 2-chloro-*s*-triazine metabolism in plants.

Although these metabolic pathways were discussed in earlier reviews (*1,2,4*), the recent identification of many additional metabolites of atrazine and simazine in more mature plants treated under field conditions made it desirable to publish an updated review on 2-chloro-*s*-triazine metabolism in plants. Many of the recently identified metabolites are clearly produced from the GSH pathway, but the origin of several of these metabolites is less clear. They may have been produced either from the GSH pathway or from a fourth competing pathway.

Metabolism of the *N*-Alkyl Side-Chains

Much of the research on the metabolism of the *N*-alkyl side-chains of the *s*-triazines was conducted from 1960-1978 (*1,2,4*). The 2-chloro-*s*-triazines such as atrazine and simazine are metabolized partially by *N*-dealkylation in many plants including corn, sorghum, cotton, soybean, wheat, sugarcane, pea, citrus, black walnut, yellow poplar, and Canada Thistle (*Crisium arvense* L.) (*1-7*). In pea and sorghum, *N*-dealkylation of atrazine occurs more readily with the ethyl side-chain than with the isopropyl side-chain and the concentration of desethyl atrazine (I) can be twice the concentration of desisopropyl atrazine (II) or atrazine (*2*) (Figure 2). *N*-Dealkylation of one side-chain results in partial loss of phytotoxicity while *N*-dealkylation of both side-chains results in nearly complete loss of phytotoxicity (*2,8*). *N*-Dealkylation proceeds primarily to mono-*N*-dealkylated metabolites in some moderately susceptible plant species while *N*-dealkylation of both side-chains becomes more important in some more tolerant species. *N*-Dealkylation is the primary route of simazine metabolism

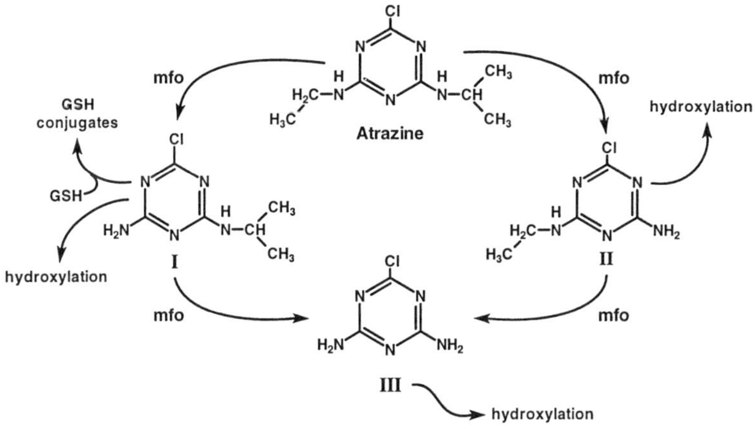

Figure 2. *N*-Dealkylation in 2-chloro-*s*-triazine metabolism in plants.

in *Lolium rigidum*, but large differences in *N*-dealkylation exist between a normal susceptible biotype and a biotype that evolved tolerance as a result of repeated herbicide exposure (*9*). Mono-*N*-dealkylated simazine accounted for 20% of the residue in a susceptible *L. rigidum* biotype versus 34% in a tolerant biotype after 12 hr of exposure to simazine. After 48 hr of exposure, di-*N*-dealkylated simazine accounted for 6.7% of the residue in the susceptible biotype versus 41% in the tolerant biotype (*9*).

The enzymes responsible for *N*-dealkylation of the *s*-triazine herbicides have not been extensively studied, but they appear to be cytochrome P450 monooxygenases similar to those that catalyze the *N*-dealkylation of the methylurea herbicides (*10*). Aminobenzotriazole, an inhibitor of cytochrome P450 monooxygenase activity, inhibits the *N*-dealkylation of simazine in a tolerant biotype of *Lolium rigidum* and also synergizes simazine toxicity (*9*). Plants, animals, and microorganisms all appear to be capable of metabolizing the 2-chloro-4,6-bis(alkylamino)-*s*-triazines by *N*-dealkylation (*2*).

Both mono- and di- *N*-dealkylated *s*-triazine metabolites can be further metabolized by hydroxylation at the 2-position of the triazine ring and some mono-*N*-dealkylated products can also be further metabolized by displacement of the chlorine with GSH as observed in the metabolism of atrazine in sorghum (*1,3*). Thus, products from the *N*-dealkylation pathway can enter either of the other two major pathways of *s*-triazine metabolism in plants (Figure 2).

Cyanazine (2-chloro-4-ethylamino-6-[2-methylpropanenitrile]-*s*-triazine) contains a nitrile group in one of the alkylamino side-chains and its metabolism may be somewhat atypical of the 2-chloro-*s*-triazine herbicides. It is metabolized by *N*-dealkylation of the ethylamino side-chain followed by metabolism of the nitrile group to a carboxylic acid group (*11-13*). 2-Methoxy- and 2-methylthio-*s*-triazine herbicides such as prometone and pyrometryn can also be metabolized by *N*-dealkylation while those with more complex *N*-alkyl side chains can be metabolized by hydroxylation of the methyl or methylene groups (*1,14*).

Hydrolytic Dehalogenation

Hydrolytic dehalogenation was one of the first reactions to be reported in the metabolism of the 2-chloro-*s*-triazines in plants (*1,15*) (Figure 1). It results in the loss of phytotoxicity and it is regarded as a detoxification mechanism (*2,16*). Hydrolytic dehalogenation is observed during 2-chloro-*s*-triazine metabolism in highly tolerant corn and *Coix Lacryma-jobi* and in susceptible wheat and rye, but it is unimportant in other plants such as tolerant sorghum and intermediately susceptible oat, barley and soybean (2). Hydrolytic dehalogenation has also been reported in the metabolism of the 2-chloro-*s*-triazines in wild cane (*sorghum bicolor*), spruce, potato, yellow poplar (*Liriodendron tulipifera* L.) and various *Setaria* and *Panicum* species (*1,2,4,5,12,17-20*).

In plants, the hydrolytic dehalogenation of the 2-chloro-*s*-triazines is catalyzed by 2,4-dihydroxy-7-methoxy-1,4(2*H*)benzoxazin-3-*H*(4*H*)-one, structurally related compounds (benzoxazinones), and their 2-glucosides (*1,2,4,21,22*). The rate of

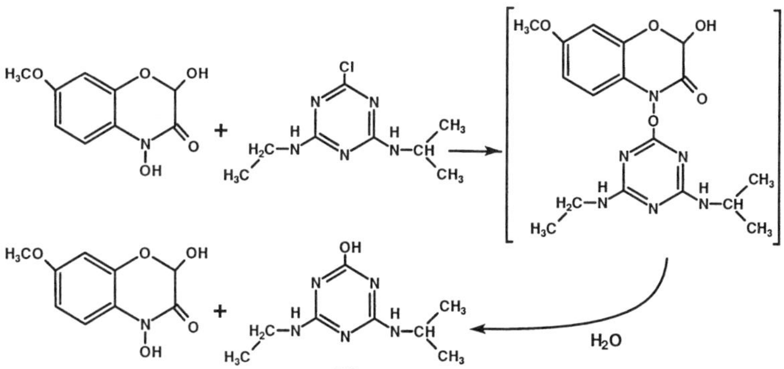

Figure 3. Benzoxazinones in the catalysis of hydrolytic dehalogation of 2-chloro-*s*-triazines in plants.

hydrolysis is positively correlated to benzoxazinone levels. The catalytic mechanism appears to involve an ether intermediate that decomposes to the hydroxylated triazine and benzoxazinone (*1,2*) (Figure 3). In some *Pseudomonas* species of bacteria, hydrolytic dehalogenation of the 2-chloro-*s*-triazines is catalyzed by an enzyme, but a comparable enzymatic reaction does not appear to have been established in plants (*23*).

The importance of hydrolytic dehalogenation to 2-chloro-*s*-triazine metabolism in corn depends upon the route of entry of the herbicide into the plant (*2*). Upon entry through the roots, from 38% to 65% of the herbicide (atrazine) is metabolized by hydrolytic dehalogenation while 10% to 25% is metabolized by GSH conjugation. In contrast, upon entry through the leaves only 10 to 25% is metabolized by hydrolytic dehalogenation and 65% to 85% is metabolized by GSH conjugation (*24,25*). Following either route of entry, mono-*N*-dealkylated 2-chloro-*s*-triazines generally account for less than 10% of the total radioactive residue (TRR) in corn, but *N*-dealkylated products from the GSH pathway and the hydrolytic dehalogenation pathway may also be present. The importance of the route of entry to metabolism is related to three factors: (a) the GST enzymes appear to be capable of a higher rate of catalysis than the benzoxazinones, (b) the GSTs required for 2-chloro-*s*-triazine metabolism are present only in the foliar tissue, and (c) the benzoxazinones are present in both the foliar and root tissues (*2,26*).

Products from the hydrolytic dehalogenation pathway include the 2-hydroxy-4,6-bis(alkylamino)-*s*-triazines (IV) as well as mono- and di- *N*-dealkylated 2-hydroxy-*s*-triazines (V, VI, and ammeline) (Figure 4). The concentration of these metabolites varies widely depending upon the species, tissue and length of treatment. In corn, the 2-hydroxy-*s*-triazine metabolites typically account for 25-60% of the herbicide residue (*2,25*), but in species such as sorghum and sugarcane where hydroxylation is less important, hydroxylated metabolites account for about 10% of the residue (*3,25*). In the early stages of metabolism in corn the simple 2-hydroxy-4,6-bis(alkylamino)-*s*-triazines predominate, accounting for 32% of the residue (*2*), but during the latter stages of metabolism the most abundant hydroxylated metabolites are the 2-hydroxy-4-amino-6-alkylamino-*s*-triazines (*25*). The most abundant residue in corn 30 days after treatment with atrazine was 2-hydroxy-4-amino-6-isopropylamino-*s*-triazine (17%) and after treatment with simazine it was 2-hydroxy-4-amino-6-ethylamino-*s*-triazine (28%) (*25*). The 2-hydroxy-4,6-bis(alkylamino)-*s*-triazine metabolites accounted for only 1-2% of the residue and ammeline for only 1 to 5% of the residue 30 days after treatment (*25*).

Both mono-*N*-dealkylated 2-hydroxy-*s*-triazine metabolites were formed when sorghum was treated with hydroxyatrazine and both mono-*N*-dealkylated 2-hydroxy-*s*-triazine metabolites and hydroxyatrazine were also formed when protein-free extracts of corn were treated with atrazine and the two 2-chloro-mono-*N*-dealkylated metabolites of atrazine; therefore, *N*-dealkylated 2-hydroxy-*s*-triazine metabolites can be formed either by hydrolytic dehalogenation followed by *N*-dealkylation or by *N*-dealkylation followed by hydrolytic dehalogenation (*2*) (Figure 4). Small amounts of 2-hydroxy-4,6-bis(amino)-*s*-triazine (ammeline) are produced in *Coix lacryma-jobi* and corn treated with atrazine, but signficant amounts of this metabolite were not detected 28 days after treating corn with simazine under conditions in which 60% of the extractable residue was 2-hydroxy-4,6-bis(ethylamino)-*s*-triazine (*2*). Ammeline appears to be formed by the slow *N*-dealkylation of the 2-hydroxy-4-amino-6-

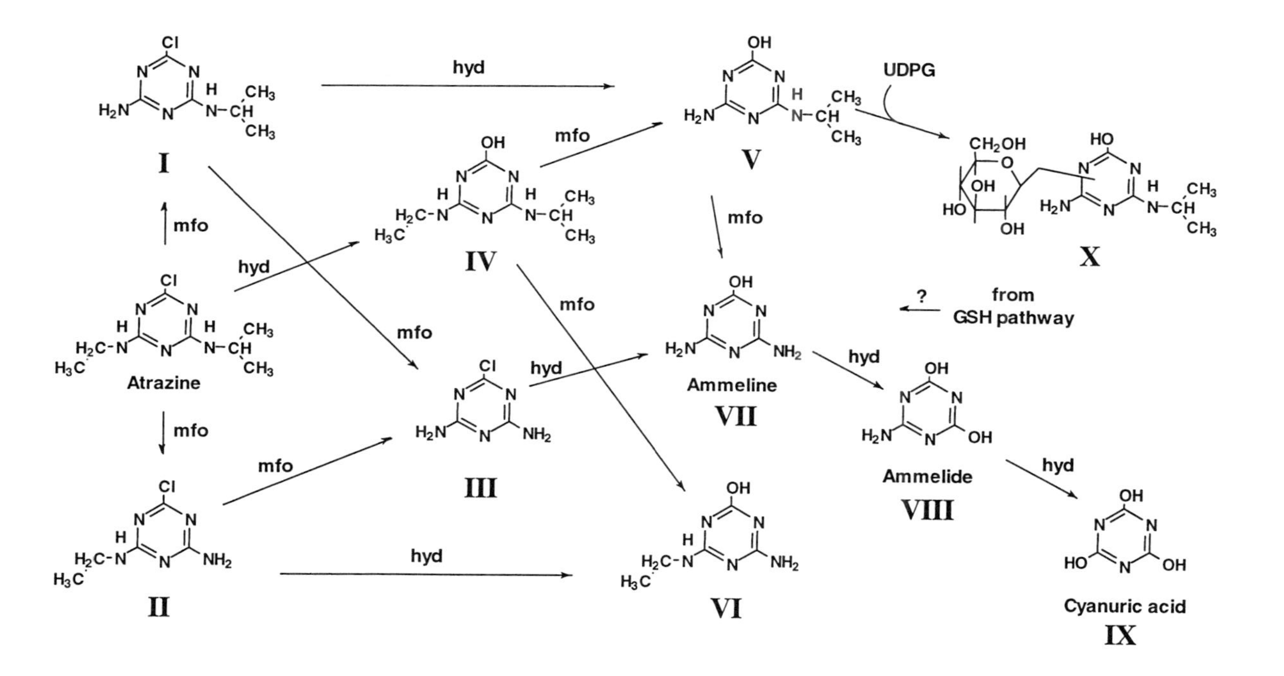

Figure 4. Interaction of the hydrolytic dehalogation and *N*-dealkylation pathways in 2-chloro-*s*-triazine metabolism in plants.

alkylamino-*s*-triazines and perhaps also by the hydroxylation of 2-chloro-4,6-bis-(amino)-*s*-triazine (Figure 4). Ammeline and 2-hydroxy-4-amino-6-isopropylamino-*s*-triazine were also observed in atrazine metabolism in sorghum, a species in which hydrolytic dehalogenation was not thought to occur at a significant rate (*2,3*). In sorghum, it is not clear whether the 2-hydroxy-*s*-triazines are formed by hydrolytic dehalogenation or by metabolism of GSH conjugates as is thought to occur during propachlor metabolism in corn (*27*).

Ammelide and cyanuric acid were reported as minor residues in several plant species following extended exposure to 2-chloro-*s*-triazine herbicides or their hydroxylated metabolites (*1,28*), but they were not reported as metabolites in a 28-day metabolism study of simazine in corn (*2*) or a 30-day metabolism study of atrazine in sorghum (*3*). Although ammeline metabolism to ammelide and cynauric acid has not been intensively studied in plants, many bacterial species are capable of these metabolic transformations (*29*). In several *Pseudomonas* species, only a series of hydrolytic enzymes appear to be required to sequentially metabolize ammeline to ammelide, cyanuric acid, biuret, urea, CO_2 and ammonia (*30,31*). It is possible that ammelide, cynauric, and $^{14}CO_2$ detected during the metabolism of [^{14}C-ring]2-chloro-*s*-triazines in plant systems could actually have been produced by bacteria.

The 2-hydroxy-*s*-triazine metabolites exist in the tautomeric 2-oxo-4,6-bis(alkylamino)-1,2-dihydro-*s*-triazine form (*1*) and it is perhaps for this reason that *O*-glucosides of the 2-hydroxy-4,6-bis(alkylamino)-*s*-triazines have not been reported commonly as metabolites. Recently, a glucoside of 2-hydroxy-4-amino-6-isopropyl-amino-*s*-triazine (X) was characterized as a metabolite of atrazine in sugarcane (*25*) (Figure 4). Metabolite X accounted for 0.9% of the total radioactive residue (TRR) while the corresponding aglycone was present as 8.5% of the TRR. Positive and negative ionization electropsray LC/MS/MS indicated that metabolite X was a glucoside of 2-hydroxy-4-amino-6-isopropylamino-*s*-triazine. The positive ionization, electrospray, collision-induced mass spectrum confirmed the M + 1 molecular ion (m/z 352) and showed that ions corresponding to the triazine aglycone m/z 170 (100%) and the ammeline ion m/z 128 (7%) were derived from the molecular ion. Metabolite X was resistant to hydrolysis by β-glucosidase, but it was hydrolyzed by acid. The data are more consistent with an *N*- than an *O*-glucoside, but either structure is possible based on these data.

The metabolism of the 2-chloro-*s*-triazines in plants by hydroxylation and by the interaction of the hydroxylation pathway with the *N*-dealkylation and GSH conjugation pathways is summarized in Figure 4. The 2-methoxy-*s*-triazine herbicides are also susceptible to benzoxazinone-catalyzed hydrolysis, but at a rate less than 20% of that observed with the 2-chloro-*s*-triazines (*1*). The 2-methylthio-*s*-triazine herbicides are resistant to hydrolysis by benzoxazinone catalysis, but they can be metabolized to 2-hydroxy-*s*-triazines through sulfoxide or sulfone intermediates (*1,14*).

Glutathione Conjugation

Atrazine and related 2-chloro-*s*-triazine herbicides are metabolized rapidly by conjugation with GSH in certain *s*-triazine tolerant plants (*32,33*) and in a potato cell suspension culture (*34*) (Figure 5). The reaction is catalyzed by glutathione S-

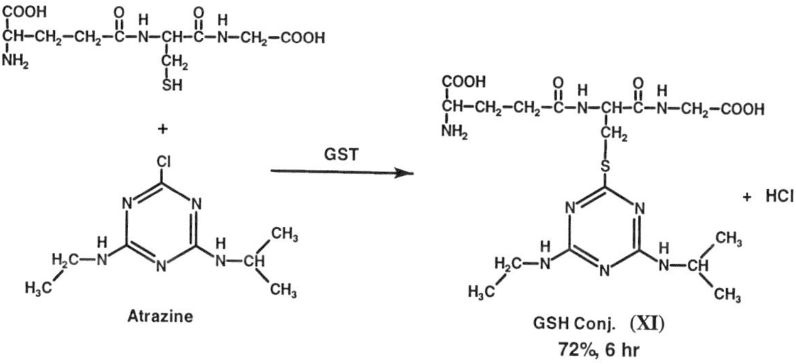

Figure 5. GSH conjugation of a 2-chloro-*s*-triazine herbicide (atrazine).

transferase (GST) enzymes present in the leaves of *s*-triazine tolerant plants such as corn, sorghum, sugarcane, Sudan grass, and *Sorghum halpenese* (*26*). Most plants contain GST enzymes (*35*), but GST enzymes capable of utilizing 2-chloro-*s*-triazines as substrates are absent from 2-chloro-*s*-triazine susceptible plants such as pea, oats, wheat, barley and *Amaranthus retroflexus* (*26*).

The relative importance of GSH conjugation and hydrolytic dehalogenation to 2-chloro-*s*-triazine tolerance and selectivity in common crop and weed species has been investigated. A mutant corn line deficient in benzoxazinone and ineffective in atrazine metabolism by hydrolytic dehalogenation was only slightly less tolerant to atrazine than other common corn lines (*21,22*). Hydrolytic dehalogenation was observed in both an atrazine tolerant and an atrazine susceptible inbred corn line, but only the tolerant line metabolized atrazine by conjugation with GSH (*24*). Atrazine tolerance in sorghum was clearly related to rapid GSH conjugation while moderate susceptibility in pea was related to a slow rate of *N*-dealkylation (*36*). It was concluded that while hydrolytic dehalogenation plays a role in the tolerance of corn to the 2-chloro-*s*-triazines, metabolism by GSH conjugation is necessary for the normal level of tolerance observed in corn, sorghum, and sugarcane (*2*). A biotype of velvet leaf (*Abutilon theophrasti*) with a 10-fold higher tolerance to atrazine compared to a more common wild-type has evolved as a result of repeated exposure to atrazine. The increased tolerance is due to the over-production of 2 constitutive GST isozymes that results in a 2- to 9-fold higher rate of GSH conjugation of atrazine (*37*).

GSH conjugation is very rapid in the metabolism of the 2-chloro-*s*-triazines in tolerant plants. About 70% of the ^{14}C-atrazine is metabolized to GSH-related conjugates within 6 hr of treatment of corn, sorghum, or sugarcane leaves with 40 nmole ^{14}C-atrazine/gram tissue. Most of the residual ^{14}C is atrazine, but low levels of *N*-dealkylated and/or hydroxylated metabolites may also be present (*33*). Similar results were observed with other 2-chloro-*s*-triazines and their metabolites that had undergone mono-*N*-dealkylation (*33*). The 2-methoxy- and 2-methylthio-*s*-triazine herbicides are not effective substrates for the GST enzyme from corn (*26*), but they are slowly metabolized to water-soluble products in excised leaves of corn sorghum and sugarcane (*33*). The methylthio-*s*-triazine herbicides appear to be partially metabolized to GSH conjugates after an initial oxidation reaction (*14*) in a manner similar to the metabolism of metribuzin to a homo-glutathione conjugate in soybean (*38*).

The GSH conjugates of the *s*-triazines are rapidly and sequentially catabolized to γ-glutamylcysteine and *S*-cysteine conjugates in corn, sorghum, and sugarcane (*3,33*) (Figure 6). This route of catabolism is also utilized in the metabolism of GSH

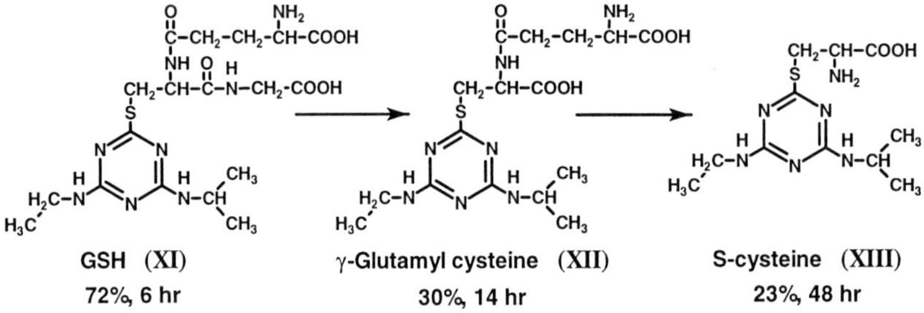

GSH (XI) 72%, 6 hr → γ-Glutamyl cysteine (XII) 30%, 14 hr → S-cysteine (XIII) 23%, 48 hr

Figure 6. Catabolism of a GSH conjugate to an *S*-cysteine conjugate.

conjugates of other pesticides such as PCNB, fluorodifen, and the chloroacetamides in other plants including peanut, corn, sorghum, and spruce (*35,39,40*). The rate of catabolism appears to vary with the plant (*33*). In corn, 83% of the GSH conjugate of atrazine was catabolized to the γ-glutamylcysteine conjugate within 6 hr.

Cysteine conjugates are frequently the branch-point in the metabolism of pesticide GSH conjugates and most metabolites derived from GSH conjugates in plants are due to different reactions having occurred at this or at later points in the metabolic pathway (*35*). This is also true in the metabolism of the GSH conjugates of the *s*-triazines. The various reactions commonly observed in the metabolism of *S*-cysteine conjugates of pesticides in plants are shown (Figure 7).

S-(*N*-Malonyl)cysteine conjugates, produced by the action of malonyl-CoA transferases, are among the common metabolites produced from GSH conjugates of pesticides in plants (*39,41*) (Figure 8). The *S*-(*N*-malonyl)cysteine conjugates can be terminal metabolites or they can undergo further metabolism, for example by oxidation of the sulfide to the sulfoxide (*27,39,41*). *S*-(4-Ethylamino-6-isopropylamino-*s*-triazinyl-2)-*N*-malonylcysteine was recently isolated as a minor metabolite of atrazine in sugarcane where it accounted for approximately 0.1% of the total radioactive residue (*25*) (Figure 8). It was identified by positive and negative ionization electrospray mass spectrometry and by collision-induced dissociation of the molecular ion. The M + 1 base peak of the positive ionization spectrum fragmented to diagnostic ions at m/z 343, 301, and 214.

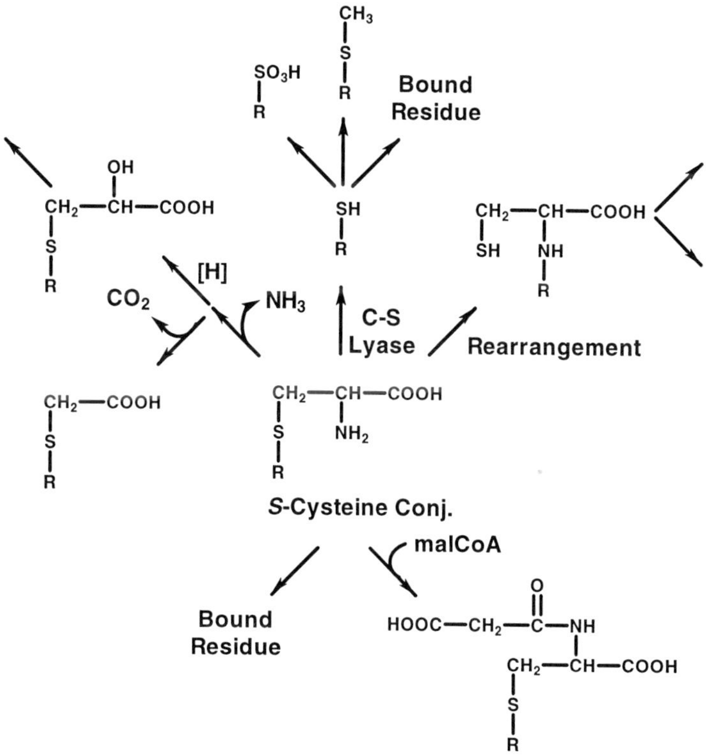

Figure 7. *S*-Cysteine conjugates are a major branch-point in the metabolism of GSH conjugates of pesticides in plants.

Two glucose-thiolactic acid conjugates were isolated recently as metabolites of atrazine in sugarcane: *S*-(4-ethylamino-6-isopropylamino-*s*-triazinyl)-2-(*O*-β-D-glucosyl)-3-thiolactic acid (11.1% TRR) and *S*-(4-amino-6-isopropylamino-*s*-triazinyl)-2-(*O*-β-D-glucosyl)-3-thiolactic acid) (3.6% TRR) (25) (Figure 9). They were identified by positive and negative ionization LC/MS/MS and by hydrolysis. The mass spectra were characterized by intense molecular ions and by diagnostic ion fragments derived from the molecular ion. These metabolites are probably formed from the corresponding GSH conjugates of atrazine and desethyl atrazine *via* *S*-cysteine conjugate intermediates (Figure 10). *S*-Thiolactic acid conjugates are

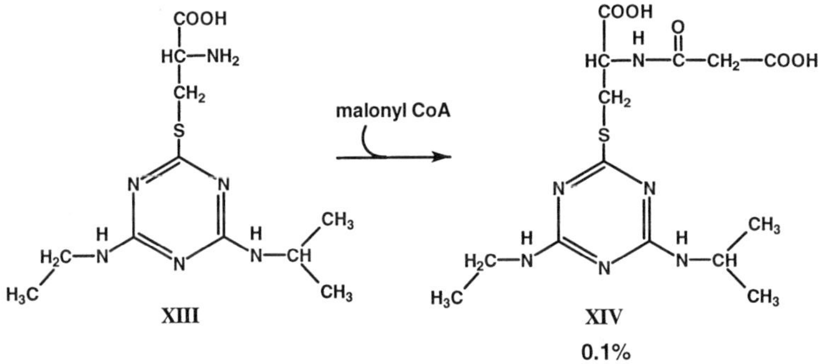

malonyl CoA

XIII

XIV

0.1%

Figure 8. Metabolism of an *S*-cysteine conjugate to a malonylcysteine conjugage.

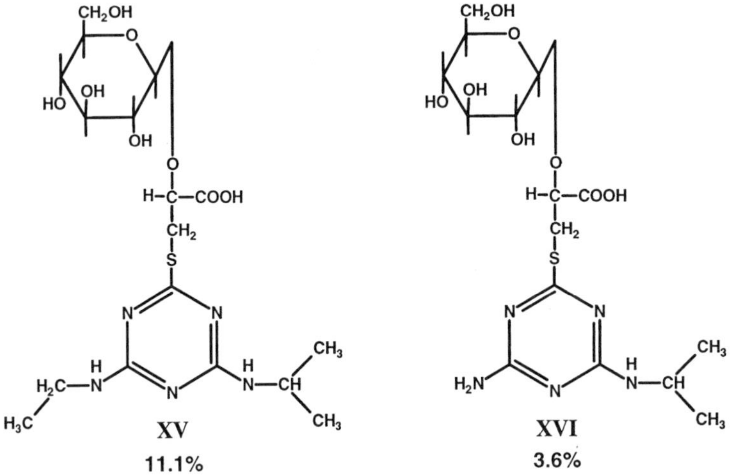

XV

11.1%

XVI

3.6%

Figure 9. Glucose-thiolactic acid conjugates of atrazine from sugarcane.

commonly formed in the metabolism of *S*-cysteine conjugates of pesticides in various plant species and they can represent a major route of metabolism of the *S*-cysteine conjugates (*35*). As indicated, they are probably formed by transamination of the *S*-cysteine conjugates to *S*-thiopyruvate intermediates that are reduced to the corresponding *S*-thiolactic acid conjugates. The *S*-thiolactic acid conjugates can exist

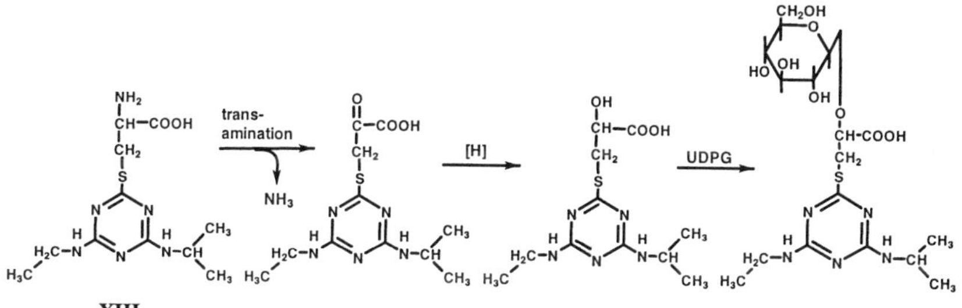

Figure 10. Proposed pathway for the metabolism of an *S*-cysteine conjugate of atrazine to a glucose-thiolactic acid conjugate.

in the free form as observed in the metabolism of PCNB in peanut (*42*), they can exist with the hydroxyl group esterified with malonic acid as observed in metazachlor metabolism in corn (*35,43*), or with the hydroxy group glucosylated as in the metabolism of fluorodifen in spruce (*44*). In corn, the *S*-(*O*-malonyl)thiolactic acid conjugate of metazachlor herbicide was slowly metabolized to the *S*-sulfinyl-lactic acid conjugate which appeared to be a more stable product. It is likely that additional forms of these *S*-thiolactic acid conjugates of atrazine/desethyl atrazine may be present in sugarcane.

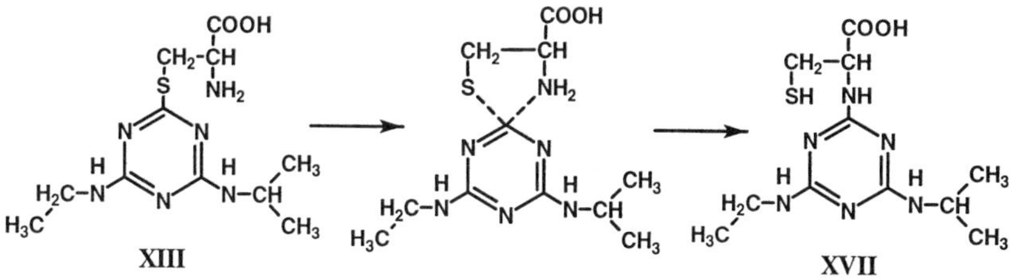

Figure 11. Rearrangement of the *S*-cysteine conjugate of atrazine.

The *S*-cysteine conjugate of atrazine undergoes a rapid nonenzymatic rearrangement to the *N*-cysteine conjugate *in vitro* at pH 7.5 (*2*). The rearrangement probably involves a 5-membered cyclic intermediate in which the sulfur of cysteine is displaced by the amine (Figure 11). The corresponding disulfide dimer of the *N*-

cysteine conjugate was identified as a metabolite of atrazine in sorghum, accounting for 4% of the TRR (*8*). An *N*-cysteine conjugate of 2-methylthio-4-ethylamino-6-(1',2'-dimethylpropyl)amino-*s*-triazine was also produced in rice from the GSH pathway and appears to be one of the few other cases where an *N*-cysteine conjugate has been observed from the GSH pathway (*14*). However, other *S*-cysteine conjugates such as *S*-(2,4-dinitrophenyl)cysteine undergo this rearrangement *in vitro* (*40*).

A lanthionine conjugate was one of the most abundant metabolites of atrazine in sorghum and accounted for approximately 21% of the TRR 10 days following treatment under greenhouse conditions (*3*). The mechanism of formation of this metabolite is not completely understood, but it may involve the reaction of the *N*-cysteine conjugate with cysteine followed by the elimination of sulfur (Figure 12). Recently, several metabolites of atrazine that produced LC/MS or LC/MS/MS spectra that were identical to that of the lanthionine conjugate of atrazine were isolated from sugarcane that had been treated with atrazine under field conditions (*25*). Since there

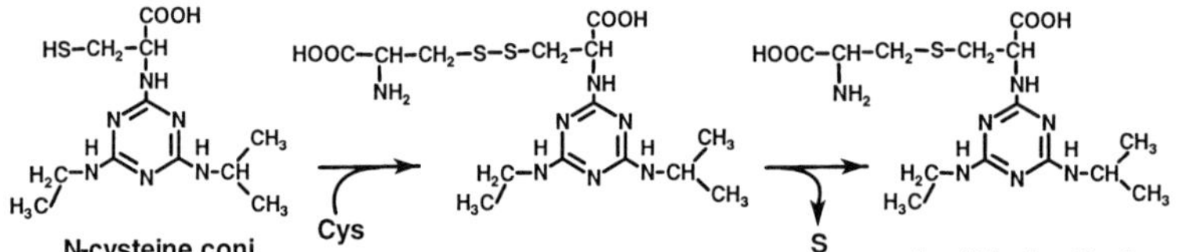

N-cysteine conj.

XVII

Cys

S

Lanthionine Conj.

21%, 10 days

XVIII

Figure 12. Proposed pathway for formation of the *N*-lanthionine conjugate.

are two asymmetrical centers in lanthionine, four diastereo isomers are possible. The various fractions that produced mass spectra identical to that of the lanthionine conjugate may have been different diastereo isomers of the lanthionine conjugate or labile complexes that degraded to the lanthionine conjugate during further purification.

The lanthionine conjugate of atrazine is further metabolized in sorghum. It declined from 21% of the TRR after 10 days to only 10% of the TRR after 30 days (*3*). Five additional metabolites of atrazine that appear to have been derived from the lanthionine conjugate were recently identified from extracts of sugarcane and sorghum by LC/MS/MS and by hydrolysis (*25*) (Figure 13). One of these metabolites, isolated from sorghum, was characterized as the sulfoxide of the lanthionine conjugate (XIX). It was no doubt produced by simple oxidation of the lanthionine conjugate. An *N*-cysteic acid conjugate (XXIII) was identified as a metabolite of atrazine in sugarcane (0.6% TRR). This metabolite may have been formed from the lanthionine conjugate of atrazine by the action of C-S lyase and oxidative enzymes or it may have been formed more directly from the *N*-cysteine conjugate of atrazine. The role of C-S lyase enzymes has been established in the metabolism of the

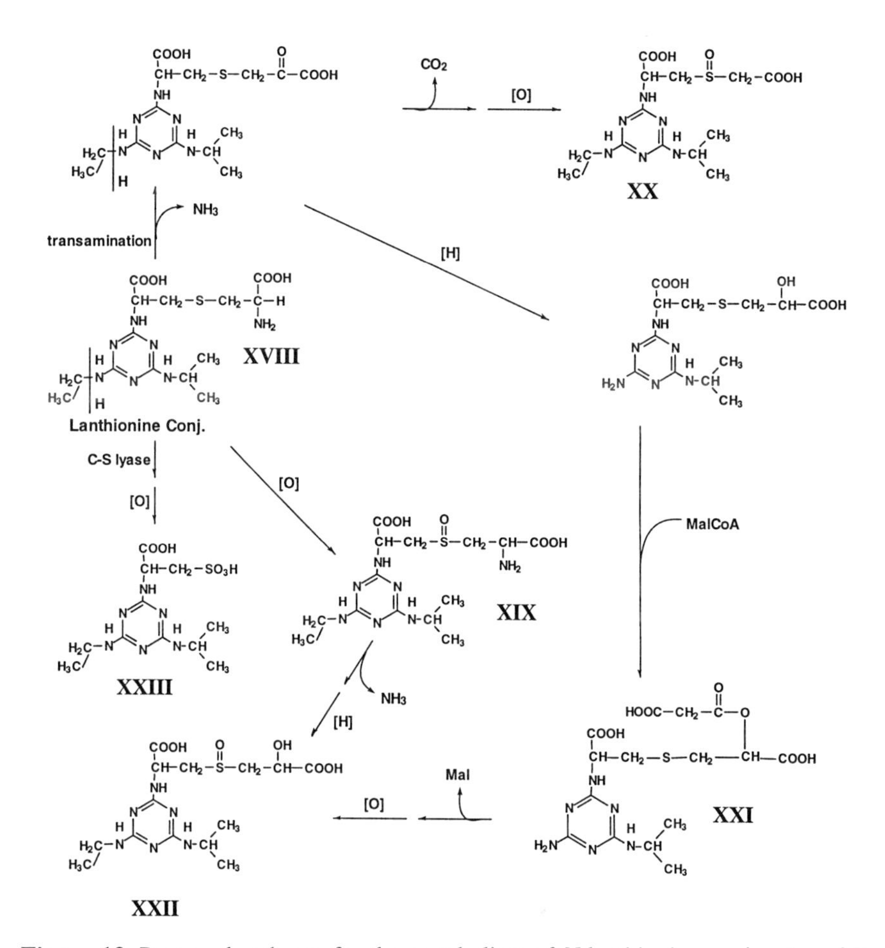

Figure 13. Proposed pathway for the metabolism of *N*-lanthionine conjugates of 2-chloro-*s*-triazines in plants. Only metabolites designated with a Roman number were actually identified by mass spectrometry. Reaction arrows indicate probable reaction sequences relating to the amino acid sidechain and do not take into account whether a metabolite was identified as a conjugate of atrazine or desethylatrazine.

cysteine conjugates of pesticides in onion and spruce (*45-46*). Each of the remaining three metabolites (XX, XXI, XXII) in Figure 13 was probably formed from a lanthionine conjugate following transamination of the lanthionine conjugate to the corresponding pyruvate intermediate. This is similar to the transamination reaction in the metabolism of the *S*-cysteine conjugate of atrazine (Figure 10). Metabolite XX, 2-[*N*-(4-ethylamino-6-isopropylamino-*s*-triazinyl-2)-cysteinyl-*S*]-acetic acid-*S*-oxide (1.3% TRR) was identified from sugarcane (*25*). It is probably formed from a pyruvate intermediate following decarboxylation and oxidation (Figure 13). This is similar to the formation of the *S*-thioacetic acid conjugate observed in the metabolism of PCNB in peanut (*42*). Metabolites XXI [3-(*N*-{4--amino-6-isopropylamino-*s*-triazinyl-2}-cysteinyl-*S*)-lactic acid-*S*-oxide] (0.5% TRR) and XXII [3-(*N*-{4-ethylamino-6-isopropylamino-*s*-triazinyl-2}-cysteinyl-*S*)-O-malonyl-lactic acid] (0.9% TRR) were isolated from sugarcane and also appear to have been derived from pyruvate intermediates (*25*) (Figure 13). Metabolite XXI was formed by metabolism of the GSH conjugate of desethyl atrazine while metabolite XXII was apparently formed by metabolism of the atrazine GSH conjugate. The formation of these metabolites is similar to the formation of the *S*-(*O*-malonyl)thiolactic acid conjugate from the *S*-cysteine conjugate of atrazine as discussed above or the formation of the *S*-(*O*-malonyl)-thiolactic acid conjugate of metazachlor from the *S*-cysteine conjugate (*43*).

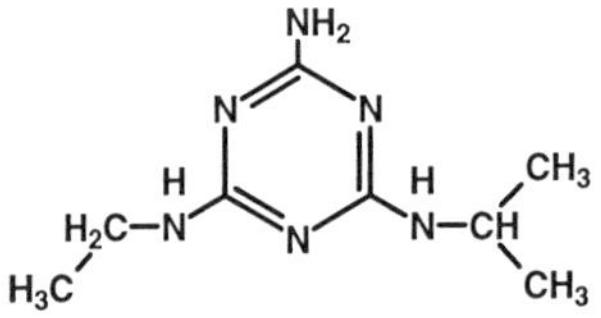

1.0%

XXIV

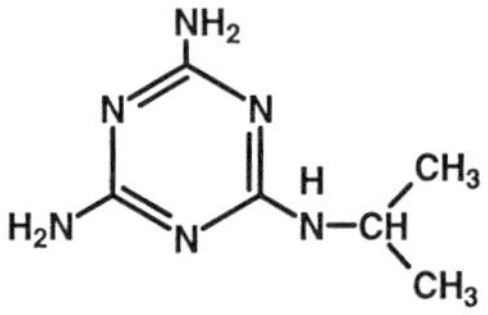

2.2%

XXV

Figure 14. 2-Amino-*s*-triazine metabolites of atrazine identified from sugarcane.

Amination Pathway

Two metabolites of atrazine recently isolated from sugarcane, 2-amino-4-ethylamino-6-isopropylamino-*s*-triazine (XXIV) and 2,4-diamino-6-isopropylamino-*s*-triazine (XXV), accounted for 1% and 2.2% of the TRR, respectively (25) (Figure 14). They were identified by comparison of their GC retention times and mass spectra with those of synthetic standards. Analogous metabolites with an amino group in the 2-position were isolated and identified from 2-methylthio-4-ethylamino-6-(1',2'-dimethylpropyl)amino-*s*-triazine (dimethametryn) metabolism in paddy rice (*14*). Other metabolites of dimethametryn included a GSH conjugate and *S*- and *N*-cysteine conjugates. When rice was treated with the *N*-lanthionine conjugate of dimethametryn, the corresponding 2-amino derivative was produced. It was proposed that the 2-amino derivatives were formed by oxidative deamination of the corresponding *N*-cysteine conjugates (*14*). The possibility should also be considered that

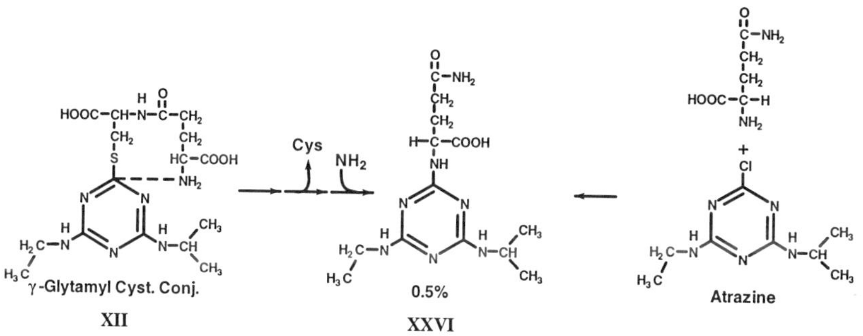

Figure 15. Possible mechanisms of *N*-glutamine conjugate formation of atrazine (characterized from sorghum).

these metabolites might be formed from the GSH pathway by nucelophilic displacement of a sulfide, sulfoxide, or sulfone conjugate by an amine derivative or ammonia or by a nucleophilic displacement of chloride from the herbicide by another amine derivative or ammonia. Trimethylamine can react with the 2-chloro-*s*-triazines to form quaternary nitrogen derivatives that undergo displacement reactions more readily than the 2-chloro-*s*-triazines; therefore, an enzyme or low molecular weight catalyst could be involved in such transformations (*1*).

Two *N*-amino acid conjugates were characterized recently as metabolites of the 2-chloro-*s*-triazines in plants: *N*-(4-ethylamino-6-isopropylamino-*s*-triazinyl-2)-γ-glutamine (XXVI) and (*N*-[4,6-diamino-*s*-triazinyl-2]-proline) (XXVII) (*25*) (Figures 15 and 16). *N*-(4-Ethylamino-6-isopropylamino-*s*-triazinyl-2)-γ-glutamine (XXVI) was isolated as a minor metabolite of atrazine from sorghum (0.2% TRR). The positive ionization electrospray LC/MS/MS spectrum of (XXVI) was characterized by an intense molecular ion and ion fragments consistent with this structure. It was hydrolyzed to hydroxy-atrazine in 87% yield. Metabolite XXVI could be formed from the GSH pathway by rearrangement of the *S*-γ-glutamylcysteine conjugate to an *N*-γ-glutamyl-cysteine conjugate followed by peptidase and amination reactions to yield that final metabolite (Figure 15). The proposed rearrangement would be comparable to that observed with the *S*-cysteine conjugate (Figure 11). Alternatively, metabolite XXVI could be formed by a direct attack of glutamine on atrazine or an activated form of atrazine (Figure 15).

Metabolite XXVII (*N*-[4,6-diamino-*s*-triazinyl-2]-proline) was recently characterized as a metabolite of simazine present in citrus and apple (*25*) (Figure 16).

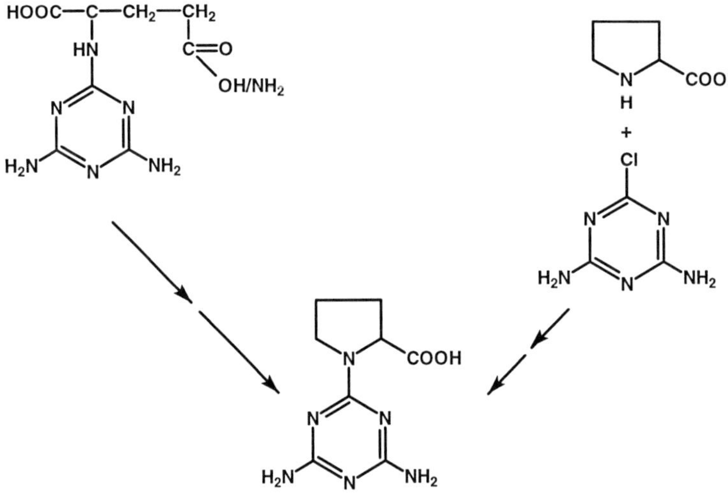

Figure 16. Possible pathways for the formation of an *N*-proline conjugate characterized as a metabolite of simazine from orange.

Metabolite XXVII was present in orange juice at 0.079 ppm and accounted for 75% of the TRR. It was isolated from orange juice and characterized by high resolution DCI/MS of both XXVII and the methyl ester of XXVII. It was also characterized by MS/MS. Two likely routes of formation include ring closure of a glutamine conjugate of 4,6-diamino-*s*-triazine or by the attack of proline on either 2-chloro-4,6-diamino-*s*-triazine or an activated form of this metabolite (Figure 16). Amino acid, peptide, and protein conjugates have also been reported to occur in the metabolism of atrazine and simazine in tea and citrus plants (*47,48*).

Summary and Conclusions

The metabolism of the 2-chloro-*s*-triazine herbicides in plants by the *N*-dealkylation and hydrolytic dehalogenation pathways is summarized in Figure 4. All metabolites except the glucoside (X) have been described previously (*1-4,25,35*). Metabolism by the GSH pathway and a possible amination pathway is summarized in Figure 17. Except for the *N*-proline conjugate which was produced from simazine metabolism in citrus and apple, all metabolites described in Figure 17 were produced from atrazine metabolism in sorghum, sugarcane, and corn (*3,25,33*).

N-Dealkylation occurs in the metabolism of most 2-chloro-*s*-triazines in most plants and it is especially important in plants that lack the hydrolytic dehalogenation or GSH conjugation pathways (*1,2,4*). The *s*-triazine herbicides with highly branched *N*-alkyl side-chains can be metabolized by hydroxylation rather than by *N*-dealkylation of the side-chains (*1,2,4,14*). The mono-*N*-dealkylated metabolites can be metabolized further by GSH conjugation or hydroxylation. There is little evidence that 2-chloro-4,6-bis(amino)-*s*-triazine is metabolized in plants by GSH conjugation, but it is presumed that it can be metabolized by hydrolytic dehalogenation (Figure 4).

Hydrolytic dehalogentation of the 2-chloro-*s*-triazines is usually catalyzed by benzoxazinones in species such as corn, but the 2-hydroxy-*s*-triazine metabolites may arise from the metabolism of GSH conjugates in some species (*1,2,4*) (Figure 4). Hydroxylation can occur before or after *N*-dealkylation (*2*). The mechanism by which ammeline is converted to ammelide and cyanuric acid in plants has not been extensively studied. There is little evidence that the hydroxylated triazine metabolites are readily metabolized to *O*-glucosides, but a glucoside of 2-hydroxy-4-amino-6-isopropylamino-*s*-triazine has now been identified. It is uncertain whether it is an *N*- or an *O*- glucoside (*25*).

GSH conjugation is important in the metabolism of the 2-chloro-*s*-triazines in highly tolerant crops such as corn, sorghum, and sugarcane. Atrazine can enter the GSH pathway directly or after it has been metabolized to desethylatrazine. The GSH conjugates of both atrazine and desethylatrazine appear to be metabolized by parallel pathways to the last products derived from the *S*-cysteine and *N*-cysteine conjugates (*3,33,25*) (Figure 17). It has also been proposed that GSH conjugates may undergo *N*-dealkylation after GSH conjugation (*14*). The rate of GSH conjugation varies significantly depending upon the 2-chloro-*s*-triazine (*26,33*). Therefore, the relative importance of GSH conjugation, hydroxylation, *N*-dealkylation and/or amination may depend upon both the *s*-triazine and the plant. The metabolism of the 2-chloro-*s*-triazines by the GSH pathway is more complex than that of some pesticides because: (a) the 2-chloro-*s*-triazines can be metabolized to GSH conjugates directly or (b) after at least one *N*-dealkylation reaction, and (c) the *S*-cysteine conjugates undergo an uncommon rearrangement to *N*-cysteine conjugates which are further metabolized by several competing reactions.

Many of the metabolites of atrazine in Figure 17 have not been previously identified; however, all but four of these products are consistent with products observed in the metabolism of other pesticide GSH conjugates in plants (*35*). The remaining four metabolites (the *N*-glutamine and *N*-proline conjugates and the two 2-amino-*s*-triazine metabolites) may have been produced from the GSH pathway or by an independent amination pathway.

Bound residues frequently account for a high percent of the TRR in crops

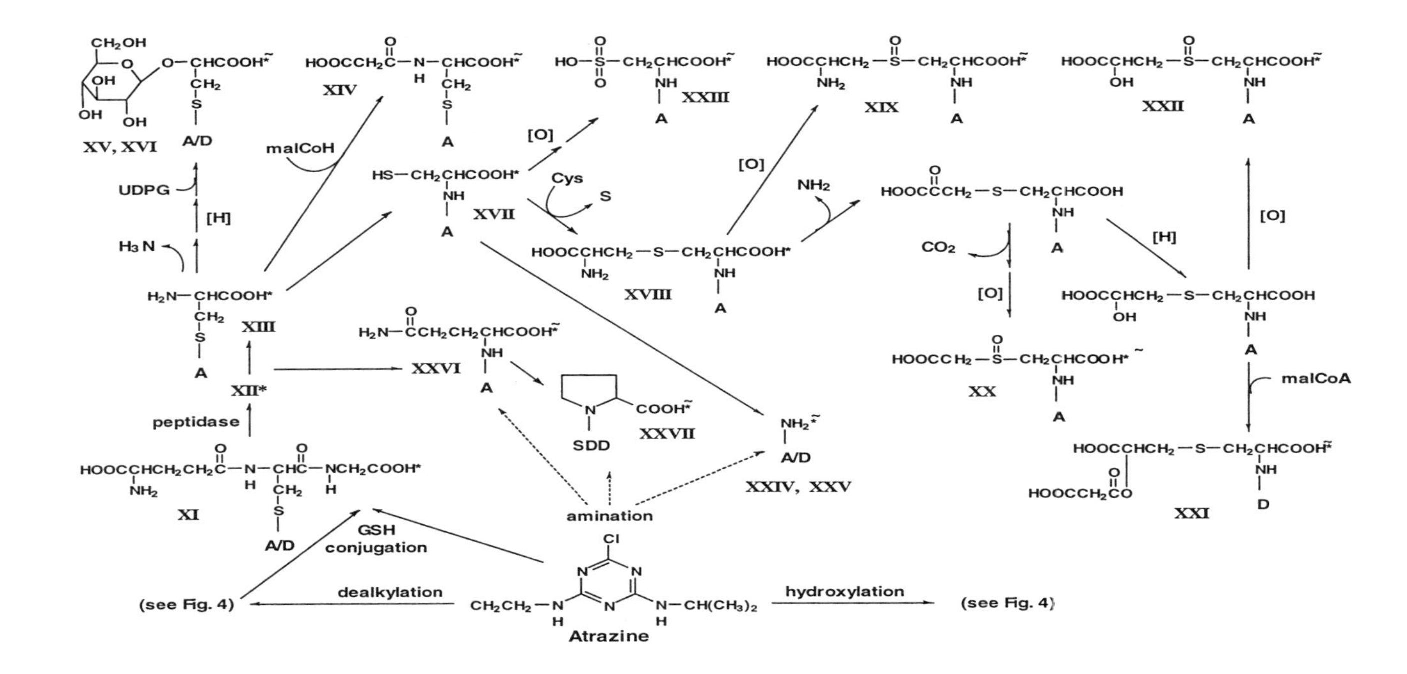

Figure 17. Proposed pathway for the metabolism of atrazine and related 2-chloro-*s*-triazines in plants by GSH conjugation and/or by amination. A = conjugates of 4-ethylamino-6-isopropylamino-*s*-triazine, D = conjugates of 4-amino-6-isopropylamino-*s*-triazine and SDD = conjugate of 4,6-diamino-*s*-triazine. Dashed arrows indicate a speculative pathway. Metabolites not previously reported are indicated by ~, those identified by mass spectrometry or other rigerous methods are indicated by *, all other metabolites are speculated intermediates.

treated with the 2-chloro-*s*-triazine herbicides (*1,3*). Some bound residues may be formed from the GSH pathway, but the mechanism of this transformation is uncertain (*1*). Bound residues are also formed from pesticides in plants where the GSH pathway does not play an important role and there may be several mechanisms by which the bound residues are formed. Ammeline and products derived from ammeline do not appear to be precursors of bound residues (*1*). The bound residues of atrazine from corn did not appear to be biologically available to sheep and rat (*49*).

Approximately 28 metabolites of atrazine have been identified in various plant metabolism studies: approximately 15 from the GSH pathway, 10 from the *N*-dealkylation and hydrolytic dehalogenation pathways, and 3 more that may be produced from the GSH pathway or perhaps by a separate process. Based upon the pathways shown in Figures 4 and 17, 48 metabolites of atrazine would be expected. One additional metabolite, a proline conjugate, has been characterized from metabolism studies of simazine in citrus.

Literature Cited

1. Esser, H.O.; Dupuis, G.; Ebert, E.; Vogel, C.; Macro, G.J. In *s-Triazines*; Kearney, P.C. and Kaufman, D.D., Ed.; Herbicides Chemistry, Degradation, and Mode of Action; Marcel Dekker, Inc.:New York, **1975** Vol. 1, 2nd edition; pp 129-208.
2. Shimabukuro, R.H.; Lamoureux, G.L.; Frear, D.S.; Bakke, J.E. In *Metabolism of s-triazines and its signficance in biological systems*; Tahori, A.S., Ed.; Pesticide Terminal Residues (Supplement to Pure and Applied Chemistry); Butterworth & Co.: London, **1971**, pp 323-342.
3. Lamoureux, G.L.; Stafford, L.E.; Shimabukuro, R.H.; Zaylskie, R.H. *J. Agric. Food Chem. 21.* **1973,** *vol 21*, 1020-1030.
4. Naylor, A.W. In *Herbicide Metabolism in Plants*; Audus, L.J., Ed.; Herbicides Physiology, Biochemistry, Ecology: Academic Press: New York, NY, **1975**, Vol. 1; pp 397-427.
5. Jordan, L.S.; Jolliffe, V.A. *Pestic. Sci.* **1973**, *4*, 467-472.
6. Wichman, J.R; Byrnes, W.R. *Weed Sci.* **1975**, *23*, 448-453.
7. Burt, G.W. *Weed Sci.* **1974,** *22*, 116-119.
8. Shimabukuro, R.H.; Walsh, W.C.; Lamoureux, G.L.; Stafford, L.E. *J. Agric. Food chem.* **1973,** *21*, 1031-1036.
9. Burnet, M.W.M.; Loveys, B.R.; Holtum, J.A.M.; Powles, S.B. *Pestic. Biochem. Physiol.* **1993,** *46*, 207-218.
10. Frear, D.S.; Swanson, H.R.; Tanaka, F.S. *Phytochem.* **1966,** *8*, 2157.
11. Benyon, K.I.; Stoydin, G.; Wright, A.N. *Pestic. Sci.* **1972,** *3*, 293-306.
12. Benyon, K.I.; Stoydin, G.; Wright, A.N. *Pestic. Sci.* **1972,** *4*, 379-387.
13. Benyon, K.I.; Stoydin, G.; Wright, A.N. *Pestic. Sci.* **1972,** *2*, 153.
14. Mayer, P., Kriemler, H.-P., and Laanio, T.L. *Agric. Biol. Chem.* **1981**, *45*, 361-368.
15. Roth, W. *C.r. hebd. Se'anc. Acad. Sci. Paris.* **1957,** *245*, 942-944.

16. Gysin, H.; Knusli, E. *Advan. Pest Control Res.* **1960**, *3*, 289.
17. Shimabukuro, R.H.; Swanson, H.R.; Walsh, W.C. *Plant Physiol.* **1970,** *46*, 103-107.
18. Lund-Hoie, K. *Weed Res.* **1969**, *9*, 142-147.
19. Thompson, Lafayette, Jr. *Weed Sci.* **1972**, *20*, 153-155.
20. Thompson, Lafayette, Jr. *Weed Sci.* **1972**, *20*, 584-587.
21. Hamilton, R.H. *Weeds.* **1963-1964**, 27-30.
22. Hamilton, R.H. *J. Agric. Food Chem.* **1964**, *12*, 14-17.
23. De Souza, M.L.; Wackett, L.P.; Boundy Mills, K.L.; Mendelbaum, R.T.; Sadowsky, M.J. *Appl. Environ. Microbiol.* **1995**, *61*, 3373-3378.
24. Shimabukuro, R.H.; Frear, D.S.; Swanson, H.R.; Walsh, W.C. *Plant Physiol.* **1971**, *47*, 10-14.
25. Ciba-Geigy, *unpublished*, **1996**.
26. Frear, D.S.; Swanson, H.R. *Phytochem.* **1970**, *9*, 2123-2132.
27. Lamoureux, G.L.; Rusness, D.G. *Pestic. Biochem. Physiol.* **1989**, *34*, 187-204.
28. Montgomery, M.L.; Botsford, D.L.; Freed, V.H. *J. Agric. Food Chem.* **1969**, *17*, 1241-1243.
29. Zeyer, J.; Bodmer, J.; Huetter, R. *Zentralbl., Bakteriol., Mikrobiol. Hyg. Abt.* **1981**, *1*, 289-298.
30. Jutzi, K.; Cook, A.; Huetter, R. *Biochem. J.* **1982**, *208*, 679-684.
31. Cook, A.; Beilstein, P.; Grossenbacher, H; Huetter, R. *Biochem. J.* **1985**, *231*, 25-30.
32. Lamoureux, G.L.; Shimabukuro, R.H.; Swanson, H.R.; Frear, D.S. *J. Agric. Food Chem.* **1970**, *18*, 81-86.
33. Lamoureux, G.L.; Stafford, L.E.; Shimabukuro, R.H. *J. Agric. Food Chem.* **1972**, *20*, 1004-1010.
34. Edwards, R.; Owen, W.J. *Pestic. Biochem. Physiol.* **1989**, *34*:246-254.
35. Lamoureux, G.L.; Rusness, D.G. In *Glutathione in the metabolism and detoxification of xenobiotics in plants;* De Kok, L.J.; Stulen, I.; Rennenberg, H.; Brunold, C.; Rauser, W.E., Eds.; Sulfur Nutrition and Assimilation in Higher Plants; SPB Academic Publishing: The Hague, **1993**; pp 221-237.
36. Shimabukuro, R.H.; Swanson, H.R. *J. Agric. Food Chem.* **1969**, *17*, 199.
37. Gronwald, J.W.; Andersen, R.N.; Yee, C. *Pestic. Biochem. Physiol.* **1989**, *34*, 84-294.
38. Frear, D.S.; Swanson, H.R.; Mansager, E.R. *Pestic. Biochem. Physiol.* **1985**, *23*, 56-65.
39. Lamoureux, G.L.; Rusness, D.G. In *Catabolism of glutathione conjugates of pesticides in higher plants;* Rosen, J.D., Magee, P.S., and Casida, J.E., Eds.; Sulfur in pesticide Action and Metabolism, ACS Symposium Series 158; American Chemical Society, Washington, DC, **1981**, pp. 135-164.
40. Lamoureux, G.L.; Rusness, D.G. In *The role of glutathione and glutathione-s-* transferases in pesticide metabolism, selectivity and mode of action in plants

and insects; Dolphin, D.; Poulson, R.; and Avramovic, O., Eds.; *Coenzymes and Cofactors Volume III, Glutathione: Chemical, Biochemical and Medical Aspects, Part B*: John Wiley and Sons, New York, **1989**, pp 154-196.

41. Lamoureux, G.L.; Rusness, D.G. In *Malonylcysteine conjugates as end-products of glutathione conjugate metabolism in plants;* Miyamoto, J.; Kearney, P.C., Eds.; Pesticide Chemistry Human Welfare and the Environment Vol 3 Mode of Action, Metabolism and Toxicology; Pergamon Press: New York, **1983**, pp 295-300.
42. Rusness, D.G.; Lamoureux, G.L. *J. Agric. Food Chem.* **1980**, *28*, 1070-1077.
43. Khalifa, M.A.; Lamoureux, G.L. In *The effect of BAS 145 138 safener on the secondary/tertiary steps of propachlor metabolism in corn.* Abstract, 7th International Congress of Pesticide Chemistry (IUPAC), Hamburg, Germany, **1990**.
44. Lamoureux, G.L.; Rusness, D.G.; Schroder, P.; Rennenberg, H. *Pestic. Biochem. Physiol.* **1991**, *39*, 291-301.
45. Lamoureux, G.L.; Rusness, D.G. *Pestic. Biochem. Physiol.* **1980**, *14*, 50-61.
46. Lamoureux, G.L.; Rusness, D.G.; Schroder, P. *Pestic. Biochem. Physiol.* **1993**, *47*, 8-20.
47. Kakhniashvili, Kh.A.; Durmishidze, S.V.; Gigauri, M.Sh. *Agrokhimiya* **1987**, *1*, 89-93 (*Chemical Abstracts 106*: 170986).
48. Kakhniashvili, Kh.A.; Durmishidze, S.V.; Gigauri, M.Sh. *Fiziol. Rast. (Moscow)* **1989**, *36*, 99-106 (*Chemical Abstracts 110*: 149757).
49. Bakke, J.E.; Shimabukuro, R.H.; Davison, K.L.; Lamoureux, G.L. *Chemosphere*, **1972**, *1*, 21-24.

Chapter 7

Atrazine Hydrolysis by a Bacterial Enzyme

Lawrence P. Wackett[1,2], Michael J. Sadowsky[2,3], Mervyn de Souza[1,2], and Raphi T. Mandelbaum[4]

[1]**Department of Biochemistry and Bioprocess Technology Institute,**
[2]**Center for Biodegradation Research and Informatics and** [3]**Department of Soil, Water and Climate, University of Minnesota, St. Paul, MN 55108**
[4]**Volcani Research Institute, Bet-Dagan 50250, Israel**

ABSTRACT

Atrazine, 2-chloro-4-(ethylamino)-6-(isopropylamino)-1,3,5-triazine, is metabolized relatively slowly in natural soils and waters by resident microorganisms. Recently, several atrazine-degrading bacterial pure cultures were isolated and the molecular basis of bacterial atrazine metabolism is now beginning to be revealed. *Pseudomonas* sp. strain ADP was isolated from a herbicide spill site for its ability to use atrazine as the sole source of nitrogen for growth. Atrazine metabolism also liberated the triazine ring carbon atoms as carbon dioxide. Hydroxyatrazine was detected transiently in the growth medium during the course of atrazine metabolism. Previously, hydroxyatrazine was proposed to be derived solely from abiotic hydrolysis catalyzed by soil organic matter and clays. The gene encoding the enzymatic hydrolysis of atrazine by *Pseudomonas* sp. ADP was cloned and expressed in *Escherichia coli.* Cell-free atrazine hydrolysis activity in the recombinant *E. coli* strain was determined by high pressure liquid chromatography. The enzyme, atrazine chlorohydrolase, was purified to homogeneity using ammonium sulfate precipitation and ion exchange chromatography. The purified chlorohydrolase showed a single band on denaturing sodium dodecyl sulfate-polyacrylamide gel electrophoresis corresponding to a subunit molecular weight of 60,000. Gene sequencing data yielded a molecular weight of 52,421. Gel filtration chromatography indicated a holoenzyme molecular weight of 240,000 consistent with an α_4 or α_5 subunit stoichometry. In [^{18}O]-H_2O, atrazine chlorohydrolase yielded [^{18}O]-hydroxyatrazine quantitatively. In control experiments incubated and analyzed under the same conditions, [^{18}O] from H_2O did not exchange into hydroxyatrazine. These data are consistent with enzymatic hydrolysis of atrazine. Other bacteria were also demonstrated to catalyze atrazine hydrolysis, suggesting this biologically-mediated reaction is widespread in soil and water.

BACKGROUND KNOWLEDGE

Atrazine is broadly applied to soils for weed control and shows significant persistence under most conditions. It is somewhat mobile in soils and, thus, found in groundwater. This has elevated people's interest in studying atrazine biodegradation, which is dependent on the metabolic activities of microorganisms. Better understanding of these may lead to management systems which will reduce the atrazine levels found in ground and surface water.

Microorganisms are the primary agents responsible for recycling the Earth's organic matter, both natural products and synthetic commercial chemicals. In total, approximately eight million organic compounds exist, many are biodegradable via microbial enzymatic transformation, but detailed information on the biodegradation of most organic compounds is lacking. However, there is increasing knowledge of how different organic functional groups are transformed by microorganisms and this will aid efforts for predicting the biodegradability of organic compounds which have not yet been investigated experimentally. This information is currently being highlighted by the University of Minnesota Biocatalysis/Biodegradation Database that is freely accessible via the World Wide Web (1). The database contains information on atrazine biodegradation as an example of the much larger class of *s*-triazine compounds.

Atrazine can theoretically be metabolized via dealkylation, deamination, dechlorination and/or ring cleavage reactions (Figure 1). Based on studies with soils (2) and a pure culture (3), it has been proposed that microbes oxidatively dealkylate the ethyl and isopropyl substituents of atrazine and that environmentally observed hydroxyatrazine derives from abiotic hydrolysis. Ring cleavage of atrazine has been suggested, but a definitive identification of metabolic intermediates is lacking. Studies have been hindered, until very recently, by a lack of bacterial pure cultures that metabolize atrazine. Microorganisms capable of metabolizing less heavily substituted *s*-triazines were obtained more readily (4,5). Atrazine-mineralizing pure cultures were isolated and described in 1995 from a bioreactor in Switzerland (6,7), an agricultural soil in Ohio (8), and a herbicide spill site in Minnesota (9). The latter one will be described here.

Hydroxyatrazine has been observed in soils (2), plants (10) and mammals (11), but this hydrolysis product of atrazine has not been typically attributed to microbial metabolism. Hydroxyatrazine is not significantly herbicidal and is unregulated because it has no known negative impact on mammalian health. Furthermore, hydroxyatrazine is much more strongly sorbed to soils and, thus, is much less prone to leach into groundwater. In this context, microbial metabolism of atrazine to yield hydroxyatrazine would constitute the ideal pathway. Hydroxyatrazine is thought to be biodegradable, suggesting that hydroxyatrazine will not accumulate in the environment.

Pseudomonas sp. ADP. *Pseudomonas* sp. ADP was isolated by enrichment culture with atrazine serving as the sole source of nitrogen (9,12). It was obtained from soil at an abandoned agrochemical dealership in which atrazine had been repeatedly spilled as a result of atrazine distribution activity. We have measured atrazine concentrations as high as 40,000 ppm at such sites, and this situation likely provides for strong selective pressure for the evolution of atrazine-metabolizing bacteria. Bacterial isolates that could metabolize atrazine were identified using an agar plate assay containing a carbon source(s) and with atrazine as the nitrogen source and visual indicator. Atrazine was present at 500 ppm, significantly above its limit of solubility, and formed an opaque background in the agar. Bacteria capable of metabolizing atrazine yielded a halo of atrazine-clearing surrounding the colony and thus, were readily differentiated from the preponderance of atrazine non-degrading bacteria.

The fastest growing isolated bacterium was subjected to taxonomic identification and characterized with respect to its metabolism of atrazine. It was a gram-negative polarly flagellated organism that was definitively identified as a

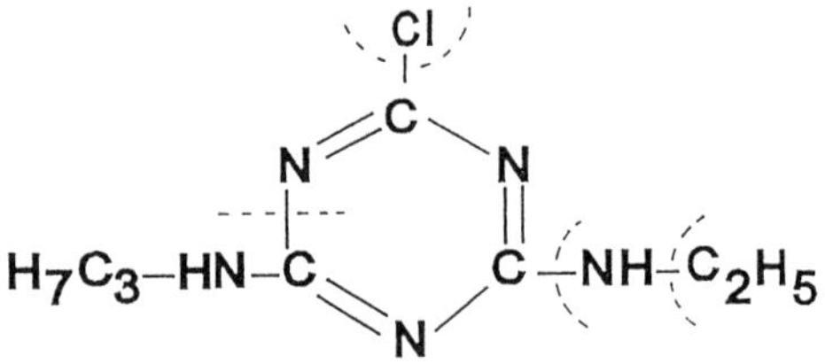

Figure 1. Structure of atrazine with dashed lines showing the points of bond cleavage during microbial metabolism.

Pseudomonas sp. by an array of biochemical tests. It was designated as *Pseudomonas* sp. strain ADP because it was sufficiently distinct from any known species. Strain ADP rapidly metabolized atrazine in excess of its requirement for nitrogen. Cell suspension in aqueous media cleared >99% of a 2000 ppm suspension of atrazine in less than 30 minutes. In soil tests, aged atrazine was removed most readily with a combined application of *Pseudomonas* sp. ADP and sodium citrate (9). The latter compound had previously been shown to serve as the sole carbon source for growth, and it supported excellent metabolism of atrazine.

The molecular basis of atrazine metabolism was further investigated using defined media conditions and [^{14}C]-atrazine to follow the fate of atrazine carbon atoms. In studies using growth-limiting atrazine concentrations, all 5 nitrogen atoms were liberated to support growth (13). Atrazine ring carbon atoms are released as carbon dioxide. During a kinetic course of atrazine metabolism, an organic solvent insoluble intermediate(s) accumulated to a steady-state level and then disappeared from bacterial cultures. One intermediate was determined to be hydroxyatrazine based on HPLC retention time, migration on thin layer chromatograms, and mass spectrometry. These data suggested that hydroxyatrazine might be the initial metabolite during atrazine transformation by *Pseudomonas* sp. ADP.

Molecular basis of hydroxyatrazine formation. Over one dozen papers had suggested that environmental hydroxyatrazine originated from abiotic reactions (for example, see 14-21) and this necessitated a rigorous examination of the mechanism of its formation in cultures of *Pseudomonas* sp. ADP. The hypothesis that hydroxyatrazine is the first intermediate in a metabolic pathway for atrazine requires that a specific gene(s) and enzyme exist to convert atrazine to hydroxyatrazine. This was tested by molecular cloning of total genomic DNA from *Pseudomonas* sp. ADP into *Escherichia coli* and then screening for atrazine metabolism (22). A recombinant *E. coli* that metabolized atrazine was identified using the atrazine agar plate assay described above. The same assay was used for subcloning experiments that gave rise to the isolation of a 1.9 kb DNA fragment on plasmid pMD4. This *E. coli* clone metabolized atrazine to hydroxyatrazine.

Atrazine Chlorohydrolase. *E. coli* (pMD4) was grown on a large scale and cell-free protein extracts were obtained that produced hydroxyatrazine. The product was determined using an HPLC assay. The crude protein extract was first fractionated by adding ammonium sulfate, with stirring, to 20% (w/v). The precipitate was harvested by centrifugation and shown to contain the enzymatic activity for hydroxyatrazine production. The partially purified protein thus obtained was applied to a Q-20 anion exchange column (Bio-Rad). The activity was retained by the column and was subsequently eluted with a 0 - 0.5 M gradient of KCl in 25 mM MOPS buffer, pH 6.9.

The highly purified protein was shown to be homogeneous as evidenced by the presence of a single polypeptide on denaturing sodium dodecyl sulfate-polyacrylamide gel electrophoresis (SDS-PAGE). The polypeptide migrated consistent with a subunit molecular weight of 60,000. A single protein was also observed by gel filtration chromatography. It showed an apparent molecular weight of 240,000, suggestive of an α_4 subunit stoichiometry.

Independently, the gene encoding atrazine chlorohydrolase was sequenced and the protein primary structure was derived from this. The translated protein is predicted to have a molecular weight of 52,421. Taken with the gel filtration data, the subunit stoichiometry could be 4 or 5.

Confirmation of enzyme-catalyzed hydrolysis. There are precedents for the biological replacement of a chlorine substituent with a hydroxyl group in which the oxygen substitutent derives from (a) water or (b) molecular oxygen. The general mechanism of hydroxylation catalyzed by the *Pseudomonas* enzyme was determined using [^{18}O]-H_2O. The product, analyzed by mass spectrometry, was [^{18}O]-hydroxyatrazine. A small amount of [^{16}O]-hydroxyatrazine was observed but that was consistent with the amount of [^{16}O]-H_2O in the reaction mixture. A control experiment with [^{16}O]-hydroxyatrazine and [^{18}O]-H_2O showed that water did not exchange into hydroxyatrazine spontaneously under the conditions of the experiment.

Conclusions. More than a dozen different bacterial consortia, obtained from unique soils, were shown to yield hydroxyatrazine (23). The gene eneocding atrazine chlorohydrolase has been demonstrated in other atrazine-metabolizing bacteria (13). These data, in concert with studies on *Pseudomonas* sp. ADP, strongly suggest that bacterial atrazine hydrolysis is widespread in the environment. Furthermore, atrazine chlorohydrolase may have applicability for engineering atrazine biodegradation in soils and waters. Both *in vivo* gene expression in various organisms and *in vitro* enzyme applications are currently under study.

Acknowledgments. This research was supported by research grants from Ciba-Geigy and Grant No. 2394-93 from BARD, the United States-Israel Binational Agricultural Research and Development fund, administered in the U.S. by U.S.D.A. Grant No. USDA/94-34339-1122.

Literature Cited:

1. **Ellis, L.B.M. and L.P. Wackett.** 1995. *Soc. Ind. Microbiol. News* **45:**167-173.
2. **Erickson, E.L., and K.H. Lee.** 1989. *Critical Rev. Environ. Cont.* **19:**1-13.
3. **Behki, R.M., and S.U. Kahn.** 1986. *J. Agric. Food Chem.* **34:**746-749.
4. **Cook, A.M., and R. Hutter.** 1981. *J. Agric. Food Chem.* **29:**1135-1143.
5. **Cook, A.M.** 1987. Biodegradation of *s*-triazine xenobiotics. *FEMS Microbiol. Rev.* **46:**93-116.
6. **Stucki, G., C.W. Yu, T. Baumgartner, and J.F. Gonsalez-Valero.** 1995. *Water Res.* **1:**291-296.
7. **Yanze-Kontchou, C., and N. Gschwind.** 1994. *Appl. Environ. Microbiol.* **60:**4297-4302.
8. **Radosevich, M., S.J. Traina, Y.L. Hao, and O.H. Tuovinen.** 1995. *Appl. Environ. Microbiol.* **61:**297-302.
9. **Mandelbaum, R.T., D.L. Allan, and L.P. Wackett.** 1995. *Appl. Environ. Microbiol.* **61:**1451-1457.
10. **Shimabukuro, R.H.** 1967. *Plant Physiol.* **43:**1925-1930.
11. **Bakke, J.E., J.D. Larson and C.E. Price.** 1972. *J. Agr. Food Chem.* **20:**602-607.
12. **Mandelbaum, R.T., L.P. Wackett, and D.L. Allan.** 1993. *Appl. Environ. Microbiol.* **59:**1695-1701.
13. **deSouza, M.L.,** unpublished data.

14. **Armstrong, D.E., G. Chesters and R.F. Harris.** 1967. *Soil Sci. Soc. Am. Proc.* **31**:61-66.
15. **Armstrong, D.E. and G. Chesters.** 1968. *Environ. Sci. Technol.* **2**:683-689.
16. **Zimdahl, R.L., V.H. Freed, M.L. Montgomery and W.R. Furtick.** 1970. *Weed Res.* **10**:18-26.
17. **Li, G.C., and G.T. Felbeck.** 1972. *Soil Sci. Soc.* **114**:201-209.
18. **Nearpass, D.C.** 1972. *Soil Sci. Soc. Am. Proc.* **36**:606-610.
19. **Skipper, H.D. and V.V. Volk.** 1972. *Weed Sci.* **20**:344-347.
20. **Fernandez-Quintanilla, C.M., A. Cole and F.W. Slife.** 1981. *Proc. EWRS Symp.* pp. 301-308.
21. **Adams, C.D., and S.J. Randtke.** 1992. *Environ. Sci. Technol.* **26**:2218-2227.
22. **de Souza, M.L., L.P. Wackett, K.L. Boundy-Mills, R.T. Mandelbaum, and M.J. Sadowsky.** 1995. *Appl. Environ. Microbiol.* **61**:3373-3378.
23. **Mandelbaum, R.T., L.P. Wackett and D.L. Allan.** 1993. *Environ. Sci. Technol.* **27**:1943-1946.

Chapter 8

Genetics of Atrazine Degradation in *Pseudomonas* sp. Strain ADP

M. J. Sadowsky[1,2], L. P. Wackett[2-4], M. L. de Souza[2,3], K. L. Boundy-Mills[1-3], and R. T. Mandelbaum[5]

[1]Department of Soil, Water, and Climate and Department of Microbiology, [2]Center for Biodegradations Research and Informatics, [3]Department of Biochemistry, and [4]Biological Process Technology Institute, University of Minnesota, St. Paul, MN 55108
[5]Volcani Research Institute, Bet-Dagan 50250, Israel

A 21.5-kilobase *Eco*RI genomic DNA fragment from *Pseudomonas* sp. strain ADP, designated pMD1, was shown to encode atrazine degradation activity in *Escherichia coli* DH5α. Atrazine degradation was demonstrated by a zone-clearing assay on agar medium containing crystalline atrazine and by chromatographic methods. A gene conferring the atrazine clearing phenotype was subsequently subcloned as a 1.9 kb *Ava*I fragment in pACYC184, designated pMD4, and was expressed in *E. coli*. Random *Tn*5 mutagenesis established that the 1.9 kb *Ava*I fragment was essential for atrazine dechlorination activity. *E. coli* containing pMD4 degraded atrazine and accumulated hydroxyatrazine. A 0.6 kb *Apa*I-*Pst*I fragment from pMD4, containing the putative atrazine chlorohydrolase gene, hybridized to DNA from atrazine-degrading bacteria isolated in Switzerland and Louisiana (USA). Sequence data indicated that a single open reading frame of 1419 nucleotides, *atz*A, encoded atrazine dechlorination activity. The *Atz*A, a polypeptide of 473 amino acids, had significant amino acid identity (41%) with *Trz*A, a protein from *Rhodococcus corallinus*, whose preferred substrate is melamine. These data suggest that genes encoding atrazine hydrolysis to hydroxyatrazine are widespread in nature and contribute to the formation of hydroxyatrazine in soil, a reaction previously attributed to abiotic processes.

Atrazine [6-chloro-N-ethyl-N'-(1-methylethyl-1,3,5-triazine-2,4-diamine)] is a widely used herbicide for the control of broad-leaf and particular grass weeds. It is the predominant member of a broad class of *s*-triazine herbicides which are used to control weeds in corn, sorghum and other crops. Atrazine is relatively persistent in soils with an average half-life ranging from 4 to 57 weeks (*1,2*). Several studies concerning the environmental fate of atrazine have shown that atrazine is transformed relatively slowly in the environment (*3,4*). It has been estimated that approximately 800 million pounds of atrazine was used in the United States from 1980-1990 (*5*). Because of its widespread use over the last thirty years, for both selective and nonselective weed control (*6*), atrazine and other *s*-triazine derivatives have been detected in soils as well as in ground and surface water in several countries (*7-13*). Atrazine mobility has directly led to the contamination of groundwater (*8,14*). This has prompted many researchers to look for microorganisms that have the ability to degrade atrazine in soils and water.

Atrazine Degradation

Degradation of atrazine in soils may occur by both biotic and abiotic processes. Several environmental factors, including soil pH (*15-17*), temperature (*18*), and moisture (*19*), tillage practices (*17,20-22*) and other soil properties (*23*) have been shown to influence atrazine persistence in soils. Degradation by microorganisms (*24-33*) reduces atrazine concentrations in soil. However, relatively few soil microorganisms that mineralize atrazine have been described.

Biodegradation of *s*-triazine compounds can occur by N-dealkylation and dechlorination processes (*26*). The *s*-triazine compounds lacking bulky side group substituents are degraded relatively rapidly in soils, due to bacterial-mediated dechlorination reactions (*29,34*). An enzyme from *Rhodococcus corallinus* has been identified which catalyzes dechlorination of some chloro-*s*-triazine compounds, but it is inactive with atrazine (*30*). Dealkylation reactions have been suggested to be the first metabolic step in the biodegradation of atrazine (*26,33,35-37*). Earlier studies suggested that atrazine dechlorination reactions in soils occurred by strictly chemical processes (*26,38*) or occurred slowly due to the action of a limited number of soil fungi (*39*). More recently, dechlorination of atrazine has also been shown to occur (*25,27,40*). For example, *Pseudomonas* sp. strain ADP rapidly dechlorinates atrazine to hydroxyatrazine [4-amino-6-[(1-methylethyl)amino]-1,3,5-triazine-2(1H)-one](Figure 1) by an enzymatic hydrolytic reaction mediated by a bacterial halidohydrolase (*25*). Thus, microorganisms have developed several biochemical mechanisms for degrading atrazine and related *s*-triazine compounds.

Atrazine Degrading Microorganisms

Numerous reports have documented the occurrence of *s*-triazine-degrading microorganisms (*24,26-29,31-33,35,37,41-45*). A majority of the organisms described, however, fail to mineralize atrazine (*24,42*). While earlier studies reported atrazine degradation only by mixed microbial consortia, more recent reports indicate that several isolated bacterial strains can degrade atrazine. Mandelbaum *et al.* (*14*) reported the isolation of a pure bacterial culture, identified as *Pseudomonas* sp. strain ADP, which degraded a high concentration of atrazine (>1,000 μg/ml) under growth and non-growth conditions. *Pseudomonas* sp. strain ADP used atrazine as a sole source of nitrogen for growth and the organism completely mineralized the *s*-triazine ring of atrazine under aerobic growth conditions. About 80% of the added atrazine was degraded within 15 hr of incubation and 100% was mineralized by 25 hr. Recently,

Radosevich *et al.* (*37*) reported the isolation of pure bacterial culture which can use atrazine as sole N and/or C sources to support microbial growth under aerobic conditions. In addition, Yanze-Kontchou and Gschwind (*33*) isolated a *Pseudomonas* strain which is capable of using atrazine as a sole carbon and energy source. Approximately 50% of the added atrazine was mineralized during a 50 day incubation period. The strain, YAYA6 had a doubling time of about 11 hr, and is proposed to degrade atrazine by two pathways: using a dechlorination reaction to yield hydroxyatrazine or by using a N-dealkylation reaction with the formation of desethylatrazine or deisopropylatrazine. The unclassified strain reported by Radosovich and coworkers (*37*) used atrazine as a sole N or C source for growth and mineralized between 40 and 50% of added atrazine. Growth, however, was very moderate and there was no change in absorbance following growth on atrazine. More recently, Bouquard *et al.* (*40*) reported that a *Rhizobium* sp. strain, PATR, has the ability to metabolized atrazine to hydroxyatrazine, via a dechlorination reaction.

Genetics of Atrazine Degradation

There are several reasons to study the genetics of atrazine degradation by microorganisms. On a broad scale, genetic studies are done to: 1) gain an understanding of the biological pathways involved in degradative processes, 2) dissect the underlying biochemistry involved in atrazine degradation, 3) understand the evolution of a "newly evolved" microbial pathway, 4) produce gene probes for ecological and phylogenetic analyses, 5) produce large quantities of biodegradation intermediates, 6) produce large amounts of enzymes for biochemical analysis and bioremediation purposes, and 7) aid in the construction of superior biodegradative organisms.

Little information is available concerning the genes and enzymes involved in the metabolism of atrazine and other *s*-triazine compounds. An inducible set of genes that encode the enzymes for melamine (1,3,5-triazine-2,4,6-triamine) metabolism were isolated from *Pseudomonas* sp. strain NRRL B-12227 (*46,47*). While NRRL B-12227 did not degrade atrazine, it degraded melamine in a six step pathway and used the intermediates as a sole N source. NRRLB-1227 also degraded other *s*-triazines including N-isopropylammeline, N-ethylammeline, ammelide, and cyanuric acid. Three of the genes involved in the melamine degradation pathway, *trz*B, *trz*C, and *trz*D, have been cloned. Similar degradative genes have been isolated from *Pseudomonas* sp. strain NRRL B-12228 and *Klebsiella pneumoniae* strain 99 (*46,47*).

Recently, two reports have documented the isolation of genes encoding atrazine degradation activity from *Rhodococcus* sp. strains (*31,32*). In *Rhodococcus* sp. strain TE1, *N*-dealkylation of atrazine was mediated by a cytochrome P-450 (*31*), encoded by a single gene, *atr*A (*32*). The cloned DNA region also contains a gene for the degradation of the herbicide EPTC (*S*-ethyl-dipropylthiocarbamate), *ept*A. The *atr*A gene was not expressed in *E. coli*, only in *Rhodococcus*. Another *Rhodococcus* strain, *R. corallinus* NRRL B-15544R, has the ability to dechlorinate the *s*-triazines desethylsimazine and desethylatrazine

(*30*). The strain, however, does not degrade atrazine or simazine. The gene responsible for the dechlorination/deamination has been sequenced and is termed *trz*A (Steffens and Mulbry, Unpublished, GenBank and *48*). However, recombinant *Rhodococcus* strains expressing both the *atr*A and *trz*A genes have been shown to transform atrazine to N-isopropylammelide and N-ethylammelide (*48*).

Strategies used to identify genes involved in atrazine degradation include transposon mutagenesis and complementation analysis. We have recently used both of these approaches to isolate and characterize a gene region encoding atrazine degradation activity from *Pseudomonas* sp. strain ADP (*25,49*). A 21.5-kilobase *Eco*RI genomic DNA fragment, designated pMD1, was shown to encode atrazine degradation activity in *E. coli* DH5α (Figure 2). Atrazine degradation was demonstrated by a zone-clearing assay on agar medium containing crystalline atrazine and by chromatographic methods. A gene conferring the atrazine clearing phenotype was subsequently subcloned as a 1.9 kb *Ava*I fragment in pACYC184, designated pMD4, and was expressed in *E. coli*. This result and random *Tn*5 mutagenesis established that the 1.9 kb *Ava*I fragment was essential for atrazine dechlorination. High pressure liquid and thin layer chromatographic analyses established that *E. coli* containing pMD4 degraded atrazine and accumulated hydroxyatrazine (Figure 1). Hydroxyatrazine was detected only transiently in *E. coli* containing pMD1. A 0.6 kb *Apa*I-*Pst*I fragment from pMD4, containing the atrazine chlorohydrolase gene, hybridized to DNA from atrazine-degrading bacteria isolated in Switzerland and Louisiana (USA). *E. coli* and *Pseudomonas* strains containing pMD1 expressed atrazine degradation activity, indicating that the enzyme can function in genetically diverse backgrounds (de Souza et al., unpublished).

Sequence data for the pMD4 gene region encoding atrazine transformation ability indicated that a single open reading frame of 1419 nucleotides, *atz*A, encodes atrazine dechlorination activity. AtzA, a polypeptide of 473 amino acids, had significant amino acid identity (41%) with TrzA, a dechlorinating enzyme from *Rhodococcus corallinus*, which has melamine as its preferred substrate (*49*). More recently, we have isolated, cloned, and sequenced the second gene in the atrazine degradation pathway, atzB, from *Pseudomonas* sp. strain ADP. The atzB gene encodes a 481 amino acid polypeptide that transforms hydroxyatrazine to N-isopropylammelide [2,4-dihydroxy-6-(isopropylamino)-s-triazine] (*50*).

Taken together, our results indicate that *atz*A is a relatively small gene which produces a protein product (atrazine chlorohydrolase) that has the ability to transform atrazine to hydroxyatrazine. Consequently, *atz*A is an ideal candidate for use in engineering bacteria and plants to metabolize atrazine to hydroxyatrazine, thereby remediating herbicide-containing soils.

Acknowledgments

This work was supported, in part, by a grant from Novartis Crop Protection, Inc. and by grant 94-34339-1122 from the United States Department of Agriculture-BARD program.

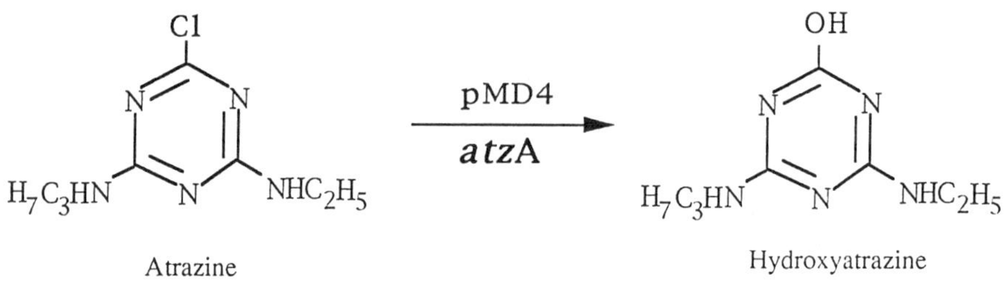

Atrazine $\xrightarrow[\textit{atzA}]{\text{pMD4}}$ Hydroxyatrazine

Figure 1. Partial pathway for atrazine degradation in *Pseudomonas* sp. strain ADP. The first step is encoded by a gene, *atz*A, located on pMD4 and generates hydroxyatrazine which is subsequently metabolized to carbon dioxide and ammonia (27,28).
(Adapted with permission from references 27 and 28. Copyright 1995 and 1993 American Society for Microbiology.)

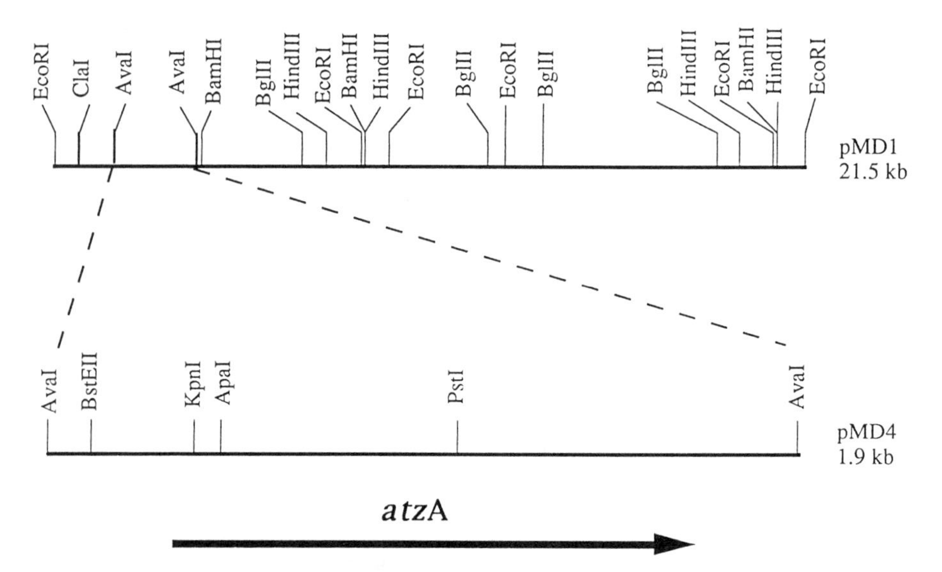

Figure 2. Physical relationship of clones expressing atrazine degradation ability. Cosmid pMD1 is a 21.5 kb *Eco*RI fragment in pLAFR3. Plasmid pMD4 is a 1.9 kb *Ava*I fragment cloned into pACYC184. Both clones express atrazine degrading activity in *Escherichia coli* DH5α. The ORF designated *atz*A is indicated by the arrow. Adapted from (25).
(Adapted with permission from reference 25. Copyright 1995 American Society for Microbiology.)

Literature Cited

1. Anderson, P. C.; Georgeson, M. *Genome.* **1989,** *31,*994-999.
2. Cohen, S.; Creager, S.; Carsel, R.; Enfierld, C. In *Treatment and disposal of pesticide waste*; R. F. Krueger and J. N. Seiber (Ed.); American Chemical Society: Washington, D.C., 1984; pp. 297-325.
3. Baker, F. W. G. In *Tropical grassy weeds*; F. W. Baker and P. J. Terry (Eds.); CAB International: Wallingford, England, 1991; pp. 96-105.
4. Erickson, E. L.; Lee, K. H. *Crit. Rev. Environ. Contam.* **1989,** *19,*1-3.
5. Gianessi, L. P. *Resources.* **1987,** *89,*1-4.
6. Belluck, D. A.; Benjamin, S. L.; Dawson, T. In *Pesticide transformation products: fate and significance in the environment*; L. Somasundaram and J. R. Coats (Eds.); American Chemical Society: Washington, D.C., 1991. pp. 254-273.
7. Baker, B. D.; Richards, P. R. In *Rational Readings on Environmental Concerns;* J. H. Lehr, (Ed.); Van-Nostrand Reinhold: New York, N.Y., 1992, pp. 84-97.
8. Belluck, D. A.; Benjamin, S. L.; Dawson, T. In *Pesticide transformation products: fate and significance in the environment;* L. Somasundaram and J. R. Coats (Eds.); American Chemical Society: Washington, D.C., 1991. pp. 254-273.
9. Eisler, R. *Atrazine hazards to fish, wildlife, and invertebrates.* A Synoptic Review; Contaminant Hazard Review report 18; Fish and Wildlife Service, U.S. Department of the Interior: Washington, D.C., 1989.
10. Kello, D. *Food Additives & Contaminants.* **1989,** 6 Suppl. *1,*S79-85
11. Koplin, D. W.; Kalkhoh, S. *J. Environ. Sci. Technol.* **1993,** *27,*134-139.
12. Parsons, D. W.; Witt, J. M. *Pesticides in ground water in the Unites States of America;* a report of a 1988 survey of state lead agencies. Oregon State Univ. Agri. Chem. Dept.: Corvalis, OR, 1988.
13. Pick, F. E.; van Dyk, L. P.; Botha, E. *Chemosphere.* **1992,** *25,*335-341.
14. Mandelbaum, R. T.; Allan, D. L.; Wackett, L. P. *Appl. Environ. Microbiol.* **1995,** *61,*1451-1457.
15. Best, J. A.; Weber, J. B. *Weed Sci.* **1974,** *22,*364-373.
16. Hiltbold, A. E.; Buchanan, G. A. *Weed Sci.* **1977,** *25,*515-520.
17. Kells, J. J.; Rieck, C. E.; Blevens, R. L.; Muir, W. M. *Weed Sci.* **1980,** *28,*101-104.
18. McCormick, L. L.; Hiltbold, A. E. *Weeds.* **1966,** *14,*77-82.
19. Roeth, F. W.; Lavy, T. L.; Burnside, O. C. *Weed Sci.* **1968,** *17,*202-205.
20. Bauman, T. T.; Ross, M. A. *Weed Sci.* **1983,** *41,*423-426.
21. Burnside, O. C.; Wicks, G. A. *Weed Sci.* **1980,** *28,*661-666.
22. Pawlak, J. A.; Kells, J. J.; Barrett, M.; Meggit, W. F. *Weed Technol.* **1987,** *1,*140-144.
23. LeBarron, H. M. *Residue Rev.* **1970,** *11,*119-140.
24. Cook, A. M. *FEMS Microbiol. Rev. 1987, 46,*93-116.
25. de Souza, M. L.; Wackett, L. P.; Boundy-Mills, K. L.; Mandelbaum, R. T.; Sadowsky. Appl. *Environ. Microbiol.* **1995,** *61,*3373-3378.

26. Erickson, E. L.; Lee, K. H. *Crit. Rev. Environ. Contam.* **1989,** *19,*1-3.
27. Mandelbaum, R. T.; Allan, D. L.; Wackett, L. P. *Appl. Environ. Microbiol.* **1995,** *61,*1451-1457.
28. Mandelbaum, R. T.; Wackett, L. P.; Allan, D. L. *Appl. Environ. Microbiol.* **1993,** *59,*1695-1701.
29. Mandelbaum, R. T.; Wackett, L. P.; Allan, D. L. *Environ. Sci. Technol.* **1993,** *27,*1943-1946.
30. Mulbry, W. W. *Appl. Environ. Microbiol.* **1994,** *60,*613-618.
31. Nagy, I.; Compernolle, F.; Ghys, K.; Vanderleyden, J.; de Mot, R. *Appl. Environ. Microbiol.* **1995,** *61,*2056-2060.
32. Shao, Z.; Behki, R. *Appl. Environ. Microbiol.* **1995,** *61,*2056-2060.
33. Yanze-Kontchou, C.; Gschwind, N. *Appl. Environ. Microbiol.* **1994,** *60,*4297-4302.
34. Cook, A. M.; Hutter, R. *J. Agric. Food Chem.* **1981,** *29,*1135-1143.
35. Behki, R. M.; Kahn, S. U. *J. Agric. Food Chem.* **1986,** *34,*746-749.
36. McMahon, P. B.; Chapelle, F. H.; Jagucki, M. L. *Environ. Sci. Technol.* **1992,** *26,*1556-1559.
37. Radosevich, M.; Traina, S. J.; Hao, Y. L.; Tuovinen, O. H. *Appl. Environ. Microbiol.* **1995,** *61,*297-302.
38. Armstrong, D. E.; Chesters, G. *Environ. Sci. Technol.* **1968,** *2,*683-689.
39. Kaufman, D. D.; Blake, *J. Soil Biol. Biochem.* **1970,** *2,*73-80.
40. Bouquard, C.; Ouazzani, J.; Prome, J. C.; Michelbriand, Y.; Plesiat, P. *Appl. Environ. Microbiol.* **1997,** *63,*862-866.
41. Behki, R.; Topp, E.; Dick, E.; Germon, P. *Appl. Environ. Microbiol.* **1993,** *59,*1955-1959.
42. Cook, A. M.; Hutter, R. *J. Agric. Food Chem.* **1981,** *29,*1135-1143.
43. Giardina, M. C.; Giardi, M. T.; Filacchioni, G. *Agric. Biol. Chem.* **1980,** *44,*2067-2072.
44. Masaphy, S.; Levanon, D.; Henis, Y. *Bioresource Technol.* **1996,** *56,*207-214.
45. Korpraditskul, R.; Katayama, A.; Kuwatasuka, S. *J. Pest. Sci.,* **1993,** *18,*293-298.
46. Eaton, R. W.; Karns, J. S. *J. Bacteriol.* **1991,** *173,*1215-1222.
47. Eaton, R. W.; Karns, J. S. *J. Bacteriol.* **1991,** *173,*1363-1366.
48. Shao, Z. Q.; Seefens, W.; Mulbry, W.; Behki, R. M. *J. Bacteriol.* **1995,** *177,*5748-5755.
49. de Souza, M. L.; Sadowsky, M. J.; Wackett, L. P. *J. Bacteriol.* **1996,** *178,*4894-4900.
50. Boundy-Mills, K. L.; de Souza, M. L.; Wackett, L. P.; Mandelbaum, R.; Sadowsky, M. J. *Applied Environ. Microbiol.* **1997,** *63,*916-923.

Chapter 9

Metabolism of Selected (*s*)-Triazines in Animals

Jinn Wu[1], Robert A. Robinson[1], and Bruce Simoneaux[2]

[1]Xenobiotic Laboratories, Inc., 107 Morgan Lane, Plainsboro, NJ 08536
[2]Novartis Crop Protection, Inc., P.O. Box 18300, Greensboro, NC 27419

The metabolism of (*s*)-triazines in animals has been extensively studied over the past twenty-five or more years. Triazines substituted with alkylamines (ethylamino and/or isopropylamino) at the C_2 and C_4 positions and Cl, SCH_3, or OH groups at the C_6 position encompass the major classes of parent compounds investigated. Extensive metabolism of these compounds has been observed in animals. N-dealkylation of the side-chains through oxidative intermediates was the major observed biotransformation. Oxidation of the alkyl groups produces primary alcohols and carboxylic acids. Conjugation of the alkanols with sulfate and glucuronic acid appears to be a minor pathway. Hydrolysis at the chlorine position to a corresponding hydroxy derivative was observed to a small extent but probably occurred as an artifact of the isolation or chromatographic technique employed. Conjugation of the chloro- and thiomethyl-groups with glutathione was the next most important biotransformation observed leading to formation of cysteine conjugates, mercapturates, sulfides, disulfides, and sulfoxides. Hydroxy-*s*-triazines were mostly stable in animal systems but did metabolize to N-dealkylated and ring N-methylated products.

Atrazine (2-chloro-4-ethylamino-6-isopropylamino-*s*-triazine) and simazine (2-chloro-4,6-bis-ethylamino-*s*-triazine) are two of the most widely used (*s*)-triazine herbicides in the United States for control of many broadleaf and grass weeds in a variety of agronomic crops. Ametryn (2-ethylamino-4-isopropylamino-6-methylthio-*s*-triazine) is a member of the thiomethyl-*s*-triazine group of herbicides that is registered for use on bananas, corn, pineapple, sugarcane, and non-crop areas. Hydroxyatrazine (2-hydroxy-4-ethylamino-6-isopropylamino-*s*-triazine) is a major soil and plant metabolite of both atrazine and ametryn and is representative of the class of metabolites known as hydroxy-*s*-triazines. The metabolism of these selected (*s*)-triazine compounds in rats, goats, and hens will be discussed.

Metabolism of Selected (*S*)-Triazines in Rats

The metabolism of atrazine and simazine in the rat has been studied extensively by various researchers since the 1960's. It is not the purpose of this review to summarize all of the available data on this topic, but rather to focus upon the more recent studies that give the clearest picture of absorption, excretion, and metabolism of (*s*)-triazines in rats.

In a recent study (*1*) of the absorption, distribution, degradation, and excretion of ^{14}C-atrazine in the Sprague-Dawley rat, single oral doses of 1.0 mg/kg and 100 mg/kg were administered by oral gavage to several groups of male and female animals. One group consisted of only male rats (1.0 mg/kg) that were cannulated to facilitate bile collection.

Absorption from the gastro-intestinal tract into the general circulation at the low dose was rapid, with the maximum whole blood concentration of radioactivity observed at 2 hours after dosing. At the high dose level, maximum whole blood concentration was reached at 24 hours.

For both the low dose and high dose rats, the highest levels of radioactive residues were in the kidney, liver and red blood cells. Other tissue residues were low and there was no evidence of accumulation.

Elimination rates were dependent upon the blood content of the tissues. The residues of radioactivity in red blood cells are probably associated with binding of residues to rat hemoglobin which is known to bind atrazine and other *s*-triazines. This binding is specific to rodent and chicken and appears irrelevant in other animal species. Hamboeck *et al* (*2*) studied the *in vitro* binding of *s*-triazines to hemoglobins in whole blood of several animal species. They concluded that chloro-*s*-triazines and alkyl-*s*-thiotriazines *per se* did not appreciably bind to erythrocytes, but that the sulfoxide metabolites derived from chloro-*s*-triazines and alkylthio-*s*-triazines do bind covalently to the sulfhydryl group of cysteine β-125 in rodent and chicken hemoglobin. It was concluded that hemoglobins from species other than rodents and chickens do not react with the sulfoxide metabolites.

Approximately 66% of the administered dose was recovered in the urine over the 0-168 hour period following dosing of the high level rats. A further *ca* 20% of the dose was recovered in the feces from the high dose animals over the same period. Low dose cannulated male rats excreted 65.7% of the dose in urine, 7.3% in bile and an additional 15.7% in tissues not including the dissected gastro-intestinal tract. Elimination was rapid regardless of dose and half-lives were estimated to be between 8 and 24 hours. The amount absorbed represented a mean of *ca* 88% of the administered dose at the low level.

The urine of high dosed male rats revealed a complex pattern of *ca* 26 metabolite fractions when analyzed by 2-D TLC. The major metabolite fraction present in urine was identified as 2-chloro-4,6-diamino-*s*-triazine (26% of dose) and a minor amount of 2-chloro-4-ethylamino-6-amino-*s*-triazine. There was also evidence for the presence of 2-acetylcysteinyl-4,6-diamino-*s*-triazine and 2-chloro-

4-amino-6-isopropylamino-*s*-triazine. The metabolic pattern in urine from the low-dosed bile duct cannulated rats was qualitatively similar to the high dosed animals.

About 78% of the radioactivity present in the feces of high dosed male rats was extractable with neutral solvents. Two-dimensional TLC of the feces extract showed a less complex pattern, but qualitatively similar to that found in urine. It consisted of *ca* 12 metabolite fractions each accounting from 0.1-2.3% of the dose. Unchanged parent, and its two monodealkylated chloro-*s*-triazine metabolites together accounted for 1.68% of the dose.

The bilary metabolite pattern for the low dose male rats was qualitatively similar but less complex than those observed for urine and feces. It consisted of *ca* 9 metabolite fractions ranging from 0.1-1.6% of the dose. The major fraction corresponded to 2-chloro-4,6-diamino-*s*-triazine. Minor fractions corresponded to the monodealkylated chloro-*s*-triazine metabolites.

The major degradation pathway for atrazine in rats is stepwise N-dealkylation, resulting in the production of monodealkylated chloro-*s*-triazines and ultimately 2-chloro-4,6-diamino-*s*-triazine, which is the major metabolite.

The metabolism of simazine (2-chloro-4,4-bis(ethylamino)-*s*-triazine) in rats was studied at 0.5 mg/kg and 100 mg/kg dose levels (*3*). The route of administration, sample collection, and nature of the dosing groups were similar to the previously discussed atrazine study. In the case of the simazine study, two dosing groups (low and high) contained both sexes of cannulated animals. Other dosing groups were utilized to provide samplings of urine, feces, blood, and selected tissues at multiple time points after oral administration.

The following conclusions are drawn from this study: Independent of the sex of the animal, about 90% and 65% of an orally administered dose was absorbed at the low and high dose level, respectively. The time to reach maximum blood concentration was dose dependent. Maximum blood concentrations were achieved within 2 hours and 18 hours after administration of the low and high dose, respectively.

The routes of excretion were dose dependent, but independent of sex. At the low dose level the principal route of excretion was urine (63%), with lesser amounts in the feces (25%). The corresponding values for the high dose level were 39% (urine) and 49% (feces). Excretion was rapid as more than 95% of the radioactivity found in urine and feces was present in the 0-48 hour samples.

Within 48 hours, the low and high level cannulated animals eliminated 8%, 69%, and 4% and 4%, 41%, and 16% of the dose via bile, urine, and feces, respectively. Hence, a significant part of dose eliminated via the feces of non-cannulated rats was absorbed and reentered the intestinal tract by biliary excretion.

Analagous to the atrazine-dosed rats, the kidneys, liver, and red blood cells from simazine-dosed animals contained the highest amounts of radioactivity independent of dose and sex. Other tissue residues were low and depuration was dependent on their blood content.

The analysis of urine, feces, and bile from the simazine-dosed rats were reported by Thanei *et al* (*4*). 2-D TLC of excreta revealed a complex pattern of *ca* 20 metabolite fractions in urine. The patterns in feces (9 fractions) and bile (4 fractions) were less complex but qualitatively similar to the urine. Based upon the structures of the metabolites identified, the metabolic pathway for simazine in the rat involves stepwise oxidative dealkylation to monodesethyl simazine and finally to 2-chloro-4,6-diamino-*s*-triazine the major metabolite in urine, feces, and bile. A minor pathway involves oxidation of the ethyl side chain resulting in primary alcohols and carboxylic acids. Another minor pathway was dechlorination by glutathione followed by degradation of the glutathione conjugates to various sulfur containing metabolites such as cysteine derivatives, mercapturates, sulfides, disulfides, and sulfoxides.

The metabolism of ametryn, a thiomethyl analog of atrazine, was studied in the rat at dose levels of 0.5 mg/kg and 200 mg/kg (*5*). The absorption, excretion, tissue uptake, and disposition of ametryn in rats was similar to the previously discussed chloro-*s*-triazines (atrazine and simazine) and therefore will not be reiterated. An overview of its metabolism in rats is as follows: Ametryn was extensively metabolized to *ca* 35 different fractions present in urine and feces extracts when separated by HPLC and TLC. The thiomethyl group of ametryn is easily oxidized to the corresponding sulfoxide which makes an excellent leaving group for nucleophilic substitution with glutathione. Since atrazine and ametryn have in common ethylamino and isopropylamino side chains, once glutathione conjugation occurs with either parent compound or their dealkylated metabolites, the subsequent metabolism would be expected to be identical.

Ametryn metabolism in the rat is characterized primarily by two competing reactions, oxidative dealkylation and glutathione conjugation. Side chain oxidative intermediates were primarily observed on the isopropyl group leading to isopropanol and isopropionate derivatives and dealkylated moieties. Glutathione conjugates of ametryn and its dealkylated thiomethyl metabolites were apparently very labile and were present in excreta primarily as mercapturates. Other minor pathways included hydrolytic deamination with subsequent sulfate conjugation and degradation of ametryn to a postulated mercaptan intermediate prior to S-glucuronide formation. Some evidence for disulfide formation from the mercaptan of parent compound was also observed in the urine of high dosed animals from both sexes. A detailed metabolic pathway for ametryn in the rat including all of the identified products and proposed intermediates is presented in Figure 1. Most of the degradative processes operative in this pathway would also be applicable to other chloro-*s*-triazine and thiomethyl-*s*-triazine analogs in the rat.

Hydroxyatrazine is a major plant and soil metabolite of atrazine and is representative of the metabolite class known as hydroxy-*s*-triazines. A recent metabolism study (*6*) was conducted in support of a rat chronic feeding/oncogenicity study. Rats of both sexes were dosed orally at 0.05 mg/kg and 36.9 mg/kg. Approximately 168 hours after administration of the single dose, the rats were sacrificed and selected tissues excised. Urine and feces were collected periodically over the 168 hour test period.

The low dose group of animals average excretion (% of dose) amounted to 82.7% in urine and 5.1% in feces. Corresponding values for the high dose group was 16.3% in urine and 87.7% in feces. Because of the limited solubility of hydroxyatrazine in water, the high percentage of the dose excreted in feces by the high dose animals probably reflects a reduced capacity for absorption of the dose.

Hydroxyatrazine has a much less complex metabolic pattern in urine and feces than those observed for the chloro- and thiomethyl-*s*-triazines investigated. HPLC of the low and high dose urine resolved mostly intact parent compound and four metabolite fractions. HPLC of the low dose feces extracts resolved parent compound and three metabolite fractions. The high dose feces extracts only contained unaltered parent compound. In addition to hydroxyatrazine, the urine and feces extracts contained 2-hydroxy-4-amino-6-isopropylamino-*s*-triazine and 2-amino-4-ethylamino-6-hydroxy-*s*-triazine. It was postulated that low dose feces extracts contained 2-hydroxy-4-etheylamino-6-isopropylamino-*s*-triazine based upon limited mass spectral data.

Metabolism of Selected (S)-Triazines in Goats and Hens

The metabolism of atrazine, simazine, and ametryn in lactating animals and chickens has been the subject of extensive investigations in recent years. Three overview documents (*7-9*) were prepared to summarize the metabolism of atrazine and simazine in large animals treated at exaggerated rates in order to detect and describe the nature of the residues. A summary of these findings are as follows:

The chloro-*s*-triazine and thiomethyl *s*-triazine compounds investigated have similar metabolic fates in large animals. These studies were conducted at levels ranging from 5 ppm to 100 ppm equivalent to parent compound in the feed. Exaggerated feeding levels were utilized in order to produce enough residues in edible commodities to facilitate identification. Typically, most of the radioactive dose was excreted rapidly by lactating animals in urine (~70%) and feces (~20%) or predominantly in excreta (~90%) by hens. The metabolic profiles in urine (~20 fractions) were much more complex than those observed in feces (~10 fractions) for lactating animals. Excreta profiles obtained from hens were also complex. These profiles resembled those obtained previously from the rat. The individual metabolites can be traced to two predominant degradative processes: cleavage of side-chain alkyl groups to give mostly simple chloro-*s*-triazine and thiomethyl-*s*-triazine metabolites and conjugation of chloro-*s*-triazine and thiomethyl-*s*-triazine moieties with glutathione leading to formation of cysteine conjugates, mercapturates, mercaptans, sulfides, and disulfides (Figure 2).

CONCLUSION

The metabolism of chloro-*s*-triazines and thiomethyl-*s*-triazines in animals have been extensively studied over the past several years and the nature of residues is well understood. The major biotransformations of the chloro-*s*-triazines involves N-dealkylation of the side-chains. The major biotransformations of the thiomethyl-*s*-triazines involves both N-dealkylation of the side-chains and conjugation with glutathione to form various mercapturates.

The metabolic pathways in animals treated with chloro- and thiomethyl-*s*-triazines, and hydroxyatrazine are shown in Figures 1, 2 and 3.

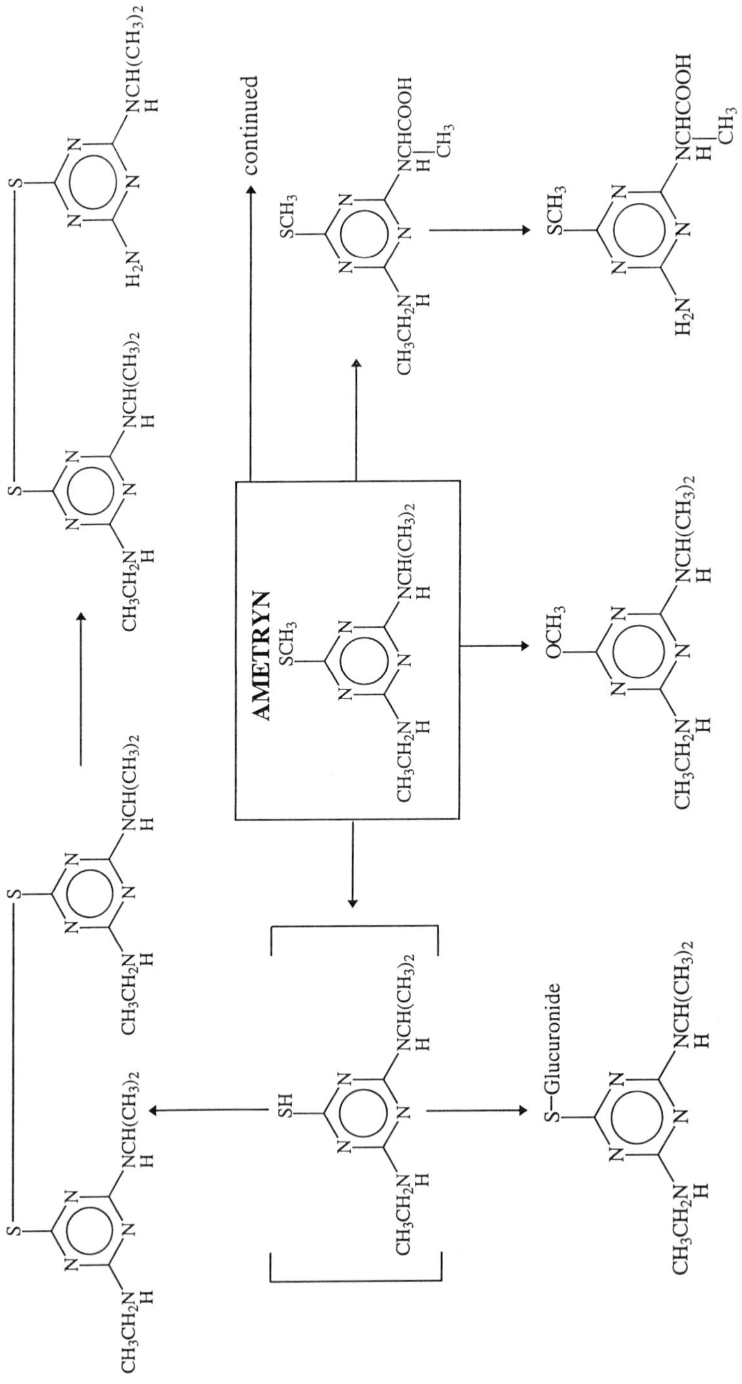

S — S

CH_3CH_2NH · N · N · N · $NHCH(CH_3)_2$ CH_3CH_2NH · N · N · N · $NHCH(CH_3)_2$

→

S — S

CH_3CH_2NH · N · N · N · $NHCH(CH_3)_2$ H_2N · N · N · N · $NHCH(CH_3)_2$

AMETRYN

SCH_3

CH_3CH_2NH · N · N · N · $NHCH(CH_3)_2$

→ continued

[SH

CH_3CH_2NH · N · N · N · $NHCH(CH_3)_2$]

SCH_3

CH_3CH_2NH · N · N · N · $NHCH(CH_3)COOH$

S–Glucuronide

CH_3CH_2NH · N · N · N · $NHCH(CH_3)_2$

OCH_3

CH_3CH_2NH · N · N · N · $NHCH(CH_3)_2$

SCH_3

H_2N · N · N · N · $NHCH(CH_3)COOH$

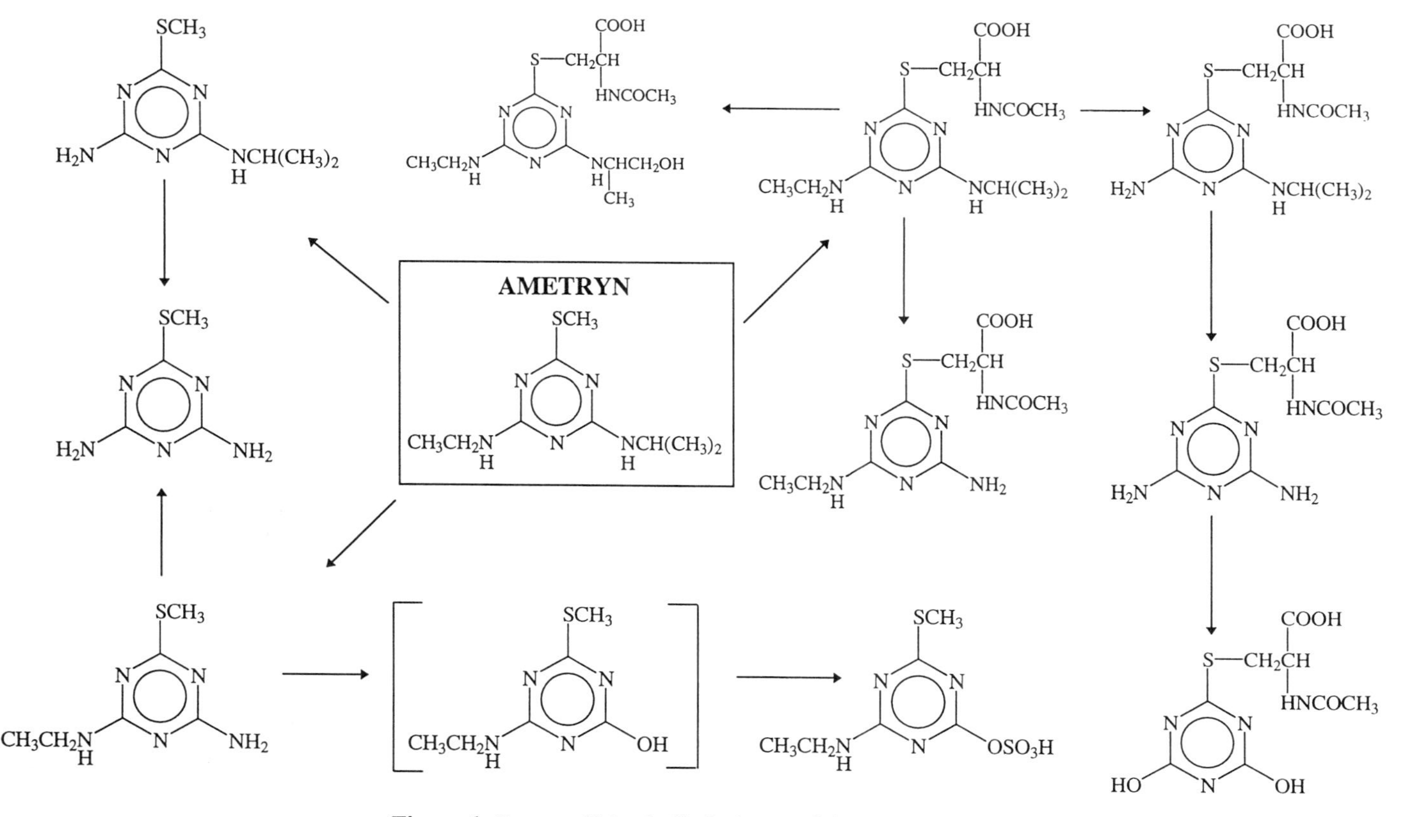

Figure 1. Proposed Metabolic Pathway of Ametryn in Rats

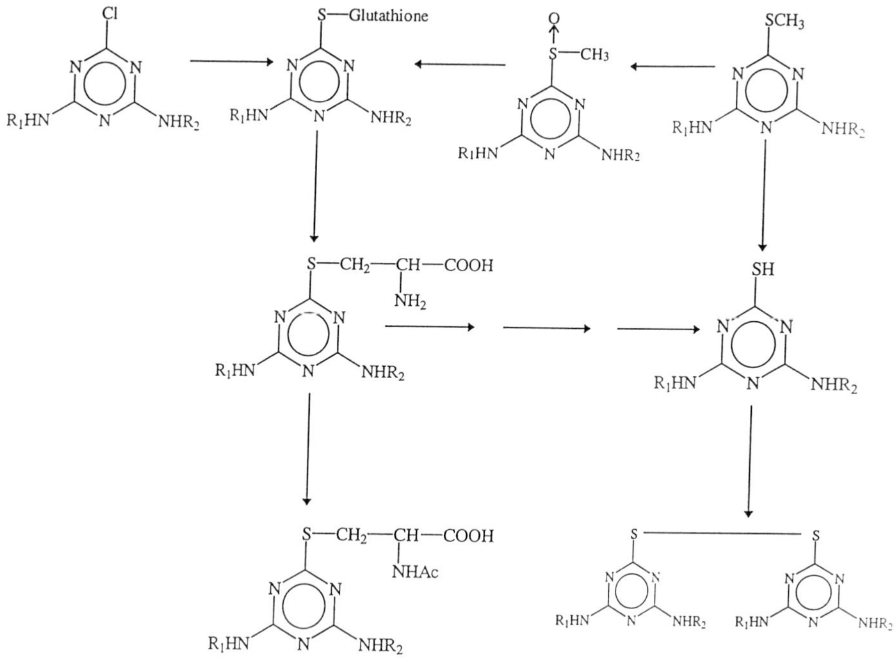

where R_1 = H or C_2H_5 and R_2 = H, C_2H_5 or C_3H_7

Figure 2. General Metabolic Pathway of Chloro- and Thiomethyl-(S)-Triazines in Animals

LITERATURE CITED

1. Paul, H. J.; Dunsire, J. P.; Hedley, D. Inveresk Research International Report No. 9523, *"The Absorption, Distribution, Degradation and Excretion of [U-^{14}C] Triazine G 30027 in the Rat,"* 1993.
2. Hamboeck, H.; Fischer, R.W.; Di Iorio, K.W. Winterhalter *Molecular Pharmacology*. **1981**, *20*, pp.579-584.
3. Johnston, A.M., Clydesdale, K., Somers, K., Speirs, G. C. Inversek Research International Report No. 8744, "The Absorption, Distribution and Excretion of [U^{14}C] Triazine G 27692 in the Rat," **1992**.
4. Thanei, P., Ciba-Geigy Report No. 7/92, "The Metabolic Profiles in Urine, Bile and Feces of Rats after Administration of [U-^{14}C] Triazine G 27692," **1992**.
5. Wu, Diana, Xenobiotic Laboratories, Inc. Report No. RPT 0022, "Analysis, Quantitation and Structure Elucidation of Metabolites in Urine and Feces from the Rat Dosed with ^{14}C-Ametryn," **1990**.

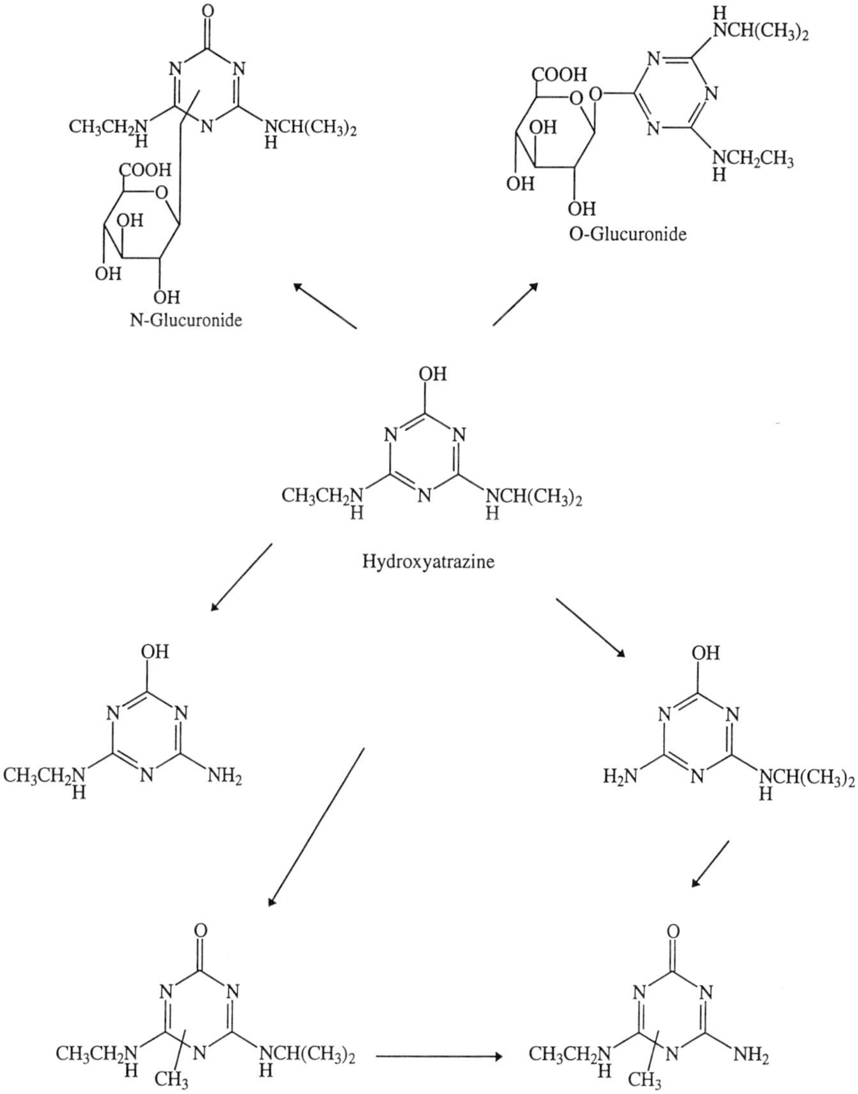

Figure 3. Proposed Metabolic Pathway of Hydroxyatrazine in Animals

6. Tortora, N., Ciba-Geigy Report No. F-00169, "Metabolism of [Triazine-^{14}C] Hydroxy-Atrazine (G-34048) in Rats, **1994**.
7. Simoneaux, B., Ciba-Geigy Report No. ABR-87112, "Nature of Atrazine Metabolism in Poultry and Ruminants," **1987**.
8. Thede, B., Simoneaux, B., Ciba-Geigy Report No. ABR-88050, "Nature of Simazine Metabolism in Poultry and Ruminants, **1988**.
9. Thede, B., Ciba-Geigy Report No. ABR-89053, "Nature of Atrazine Residues in Animals," **1989**.

Chapter 10

Magnitude and Nature of (*s*)-Triazine Residues in Foodstuffs as Predicted from Radiolabeled Studies on Selected Animals and Plants

Bruce J. Simoneaux[1], Dennis S. Hackett[1], Leslie D. Bray[1], and Fred Thalaker[2]

[1]Agricultural Division, Ciba-Geigy Corporation, P.O. Box 18300, Greensboro, NC 27419
[2]Metabolism and Environmental Fate, Corning-Hazelton P.O. Box 7545, Madison, WI 53707–7545

Chloro-*s*-triazines, hydroxy-*s*-triazines, amino-*s*-triazines, and conjugates of chloro-*s*-triazines derived from the glutathione pathway represent the major classes of residues identified in plants treated with ^{14}C-labeled atrazine and simazine. The principal registered crops for these herbicides were grown in field plots to determine the magnitude and nature of these residues at various harvest intervals. The distribution and identification of metabolites were determined in agricultural commodities to assess any potential dietary exposure to humans as a result of food consumption. An overall dietary exposure assessment based upon a new three level-dairy feeding study and magnitude of (*s*)-triazine residues in foodstuffs was calculated.

Atrazine, 2-chloro-4-ethylamino-6-isopropylamino-*s*-triazine, and simazine, 2-chloro-4,6-bis-ethylamino-*s*-triazine are representative chloro-*s*-triazine herbicides used extensively for weed control in corn, citrus, grapes, sorghum, sugarcane, and several other fruit and nut crops. Atrazine can be applied either pre- or post-emergence to corn and sorghum while simazine can only be applied pre-emergence to corn. All studies on crop uptake and metabolism were conducted at the maximum registered or an exaggerated use rates at the time the studies were initiated. In 1992, atrazine use rates were lowered on corn and sorghum to a single maximum rate of 2.0 lbs a.i./A and a maximum of 2.5 lbs a.i./A per calendar year. The nature and magnitude of terminal metabolic products was determined in the raw agricultural commodities (RACs) in order to assess any potential dietary risk to humans as a result of food consumption either directly or indirectly.

Dietary exposure was used in conjunction with toxicology information to determine the overall risk to various sub-populations in man. A dietary exposure assessment was conducted for the U.S. population and subgroups using anticipated residues in treated foods and animal feeds.

Nature and Magnitude of the Residue(s) in Plants

The pathway of atrazine metabolism in corn, sorghum, and sugarcane is complex and has been thoroughly discussed in another chapter of this symposium series book. Simazine follows similar metabolic processes in corn and citrus. The primary metabolic transformations of these chloro-*s*-triazines in plants are the result of three competing reactions: N-dealkylation of the triazine ring side-chains, hydrolytic dehalogenation, or nucleophilic displacement of the chlorine atom with glutathione (GSH). These metabolites can be grouped into four classes: chloro-*s*-triazines, hydroxy-*s*-triazines, amino-*s*-triazines, and conjugates. Some of the terminal metabolites in plants are illustrated in the glossary. The Phase I metabolites differ in the nature of the group attached at the 2-position of the triazine ring (Cl, OH, or NH_2). Phase II metabolites differ in the type of natural product attached to the triazine ring (amino acid, peptide, or sugar).

When atrazine was applied as a post-emergence spray at exaggerated use rates to corn and sorghum (*1*) (3 lbs a.i./A) in small field plots at three different geographic locations (Illinois, Mississippi, and New York), total radioactive residues (TRR) averaged less than 1.5 ppm for corn forage and 3.0 ppm or less for sorghum forage (Table I). Total residues declined rapidly with time after post-emergence treatment with some evidence of concentration in fodder samples due to desiccation of the plants. Total radioactive residues in grain samples averaged 0.05 ppm for corn and 0.184 ppm for sorghum. The combined Phase I metabolites accounted for 25% to 50% of the TRR in early forage samples and less than 10% of the TRR in grain samples. Atrazine and its dealkylated chloro-*s*-triazine metabolites, desethylatrazine (6-chloro-N-(1-methylethyl)-1,3,5-triazine-2,4-diamine), desisopropylatrazine (6-chloro-N-ethyl-1,3,5-triazine-2,4-diamine), and diaminochloro-*s*-triazine (6-chloro-1,3,5-triazine-2,4-diamine), accounted for less than 20% of the early forage residues and only 2% or less of the grain residues. Individual radiolabeled chloro-*s*-triazines could not be measured directly in grain extracts but were estimated to be equal to the amount of organic-soluble radioactivity present after partitioning the aqueous solubles with chloroform. The actual content of chloro-*s*-triazines in grain is likely to be less than this estimate predicts because residues of hydroxy-*s*-triazine and amino-*s*-triazine metabolites can be found in this fraction as well. Hydroxy-*s*-triazines (hydroxyatrazine (4-(ethylamino)-6-[(1-methylethyl)amino]-1,3,5-triazine-2(1H)-one), desethylhydroxy-atrazine (4-amino-6-[(1-methylethyl)amino]-1,3,5-triazine-2(1H)-one), desethylhydroxysimazine (4-amino-6-(ethylamino)-1,3,5-triazin-2(1H)-one, and ammeline (4,6-diamino-1,3,5-triazin-2(1H)-one), accounted for approximately two to four times as much of the TRR in corn as was found in sorghum. The difference can be accounted for by the greater potential for sorghum to conjugate through the glutathione pathway than corn. Amino-*s*-triazines (aminoatrazine (N-ethyl-N'-(1-methylethyl)-1,3,5-triazine-

2,4,6-triamine) and N-(1-methylethyl)-1,3,5-triazine-2,4,6-triamine) account for approximately 2% of the TRR in forage samples and were not detected in grain extracts. The remaining residues can be accounted for by conjugates and non-extractable radioactivity.

When simazine was applied as a pre-emergence spray at the maximum use rate to corn (*2*) (2 lb a.i./A) in Illinois, total radioactive residues were less than 0.5 ppm in forage and fodder samples and only 0.044 ppm in grain (Table II). The 30-DAT forage sample was not further analyzed because it is not common practice to feed this early forage to livestock. The combined Phase I metabolites accounted for approximately 25% to 50% of the TRR in fodder and silage samples and 10% of the grain residues. Chloro-*s*-triazines accounted for 10% or less of the silage and fodder residues and only 1% of the grain residues. Hydroxy-*s*-triazines accounted for the bulk of the Phase I metabolites (10% to 35%) in grain, fodder and silage samples. Amino-*s*-triazines accounted for less than 2% of the fodder and silage TRR. In general, pre-emergence treatment of corn with simazine results in less total residues than post-emergence treatment of corn with atrazine. The percentages of identified Phase I metabolites are approximately the same at similar stages of maturity for fodder and grain samples.

Sugarcane (*3*) was treated four times with ^{14}C-atrazine for a total of 10 lbs a.i./A in California. By plant maturity, the stripped canes contained 2.09 ppm total radioactive residues (Table III). Analysis of the cane extracts showed that 0.108 ppm were associated with chloro-*s*-triazines, 0.173 ppm with hydroxy-*s*-triazines and most of the remaining residues were complex conjugates. A comparison of residues from atrazine-treated field plots demonstrates that the amount of chloro-*s*-triazines and hydroxy-*s*-triazines in cane are very similar to the amount in cane from the radiolabeled metabolism study. When the cane grown in the residue plots was processed, refined sugar and molasses did not contain any chloro-*s*-triazine or hydroxy-*s*-triazine residues above the limit of detection (0.02 ppm for hydroxy-*s*-triazines and 0.002 ppm for chloro-*s*-triazines) except for one value of hydroxy-*s*-triazine in molasses at 0.020 ppm. For the purposes of dietary exposure assessment that will be discussed later in this chapter, the three groups of triazine residues evaluated are chloro-*s*-triazines, hydroxy-*s*-triazines, and TRR minus hydroxy-*s*-triazines (primarily conjugates and small amounts of chloro-*s*-triazines). The majority of the radiolabeled residues reside in the latter group.

Dietary Exposure Analysis

A dietary exposure assessment was conducted for atrazine and its metabolites (*4*) and the results were expressed as a percentage of the reference dose (RfD) which is based on a no effect level (NOEL) obtained from the chronic rat feeding studies and a 100-fold safety factor. The exposure assessment was divided into three residue subsets (provided below with the corresponding reference doses):

Table I. Average Values for Total Radioactive Residues in Corn and Sorghum Treated Post-Emergence at Exaggerated Use Rates and the Relative Percentages of Phase I Metabolite Classes

	CORN							
	30-Day Forage		*Silage*		*Fodder*		*Grain*	
PHI (Days)	30		79		112		112	
Metabolite Class	%	PPM	%	PPM	%	PPM	%	PPM
TRR	100	1.333	100	0.623	100	1.403	100	0.0500
Chloro-*s*-triazines	16.0	0.212	1.2	0.008	0.7	0.009	1.0	0.0005
Hydroxy-*s*-triazines	34.8	0.464	29.3	0.182	20.9	0.293	8.9	0.0040
Amino-*s*-triazines	1.7	0.022	1.8	0.011	2.6	0.036	ND	<0.0010
Total Phase I	52.4	0.698	32.3	0.201	24.2	0.308	9.9	0.0045
	SORGHUM							
	30-Day Forage		*Silage*		*Fodder*		*Grain*	
PHI (Days)	30		87		126		126	
Metabolite Class	%	PPM	%	PPM	%	PPM	%	PPM
TRR	100	3.036	100	0.861	100	0.789	100	0.184
Chloro-*s*-triazines	18.9	0.574	3.0	0.026	2.8	0.022	2.0	0.004
Hydroxy-*s*-triazines	6.9	0.209	6.3	0.055	7.4	0.058	4.1	0.008
Amino-*s*-triazines	2.0	0.061	2.1	0.018	2.0	0.016	ND	<0.001
Total Phase I	27.8	0.844	11.4	0.099	12.2	0.096	6.1	0.012

Table II. Total Radioactive Residues in Simazine Treated at Maximum Use Rate Pre-Emergence Corn and the Relative Percentages of Phase I Metabolite Classes

	30-Day Forage		*Silage*		*Fodder*		*Grain*	
PHI (Days)		30		120		162		162
Metabolite Class	%	PPM	%	PPM	%	PPM	%	PPM
	100	0.158	100	0.209	100	0.493	100	0.044
Chloro-*s*-triazines	NA		8.4	0.018	3.7	0.018	1.2	0.0005
Hydroxy-*s*-triazines	NA		35.2	0.074	18.5	0.091	10.2	0.004
Amino-*s*-triazines-	NA		1.0	0.002	2.0	0.010	ND	<0.001
Total Phase I	NA		44.6	0.094	24.2	0.119	11.4	0.0045

Table III. Comparison of Triazine Residues in Sugarcane Treated at lb. ai/A Raw Agricultural Commodities from Metabolism and Residue Field Trials

	Radiolabeled Metabolism Study (PPM)		*Field Residue Study (PPM)*	
Metabolite Class	Harvest Cane	Mature Forage	Refined Sugar	Molasses
TRR	2.093			
Chloro-*s*-triazines	0.108	0.14	<0.002	<0.002
Hydroxy-*s*-triazines	0.173	0.10	<0.020	0.020
TRR Minus Hydroxy-*s*-triazines	1.920	1.20[a]	0.24[a]	0.24[a]

[a]Estimated residue based on proportionality factor derived from metabolism study

1) combined parent atrazine plus chloro-*s*-triazine metabolites (atrazine RfD = 0.035 mg/kg body weight/day)
2) combined free hydroxy-*s*-triazine metabolites (% RfD = 0.01 mg/kg body weight/day for hydroxy-*s*-atrazine)
3) total triazine residues minus combined free hydroxy-*s*-triazines [estimates conjugates plus chloro-*s*-triazines (RfD = 0.035 mg/kg body weight/day for atrazine)]

The dietary exposure analysis was conducted using the Technical Assessment Systems Inc. (TAS) Exposure 1 program which utilizes data from the USDA 1977-78 nationwide food consumption survey. The TAS Exposure 1 program was used to estimate the mean chronic exposure to food constituents comprising the diets of the average U.S. population and population subgroups. To assess residue levels in meat, milk, poultry and eggs, a representative dairy cattle diet was developed by Dr. Jim Spain of the Animal Sciences Center at the University of Missouri (*5*). This diet, containing atrazine-treated feed commodities was developed to provide adequate nutrition to lactating dairy cattle. Anticipated residues in cattle feed constituents were adjusted for the percent of crops treated with atrazine in the U.S. (market share data) to provide an exposure estimate on a national basis (48 states) (*6*). All estimated residues in the cattle feed constituents were corrected for percent dry weight. In addition, the residue contribution from pre-emergence and post-emergence application of atrazine to corn and sorghum were adjusted accordingly for the percent of total acreage treated pre- and post-emergence in the U.S (*6*).

This dietary exposure analysis utilized anticipated residue data from both pre-emergence and post-emergence field trials and radiolabeled field studies on corn and sorghum (*1, 7-10*), rotational crops (*11*), sugarcane (*3*), and a 3-level dairy cow feeding study (*12*). The levels of chloro-*s*-triazine residues (atrazine, desethylatrazine (6-chloro-N-(1-methylethyl)-1,3,5-triazine-2,4-diamine), desisopropylatrazine (6-chloro-N-ethyl-1,3,5-triazine-2,4-diamine), and diaminochloro-*s*-triazine (6-chloro-1,3,5-triazine-2,4-diamine), were measured directly from field residue samples. The chloro-*s*-triazines were summed for each commodity at each site and the results averaged. The hydroxy-*s*-triazine residues (hydroxyatrazine (4-(ethylamino)-6-[(1-methylethyl)amino]-1,3,5-triazine-2(1H)-one); desethylhydroxy-atrazine (4-amino-6-[(1-methylethyl)amino]-1,3,5-triazine-2(1H)-one); desethylhydroxysimazine (4-amino-6-(ethylamino)-1,3,5-triazin-2(1H)-one); and 4,6-diamino-1,3,5-triazin-2(1H)-one) and the total triazine residues were measured from ^{14}C-metabolism studies. For each commodity (except sugarcane, molasses and animal substrates), the total triazine residues minus hydroxy-*s*-triazines were determined by subtracting the total ppm of combined free hydroxy-*s*-triazines from the total radioactive residue (TRR). The magnitude of hydroxy-*s*-triazines and total triazines (and total triazines minus hydroxy-*s*-triazines) in the residue field trials were estimated from ^{14}C-metabolism studies using proportionality factors. A description of the data used in the exposure estimate is provided below.

Because of the difficulty associated with processing radioactive-treated substrates, processing data for sugarcane are limited to chloro-*s*-triazines and hydroxy-*s*-triazines (*13,14*). In addition, chloro-*s*-triazine data only were available for meat, milk, and eggs so the amount of hydroxy-*s*-triazines and total triazines minus the hydroxy contribution were estimated.

In order to estimate the total triazines minus the hydroxy-*s*-triazine contribution in sugarcane and molasses, a proportionality factor was derived from other commodities using the ratio of total triazines minus hydroxy-*s*-triazines/total hydroxy-*s*-triazines. This ratio was calculated for other substrates and the ratios were averaged. This proportionality factor was then multiplied by the total amount of hydroxy-*s*-triazines in refined sugar and molasses in order to conservatively estimate the total triazine minus hydroxy contribution. No chloro- or hydroxy-*s*-triazine residues were detected in refined sugar generated from eight sugarcane trials conducted in 1993 (*13,14*). All non-detects were evaluated as one-half the limit of quantitation (1/2 LOQ = <0.02 ppm). Similarly, no chloro-*s*-triazine residues were detected in molasses and only one detect of hydroxy-*s*-atrazine was found at the LOQ (0.02 ppm). The proportionality factor was multiplied by the LOQ value of 0.02 ppm for both sugar and molasses for the amount of total triazines minus hydroxy-*s*-triazines. A comparison between tolerances and anticipated residue levels is provided in Table IV.

The major residue contributors to cattle diet are corn silage and sorghum forage. In order to realistically assess anticipated residues, field trials were conducted with corn and sorghum at the current maximum label use rate for pre-emergence application (2 lb a.i./A), as well as applications made pre-emergence at 0.5 lb a.i./A followed by 2 lb a.i./A post-emergence (*9,10*). These field trials afforded a geographical distribution of the magnitude of chloro-*s*-triazine residues at the maximum allowable use rates. The majority of U.S. atrazine use on corn and sorghum is pre-emergence (75.5% and 73.3%, respectively), with post-emergence application accounting for the remainder (*6*). The average residue contributions from the pre- and post-emergence residue field trials were adjusted by the percent of total use practice in the U.S. Hydroxy-*s*-triazine and

Table IV. Summary of Tolerances and Anticipated Residues from Maximum or Exaggerated Use Rates and Short Pre-Harvest Intervals in the Raw Agricultural Commodities Used in the Dietary Exposure Assessment of Atrazine

Commodity	*Tolerance (ppm)*	*Chloro-s-Triazines (ppm)*	*Hydroxy-s-Triazines (ppm)*	*Total Triazines Minus Hydroxy-s-triazines (ppm)*
Corn Silage	15	0.035044	0.598157	1.5371
Corn Grain	0.25	0.000275	0.001553	0.03270
Sorghum Forage	15	0.22740	0.185238	3.465
Sorghum Grain	0.25	0.000962	0.00295	0.06671
Refined Sugar	0.25	<0.002	<0.02	0.24*
Molasses	1.5	<0.1	0.02	0.24*
Wheat Grain	0.25	0.002	0.002	0.052
Wheat Straw	5	0.051	0.019	0.292
Macadamia Nuts	0.25	0.1	0.1	0.1
Guava	0.05	0.01	0.01	0.01

*Estimated value as described in text.

total triazine (TRR) data for corn and sorghum were generated from both pre- and post-emergence ^{14}C-metabolism studies. These data were used to predict hydroxy-*s*-triazine and total triazine levels in the residue field trials. The average concentrations of hydroxy-*s*-triazines and the total triazines minus hydroxy-*s*-triazines (predicted from the radiolabeled studies) were also adjusted for the percent of total use practice in the U.S.

In order to predict the pre-emergence contribution of hydroxy-*s*-triazines and total triazines minus hydroxy-*s*-triazines in the residue field trials, a ^{14}C-metabolism study was conducted by pre-emergence application to corn and sorghum at the current 2 lb a.i./A use rate (*7,8*). This study was performed at one site and provided ratios of the amount of hydroxy-*s*-triazines and total triazines present in relation to the amount of chloro-*s*-triazine residues. Using these ratios, it was possible to predict the amount of hydroxy-*s*-triazine and total triazines present in the field studies based on the amount of chloro-*s*-triazines present in the field samples. For the post-emergence residue contribution of hydroxy-*s*-triazines and the total triazines, ratios were generated from ^{14}C-metabolism studies conducted by post-emergence application of atrazine at an exaggerated rate of 3 lb a.i./A at three sites (*1*). Although these data overestimated anticipated residues expected from the current label's maximum post application rate of 2.0 lb a.i./A, these data were appropriate to generate ratios to predict hydroxy-*s*-triazines and total triazines in the field trials.

Levels of chloro-*s*-triazines in milk were estimated from a three-level feeding study using ^{14}C-labeled atrazine (*12*). Levels of chloro-*s*-triazines in beef tissues and poultry tissues plus eggs were estimated from three-level dairy cattle and poultry feeding studies (*15,16*). Residue levels of hydroxy-*s*-triazines in cattle tissues and milk were estimated from a ^{14}C-hydroxy-*s*-atrazine feeding study in lactating goats (*17*). Hydroxy-*s*-triazine residues in poultry tissues and eggs were estimated from a ^{14}C-atrazine biosynthesized metabolites poultry feeding study (*18*). Total triazine levels in cattle tissues, milk, poultry tissues, and eggs were estimated from feeding studies conducted with ^{14}C-atrazine biosynthesized metabolites in goats and poultry (*18,19*) and as summarized by the EPA (*20*).

The total dietary exposure was compared to the reference dose for atrazine (chloro-*s*-triazines and total residues minus the hydroxy-*s*-triazine component) and hydroxy-*s*-atrazine (hydroxy-*s*-triazines). The dietary exposure assessment for the three residue subsets and the infant and children population subgroups is provided below in Table V.

Table V. Dietary Exposure Assessment

Most Sensitive Population Subgroup and Exposure Scenario	*Dietary Exposure mg/kg body wt/day*	*% of Reference Dose*
Total Chloro-*s*-Triazines		
(RfD = 0.035 mg/kg/day)		
U.S. Pop, 48 states, All Seasons	0.000007	0.02
Non-nursing infants (<1 year old)	0.000031	0.09
Children (1 - 6 years)	0.000019	0.05
Total Free Hydroxy-*s*-Triazines		
(RfD = 0.01 mg/kg/day)		
U.S. Pop, 48 states, All Seasons	0.000039	0.39
Non-nursing infants (<1 year old)	0.000162	1.62
Total Triazines minus Hydroxy-*s*-Triazines		
(RfD = 0.035 mg/kg/day)		
U.S. Pop, 48 states, All Seasons	0.000229	0.65
Children (1 - 6 years)	0.000583	1.67

The threshold estimates of dietary exposure non-nursing infants (<1 year old) and children (1-6 years) with an exposure of 0.09% and 0.05% of the atrazine RfD, respectively, for chloro-*s*-triazine residues. The theoretical estimates of hydroxy-*s*-triazine exposure for non-nursing infants (<1 year old) were 1.62% of the RfD for hydroxy-*s*-atrazine. For the total triazine minus hydroxy-*s*-triazine residue subset, the most sensitive sub-population was children (1-6 years) had a theoretical estimated exposure of 1.67% of the RfD for atrazine. It is evident that the total dietary exposure in each case represents only a very small percentage of the reference dose which is based on a no effect level and a wide margin of safety.

A dietary exposure assessment was also conducted on simazine (*21*). As with atrazine, this exposure assessment included three residue subsets (chloro-*s*-triazines, hydroxy-*s*-triazines and total triazines minus hydroxy-*s*-triazines). This chronic dietary exposure analysis utilized residue data from radiolabeled and residue field trials on corn (*2*), citrus (*22*), grapes (*23*) and apples (*24*). The ^{14}C-metabolism data were used to predict the hydroxy-*s*-triazines and total triazines (and total triazines minus hydroxy-*s*-triazines) generated in the field residue trials, in which only chloro-*s*-triazines were measured. Data for all commodities were not available so surrogate data were used from the same or similar crop groups.

The anticipated residues were adjusted for percent market share to provide estimates on a national basis and all feed item values were corrected for percent dry weight (*6*).

The dietary burden to cattle was evaluated using a cattle diet developed by Dr. Jim Spain of the Animal Sciences Center at the University of Missouri (*5*). The diet used in this exposure assessment was developed to provide adequate nutrition to lactating dairy cattle and also to maximize the quantity and number of feed items treated with simazine. The dietary transfer of residues from meat, milk, poultry and eggs (chloro-*s*-triazines only) was calculated using transfer factors (slopes) obtained in ruminant and poultry three-level feeding studies (*25,26*). The transfer of hydroxy-*s*-triazines and total triazines minus hydroxy-*s*-triazines in poultry meat and eggs was estimated from total triazine levels measured in poultry metabolism studies conducted using exaggerated dose levels. Use of these values gives a conservative estimate for each of the two residue subsets (hydroxy-*s*-triazines and total triazines minus the hydroxy-*s*-triazine contribution) since the tissue levels (obtained by combustion analysis) represent TRR.

For simazine plus chloro metabolites and the total triazines minus hydroxy-*s*-triazine contribution, the EPA-approved reference dose of 0.005 mg/kg body weight/day was used based on the rat chronic NOEL of 0.05 mg/kg body weight/day and a one hundred-fold safety factor. The hydroxy-*s*-triazine residue contribution was evaluated utilizing the reference dose for hydroxy-*s*-atrazine (0.01 mg/kg/body weight/day) based on the rat chronic NOEL for hydroxy-*s*-atrazine and a one hundred-fold safety factor. It was assumed that the reference dose for hydroxy-*s*-atrazine would serve as an acceptable surrogate for hydroxy-*s*-simazine due to the structural similarity between these chemicals. The results of the dietary exposure assessment for the U.S. population and infants and children sub-populations are summarized in Table VI below.

Table VI. Simazine Dietary Exposure Assessment

Most Sensitive Population Subgroup and Exposure Scenario	*Dietary Exposure mg/kg body wt/day*	*% of Reference Dose*
Total Chloro-*s*-Triazines		
(RfD = 0.005 mg/kg/day)		
U.S. Pop., 48 States, All Seasons	0.0000012	0.02
Non-nursing Infants (< 1 year old)	0.0000048	0.10
Total Free Hydroxy-*s*-Triazines		
(RfD = 0.01 mg/kg/day)		
U.S. Pop., 48 States, All Seasons	0.000047	0.47
Children (1 - 6 years)	0.000132	1.32
Total Triazines Minus Hydroxy-*s*-Triazines		
(RfD = 0.005 mg/kg/day)		
U.S. Pop., 48 States, All Seasons	0.000093	1.87
Children (1 - 6 years)	0.000260	5.21

The theoretical estimated dietary exposure to chloro-*s*-triazine residues for non-nursing infants (<1 year old) represented 0.10% of the simazine RfD. The theoretical exposure estimates for hydroxy-*s*-triazine children (1-6 years) were 1.32% of the RfD for hydroxy-*s*-atrazine. For the total triazine minus hydroxy-*s*-triazine residue subset, children (1-6 years) had a theoretical estimated exposure of 5.21% of the RfD for simazine. As with atrazine, it is evident that the total dietary exposure for simazine represents a very small part of the reference dose, which is based on a no-effect level and a wide margin of safety.

Literature Cited

1. John D. Larson, HWI-6117-178, "14C-Atrazine: Nature of the Residue in Corn and Sorghum."
2. John D. Larson, HWI 6117-210, "^{14}C-Simazine: Nature of the Residue in Corn."
3. John D. Larson, HWI 6117-181, Ciba Protocol No. 95-91, "^{14}C-Atrazine: Nature of the Residue in Sugarcane."
4. Leslie D. Bray, ABR-97064, "Revised dietary Assessment for Atrazine Including Two Exposure Scenarios."
5. Leslie D. Bray, ABR-97067, "Rationale for the Dairy Cattle Diet Utilized in the Revised Dietary Exposure assessment for Atrazine and Simazine."
6. Communication from Maritz Marketing Research Inc., St. Louis County, MO, and Doane Marketing Research Inc., St. Louis, MO to Ciba Crop Protection.
7. F. W. Thalacker and S. G. Ash, CHW 6117-335, "^{14}C-Atrazine: Nature and Magnitude of the Hydroxytriazine and Chloro-*s*-triazine Residues In Corn Following a Pre-Emergence Application at 2 lb ai/A."
8. F. W. Thalacker and S. G. Ash, CHW 6117-337, "^{14}C-Atrazine: Nature and Magnitude of the Hydroxy-*s*-triazine and Chloro-*s*-triazine Residues In Sorghum Following a Pre-Emergence Application at 2 lb ai/A."
9. S. E. Boyette, ABR-96087, "Atrazine - Magnitude of the Residues In or On Corn."
10. S. E. Boyette, ABR-96088, "Atrazine - Magnitude of the Residues In or On Grain Sorghum."
11. John D. Larson, HWI 6117-183, "Uptake and Metabolism of Atrazine in Field Rotational Crops Following Corn and Sorghum Treated at a Rate of 3.0 lb AI/Acre."
12. Frederic W. Thalacker, CHW 6117-325, "Determination of Transfer Rate and Nature of the Residue(s) in Milk from ^{14}C-Atrazine Treated Cows."
13. V. G. Novak, F. B. Selman and R. A. Kahrs, ABR-93044, "Atrazine - Magnitude of Residues in or on Sugarcane and Processed Fractions following Applications of AAtrex Nine-O or AAtrex 4L."
14. R. E. M. Wurz, ABR-93044, Amendment 1, "Atrazine - Magnitude of Residues in or on Sugarcane and Processed Fractions Following Applications of AAtrex Nine-O or AAtrex 4L."
15. T. R. Bade, ABR-87060, "Residues of Atrazine and its Chlorometabolites in Dairy Tissues and Milk Following Administration of Atrazine."

16. M. W. Cheung, ABR-87102, "Magnitude of Residues of Atrazine and its Chlorometabolites in Poultry Tissues and Eggs from Laying Hens Fed Atrazine in their Diet."
17. W. J. Tortora, EHC F-00123, "Metabolism of [Triazine-^{14}C]-Hydroxy-Atrazine in Lactating Goats."
18. J. Emrani, ABR-89006, "Fate of Corn Biosynthesized Metabolites of -^{14}C-Atrazine in Chickens."
19. J. E. Cassidy, GAAC-71021, "Metabolism of Atrazine Metabolites in Corn by Goats."
20. Letter from Walt Waldrop (EPA) to Thomas Parshley (Ciba), January 15, 1991, "Atrazine Special Review - Metabolism, revised anticipated residues."
21. L. D. Bray, ABR-96093, "Updated Dietary Risk Exposure Assessment for Simazine."
22. Thomas J. Burnett, Pan Ag No. 92115, "^{14}C-Simazine: Nature of the Residue in Citrus."
23. John D. Larson, HWI 6463-106, "^{14}C-Simazine: Nature of the Residue in Grapes."
24. John D. Larson, HWI 6463-104, "^{14}C-Simazine: Nature of the Residue in Apples."
25. M. Cheung, ABR-88036, "Residues of Simazine and Its Chlorometabolites in Dairy Tissues and Milk Following Administration of Simazine."
26. M. Cheung, ABR-88025, "Residues of Simazine and its Chlorometabolites in Poultry Tissues and Eggs From Laying Hens Following Administration of Simazine.

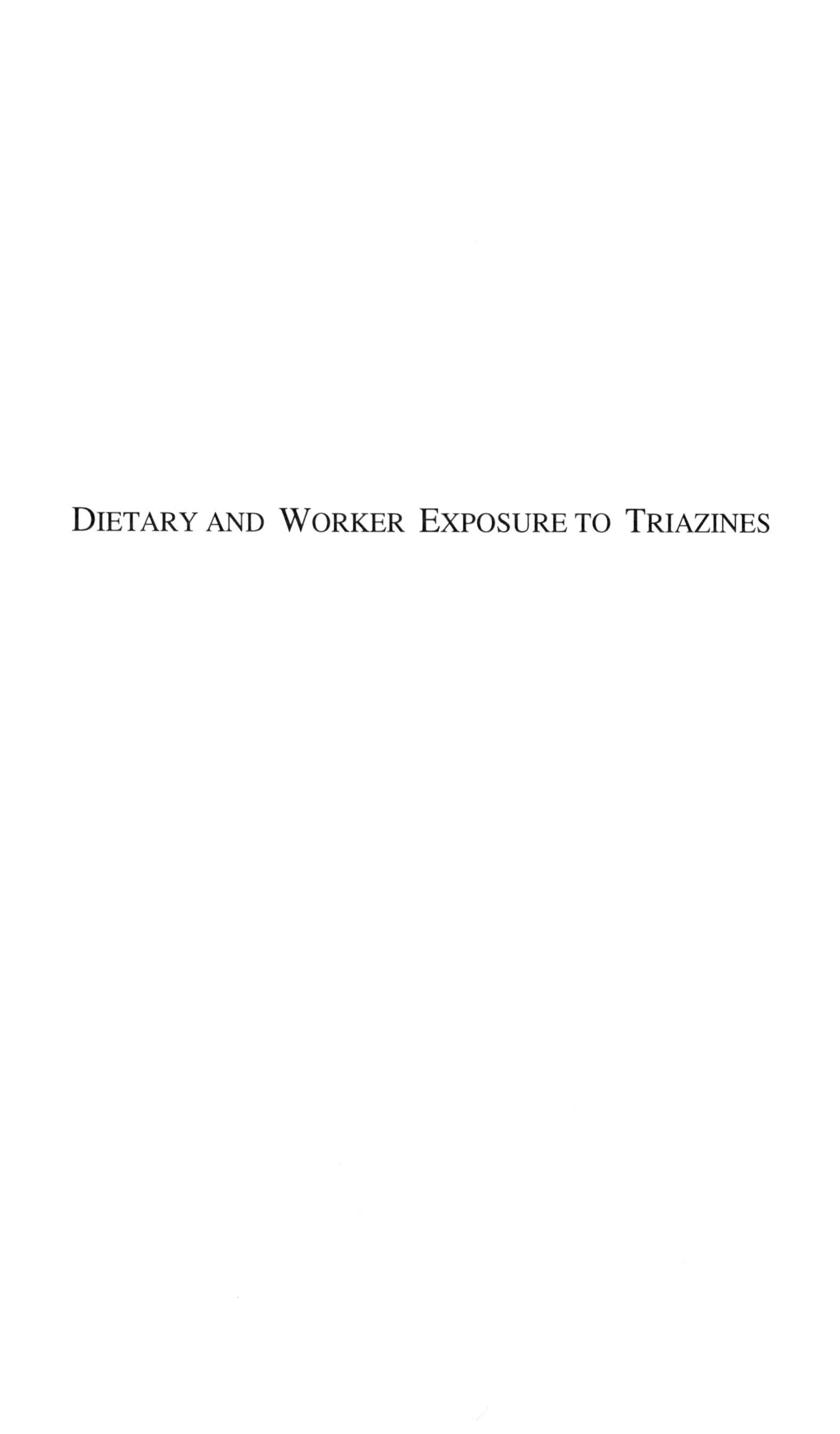

DIETARY AND WORKER EXPOSURE TO TRIAZINES

Chapter 11

Pesticide Residues in Processed Foods: Not a Food Safety Concern

E. R. Elkins, R. S. Lyon, and R. Jarman

National Food Processors Association, 1401 New York Avenue, Suite 400, Washington, DC 20005

Residue analyses results indicate that residues of triazine herbicides are not a source of concern to the processed foods industry. As with all properly applied pesticides, no food safety concerns are warranted. A close examination of the FDA Residue Monitoring program data show no residues of cyanazine, atrazine, simazine or ametryn in 1992, 1993, or 1994 representing 76,973 samples. The NFPA database (6563 values) contains one positive value (0.04 ppm) for simazine in corn and one positive value for atrazine in wheat rough (0.05 ppm). Several general pesticide issues which continue to confront the industry include inadvertent spray residues, residue concentration in soil, illegal or unintentional use of unregistered or canceled materials, consumer expectations for residue-free products, pesticide use by foreign product sources, and pesticide standards in countries to which we export. These industry concerns are discussed.

The National Food Processors Association (NFPA) has long been interested in the issue of pesticide residues on processed foods, or foods ready to eat. In 1960, we developed the NFPA Protective Screen Program. The objective of this program is the prevention of illegal or unnecessary residues in processed foods. The program, a set of detailed recommendations that have evolved from more than 25 years of experience in the operation of active programs that are helpful in the prevention of illegal or unnecessary residues, is published annually (*1*).

NFPA has been involved in research on pesticide residue chemistry, the influence of food processing operations that remove pesticide residues from foods, and the development and improvement of analytical methods for detection and quantitation of pesticide residues for more than forty years. Over the past eight years we have built and maintained a substantial pesticide residue database that includes over a million records, many on processed foods. We also maintain the capability of a risk assessment program using the EPA Dietary Risk Evaluation System (DRES) software. This research and the DRES program provides NFPA a sound basis for evaluation of current issues dealing with pesticide residues in food, and based on this experience, we can say that pesticide residues in processed foods are not a food safety concern. This includes the triazine herbicides; the subject of this book.

The triazine herbicides have a long history. This class of herbicides includes atrazine, simazine, and cyanazine. Both cyanazine and atrazine control weeds by interrupting photosynthesis in susceptible weed species. Atrazine was first introduced in 1958 and provided farmers with the first effective alternative to cultivation and 2,4-D for the control of grasses and other weeds. Atrazine quickly became the leading herbicide applied to corn. Corn is both the largest and the highest value crop grown in the United States. According to USDA, farmers in the US produced 6.3 billion bushels of corn in 1993 with a farm-gate value of $16.6 billion. Corn production has increased significantly over the past forty years while the number of planted acres has declined, largely due to the increased productivity of the acres cultivated . Cyanazine was introduced to the corn market in 1971 and provides weed control similar to atrazine, but without limiting rotational crop options. Atrazine accounted for 55% of all the pounds of herbicides applied to corn in 1992, according to USDA. Most of these herbicides are used in combination with each other. Farmers tend to use herbicides that provide the most cost-effective weed control consistent with tillage practices and local weed pressures.

Dietary Exposure-Processed Foods

As we discuss, the commercial operations that remove pesticides from food keep in mind that the triazine herbicides are selective, pre-emergence and post-emergence herbicides and one would not expect them to be present in foods.

In general, recovery of the edible portion of a vegetable or fruit may involve husking, peeling, shelling or coring operations which effectively remove most of the pesticide with the discarded portions of the plant. Peas and corn are examples of foods in which pesticides seldom, if ever, come in contact with the edible portion.

To illustrate the effect of commercial processing on pesticide residues on food we have selected tomatoes. Figures 1, 2, and 3 are flow charts showing unit operations in commercial processing that remove pesticides from tomatoes. These procedures are required in the processing of tomatoes and were not specifically designed to remove pesticide residues.

Tomatoes are received in bulk trailer carriers and are usually processed the day received. The washing procedures are started immediately while the tomatoes are still in the bulk trailers or trucks. Tomato processing is very water-intensive using approximately 1000 gal. of water per ton of raw tomatoes. The raw tomatoes are moved by conveyor, while still being washed, to sorting tables where any rotten or unusable fruit is removed. Tomatoes are peeled by

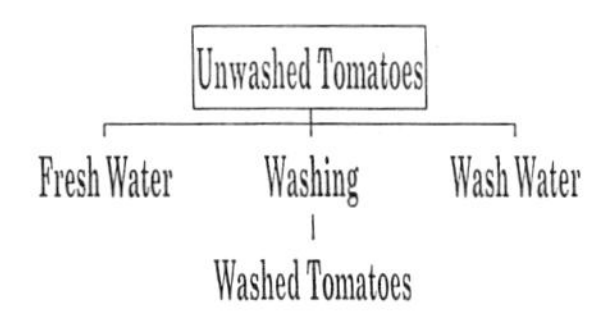

FIG. 1-Flow Chart for Washed Tomatoes

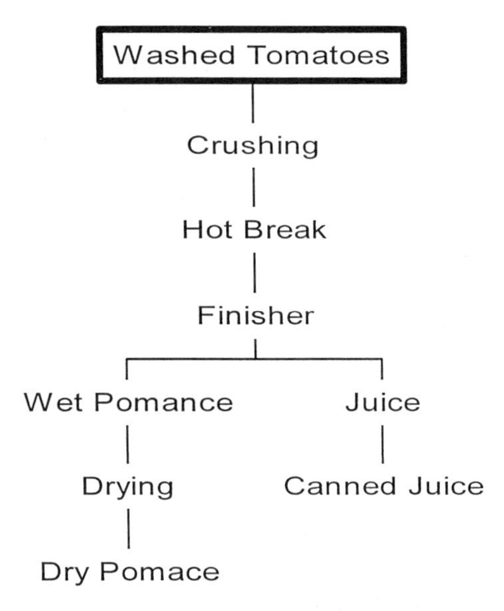

FIG. 2-Flow Chart for Canned Tomato Juice

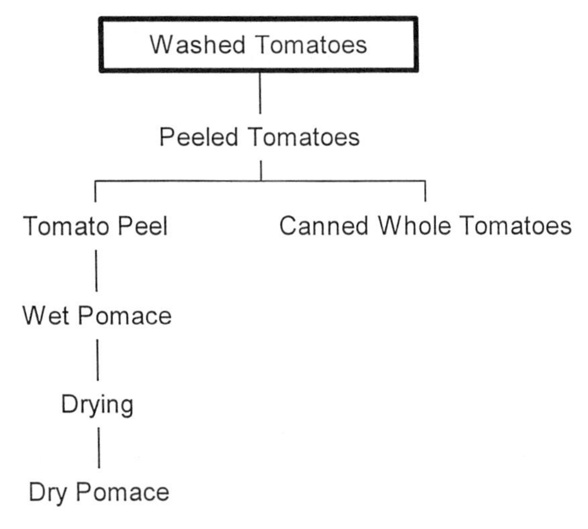

FIG. 3-Flow Chart for Canned Whole Tomatoes

either hot lye or steam as the washing procedure continues. After peeling, the tomatoes are subjected to additional sorting to remove unusable fruit. The finished products, whole tomatoes, tomato juice, tomato sauce, catsup, etc., are canned and retorted. The effects of processing on the residues of malathion and carbaryl in tomatoes have been reported (*2*). Washing, blanching and canning removed 99% of both residues from tomatoes.

As with other processed foods, residues of triazine herbicides are not a source of concern and as with all properly applied pesticides, no food safety concerns are warranted. The NFPA pesticide residue database confirms this statement. The database includes residue data from the processed foods industry, FDA and USDA. Residue data on infant foods and adult foods eaten by infants that were obtained by USDA APHIS for FDA (*3*) are also a part of our database on pesticide residues. The data in this database substantiates the fact that pesticide residues in foods ready-to-eat are rare, and if found at all are at, or close to the quantitation limit of the method used. A close examination of the FDA database shows no residues of cyanazine, atrazine, simazine or ametryn in 1992, 1993, or 1994 representing 76,973 samples. The NFPA database (6563 processed and raw values) contains one positive value (0.04 ppm) for simazine in raw, unprocessed corn and one positive value for atrazine in unprocessed wheat rough (0.05 ppm). The USDA residue program does not test for these herbicides. The available residue data clearly indicate that triazine herbicides do not present a food safety issue.

Pesticide Issues of Concern

While it is clear that pesticides in general and triazines in particular are not a food safety problem, legal and consumer demands raise a number of concerns for food processors. These concerns are not necessarily unique to food processors but remain as issues that can, and in some instances, have serious implications for the food processing industry.

Loss of Pesticides. Pesticides are important and in many situations essential tools for producing a safe, abundant food supply. However, in a growing number of instances, pesticide uses on fruits and vegetables are being lost. Public health, environmental, and/or worker safety concerns result in the restriction or elimination of a pesticide's use. In addition, pesticide manufacturers/registrants find the costs of maintaining the registration of some products, particularly if their uses are significantly restricted, are not justified by the market for those products. The loss of these "minor use" pesticides can have a major impact on fruit and vegetable production.

Pesticide uses are also affected by federal environmental statutes. The Clean Air Act Amendments of 1990 (CAAA), for example, establishes an inflexible standard for eliminating the use and production of chemicals in the United States that exceed the statuary limit for ozone depleting potential. Under the CAAA, methyl bromide production and use in the United States is slated to end in 2001. Unless efforts to change this outcome are successful, U.S.

agriculture and food processors will lose an extremely important pest control tool, U.S. producers will be placed at a competitive disadvantage with foreign producers, and exports of U.S. agricultural products may be significantly affected. The loss of methyl bromide is particularly problematic in that the availability of effective, practical alternatives is by no means certain.

The application of the Federal Food, Drug, and Cosmetic Act (FFDCA) in evaluating the safety of pesticide residues in food directly affects the availability of pesticides. The strict interpretation of the Delaney clause of the FFDCA was a factor in accelerating the loss of pesticide uses. In Les v. Reilly the Ninth Circuit U.S. Court of Appeals ruled that the Delaney clause established a zero risk standard for pesticides that EPA determined induce cancer in humans or animals and concentrate in processed foods or animal feeds or are applied directly to processed foods. Passage of the Food Quality Protection Act of 1996 (FQPA) removed the application of the Delaney clause from pesticide registration decisions. However, FQPA raises other issues that may impact the availability of minor use pesticides. The actual significance of FQPA to the availability of pesticides will be determined as the U.S. Environmental Protection Agency (EPA) applies the new mandated requirements of the statute.

Food processors' concerns over the loss of pesticide uses are for the most part the same as those of food producers. These include the reduced ability to deal with and the potential for increased pest resistance, the lack of acceptable alternative crop and product protection tools, limitations to the development and implementation of effective integrated pest management programs, increased production costs without true public health or environmental improvements, and the potential that the pesticides used to replace eliminated pesticides will cause more problems than they solve.

Illegal Residues. Government and industry data show that illegal pesticide residues in either raw agricultural commodities or processed food products are rare, and more importantly do not pose an actual food safety concern in most instances. However, the presence of an illegal residue can subject a processed food product to government enforcement actions, large scale recalls by the processing company, and loss of consumer confidence in the product. With product loyalty and the investment in processing and introducing a product in the market place at stake, the presence of an illegal residue is a major concern for food processors. The problem food processors face is that an illegal residue can come from a number of sources.

The most obvious, and fortunately, the least likely source of an illegal residue is the illegal use of a pesticide during crop production or in post-harvest applications. Food processors' application of the NFPA protective screen program or variations of it, is focused primarily on this source of illegal residues. Through close coordination with growers, product specifications, spray histories, and as a last resort, analytical monitoring, food processors are able to confirm legitimate pesticide applications are the norm. Appropriate and legal pesticide applications, however, can result in illegal pesticide residues that are more difficult to control. Previous legal use of persistent pesticides that have been removed from domestic agricultural applications such as DDT and aldrin may leave soil residues which result in detectable residues in raw agricultural commodities. Also, crop rotations and changes in cropping over time may result in illegal pesticide residues that come from previous legal and totally justified pesticide applications. In some circumstances this source of illegal pesticide residues must be addressed with changes in the registration of particular pesticides.

The international movement of food products, both raw commodities and finished products, present an increasingly complex array of problems. Despite growing efforts to harmonize pesticide residue tolerances, such as the ongoing programs among the United States, Canada, and Mexico, international uniformity in allowable pesticide uses and legal tolerances is far from complete. Raw agricultural commodities legally treated with a pesticide in an exporting country may not satisfy domestic pesticide standards. There are numerous examples of raw commodities being denied entry to the United States or entry being significantly delayed. Food processors' ability to identify and address potential problem pesticides on imported commodities is greatly affected by the degree of communication and oversight possible with foreign suppliers. The converse situation exists in exporting processed products. Food processors must pay particular attention to the requirements of the countries to which products are exported. Situations can develop virtually overnight that result in an exported product being prohibited from a foreign market.

Another potential source of illegal residues of concern to food processors results from unintentional deposition of pesticides. Spray drift, natural processes, human error, and cross contamination of pesticides themselves are among the possible situations that may lead to the presence of an illegal residue. The ability to detect lower and lower levels of pesticide residues make these unintended and innocent sources of illegal residues even more problematic.
It is to the credit of the domestic and international food production industry that the actual occurrence of illegal pesticide residues, particularly those that might be a real health issue, is minimal. However, experience has shown that illegal residues justify food processors attention and concern.

Consumer Expectations. A fundamental concern faced by food processors is consumer perception and response to the safety and quality of their food products. Despite all the evidence that shows pesticide residues are not a food safety problem and the checks and balances throughout the food regulatory and production system, potential consumer doubt about their food is a major concern for the food producer. As the so-called "alar incident" demonstrated, consumer perception can have a profound economic impact on food processors as well as growers and a devastating impact on the credibility of our food safety system. Reports that consumers actually questioned whether apple juice needed to be treated as a hazardous waste because residues of alar might be present point to the need for a sustained and long term effort in consumer education and risk communication. However, such reports also demonstrate that food processors must be aware of consumer concerns, no matter how scientifically or legally unjustified we think they are.

Conclusions

Even though no food safety concern exists relative to triazine residues in processed or unprocessed foods there are still many issues that need to be addressed such as the loss of pesticides when no alternatives are available. Illegal residues and consumer expectations also are of concern. We need to continue developing methods for dealing with these concerns.

Literature Cited

(1) Judge, James J. The Almanac of the Canning, Freezing and Preserving Industry, Westminster, MD
(2) Elkins, E.R. JAOAC Int. 1989, 72, pp 533-535.
(3) Yess, N.J., Gunderson, E.L., & Roy, R.R. J. AOAC Int. 76, 492-507.

Chapter 12

Use of a Multiresidue Method for the Determination of Triazine Herbicides and Their Metabolites in Agricultural Products

John R. Pardue[1] and Rodney Bong[2]

[1]**Southeast Regional Laboratory for Food and Drug Admnistration, 60 Eighth Street, NE, Atlanta, GA 30309**
[2]**Minneapolis District Laboratory, Food and Drug Administration, 240 Hennepin Avenue, Minneapolis, MN 55401**

325 samples of both domestic and imported agricultural products were analyzed using a method which will determine 19 triazine herbicides and 3 metabolites. The method consisted of blending a composite portion with methanol and filtering, followed by partitioning into methylene chloride from an aqueous saline solution. The organic extract was then evaporated and the herbicides isolated using solid-phase extraction (SPE) with a strong cation exchange (SCX) cartridge prior to gas chromatography and nitrogen-phosphorus detection (NPD). The only triazine detected was simazine found in ten samples of oranges from trace levels to 0.035 ppm. Duplicate and triplicate recoveries were conducted for four selected herbicides in each of the 10 commodities examined in the surveys. These recoveries, spiked at 0.1 ppm, averaged from 82.5% to 104.6%.

Triazines are widely used chemicals applied for season long weed control as both selective and non selective herbicides (*1*). In 1987 atrazine was used at the rate of 100 million pounds and was estimated by EPA to be the most widely used pesticide (*2*).

Symmetrical triazine herbicides are six membered ring compounds containing 3 cyclic nitrogen atoms and 2 amino groups external to the ring. Substitutions at the sixth position of the ring and of the amino groups produce a large number of compounds with herbicidal activity. Substitution at the six position with -Cl yields atrazine, simazine, and propazine; substitution with $-OCH_3$ yields atraton, prometon, and simeton; and with $-SCH_3$ yields ametryn, dipropetryn, and simetryn. The amino groups attached to the ring are usually of the form $-NHC_nH_{2n+1}$ and the alkyl group attached is usually either ethyl or isopropyl or, less frequently, tertiary butyl.

Loss of one or both of the alkyl groups (C_nH_{2n+1}) produces metabolites which were studied and shown to be recovered by the method used in the surveys (*3*). The desalkyl metabolites of atrazine have been found in surface and ground water (*4,5*) and in soil (*6-9*). While these metabolites are listed as atrazine metabolites, an examination of their structures reveal that they are the same as those produced by the dealkylation of simazine and propazine. The structures and relationship between atrazine, simazine, propazine, and three desalkyl metabolites are illustrated in Figure 1.

These metabolites are potentially important as residues in food products. They are included in the tolerances in determining total chlorotriazine residues for atrazine in certain grasses and millet, and as residues for simazine in bananas and fish (*10*).

This paper describes pertinent steps of the screening method and the results of two Food and Drug Administration surveys for triazine herbicides and selected metabolites conducted in FY 1995 and FY 1996.

Materials and Method

Reagents and Materials. Solvents were pesticide grade obtained from Baxter, Burdick and Jackson Div., Muskegon, MD. Reference standards were obtained from U. S. Environmental Protection Agency, Pesticide and Industrial Chemicals Repository, Research Triangle Park, NC; Crescent Chemical Co., Hauppauge, NY; and Ultra Scientific, North Kingstown, NY. SPE tubes were Supelclean LC-SCX, 3 mL obtained from Supelco, Inc., Bellefonte, PA.

Sample Collection and Preparation. Both imported and domestically grown food samples were collected for triazine analysis in nine different Food and Drug Administration (FDA) districts or regions located throughout the United States. The first survey contained 232 samples consisting of apples, bananas, cherries, corn, grapefruit, grapes, olives, oranges, pears, and plums and were collected and analyzed during FY 1995. Because of the special interest by FDA in those foods which are consumed by children and infants, an additional 93 samples were analyzed during FY 1996. These products consisted of apples, bananas, grapes, oranges, pears, and plums.

Samples were composited at the collecting district or regional laboratory according to PAM I, Table 102-a (*11*), frozen, and sent to the Minneapolis laboratory for analysis. Compositing consisted of grinding the entire commodity after removing and discarding any stems, stalks, and/or crown tissue which might be present.

Extraction and Cleanup. A 100 g portion of each composited sample was extracted with methanol and filtered. A portion of the filtrate equivalent to 50 g of sample was diluted with water and saturated salt solution, and then, extracted twice with methylene chloride. The methylene chloride extracts were dried by passing them through sodium sulfate into a 500 mL Kuderna-Danish evaporator. The methylene chloride was concentrated on a steam bath to about 2 mL, 25 mL hexane added, and the solvent mixture reconcentrated leaving only hexane. This solution was then diluted to 10 mL with hexane prior to the cleanup step.

A one mL aliquot of the hexane solution was placed onto a previously washed Supelclean LC-SCX tube. The tube was again washed with methylene chloride and acetone to remove crop co-extractives. Any triazine compounds present were then

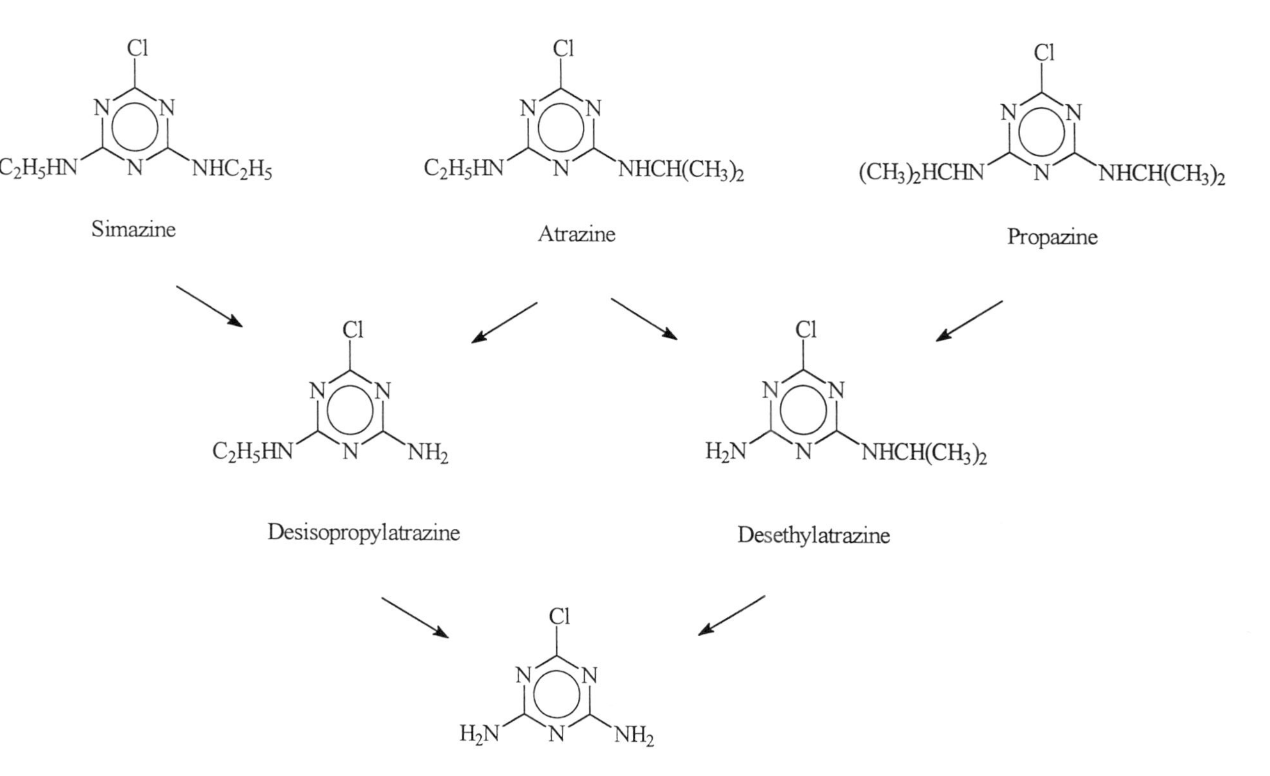

Figure 1. Structure and relationship between simazine, atrazine, propazine and their desalkyl metabolites.

eluted from the tube with a solution of 1N NH_4OH/methanol (1+ 3) into a pH 6.5 phosphate buffer. The buffer solution was extracted twice with methylene chloride, methylene chloride evaporated to dryness, and the residue dissolved in 2.0 mL acetone. This solution was then analyzed by gas chromatography.

Step by step descriptions of this method and the method used to analyze the the metabolite, diaminochloro-s-triazine, are published in the original paper (*3*). Because only low amounts (less than 0.04 ppm) of the parent herbicides and none of the mono-dealkylated metabolites were detected in any of the samples, the method to determine diaminochloro-s-triazine was not used. The basic steps of this procedure and those of the method used for the surveys are illustrated in Figure 2. The method for diaminochloro-s-triazine was developed because it was found that only a very small amount of this compound was recovered by the original procedure. The major differences in the two methods are (1) the reduction of water and removal of methanol prior to partitioning diaminochloro-s-triazine into the organic solvent and (2) substituting ethyl acetate for methylene chloride as the partitioning solvent. Both of these changes were taken because of the high solubility of the metabolite in the aqueous methanol solution. Recoveries show that diaminochloro-s-triazine remains in the methanolic layer when it is extracted with methylene chloride (*3*).

Gas Chromatographic Analysis. Analysis was done using a HP 5890A with 19234B/C nitrogen-phosphorus detector (Hewlett-Packard Co., Wilmington, DE); DB-17, 30 m X 0.53 mm with 1 μm film thickness (J&W Scientific, Folsom, CA). Temperatures (°C) were as follows: inlet, 220; detector, 220; column programmed from 150 to 230 at a rate of 4°C/min and a final hold time of 12 min. Gas flows (mL/min) were as follows: helium carrier, 15; helium auxiliary, 35; air, 90; hydrogen, 3.5. Retention times relative to atrazine for the 22 compounds tested in the original study ranged from 0.76 for melamine to 2.94 for hexazinone. With these conditions, atrazine elutes in 10 minutes.

Mass Spectral Analysis. Confirmation of detected residues was with an HP 5995 Quadrapole instrument using electron impact mass spectroscopy. Column and column conditions were the same as for the gas chromatographic analysis except a 0.25 mm DB-17 column was used.

Results and Discussion.

Triazine residues were found in 10 samples of the 325 samples examined during the surveys. Tables I and II list the number and types of samples analyzed during the two fiscal years.

All residues detected were simazine found in domestically grown oranges. One positive finding of 0.035 ppm and four trace levels (less than 0.01 ppm) were found in the 25 orange samples examined during FY 1995. One positive of 0.030 ppm and four trace levels were found in 20 orange samples analyzed during FY 1996. These samples were composited by grinding the whole oranges. They were then analyzed on the whole basis, therefore, it was not determined if the residues were on the peel, pulp, or juice. These positive findings were well below the established tolerance for simazine of 0.25 ppm in oranges (*10*).

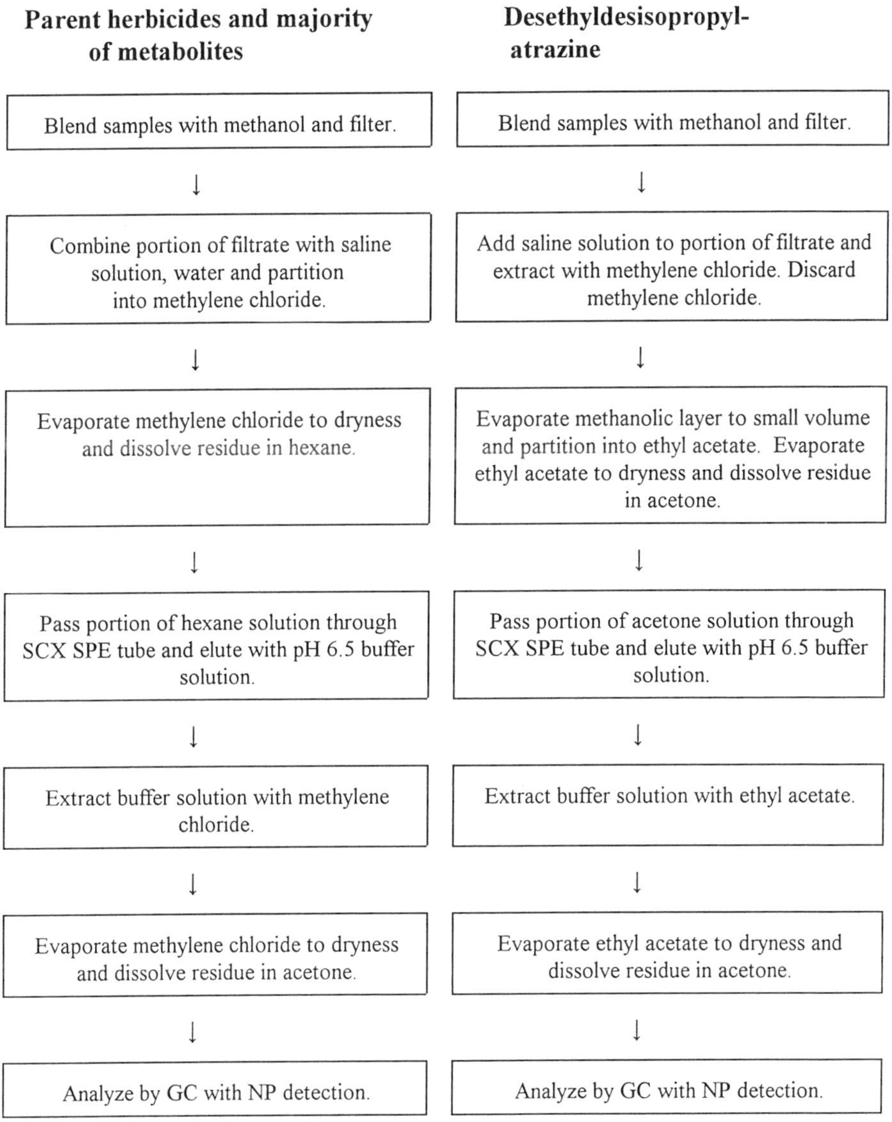

Parent herbicides and majority of metabolites	Desethyldesisopropyl-atrazine
Blend samples with methanol and filter.	Blend samples with methanol and filter.
↓	↓
Combine portion of filtrate with saline solution, water and partition into methylene chloride.	Add saline solution to portion of filtrate and extract with methylene chloride. Discard methylene chloride.
↓	↓
Evaporate methylene chloride to dryness and dissolve residue in hexane.	Evaporate methanolic layer to small volume and partition into ethyl acetate. Evaporate ethyl acetate to dryness and dissolve residue in acetone.
↓	↓
Pass portion of hexane solution through SCX SPE tube and elute with pH 6.5 buffer solution.	Pass portion of acetone solution through SCX SPE tube and elute with pH 6.5 buffer solution.
↓	↓
Extract buffer solution with methylene chloride.	Extract buffer solution with ethyl acetate.
↓	↓
Evaporate methylene chloride to dryness and dissolve residue in acetone.	Evaporate ethyl acetate to dryness and dissolve residue in acetone.
↓	↓
Analyze by GC with NP detection.	Analyze by GC with NP detection.

Figure 2. Flow diagram illustrating basic steps of method used for the determination of triazine herbicides and their mono dealkylated metabolites as compared to the method developed to determine diaminochloro-s-triazine.

Table I. Samples analyzed during FY 1995.

Commodity	Total No. Samples	No. Domestic Samples	No. Import Samples	Principal Countries of Origin for Imports
Apples	25	9	16	Chile
Bananas	25	0	25	Mexico, Ecuador, Panama
Cherries	25	25	0	
Corn	25	10	15	Thailand, Mexico, Canada
Grapefruit	12	8	4	Israel
Grapes	25	5	20	Chile, Greece
Olives	25	0	25	Spain, Greece, Italy
Oranges	25	20	5	Italy
Pears	25	19	6	Chile
Plums	20	13	7	Chile
Total	232	109	123	

Table II. Samples analyzed during FY 1996.

Commodity	Total No. Samples	No. Domestic Samples	No. Import Samples	Principal Countries of Origin for Imports
Apples	10	8	2	Chile, Canada
Bananas	20	0	20	Mexico, Ecuador, Panama
Grapes	20	6	14	Mexico, Chile, Greece
Oranges	20	18	2	China
Pears	20	10	10	Chile
Plums	3	3	0	
Total	93	45	48	

The simazine residues found at 0.035 and 0.030 ppm were confirmed by GC/MS using electron impact, selected ion monitoring. Ion ratios at mass units 201, 186, and 173 for the sample residues were matched with those of the reference standard simazine analyzed under the same conditions.

For the FY 1995 survey, each commodity was fortified twice (on separate days) at the 0.1 ppm level with four triazines. The four triazines selected for the recovery studies were three having different substitutions at the sixth position (-Cl, $-OCH_3$, $-SCH_3$) of the triazine nucleus and one of the metabolites of atrazine. Atrazine, secbumeton, and simetryn were the three herbicides chosen for the recovery studies, and desethylatrazine was the metabolite. These same four compounds were selected for single recovery determinations in apples, bananas, grapes, and oranges for the FY 1996 survey. Table III lists the cumulative recoveries which were obtained for both FY 1995 and for 1996. The recovery values for the 1996 survey are the third value in each grouping.

Table III. Combined FY 1995 & 1996 Recovery Data for Selected Triazine Herbicides

Commodity	Recovery, %							
	Desethylatrazine		Atrazine		Secbumeton		Simetryn	
Apples	90.6	92.5	104.7	102.6	103.1	109.3	94.1	102.0
	70.1		94.8		115.1		102.9	
Bananas	88.7	89.6	98.3	99.1	100.2	110.0	97.1	102.2
	94.0		112.2		115.1		110.8	
Cherries	84.3	80.1	102.6	92.2	113.1	98.7	110.8	98.0
Corn	104.7	87.9	103.5	104.3	110.3	110.3	102.0	101.0
Grapefruit	82.1	60.8	101.7	93.0	107.2	98.0	101.0	91.9
Grapes	84.2	77.7	103.5	94.8	113.4	103.0	27.5	51.0
	72.7		94.8		103.0		96.5	
Olives	82.1	77.4	100.0	94.8	102.1	103.1	93.1	94.1
Oranges	60.4	86.8	97.4	100.0	96.9	99.0	71.6	94.1
	73.6		100.9		104.8		96.1	
Pears	79.6	79.6	94.8	93.0	96.9	101.0	95.1	97.1
Plums	104.7	75.5	103.5	101.7	102.0	103.9	110.3	111.3
Fortification Level (ppm)	0.106		0.115		0.115		0.102	
Mean Recovery	82.5%		99.4%		104.6%		93.8%	
Std. Dev.	11.1		4.2		5.5		19.0	
Coff. of Variation	13.5		5.1		5.3		20.3	

The statistical values shown for each of the four herbicides are based on pooled recovery data representing all commodities. Most of the recovery values were within the range reported in the original work. One notable exception was the low recovery for simetryn from grapes (27.5 and 51.0%) obtained during the 1995 survey. Recoveries for grapes were not conducted with the original work. For this reason, it was originally believed that the recoveries might be a peculiarity of the particular sample matrix and compound, and the results were not eliminated. When the recovery was repeated with the FY 1996 survey, however, a value of 96.5% was obtained, and it does not appear as if the sample matrix was the problem. The original low recoveries were probably due to unfamiliarity with the method. If these two low results are not used for the statistical evaluation of simetryn, the mean recovery is 98.8% for all commodities. The standard deviation is lowered from 19.0 to 8.6, and the coefficient of variation from 20.3 to 8.7. These results bring the calculated values for each of the four compounds within the range reported in the original work (*3*).

The recovery values and the results of these two surveys demonstrate that the methodology is excellent for determining triazine herbicides and their desalkyl metabolites in a variety of food commodities. The frequency and levels of positive findings of these herbicides in the surveys indicate that there is a low dietary exposure rate.

Literature Cited

1. *Farm Chemicals Handbook '95*; Meister, R. T.; Sine, C., Eds.; Meister Publishing Company, Willoughby, OH, 1995, Vol. 81, 32-33.
2. Ware, W. W.; *The Pesticide Book*; Thomson Publications, Fresno, CA,1989, 112.
3. Pardue, J. R. *J. AOAC Int.* **1995**,*78*, 856-862.
4. Pereira, W. E.; Rostad, C. E.; Leiker, T. J. *Anal. Chim. Acta.* **1990**, *228*, 69-75.
5. Thurman, E. M.; Goolsby, D. A.; Meyer, M. T.; Kolpin, D. W. *Environ. Sci. Technol.* **1991**, *25*, 1794-1796.
6. Steinwandter, H. *J. Anal. Chem.* **1991**, *339*, 30-33.
7. Gorder, G. W.; Dahm, P. A. *J. Agric. Food Chem.* **1981**, *29*, 629-634.
8. Karlaganis, G.; Von Arx, R. *J. Chromatogr.* **1991**, *549*, 229-236.
9. Durand, G.; Forteza, R.; Barceló, D. *Chromatographia* **1989**, *28*, 597-604.
10. *Code of Federal Regulations*, Title 40, Government Printing Office, Washington,
D. C., 1994; Section 180.
11. *Pesticide Analytical Manual*; McMahon, B. M.; Hardin, N. F.,Eds.; Volumn I, 3rd Edition, Table 201-a.

Chapter 13

An Immunochemical Approach to Estimating Worker Exposure to Atrazine

James F. Brady, JoLyn Turner, Max W. Cheung, John D. Vargo, Jennifer G. Kelly, Denise W. King, and Andrea C. Alemanni

Ciba Crop Protection, P.O. Box 18300, Greensboro, NC 27419–8300

Atrazine [2-chloro-4-(ethylamino)-6-(isopropylamino)-*s*-triazine] is a herbicide used to control annual broadleaf and grass weeds. Since a large volume of atrazine is used in several crops throughout the U.S. (*1*), Ciba Crop Protection initiated studies to quantify occupational exposure to this valuable herbicide. Estimation of exposure to pesticides such as atrazine is often accomplished through biological monitoring techniques. Development of an analytical method to complement biological monitoring is dependent upon elucidation of the metabolic profile of the test substance to the extent that a suitable biomarker has been identified. Samples should preferentially be collected in a non-invasive fashion and be of adequate volume to allow repeat analyses. The analytical method itself should be rapid, sensitive, and specific. Ideally, the method should generate minimal amounts of solvent waste because a large number of samples may be necessary to obtain a clear picture of worker exposure.

Analytes excreted in urine are frequently the biomarkers of choice because of the ease of sample collection and the large volume of urine voided daily, approximately 1.5 L (*2*). Previous attempts to quantify atrazine in urine arising from occupational exposure found only trace amounts of unaltered atrazine (*3,4*). These results effectively eliminated the parent compound as a potential urinary marker. Lucas *et al.* (*4*) concluded glutathione conjugation and mercapturate formation was a major metabolic pathway. A method was developed utilizing an enzyme-linked immunosorbent assay for atrazine mercapturate [2-(L-cysteine, N-acetyl)-4-(ethylamino)-6-(isopropylamino)-*s*-triazine] based on an antibody obtained from Karu *et al.* (*5*). This paper describes an enzyme immunoassay method for atrazine mercapturate in urine based on a commercially available immunoassay kit. This method has been applied to metabolism studies in Rhesus monkeys and to the analysis of selected human samples collected from professional pesticide applicators and mixer/loaders.

MATERIALS AND METHODS

Materials. Enzyme immunoassay analyses were performed using EnviroGard triazine plate kits from Millipore Corporation, Bedford, MA (now available through Strategic Diagnostics, 128 Sandy Drive, Newark, DE). Diol solid phase extraction cartridges (SPE) (360 mg of packing) were obtained from Waters Corporation, Milford, MA. Samples were centrifuged on Dynac II or IEC Centra CL2 centrifuges obtained through Fisher Scientific, Pittsburgh, PA. Absorbance readings were measured on an ICN MCC340/MK II microplate reader. This instrument was controlled by a Gateway 2000 386DX/33 computer. High performance liquid chromatography with mass spectrometric detection (LC\MS\MS) was conducted using a Perkin Elmer Sciex API-III+ mass spectrometer fitted with an IonSpray liquid introduction interface. The instrument was operated in the positive ion mode. The chromatograph was equipped with an ISS-200 autosampler, a model 410 gradient pump, and an SEC 4 solvent chamber from Perkin Elmer, Norwalk, Connecticut. The YMC ODS-AQ column, 250 x 4.6 mm, was purchased from YMC, Inc. of Wilmington, North Carolina. The column was maintained at 30°C by an Eppendorf column heater from Fisher Scientific, Pittsburgh, Pennsylvania. The LC system was controlled with an Apple MacIntosh Quadra 950 computer.

Methods. Sample collection. Urine samples were collected from Rhesus monkeys dosed intravenously with ^{14}C-atrazine. Human urine samples were obtained from certified professional agrochemical mixer/loaders and applicators (MLA) from Ohio, Indiana, Illinois, Iowa, and Nebraska. All samples were shipped and stored frozen until thawed for analysis.

Sample extraction for immunoassay analysis. A representative 1.0 ml subsample of each urine sample was combined with 50 µl of 4 M HCl and 0.2 g of NaCl in a 16 x 100 mm borosilicate culture tube. This solution was partitioned three times with 4 ml of 25% methylene chloride:ethyl acetate by vortexing for 15 s. Each partition was followed by centrifugation (200 x *g*) to separate the organic (top layer) and aqueous fractions. The organic extracts were combined and dried with 3 ml of hexane and approximately 2.5 g of sodium sulfate. The dry organic solution was added to a diol SPE pre-washed with 3 ml of methylene chloride. After rinsing the SPE with additional methylene chloride (3 ml), atrazine mercapturate residues were eluted in 3 ml of 0.1% ammonium hydroxide:ethanol. The eluate was brought to dryness under N_2 and re-constituted in 2.0 ml of 0.10 M Tris-HCl buffer, pH 7.5.

Enzyme immunoassay. Immunoassay analysis of the buffered extract was conducted using the EnviroGard plate kit configured in the 1 x 8 strip format. The assay was performed as previously described (*6*). The positions of all standard and sample solutions in the microtiter plate were recorded by the analyst on a plate layout sheet which mimics the 8 x 12 array of the plate. Each standard or sample solution (120 µl, analyzed in duplicate) was added to a well of an uncoated polystytrene microtiter plate (the reservoir plate). Using the reservoir plate enabled the analyst to transfer all solutions to their designated positions in the antibody coated strips to incubate for equivalent amounts of time during the inhibition phase of the assay. Eighty µl of the 120 µl reservoir was transferred using a multichannel pipette. The same amount of enzyme conjugate was added and the plate was incubated for 1 hour at room temperature with gentle agitation (approximately 90 oscillations/min). The contents of the plate were removed and all wells washed three times with distilled, de-ionized water. Substrate solution (120 µl) was added to each well and the plate incubated as described above for approximately 30 min. Color development was stopped by addition of 50 µl of 2 M sulfuric acid. The absorbance of all wells was monitored at 450 nm. The absorbances of analytical standards ranging from 0.25 to 5 parts per billion (ppb) were used to calculate a standard curve in the form of $y = m \log(x) + b$. Results were expressed as atrazine mercapturate-equivalents. Due to the two-fold dilution of the final extract prior to immunoassay analysis, the lower limit of detection for this method is 0.50 ppb.

Fortification studies. The ability of this method to quantify atrazine mercapturate residues at the 1.0 ppb level was determined by a series of fortification experiments. Urine samples (N = 12) were collected from field personnel from five states prior to these individuals being exposed to atrazine on an occupational basis during the 1995 field season. Aliquots of these samples (1.0 ml) were fortified with 1.0 ppb of atrazine mercapturate and analyzed pre- and post-fortification. The net differences in concentration were used to determine if a limit of quantitation (LOQ) could be established at 1.0 ppb.

Specificity studies. The reactivity of this method to atrazine or its metabolites found in urine subsequent to occupational exposure to atrazine was evaluated in two ways. Aliquots of distilled, deionized water were fortified with concentrations of atrazine, atrazine mercapturate, and the mercapturates of the deethyl, deisopropyl, and diamino degradates of atrazine ranging from 1000 to 0.01 ppb. These samples were analyzed as described above. Data from these experiments were used to determine the amount of each test substance that produced half the response of an unfortified sample (I_{50}). The percent reactivity of each test substance relative to atrazine mercapturate was calculated by dividing the I_{50} of atrazine mercapturate by the I_{50} of each test substance and multiplying that quotient by 100%. The unconjugated degradates were not examined because investigators had previously evaluated their reactivity to the antibodies in the EnviroGard kit (*7-9*). The limit of detection for each compound was calculated from these data according a to modification of Rodbard's method (*9*). In addition, control urine was fortified with ^{14}C-atrazine and extracted as described above. Radioactivity in the diol eluate was determined by liquid scintillation counting to assess the amount of atrazine in the fraction that would be subjected to immunoanalysis.

Immunoanalysis of monkey urine samples. Samples were analyzed as described above. When the result was greater than 5.0 ppb, the sample was re-extracted and the new extract was diluted further, typically 20-fold, and re-assayed. Each set of samples (usually ten to twelve) was accompanied by a control sample and that control fortified with two different concentrations of atrazine mercapturate, 1.0 ppb and a higher level near the suspected range of the highest residues in that set. Sample results were corrected for mean procedural recoveries less than 100%.

Sample preparation for LC/MS/MS. Composite urine samples were subjected to anion and cation exchange chromatography (data not shown) and dissolved in 25% methanol/water for injection.

LC\MS\MS analysis. Fifty microliters of each sample solution were injected. A linear solvent gradient was generated over fifteen minutes beginning with 10% methanol/water containing 0.1% acetic acid and finishing with 100% acetonitrile, also containing 0.1% acetic acid, at a flow rate of 1.0 ml/min. Atrazine mercapturate was measured by evaluating its precursor/product ion pair, 343.2/214.2 amu. Operating in the MS/MS mode was required by the presence of an interference having the same mass as the parent molecular ion.

RESULTS AND DISCUSSION

Immunological methods are ideally suited to complement biological monitoring techniques. Immunoassays can process large numbers of samples quickly allowing rapid turnaround of data. Immunoassays are frequently more sensitive than chromatographic methods in a matrix as complex as urine and may be more suited to the analysis of polar metabolites. Cheung *et al.* (*10*) for example, analyzed chlordimeform residues in urine by high performance liquid chromatography with a LOQ of 0.05 ppm. Weisskopf and Seiber (*11*) achieved detection limits of two ppb for urinary dialkyl phosphates. In contrast, Feng *et al.* (*12*) used an enzyme immunoassay to quantify as low as 0.25 ppb of alachlor mercapturate from monkey urine. Lucas *et al.* (*4*) detected 0.5 ppb of atrazine mercapturate in human urine by ELISA techniques.

The complexity of the urinary matrix is reflected in the variety of approaches different investigators have taken to immunoassay development. The "ruggedness" of the antibody preparation, or its ability to tolerate background interferences without becoming non-specifically inhibited, is perhaps the dominating factor controlling the extent of sample preparation. Feng *et al.* (*12*) successfully assayed analytical standards and Rhesus monkey urine samples diluted into pooled human control urine. Lucas *et al.* (*4*) diluted one volume of each human urine sample with at least three volumes of buffer to avoid matrix effects. In this study, preliminary method development experiments indicated that a variety of urine extracts produced strong non-specific inhibition with the antibodies in the EnviroGard kit. Thus, it became necessary to extract and isolate atrazine mercapturate from background interferences to overcome limitations imposed by the antibodies at hand.

Fortification studies. The limit of quantitation was determined in this study by fortifying several urine subsamples from commercial mixer/loader applicators (MLAS) of atrazine. Results of this experiment indicate reactivity of the assay to the addition of 1.0 ppb of atrazine mercapturate despite a range of immunoreactive components in the samples (Table I). The mean recovery of 108% indicates the method is accurate. On the other hand, the large spread of measurement depicted by the standard deviation of ± 31.8% implies the method lacks precision. The lack of precision is likely due to the large amount of an unidentified background interference in some samples, particularly those in the group containing pre-fortification values in excess of 1.5 ppb.

The utility of this method to the analysis of human samples can be assessed by reviewing the results of human procedural recovery samples (Table II). These data show excellent recoveries were obtained at the LOQ as well as at four higher concentrations in urine collected from a number of MLAs.

Table II. Analytical Results of Procedural Recovery Samples in Human Urine.

Fortification Level (ppb)	*N*	*Mean Percent Recovery ± SD*
1	15	103 ± 19
4	6	112 ± 19
5	10	102 ± 11
10	10	86 ± 16
20	4	92 ± 15

Specificity Studies. Results of the specificity studies indicate only atrazine and atrazine mercapturate were reactive among the test substances screened (Table III). Atrazine was found to have a much smaller I_{50} (0.24 ppb) than its mercapturate degradate (1.1 ppb). Although the parent herbicide can be detected at much lower levels, the experiment following ^{14}C-atrazine through the procedure demonstrated that only 0.3% of the atrazine applied was found to be carried through the extraction and cleanup steps. Therefore, unless a sample contains a high concentration of atrazine, the presence of atrazine in a sample is not likely to make a substantial contribution to the immunoassay signal. Recalling that previous investigators (*3,4*) found little parent compound in the urine, this method can be regarded as specific for atrazine mercapturate.

Rhesus monkey urine analyses. Urine samples collected from Rhesus monkeys injected with ^{14}C-atrazine were analyzed as part of a metabolism study to determine the urinary profile and excretion kinetics resulting from exposure to atrazine in a primate model. Three of the four test subjects voided nearly all of the mercapturate residues in the first twenty-four hours post-inoculation (Figure 1).

Table III. Cross-reactivity Parameters of the Atrazine Mercapturate Enzyme Immunoassay. All units are ng/ml.

Test Substance	*LOD*[1]	*LOQ*[2]	I_{50}	*Percent Reactivity Relative to Atrazine mercapturate*
Atrazine mercapturate	0.50	1.0	1.1	100
Atrazine[3]	0.50	--	0.24	458
Deethyl mercapturate	0.51	--	84	1.3
Diamino mercapturate	NR[4]	--	1000	<1.0
Deisopropyl mercapturate	NR	--	1000	<1.0
Deethylatrazine	3.7	--	19	1.0
Deisopropylatrazine	11	--	59	<1.0
Diaminoatrazine	NR	--	--	<1.0

[1] Limit of Detection (LOD), is the smallest dose that yields a response that is statistically different than the response of the zero dose.

[2] Limit of Quantitation (LOQ) was determined only for atrazine mercapturate.

[3] The I_{50} of 0.24 ppb indicates the antibodies in the Envirogard kit are much more reactive to atrazine than to its mercapturate degradate. However, this method cannot determine residues less than 0.50 ppb. Consequently, the LOD for atrazine defaults to 0.50 ppb.

[4] NR, test substance determined to be non-reactive to the antibodies used in the EnviroGard plate kit.

Only one monkey continued to eliminate significant concentrations of residues in the second day following exposure. After forty-eight hours post-treatment, only trace amounts of atrazine mercapturate were observed in the voids.

The near complete elimination of residues in three of the four primate subjects mirrors the trend found by Catennacci *et al.* (*3*) in the elimination of atrazine by factory workers. Both the present work and that of Catennacci *et al.* found that the concentration of residues in urine peaks approximately twelve hours after exposure and decreases rapidly, nearly to baseline, after twenty-four hours.

The immunoassay was shown to provide a convenient method of monitoring the overall elimination of residues as depicted by the amount of radioactivity in each of the voids (Figure 2). The mercapturate signal from monkey three provides a reasonable comparison of the overall trend from the peak of elimination in the first four hours post-treatment until the radioactivity on day three dropped to near background levels, approximately 20 DPM or less (Figure 2). Data from the field applicators showed similar trends.

Although the assay was successfully adapted to Rhesus monkey urine (Table IV), monkey urine samples produced an unusual effect when the final buffered extract was diluted more than twenty-fold as was generally required in the 0-4 through 8-12 hour interval samples. These collection intervals contained the highest concentration of residues and higher dilutions were needed to bring the

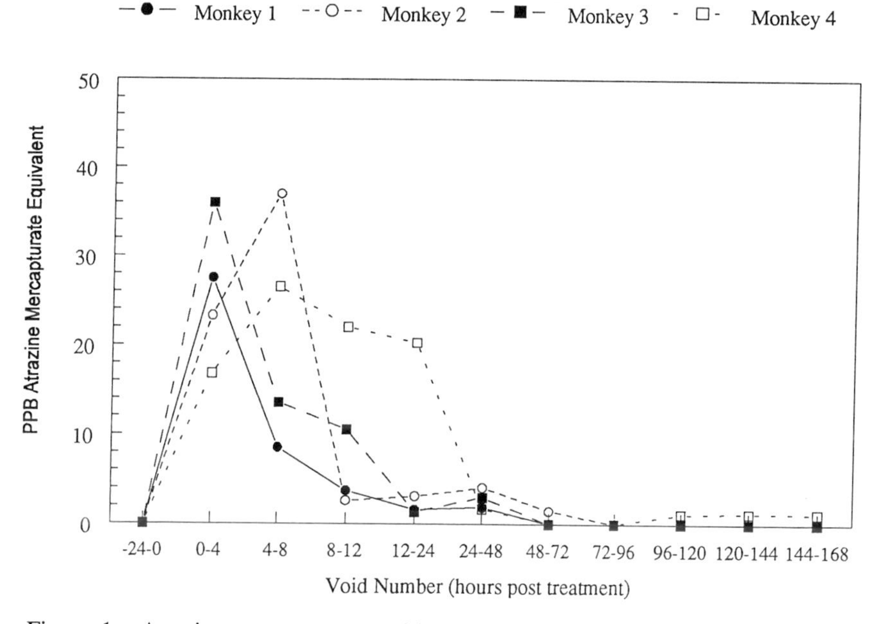

Figure 1. Atrazine mercapturate residues found in each sampling interval from each of four Rhesus monkeys dosed intravenously with ^{14}C-atrazine.

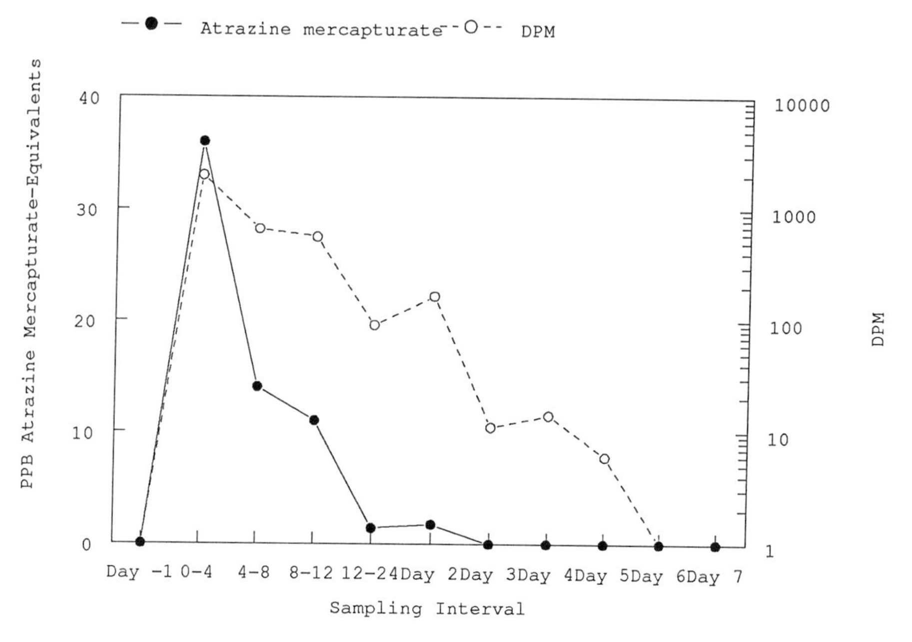

Figure 2. Comparison of immunoassay results and total radioactivity for each sampling interval from Monkey Number 3.

results into the range of measurement. Analytical results from samples diluted in this fashion lacked the precision between replicates typically observed; one value was often 200% of the other replicate. This problem was remedied by extracting less than the normal volume of sample brought to 1.0 ml with water. As little as 250 μl of sample could be extracted without effecting the extractability of atrazine mercapturate. Extracts produced in this fashion did not have to be diluted as greatly as those from the full aliquot of sample prior to immunoassay and yielded assay results consistent with those usually obtained.

Table IV. Analytical Results of Procedural Recovery Samples in Rhesus Monkey Urine.

Fortification Level (ppb)	*N*	*Mean Percent Recovery ± SD*
1.0	6	90 ± 22
50	6	87 ± 12

Although enzyme immunoassay results have been shown to mirror the overall elimination of residues, confirmation of atrazine mercapturate as the reactive moiety has proven to be problematic. LC\MS\MS analysis of composited monkey urine found an average of 16-fold less atrazine mercapturate than that determined by immunoassay (Table V). Thus, the urine contained immunoreactive species in addition to atrazine mercapturate. These unknowns displayed similar chemical behavior as atrazine mercapturate to the extent they could be carried through the extraction and preparative cleanup steps prior to immunoanalysis. The cross-reactivity profile of the antibodies in the EnviroGard kit dictates that any reactive compound must be closely structurally related to atrazine. Moreover, the need to quantitate the daughter ion as part of the LC\MS analysis was attributed to an interference having a mass identical to that of the atrazine mercapturate parent ion. Based on these observations, the unknown species may be a metabolite of atrazine having the same mass as atrazine mercapturate. Discrepancies between the concentrations of atrazine mercapturate found in composite monkey samples by immunoassay and LC\MS\MS are unresolved. Until the immunologically reactive species are more fully characterized, the immunoassay will prove most useful as an indicator of the trend of exposure.

CONCLUSION

A commercially available immunoassay kit for atrazine has been adapted for use to quantify atrazine mercapturate in urine. The assay has a range of measurement of 0.25 to 5.0 ppb and a limit of quantitation of 1.0 ppb. The assay is selective for atrazine mercapturate, but subject to interference by an unknown metabolite in monkey urine. Results of analyses of Rhesus monkey urine samples indicate the test subjects void nearly all residues within 48 hours of exposure. Procedural recovery data using human urine indicates the method is applicable to the analysis of human samples. Cross reactivity with unknown reactive metabolites or species is currently limiting this immunoassay as a quantitative tool. However, the assay

Table V. Comparison of EIA and LC\MS\MS Analyses of Composited 0-4 Hour Rhesus Monkey Urine

	Analytical Results (ppb atrazine mercapturate)		
Animal No.	*EIA*	*LC\MS\MS*	*Ratio of EIA to LC\MS\MS data*
1	43	4	10.8
2	71	4	17.8
3	65	5	13.0
4	92	4	23.0
Mean:	68	4.3	16

has been shown to provide a convenient means of monitoring the overall trend of exposure.

Literature Cited

1. U.S. EPA. Pesticides Industry Sales and Usage, 1992 and 1993 Market Estimates; U.S. Environmental Protection Agency, Office of Prevention, Pesticides and Toxic Substances; U.S. Government Printing Office: Washington, DC, 1994; EPA 733-K-94-001.
2 Liappis, N. Z. *Klin. Chem.* **1973**, *11*, pp. 279.
3. Catenacci, G.; Maroni, M.; Cottica, D.; Pozzoli, L. Bull. *Environ. Contam. Toxicol.* **1990**, *44*, pp. 1-7.
4. Lucas, A. D.; Jones, A. D.; Goodrow, M. H.; Saiz, S. G.; Blewett, C.; Seiber, J. N.; Hammock, B.D. *Chem. Res. Toxicol.* **1993**, *6*, pp. 107-116.
5. Karu, A. E.; Harrison, R. O.; Schmidt, D. J.; Clarkson, C. E.; Grassman, J.; Goodrow, M. H.; Lucas, A. D.; Hammock, B. D.; White, R .J.; Van Emon, J. M. In *Immunoassays for Monitoring Human Exposure to Toxic Chemicals*; Vanderlaan, M., Stanker, L. H., Watkins, B. E., and Roberts, D. W., Eds.; ACS Symposium Series 451; American Chemical Society: Washington, DC, 1990; pp. 59-77.
6. Brady, J. F. *et al. J. Agric. Food Chem.* **1995**, *43*, pp. 268-274.
7. Brady, J. F. Analytical method AG-568, 1990; Ciba Crop Protection, Greensboro, NC.
8. Thurman, E. M.; Meyer, M.; Pomes, M.; Perry, C. A.; Schwab, A. P. *Anal. Chem.* **1990**, *62*, pp. 2043-2048.
9. Brady, J. F. In *Immunoassay of Agrochemicals, Emerging Technologies*; Nelson, J. O., Karu, A. E. and Wong, R. B., Eds.; ACS Symposium Series 586; American Chemical Society: Washington, DC, 1995; pp. 266-287.
10. Cheung, M. W.; Kahrs, R. A.; Nixon, W. B.; Ross, J. A.; Tweedy, B. G. In *Biological Monitoring for Pesticide Exposure;* Wang, R. G. M., Franklin, C. A., Honeycutt, R. C., and Reinet, J. C., Eds.; ACS Symposium Series 382; American Chemical Society: Washington, DC, 1989; pp. 231-239.

11. Weisskopf, C. P.; Seiber, J. N. In *Biological Monitoring for Pesticide Exposure;* Wang, R. G. M., Franklin, C. A., Honeycutt, R. C., and Reinet, J. C., Eds.; ACS Symposium Series 382; American Chemical Society: Washington, DC, 1989; pp. 206-214.
12. Feng, P. C.; Sharp, C. R; Horton, S. R. *J. Agric. Food Chem.* **1994**, *42*, pp. 316-319.

Chapter 14

The Assessment of Worker Exposure to Atrazine and Simazine: A Tiered Approach

Curt Lunchick[1] and Frank Selman[2]

[1]Jellinek, Schwartz and Connolly, Inc., 1525 Wilson Boulevard, Suite 600, Arlington, VA 22209
[2]Novartis Crop Protection, P.O. Box 18300 Greensboro, NC 27419–8300

The accurate assessment of dermal and inhalation exposure received by workers mixing, loading, and applying atrazine and simazine is critical to any subsequent assessment of the potential risk to these individuals. An exposure assessment is comprised of the determination of basic exposure data and information on how workers use atrazine and simazine. The exposure assessment should also follow a tiered or iterative approach in which existing information and assumptions are used in the first or initial tier. Subsequent tiers or iterations involve identifying data gaps with the greatest possible impact on the exposure assessment and replacing the assumptions used for those data gaps with actual data. The worker exposure assessment for the Special Review of atrazine and simazine has followed a tiered approach. Estimates of worker exposure to atrazine and simazine from use on crops was reduced significantly through refinement of product use information, inclusion of human dermal absorption potential and use of updated Pesticide Handlers Exposure Database (PHED) information.

Atrazine and simazine are two *s*-triazine herbicides that the United States Environmental Protection Agency (EPA, the Agency) placed into Special Review in November, 1994 (*1*). A Special Review is a process used by EPA that permits detailed and public evaluation concerning the potential risks and benefits of pesticides.

The information obtained by the EPA through the Special Review process is used to help the Agency make a regulatory decision on the registration of the pesticide by balancing the risks and benefits associated with its use. The initiation of the Special Review implies that the Agency has concerns about potential adverse effects associated with the use of atrazine and simazine. The initiation of the Special Review does not mean that the Agency has determined that the risks resulting from the use of atrazine and simazine are unreasonable. An assessment of worker exposure will be used by EPA to determine if the risks resulting from occupational exposure are acceptable.

Development of Worker Exposure Data

The quantification of dermal and inhalation exposure to pesticides received by mixer/loaders and applicators began in the 1950's with an exposure study carried out by Batchelor and Walker with parathion (*2*). These studies involved the use of cellulose patches or passive dosimeters placed on the workers' clothing or exposed skin. Passive dosimetry methods were modified through the years with the introduction of surgical gauze patches, multi-layered dosimeters, the use of personal air samplers, and placement of patches under the clothing. The use of whole-body dosimeters, such as cotton long underwear, has become prevalent during the past 10 years.

As the number of worker exposure studies accumulated and exposure data were compared, understanding of variables affecting worker exposure increased. Wolfe, *et al.* reviewed a large number of exposure studies in 1966 and concluded that important factors affecting dermal and inhalation exposure received by workers included environmental conditions, working habits of the worker, methods of application, the application rate, and the amount of pesticide handled (*3*). This understanding, that exposure received during pesticide application is not chemical-specific, led to a formal proposal at the 187th Meeting of the American Chemical Society in St. Louis, Missouri to create a mixer/loader and applicator exposure database (*4*).

The Pesticide Handlers Exposure Database (PHED) is a computerized database of mixer/loader, applicator, and flagger exposure data that was developed in response to the proposal. The database is a joint effort of EPA, Health Canada, and the American Crop Protection Association (ACPA). PHED version 1.0 was released in June 1992 and was soon replaced by PHED version 1.01. The revised version contained edits to the database's program but no changes to the exposure data. A second version (1.1) was released in March 1995. In the period prior to the release of PHED, the Agency developed a database on exposure derived from an evaluation of published literature sources of exposure data. This database was prepared in 1985 and used by EPA to estimate exposure for pesticides such as alachlor (*5*).

A Tiered Approach to Exposure Assessments

The tiered approach to pesticide worker exposure assessment is a logical and iterative method of estimating the exposure. The concept of a tiered approach to pesticide workers exposure assessment was presented and discussed at the "Workshop on Risk Assessment for Worker Exposure to Agricultural Pesticides" held in the Netherlands in 1992 (*6*) and at a workshop on the "Methods of Pesticide Exposure Assessment" held in 1993 in Canada (*7*).

In the first-tier of such an approach to pesticide exposure assessments, existing information is used to estimate the worker's exposure. Assumptions are used to fill in data gaps. A first-tier assessment often is based on surrogate exposure data, general use information assumptions relating to the use of the pesticide on different crops, and an assumption of 100% dermal penetration in the absence of dermal penetration data derived from animal studies. The exposure data may be potential dermal exposure (PDE) based on the measurement of exposure outside the clothing. Assumptions concerning the protective value of clothing must be made to estimate the actual dermal exposure (ADE) to the skin under the clothing.

The benefit of a first-tier exposure assessment is that the exposure assessment can be quickly conducted using existing information and assumptions to fill in the data gaps. Because the assumptions generally overestimate the actual data, a first-tier assessment is expected to overestimate the actual exposures that workers receive. Therefore, if acceptable risk can be demonstrated based on a first-tier exposure assessment, more detailed information normally is not collected and the risk assessment process is concluded. If acceptable risk is not demonstrated, key assumptions used in the first-tier assessment are identified and replaced with data in the second-tier assessment.

A second-tier exposure assessment often involves using ADE data under work clothing or personal protective equipment worn during the use of the specific pesticide in question. Chemical-specific dermal penetration data derived from animal studies (or human data if available) are used to estimate the absorbed dermal dose. Detailed crop use information is also obtained. Refinements in the second-tier database yield an estimate of exposure or absorbed dose which is more accurate and often substantially less than the initial first-tier estimate.

A third-tier assessment of worker exposure or absorbed dose is conducted only if the second-tier assessment does not demonstrate acceptable risk. In a third-tier assessment the absorbed dose is often determined from a biological monitoring study conducted with workers using the pesticide in question. The absorbed dose is often determined based on workers using possible exposure mitigation methods such as closed loading systems or enclosed cab tractors.

The exposure assessment for mixer/loaders and applicators handling atrazine and simazine has followed the tiered approach to assess exposure. The Agency prepared a first-tier assessment for the Position Document 1 (PD-1) (*1*). Ciba Crop Protection (Ciba) and the Agency's Occupational and Residential Exposure Branch (OREB) have both prepared second-tiered assessments in response to the initial PD-1 assessment.

The Agency's PD-1 Assessment of Workers Exposure: Exposure Data. The initial assessment of worker exposure to atrazine and simazine that underlies the exposure and risk estimates presented in the November 1994 PD-1 was based on the pre-PHED exposure database developed in 1985. The pre-PHED exposure database was used in the initial Agency assessments of worker exposure to atrazine in January 1988 (*8*) and simazine in May 1989 (*9*).

The exposure estimates for mixer/loaders, groundboom applicators, and flaggers that are in the pre-PHED exposure database are presented in Table I.

Table I. Summary of Pre-PHED Database Exposure Estimates Used for the PD-1

Job Function	*Exposure Estimate*
Open-Pour Mixer/Loader	0.93 mg/lb a.i.
Closed-Loading Mixer/Loader	0.015 mg/lb a.i.
Open-Cab Groundboom Applicator	56.7 mg/hr/lb a.i.
Enclosed-Cab Groundboom Applicator	2.2 mg/hr/lb a.i.
Pilot	0.58 mg/hr/lb a.i.
Flagger	3.2 mg/hr/lb a.i.
Macadamia Nut Mixer/Loader	5.4 mg/lb a.i.
Hand-Held Sprayer	115 mg/hr/lb a.i.

The exposure estimates used in the pre-PHED database have several key assumptions underlying the estimates. Many of the studies in the database measured the PDE, and exposure under the clothing was assumed to be 50% of the PDE to the covered body areas. The applicator and flagger exposure estimates were expressed in mg/hr/lb a.i. PHED allows expression of exposure in mg/lb a.i., thus eliminating the time variable. With the exception of the mixer/loader replicates, the published literature studies upon which the database relied did not provide the quantity of active ingredients handled thereby eliminating the possibility of expressing exposure as mg/lb a.i. In addition, several of the groundboom applicator studies did not indicate whether the tractors had open or enclosed cabs. Data from studies that did not specify cab type were assumed to be open-cab vehicles. Finally, the database did not contain exposure data for wettable powder or dry flowable formulations. Therefore, in the PD-1, estimates for exposure to the water-dispersible granular (WDG) formulations of atrazine or simazine were not included (*1*).

Dermal Penetration. Unlike many first-tier assessments, dermal penetration data were available for atrazine and simazine. The data were obtained from dermal penetration studies conducted by Ciba in rats and submitted to the Agency for review. Therefore, the EPA's Health Effects Division (HED) did not use the 100% dermal penetration assumption that is used in the absence of dermal penetration data.

HED estimated the dermal penetration of atrazine and simazine using two scenarios:

1) the actual percentage of atrazine and simazine absorbed (2% and 1%, respectively); and 2) adding the actual dermal penetration value and the amount retained in the skin at the dose site and potentially available for absorption (26.9%

for atrazine and 32% for simazine) (*1*). The worst-case potential dermal penetration values of 26.9% for atrazine and 32% for simazine were used for the PD-1 risk assessment (*1*).

Use Information. Information on how atrazine and simazine are actually used by pesticide handlers is critical to the estimation of daily and annual exposure. As is standard practice, use information incorporated into the Agency's exposure and risk assessments was prepared by the EPA's Biological and Economic Analysis Division (BEAD) of the Office of Pesticide Programs. The use information provided to HED by BEAD for use in the PD-1 assessment is summarized in Table II.

The Agency assessment of simazine worker exposure and risk presented in the PD-1 only involved the use of simazine on corn. Simazine is a minor corn herbicide. In 1994 only 1% of the corn acreage in the 10 major corn producing states was treated with simazine (*10*). The use information prepared by BEAD for the 1989 simazine exposure assessment indicated that corn accounted for 30% of simazine use (*11*).

HED Evaluation of Its First-tier Assessment. HED evaluated the strengths and uncertainties of its first-tier assessment of worker exposure prepared for the PD-1 (*12*). The evaluation acknowledged that the occupational exposure assessment was not intended to comprehensively address all uses of atrazine and simazine. Rather, the purpose was to address the major crops and application methods that potentially resulted in the highest exposure. The HED assessment also estimated the absorbed doses of atrazine and simazine using both the actual absorbed percentage (2% and 1% for atrazine and simazine, respectively) and the potential dermal penetration estimate (26.9% and 32% for atrazine and simazine, respectively). HED recommended the use of the potential dermal penetration estimates because the assumption that pesticide remaining on the skin is eventually absorbed is consistent with this risk assessment policy.

The most important point of HED's evaluation of its worker exposure assessment is that the risk assessment was not based on PHED. HED acknowledged *"that this risk assessment is based on existing exposure assessments that were completed several years ago. No attempt has been made to investigate additional sources of data, such as the Pesticide Handlers Exposure Database, at this time. Those sources of data may significantly affect current exposure and risk assessments for the triazines"* (*12*).

The Second-Tier Assessment of Worker Exposure

Immediately following the issuance of the worker exposure and risk assessment in the PD-1, Ciba began preparations to develop a second-tier assessment of worker exposure. The initial effort was to evaluate the first-tier assessment and identify key data gaps or assumptions in the assessment where use of actual data would significantly improve the accuracy of the atrazine and simazine worker exposure assessment. Three key areas where additional data could have significant impact were identified.

These areas were:

- Replace the pre-PHED information with PHED.
- Conduct a human dermal penetration study to replace the animal dermal penetration data and assumptions, and
- Obtain newer and more detailed use information, especially on crops where existing data were limited, such as sugarcane and macadamia nuts.

All of this information could not be obtained during the 120-day response period following the issuance of the PD-1, so an interim response was prepared. Ciba's March 1995 response to the PD-1 partially met the goals established for the second-tier worker exposure assessment. The Ciba worker exposure assessment submitted to the Agency in response to the PD-1 was based on PHED version 1.01, and contained more recent and detailed use information. The final second-tier exposure assessment discussed in this paper meets the three goals and is based on exposure data from the most recent version of PHED (1.1), estimates the absorbed dose using human dermal penetration data, and contains use information from 1992 to 1995.

PHED-Based Exposure Data. PHED version 1.1 was used to estimate the dermal and inhalation exposure received during the mixing/loading and application of atrazine or simazine. PHED contains a larger number of replications from more exposure studies than did the database used in the PD-1. This provides for a more representative estimation of exposure. In addition, the exposure data were based on measurements of exposure under the workers' clothing. This eliminated the need to assume a protective value for clothing as was necessary in the first-tier assessment. PHED permits the selection of exposure data based on the quality of the laboratory and field recoveries in each study. Ciba followed OREB guidelines and selected exposure data based on the two highest grades, grades A or B.

Additional information available in PHED concerning the quantity of active ingredient handled during each replicate allows the exposure to be expressed as mg/lb a.i. for all replicates. This eliminates variability of the monitoring duration from the exposure estimates. The database used for the PD-1 estimates did not have sufficient information to permit estimating applicator and flagger exposure on a mg/lb a.i. basis. Additionally, PHED version 1.1 contains exposure data to permit the estimation of exposure resulting from handling a WDG formulation.

PHED provides the regulatory agencies and the agrochemical industry with a common exposure database. In preparation for its second-tier assessment of atrazine and simazine worker exposure, OREB (*13*) is using PHED version 1.1 to estimate worker exposure. Table III summarizes the PHED exposure estimates calculated by OREB and Ciba, and compares them with the pre-PHED exposure estimates used for the PD-1 risk assessment.

Table II. Summary of Use Information Used in the PD-1

Crop/Application Method	*Application Rate (lb a.i./A)*	*Acres Treated Per Day*	*Acres Treated Annually*
Atrazine-Corn-Grower-Groundboom	1.2	112	195
Atrazine-Corn-Commercial-Groundboom	1.2	400	6,000
Atrazine-Corn-Commercial Aerial	1.2	385	5,775
Atrazine-Sorghum-Grower-Groundboom	1.0	107	135
Atrazine-Sugarcane-Commercial-Groundboom	1.0	400	6,000
Atrazine-Sugarcane-Commercial-Aerial	1.0	440	13,200
Atrazine-Macadamia Nuts-Handspray	2.0	5.13	20.5
Simazine-Corn-Grower-Groundboom	1.1	110	195
Simazine-Corn-Commercial-Groundboom	1.1	400	6,000
Simazine-Corn-Commercial-Aerial	1.1	385	5,775

Table III. Comparison of Exposure Estimates

Uses	*PD-1 (Pre-PHED) Dermal + Inhalation Exposure*	*OREB PHED Version 1.1 Dermal Exposure*	*Ciba PHED Version 1.1 Dermal Exposure*	*Ciba PHED Version 1.1 Inhalation Exposure*
Mixer/Loader				
Liquid Open-Pour	0.93 mg/lb a.i.	0.0425 mg/lb a.i.	0.0415 mg/lb a.i.	0.0012 mg/lb a.i.
Liquid Closed-System	0.015 mg/lb a.i.	0.0086 mg/lb a.i.	0.0096 mg/lb a.i.	0.00013 mg/lb a.i.
Wettable Powder	No Estimate	0.1675 mg/lb a.i.	----[a]	----[a]
WDG	No Estimate	0.0809 mg/lb a.i.	0.0673 mg/lb a.i.	0.00077 mg/lb a.i.
Applicator				
Open-Cab Groundboom	56.7 mg/hr/lb a.i.	0.0187 mg/lb a.i.	0.0160 mg/lb a.i.	0.00063 mg/lb a.i.
Enclosed-Cab Groundboom	2.2 mg/hr/lb a.i.	0.0068 mg/lb a.i.	0.0040 mg/lb a.i.	0.000043 mg/lb a.i.
Pilot	0.58 mg/hr/lb a.i.	0.0051 mg/lb a.i.	0.0044 mg/lb a.i.	0.000018 mg/lb a.i.
Flagger	3.2 mg/hr/lb a.i.	Not Determined	0.0115 mg/lb a.i.	0.00019 mg/lb a.i.
Hand-Spray Applicator	115 mg/hr/lb a.i.	0.461 mg/lb a.i.	Not Determined	Not Determined

[a] The wettable powder formulations of atrazine and simazine have been replaced by the WDF formulation.

Two important conclusions can be reached from comparing the exposure data presented in Table III. PHED version 1.1 provides significantly lower estimates of exposure compared to the pre-PHED calculations used for the PD-1. The PHED 1.1 data eliminate the need to assume a protective value for clothing and eliminate the time variable from the exposure estimate. In addition, PHED version 1.1 subsets contain more replicates from more studies than the original database subsets.

The second important conclusion is that both OREB and Ciba have independently developed similar estimates of exposure using PHED 1.1. The use of a common database was a goal in the development of PHED. The independent development of similar exposure estimates indicates this goal has been reached, thus now permitting focus on atrazine and simazine use information and toxicology.

Refinement of the Dermal Penetration Data. As previously discussed, the dermal penetration estimates of 26.9% for atrazine and 32% for simazine used in the Agency's PD-1 risk assessment were obtained from dermal penetration studies conducted in rats. The rat dermal penetration data produced uncertainty regarding the quantity of the triazines bound to the skin at the dose site but not absorbed into the body at the time of sacrifice. Some unknown portion of the bound material could be available for absorption into the body over time with the remaining material never being absorbed as the epidermal cells with bound triazines are sloughed off and replaced. An additional source of uncertainty was the interspecies translation of dermal penetration mechanics for both atrazine and simazine.

Because there were questions regarding the applicability of the rat dermal penetration study, Ciba conducted a dermal penetration study with atrazine on human volunteers (*14*). The 8 and 80 $\mu g/cm^2$ dose levels used in the study were selected based on PHED estimates of dose levels received by workers. The atrazine dose remained on the skin for 24 hours before the dose site was washed. Urine was collected over 168 hours. The human dermal penetration study results indicate 5.6% of the 8 $\mu g/cm^2$ atrazine dose was absorbed and 1.2% of the 80 $\mu g/cm^2$ atrazine dose is absorbed.

For the second-tier assessment, the human dermal penetration study data were used (5.6% absorbed), rather than the data from the rat dermal study. The structural similarities of atrazine and simazine and the similar rat dermal penetration potentials of the two triazines justifies the use of the human atrazine dermal penetration data for simazine.

Refinement of Atrazine and Simazine Use Information. The use information presented in Table II and utilized in the Agency's first-tier (PD-1) assessment of worker exposure was evaluated by Ciba. Extensive use information was obtained from market surveys, published use information, and contacts with specific crop associations. The new information produced minor adjustments to the PD-1 use information for some crops, such as atrazine use on corn and major revisions for use patterns on other crops, such as atrazine use on macadamia nuts. Use information for simazine was obtained for crops such as caneberries and citrus that were not evaluated by the Agency in the PD-1.

Ciba used Maritz Marketing Research to obtain use information for atrazine and simazine on corn and atrazine on sorghum. An additional source of information was the 1992 U.S. Census of Agriculture (*15*). The use information confirmed atrazine's role as a major corn herbicide applied to 63% of the nations corn acreage in 1993 (*16*). Simazine remains a very minor corn herbicide and was applied to only 1.2% of the nation's corn acreage in 1993 (*16*). Atrazine is also an important herbicide in sorghum production and was applied to 61% of the nation's sorghum acreage (*16*). The average application rates and acreage treated obtained from Maritz and from the Census of Agriculture for corn and sorghum were similar to the use information cited in the PD-1.

Atrazine is a critical herbicide for sugarcane production. Atrazine use in Florida sugarcane were obtained from the Florida Sugar Cane League, Inc. and Hawaii sugarcane use was obtained from the Hawaiian Sugar Planters' Association. The survey of Florida sugarcane production indicated that aerial application decreased substantially from the 60% to 80% of atrazine applications estimated by EPA in 1987 (*17*) to 20% of atrazine applications in 1994. The surveys also confirmed the assumption that large quantities of atrazine are handled by sugarcane applicators.

The first-tier assessment of atrazine use on macadamia nuts in the PD-1 assumed that atrazine was applied by hand-held sprayers and that an individual applicator sprayed 20.5 A/yr. Macadamia nut producers and agricultural extension agents in Hawaii were contacted to ascertain how atrazine is used on macadamia nuts. Five plantations account for 66% of Hawaii's 20,100 A of macadamia nuts. The remaining acreage is in 2 to 20 A parcels. Atrazine use on macadamia nuts is primarily by two of the five plantations and is applied by groundboom equipment. A total of 8,048 lb of atrazine was applied to macadamia nuts on the two plantations. Hand-spray applications of atrazine were not used on the five plantations and accounted for less than 100 lb of atrazine used on the small orchards.

Simazine use information on citrus, blueberries, caneberries, cranberries, strawberries, and tree nurseries was submitted to EPA as part of a simazine exposure assessment (*18*). The information for these uses of simazine was obtained from the 1987 Census of Agriculture (*15*), the U.S. Forest Service (*19*), and use recommendations issued by several state extension agencies. This use information was evaluated and accepted by OREB (*20*)

A summary of the second-tier use information is provided in Table IV.

Ciba used the revised exposure estimates from PHED version 1.1, the dermal penetration value of 5.6% from the human study, and the revised use information to estimate the exposure to mixer/loaders and applications handling atrazine and simazine. The lifetime average daily dose (LADD) is the exposure estimate used in the risk assessment. The LADD is calculated using the following steps:

1) Absorbed Dose = Dermal Exposure x Dermal Penetration + Inhalation Exposure. The dermal and inhalation exposures are expressed as mg/lb a.i. and

Table IV. Summary of Use Information for Second-Tier Exposure Assessment

Crop/Application Method	*Application Rate (lb a.i./A)*	*Acres Treated Annually*	*Lbs a.i. Handled Annually*
Atrazine			
Corn-Grower-Groundboom	1.33	150	200
Corn-Commercial Groundboom	1.33	4,500	6,000
Corn-Commercial Aerial	1.33	770	1,025
Sorghum-Grower-Groundboom	1.25	150	188
Sorghum-Commercial Groundboom	1.25	600	750
Sorghum-Commercial Aerial	1.25	770	965
Sugarcane-Hawaii-Groundboom	4.1	840	3,450
Sugarcane-Florida-Groundboom	3.3 to 5.1	2,420	12,342
Macadamia Plantation-Groundboom	2.2		800
Simazine			
Corn-Grower-Groundboom (EC)	1.2	150	180
Corn-Grower-Groundboom (WDG)	1.6	150	240
Corn-Commercial-Groundboom (EC)	1.2	600	720
Corn-Commercial-Groundboom (WDG)	1.6	600	960
Corn-Commercial-Aerial (EC)	1.2	385	462
Corn-Commercial-Aerial (WDG)	1.6	385	616
Orchard-Groundboom	4.0	30	120
Blueberries-Groundboom	4.0	3.1	12.4
Caneberries-Oregon-Groundboom	4.0	3.8	15.2
Cranberries-Groundboom	4.0	10	40
Strawberries-Northwest-Groundboom	1.0	5	5.0
Tree Nurseries-Groundboom	3.0	15	45

dermal penetration is a percentage. Inhalation exposure is assumed to be 100% absorbed. Absorbed dose is also expressed as mg/lb a.i.

2) Absorbed Annual Dose = Absorbed dose x lb a.i. handled annually ÷ body weight. The body weight is 70 kg and the absorbed annual dose is expressed as mg/kg/yr.

3) LADD = Absorbed Annual Dose ÷ 365 days/yr x 35 years/70 years. The absorbed annual dose is amortized by the days in a year and an occupational exposure that occurs for 35 years during a 70-year lifespan.

These series of equations can be used to estimate the LADD for both the PD-1 first-tier exposure assessment and for the second-tier assessment. The only variation in use of these equations is where the exposure was expressed as mg/hr/lb a.i. in the first-tier assessment. In these specific instances the hours per year that triazine or simazine was applied is added to the absorbed annual dose equation.

Tables V and VI summarize the LADD for atrazine and simazine, respectively, calculated in the second-tier exposure assessment and compare the estimates to the first-tier LADD estimates (*12*).

Conclusion

The tiered approach to assessing the agricultural worker's exposure to atrazine and simazine provides a logical, iterative method of estimating the exposure for use in the risk assessment process. The U.S. Environmental Protection Agency prepared the first-tier assessment of worker exposure. This assessment was used in the PD-1 for atrazine and simazine.

Since that time, access to more extensive, and detailed, exposure data from PHED 1.1 has become available. Additionally, Ciba conducted a dermal penetration study on humans to provide a more realistic estimate of the absorption percentage used in the exposure assessment. Finally, prior to developing second-tier exposure estimates, additional, and more recent, atrazine and simazine use information was obtained. This information included additional crops, use of the water dispersible granule formulation, and significant changes in some earlier assumptions regarding application equipment and crop acreage.

The second-tier assessment is more refined, utilizes more recent information, and relies on fewer assumptions than the first-tier assessment. The second-tier assessments of atrazine and simazine worker exposure show significantly less exposure than that provided by the first tier approach. The second-tier assessment demonstrates that the LADDs for growers applying atrazine or simazine to corn are approximately 200 and 260 times lower, respectively, than the initial estimates, and that the revised estimates of LADD for commercial applicators applying atrazine to corn were approximately 35 to 150 times lower than the initial exposure estimates based on the first tier assumptions. Additionally, the clarification of application procedures for atrazine application to macadamia nuts has led to a second-tier exposure estimate that is approximately 300 times less than the first-tier estimate.

Table V. Summary and Comparison of Atrazine Second-Tier Exposure Estimates to the First-Tier PD-1 Estimates

Use Pattern	*LADD (mg/kg/day)*	
	First-Tier	*Second-Tier*
Corn		
Grower[1]-EC	4.3×10^{-3}	2.0×10^{-5}
Grower-WDG	Not Determined	2.6×10^{-5}
Commercial M/L-EC Open Pour	3.5×10^{-2}	4.1×10^{-4}
Commercial M/L-WDG	Not Determined	5.3×10^{-4}
Commercial M/L-Closed System	5.5×10^{-4}	7.8×10^{-5}
Commercial Ground Appl.- Open Cab	3.0×10^{-2}	1.8×10^{-4}
Commercial Ground Appl.- Enclosed Cab	1.1×10^{-3}	3.1×10^{-5}
Aerial M/L - EC Open Pour	Not Determined	7.1×10^{-5}
Aerial M/L - WDG	Not Determined	9.1×10^{-5}
Aerial M/L - Closed System	5.5×10^{-4}	1.3×10^{-5}
Pilot	4.3×10^{-5}	5.0×10^{-6}
Flagger	2.4×10^{-4}	1.7×10^{-5}
Sorghum		
Grower-EC	3.0×10^{-3}	1.8×10^{-5}
Grower-WDG	Not Determined	2.4×10^{-5}
Commercial M/L- EC Open Pour	Not Determined	2.4×10^{-5}
Commercial M/L- WDG	Not Determined	6.7×10^{-5}
Commercial M/L Closed System	Not Determined	1.0×10^{-5}
Commercial Ground Appl. - Open Cab	Not Determined	2.2×10^{-5}
Commercial Ground Appl. - Enclosed Cab	Not Determined	4.0×10^{-6}
Aerial M/L - EC Open Pour	Not Determined	6.7×10^{-5}
Aerial M/L - WDG	Not Determined	8.6×10^{-5}
Aerial M/L - Closed System	Not Determined	1.3×10^{-5}
Pilot	Not Determined	5.0×10^{-6}
Flagger	Not Determined	1.6×10^{-5}
Hawaii - Sugarcane		
M/L/A - EC Open Pour and Open Cab	Not Determined	3.4×10^{-4}
M/L/A - WDG Open Pour and Open Cab	Not Determined	4.5×10^{-4}
M/L/A - Closed Loading and Enclosed Cab	Not Determined	6.3×10^{-5}
Florida - Sugarcane		
M/L/A - EC Open Pour and Open Cab	5.4×10^{-2}	1.2×10^{-3}
M/L/A - WDG Open Pour and Open Cab	Not Determined	1.6×10^{-3}
M/L/A - Closed Loading and Enclosed Cab	1.4×10^{-3}	2.3×10^{-4}
Macadamia Nuts		
M/L/A Single Appl. Hand Spray	2.6×10^{-2}	Not Determined
M/L/A Groundboom EC	Not Determined	7.9×10^{-5}
M/L/A Groundboom WDG	Not Determined	1.0×10^{-4}

[1]Grower does open pour mixing/loading and open cab groundboom application
M/L = Mixer/loader, M/L/A = Combined mixer/loader/applicator
EC = Emulsifiable concentrate
WDG = Water dispersible granule

Table VI. Summary and Comparison of Simazine Second-Tier Exposure Estimates to the First-Tier PD-1 Estimates

Use Pattern	*LADD (mg/kg/day)*	
	First-Tier	*Second-Tier*
Corn		
Grower[1]-EC	4.8×10^{-3}	1.8×10^{-5}
Grower-WDG	Not Determined	3.1×10^{-5}
Commercial M/L - EC Open Pour	3.8×10^{-2}	5.0×10^{-5}
Commercial M/L - WDG	Not Determined	8.5×10^{-5}
Commercial M/L - Closed System	6.1×10^{-4}	9.0×10^{-6}
Commercial Ground Appl. - Open Cab	3.2×10^{-2}	2.2×10^{-4}
Commercial Ground Appl. - Enclosed Cab	1.2×10^{-3}	5.0×10^{-6}
Aerial M/L - EC Open Pour	Not Determined	3.2×10^{-5}
Aerial M/L - WDG	Not Determined	5.5×10^{-5}
Aerial M/L - Closed System	6.1×10^{-4}	6.0×10^{-6}
Pilot	4.8×10^{-5}	2.0×10^{-6}
Flagger	2.6×10^{-4}	8.0×10^{-6}
Citrus		
Grower - EC	Not Determined	1.2×10^{-5}
Grower - WDG	Not Determined	1.6×10^{-5}
Blueberries		
Grower - EC	Not Determined	1.0×10^{-6}
Grower - WDG	Not Determined	2.0×10^{-6}
Caneberries		
Grower - EC	Not Determined	1.0×10^{-6}
Grower - WDG	Not Determined	2.0×10^{-6}
Cranberries		
Grower - EC	Not Determined	4.0×10^{-6}
Grower - WDG	Not Determined	5.0×10^{-6}
Strawberries		
Grower - EC	Not Determined	5.0×10^{-7}
Grower - WDG	Not Determined	1.0×10^{-6}
Tree Nurseries		
Grower - EC	Not Determined	4.0×10^{-6}
Grower - WDG	Not Determined	6.0×10^{-6}

[1]Grower does open pour mixing/loading and open cab groundboom application
M/L = Mixer/loader, M/L/A = Combined mixer/loader/applicator
EC = Emulsifiable concentrate
WDG = Water dispersible granule

The second-tier exposure assessment used new information from studies, updated surveys and PHED 1.1 to accurately evaluate occupational exposure to atrazine and simazine.

Literature Cited

1. U.S. Environmental Protection Agency; *The Triazine Herbicides Atrazine, Simazine, and Cyanazine;* Position Document 1 Initiation of Special Review. 9 November 1994.
2. Batchelor, G.S.; Walker, K.C. *Health Hazards Involved in Use of Parathion in Fruit Orchards of North Central Washington. Arch Industr Hyg.* **1954**, *10*, pp. 522.
3. Wolfe, H.R.; Durham, W.F.; Armstrong, J.F. *Exposure of Workers to Pesticides. Arch Environ Health.* April **1967**. Vol *14*, pp. 622-633.
4. Hackathorn, D.R.; Eberhart, D.C. *Data-Base Proposal for Use in Predicting Mixer-Loader-Applicator Exposure. Dermal Exposure Related to Pesticide Use;* American Chemical Society Symposium Series 273; Published by the American Chemical Society: Washington, DC, 1985, pp. 341-355.
5. Lunchick, C. *et al.; The Environmental Protection Agency's Use of Biological Monitoring Data for the Special Review of Alachlor. Biological Monitoring for Pesticide Exposure;* American Chemical Society Symposium Series 382; Published by the American Chemical Society: Washington, DC, 1989, pp. 327-337.
6. Henderson, P. Th. *et al. Risk Assessment for Workers Exposure to Agricultural Pesticides.* Review of a Workshop. Ann Occup Hyg 37:5, 1993, pp. 499-507.
7. Carmichael, N.G. In *Pesticide Exposure: Overview of the Tier Approach;* Methods of Pesticide Exposure Assessment; Curry, P.B. *et al.*, Ed.; Plenum Press: New York, 1995, pp. 117-122.
8. U.S. Environmental Protection Agency, "*Exposure Assessment for Policy Group,*" Memorandum from Michael Firestone, Office of Pesticide Programs, Health Effects Division, " 6 January, 1988.
9. U.S. Environmental Protection Agency; *"Nondietary Exposure Assessment of Simazine,"* Memorandum from Curt Lunchick, Office of Pesticide Programs, Health Effects Division, 22 May, 1989.
10. U.S. Department of Agriculture; National Agricultural Statistics Service. Agricultural Chemical Usage 1994 Field Crops Summary Ag Ch 1 (95), March 1995.
11. U.S. Environmental Protection Agency; *"Transmittal of Qualitative Use Assessment of Simazine,"* Memorandum from James G. Saulmon, Office of Pesticide Programs, Biological and Economic Analysis Division, 13 January 1989.
12. U.S. Environmental Protection Agency; *"Revised Occupational and Residential Risk Assessment for the Triazines,"* Memorandum from Michael Beringer, Office of Pesticide Programs, Health Effects Division, 7 March 1994.
13. U.S. Environmental Protection Agency; *"Revised Occupational and Residential Exposure Assessment for Atrazine and Simazine,"* Memorandum from Tracy Fitzgerald, Office of Pesticide Programs, Health Effects Division. 4 October 1995.

14. *In Vivo* Percutaneous Absorption of Atrazine in Man. Ciba Crop Protection Report No. ABR-96067. October, 1997
15. Bureau of the Census, U.S. Department of Commerce, AC 92-A51, October 1994.
16. Maritz Marketing Research Use Information. Submitted with initial PD-1 response to EPA. March 1995.
17. U.S. Environmental Protection Agency; *"Use Related Exposure Data for Atrazine Exposure Assessment,"* Memorandum from Richard Petrie, Office of Pesticide Programs. Benefits and Use Division. 11 December 1987.
18. Assessment of Worker Exposure for Simazine EC, WP, and WDG Formulations Using the Pesticide Handlers Exposure Database, 4 January 1994.
19. Human Health Risk Assessment for the Use of Pesticides in USDA Forest Service Nurseries. U.S. Forest Service, U.S. Department of Agriculture, FS-412, October 1987.
20. U.S. Environmental Protection Agency; *"Exposure Assessment for Uses of Simazine EC, WP and WDG Formulations Using the Pesticide Handlers Exposure Database,"* Memorandum from Arthur O. Schloser, Office of Pesticide Programs, Health Effects Division. 15 July 1994.

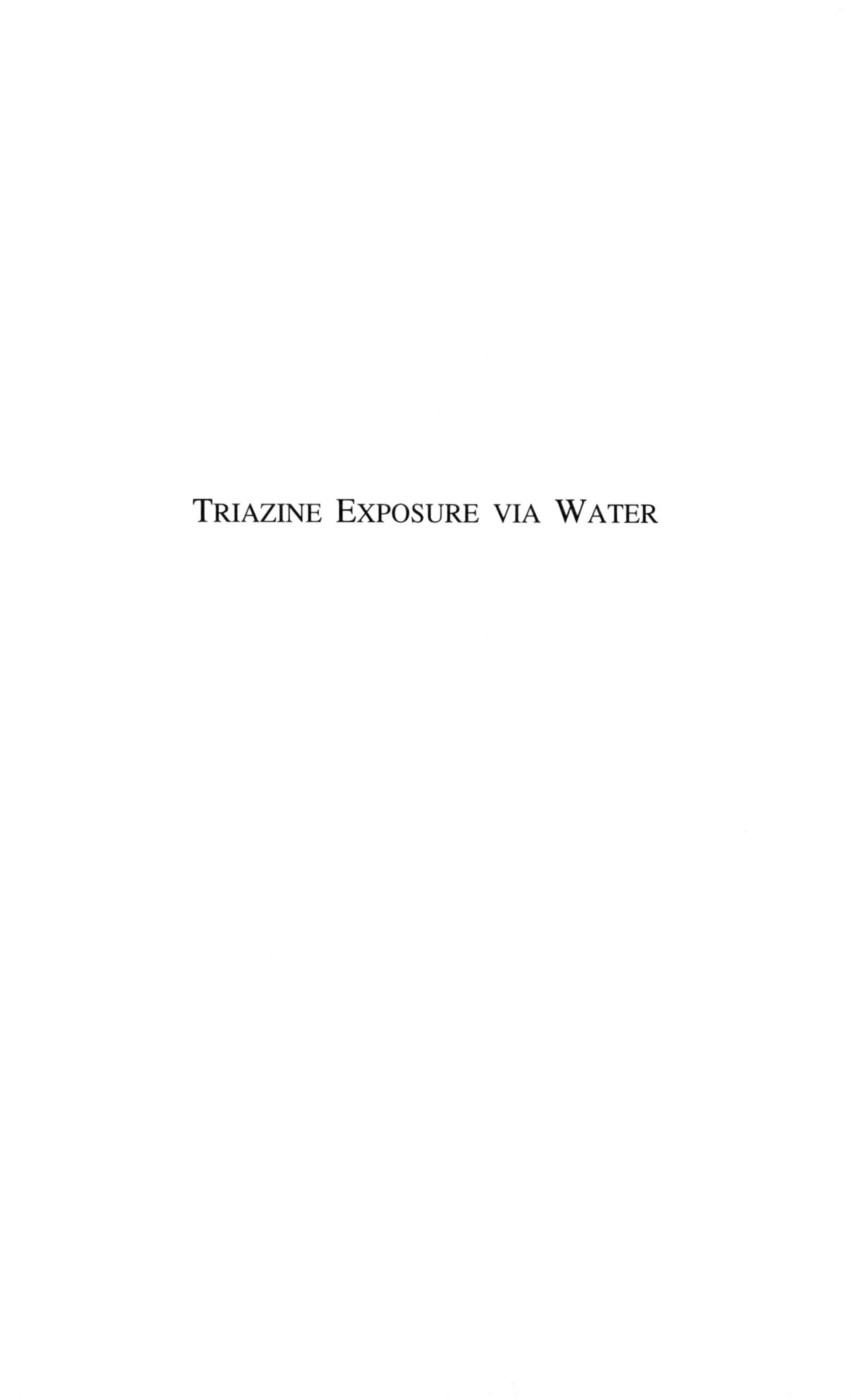

TRIAZINE EXPOSURE VIA WATER

Chapter 15

Atmospheric Transport and Deposition, an Additional Input Pathway for Atrazine to Surface Waters

Dorothea F. Rawn[1], Thor H. J. Halldorson[2], and Derek C. G. Muir[2,3]

[1]Department of Soil Science, University of Manitoba, Winnipeg, Manitoba R3T 2N2 Canada
[2]Department of Fisheries and Oceans, Freshwater Institute, 501 University Crescent, Winnipeg, Manitoba, R3T 2N6, Canada

This paper reviews the current literature observations of atrazine in precipitation and ambient air and evaluates the importance of atmospheric deposition to surface waters using measurements of atrazine in water, air and precipitation at two locations where this chemical is not used. Precipitation was found to be an important route of entry into aquatic systems, particularly during periods when extensive regional use of atrazine occurs. Gas exchange and dry deposition were estimated using established methods and gas exchange was found to be the primary dry process of atrazine deposition to surface waters.

Surface runoff losses from treated fields and groundwater contamination have been traditionally viewed as the main sources of the triazine herbicides to rivers (*1*). There has been little attention paid, until recently, to the role of atmospheric contributions to surface waters by current use herbicides, particularly in agricultural regions. In the most comprehensive study to date, Goolsby *et al.* (*2*) have measured atrazine in precipitation samples from both agricultural and non-agricultural regions of the US. Maximum atrazine concentrations in precipitation samples collected through midwestern and northeastern US were observed in May and June during 1990 and 1991 (*3*). Atrazine deposition via precipitation and gas exchange was examined in the Chesapeake Bay region (*4*), based on bulk rain and ambient air levels. From this and previous studies, Glotfelty *et al.* (*4*) concluded that seasonal high atrazine concentrations in surface waters resulted from runoff contamination, whereas atmospheric deposition was responsible for low-level, widespread contamination year round. Although the triazines have been detected in precipitation and air, there has been little research to examine dry deposition and gas exchange of these chemicals (Figure 1), or the relative importance of each of these pathways to surface waters. The objective of this paper is to review existing

[3]Current Address: National Water Research Institute, Environment Canada, Burlington, Ontario L7R 4A7 Canada

information on atrazine in precipitation and air, and to evaluate the importance of atmospheric deposition pathways.

Triazines are relatively non-volatile (Table I), which may be the reason why volatilization losses and atmospheric transport of these chemicals have not been studied as extensively as surface runoff losses. Atrazine losses in the vapor phase from treated fields in a southern California study were found to be 0.16% during the first 3 days of the experiment (*5*). Atrazine losses through volatilization from treated fields in Maryland, however, were found to be 2.4% (*6*). Surface runoff losses of pesticides are considered catastrophic if they exceed 2% of applied levels in a single event (*7*). In 1989, 1.7% of the atrazine applied in the US Midwest was estimated to have been transported to the Gulf of Mexico (*8*). Thus, overall atrazine losses via volatilization may be similar to runoff losses.

Table I: Physical/Chemical Properties related to atmospheric behaviour of triazines.

Chemical	Molecular Weight[a] $g \cdot mol^{-1}$	Water Solubility[a] $mg \cdot L^{-1}$ (T°C)	Subcooled Liquid Vapor Pressure[b] VP_L, mPa	Melting Point[a] °C	Henry's Law Constant[c] $Pa \cdot m^3/mol$
Atrazine	215.7	33 (25)	1.22×10^{-3}	174	2.52×10^{-4}
Simazine	201.7	6.2 (25)	6.23×10^{-6}	225-7	9.59×10^{-3}
Terbutryn	241.4	25 (20)	2.99×10^{-3}	104-5	2.70×10^{-3}

[a](*9*)
[b]Calculated using the relationship: $(VP_{liquid}/VP_{solid}) = \exp[6.81\,(T_{Melting\ point}/T-1)]$ (*35*)
[c](*10*)

Seasonal trends of atrazine in precipitation have been found, generally corresponding to application times (*11*). Goolsby and coworkers (*2*) found temporal patterns of atrazine in rainwater samples fit closely to the trends in water previously observed for the Mississippi River. The highest observed concentrations were detected in samples collected during application times from regions where greatest atrazine use occurred. Atrazine was found infrequently in areas where it is not used, such as Maine and parts of Michigan. When detected in these areas, it was at much lower concentrations than observed in high use regions (*2*).

Recent atrazine measurements in several European countries also show seasonal trends in precipitation and evidence of long range transport. Although atrazine use was reduced in Switzerland in 1988, detectable levels of atrazine were found in precipitation samples collected during 1988 and 1989 (*12*) (Table II). Detectable levels of atrazine were found in approximately 25% of the precipitation samples collected in Germany, both prior to, and following, the ban on atrazine use in 1991 (*13*). Atrazine was detected in precipitation samples at four locations in Northern Germany, however concentrations were not found to be significantly different between agricultural, urban and coastal sites (*14*) and no seasonal or spatial differences in atrazine concentrations were observed. Atrazine was not registered for use in Germany throughout the duration of the latter study. Maximum atrazine concentrations in rainwater samples collected near an Italian forest were 1.9 μg/L, with temporal trends reflecting local high use intervals (*15*). Combined wet and dryfall concentrations of atrazine were found at maximum levels in early spring in a rural area of France (*16*). In Paris, however, the highest observed levels were found during sampling in June, 1991 (*16*). Atrazine residues in Norwegian rain samples collected during 1993 were detectable during early to mid-May sampling times (*17*), four years following its removal from the market in Norway. In all cases, highest concentrations observed in precipitation samples reflected local or regional atrazine application.

Table II: Recent observations of atrazine in precipitation.

Location	Concentration Range (ng/L)	Duration of Study	Maximum Concentration Observed	Reference
Ontario, Canada	<10 - 445	Apr. ‘91 - Sept. ‘92	May ‘92	(*21*)
Midwestern - Eastern US	<50 - 16,000	Mar. ‘90 - Sept ‘91	June ‘90, ‘91	(2)
Iowa, US	<100 - 40,000	Oct. ‘87 - Sept. ‘90	June ‘89	(*22*)
Minnesota, US	<20 - 2,000	Mar. - Nov. ‘89 Mar. - Apr. ‘90	May ‘89, June ‘90	(*23*)
Northeastern US	<50 - 1,500	Apr. - Aug. ‘85	May ‘85	(11)
Northern Germany	<10 - 113	Mar. ‘90 - Mar. ‘92	May ‘90, ‘91	(*13*)
North Sea Germany	<3 - 140	Apr. - Jul. ‘93	May ‘93	(*14*)
France	< 5 - 140	Mar. ‘91 - Feb. ‘92	April - May ‘91	(*16*)
Switzerland	<0.1 - 600	Feb. ‘88 - Jul. ‘89	May - June ‘89	(*12*)
Italy	<100 - 1,990	May - Oct. ‘88	June ‘88	(*15*)
Norway	<10 - 86	Jun. - Sept. ‘92/‘93	May ‘93	(*17*)

Atmospheric deposition of semivolatile organic contaminants (SOCs) also occurs via dry processes which can be separated into gas exchange and particle associated deposition (Figure 1). There has been limited research to estimate relative contributions of SOCs through each of the pathways. Estimates of atrazine associated with the dryfall component were made for the Great Lakes region by Eisenreich and Strachan (*18*) using total air concentrations (vapor + particle). Particle deposition is a function of both the size and mass of a particle, in addition to nonchemical factors such as the wind speed, temperature and surface characteristics (*19*). The majority of work in this area has been performed for persistent organochlorine compounds, rather than current use herbicides. For some chemicals, relationships between concentration associated with the particles and ambient air temperature have been developed (*18*; *20*).

As far as we are aware, there have been no direct measurements of atrazine in dry deposition although several authors have measured total precipitation and dryfall using uncovered precipitation collectors (*13*; *16*; *22*). Representative measurement of dry deposition is difficult to perform. Many devices have been used to measure dryfall and have resulted in the bias toward one particle size or another. Inverted frisbee and coated flat-plate samplers were used in the determination of polychlorinated dibenzo-*p*-dioxin (PCDD) and -furan (PCDF) dry deposition fluxes (*24*). Koester and Hites (*24*) concluded that frisbee and flat plate collectors were less efficient than water surfaces, based on mass balance estimates.

Knowledge of depositional velocity (V_d) and particle surface areas (S.A.), used in the calculation of the fraction associated with the particle phase ($\varnothing$) are critical for estimating dry deposition. V_d is dependent upon the fraction of coarse or fine particles in air available for dry deposition (*25*). Estimates of particle S.A. from rural and urban regions differ by greater than one order of magnitude (*26*). Surface area estimates are used in combination with subcooled liquid vapor pressure (VP_L) to estimate the fraction of a given SOC associated with the particle phase using the Junge Pankow model (*27*).

There have been far fewer measurements of atrazine in air than in precipitation (Table III). Concentrations of atrazine in air have generally been reported to be low, or non-detectable (< 1 ng/m^3). Particle-associated or vapor phase atrazine was not detected in ambient air samples collected in Italy (*15*). This may have been a function of the high detection limit 1.6 ng/m^3 for atrazine. In Paris, air samples collected between April and June contained detectable levels of atrazine in the vapor phase (*16*) at levels (0.05 ng/m^3) very near detection limits (0.03 ng/m^3), when observed. In a Japanese study of pesticides in ambient air, atrazine was detected during both spring and summer sampling times (*28*), but, only in the vapor phase (0.2-0.32 ng/m^3), with no contribution from the particulate phase. Ambient air measurements of atrazine ranged between 0.1-20 ng/m^3 in the Wye River region, near Chesapeake Bay and varied seasonally. Five percent of the atrazine concentrations detected in ambient air was associated with the particle phase during the warmer sampling times and increased particle-bound atrazine was observed during winter sampling events (*4*). Seasonal differences in ambient air levels of atrazine were observed in the Wye River region, with winter samples containing approximately 1% of observed summer values. Alachlor and metolachlor, two other herbicides used in this region, were limited temporally to local use times in ambient air samples, whereas both atrazine and simazine were detected throughout the year (*4*). Glotfelty and coworkers (*4*) concluded that the source of high pre-application atrazine levels resulted from long range transport from southern states, where atrazine application occurred during March and April. Based on the measured atrazine concentrations in each of the wetfall, ambient air and Chesapeake Bay surface waters, atrazine deposition via gas exchange, from the atmosphere to the river was considered important (*4*).

In the following sections we examine the relative importance of atmospheric contributions of the triazine herbicide atrazine via individual atmospheric pathways based on data from two locations; one in the centre of the agricultural region of Manitoba, and the other at a remote location, within a boreal forest in northwestern Ontario.

Table III: Atrazine in ambient air.

Location	Concentration Range (ng/m^3) [where available]	Duration of Study	Fraction associated with particle phase	Reference
Ontario Canada	0.002 - 0.037	May '90 - Sept. '92	not determined	(*1*)
Midwest, US	0.008 - 20	Mar. '81 - Oct. '82	0.05 - summer 0.32 - winter	(*4*)
Japan	0.2 - 0.32	Jul. '91, Apr. '92	0	(*28*)
Italy	<1.6 (not detected)	May - Oct. '88	not determined	(*15*)
France	0.05	Mar. '91 - Feb. '92	0	(*16*)

Materials and Methods:

Watershed Description: The South Tobacco Creek Watershed (49°19'28"N, 98°21'50"W) (Figure 2) is a 70 km^2 agricultural watershed in southern Manitoba. Detailed land use information is available including weekly pesticide application information within the watershed, details of the formulation applied and rate of application, on a field by field basis. A survey of 41 landowners over the past 5 years show that there has been no atrazine application within this watershed. South Tobacco Creek, which drains the watershed ranged in dissolved organic carbon from 6.5 to 68 mg·L^{-1} and in pH from 6.7 to 8.0, during the course of the study.

Experimental Lakes Area: The lake and atmospheric samples were collected in a boreal forest in northwestern Ontario 50 km east of Kenora, ON (49°45'N, 93°47'W) (Figure 2). This location is in a remote area, at least 100 km from the nearest agricultural area and 1000 km from the US "corn belt".

Sample Collection: South Tobacco Creek Watershed: Air, precipitation and creek water were sampled over a three year period, from 1993-1995 to determine pesticide levels in each of the compartments. The sampling program was designed to collect large volumes of air, precipitation and creek water in order to determine locally used pesticides and those entering the watershed via long range transport.

Air samples were collected using a high volume (General Metalworks PS-1) PUF sampler with a GF/A glass fibre filter to trap particulate matter. Filters were prepared for use by heating for 18 hours at 265°C. PUF plugs were pre-cleaned by Soxhlet with hexane for 24 h prior to use. Approximately 350 m^3 of air was drawn through the sampling unit over a 24 h period every 6 days during the early part of the growing season, sampling commenced in June 1993. Samples were collected every 12 days later in the field season, annually. Samples were collected June-October 1993, May-November, 1994 and May-October 1995. PUFs were placed in sealed glass jars and stored at 4°C until extraction and analysis. Filters were placed in "Whirl pak" bags and stored at 4°C until extraction and analysis.

Rainfall was collected, using an automated wet-only (0.2025 m^2) sampler

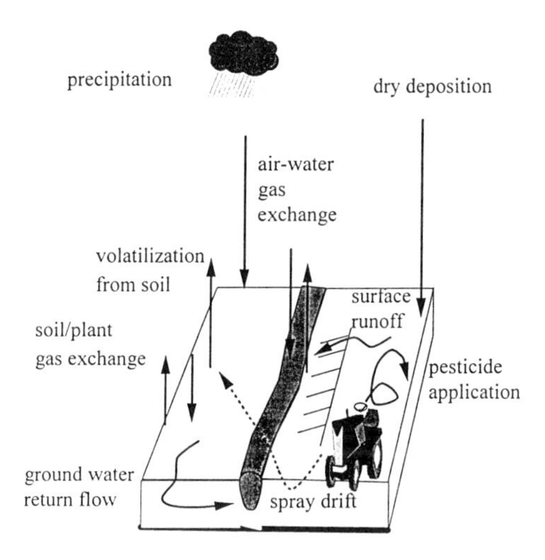

Figure 1. Atmospheric inputs and loss pathways of atrazine in agricultural watersheds.

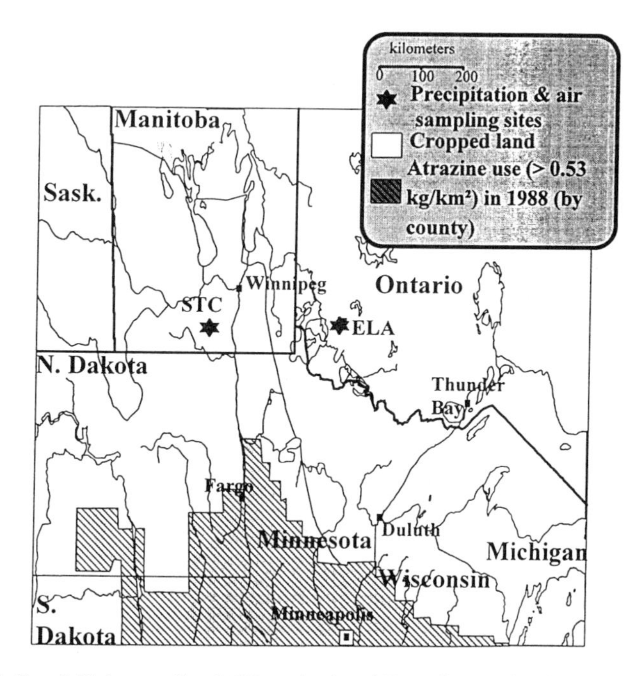

Figure 2. South Tobacco Creek Watershed and Experimental Lakes Area locations.

(Meterological Instrument Centre) in the centre of the watershed, continuously over 2 week periods in 18 L stainless steel transfer tanks. Sample volumes ranged from 3 to 18 L. Rain water was filtered through a glass wool plug at the bottom of the collection basin, for removal of large particles and dichloromethane (DCM) was used as a preservative.

Creek water was collected using a submersible pump in the centre of the creek and pumping 18 L into stainless steel containers. Samples were filtered under pressure through precleaned 1 μm GFC glass fibre filters. Samples were collected on a weekly (1994-5) and biweekly (beginning in July 1993) sampling schedule, from spring melt through freeze up.

Experimental Lakes Area: Air and precipitation samples were collected throughout the 1995 field season. Air samples were collected for a 24 h period on a 12 day cycle using a PS-1 high volume air sampler. Precipitation samples were collected on a continuous basis from May to October 1995, using an automated wet-only sampler. Sample containers were exchanged on a monthly basis through the season.

Sample Analysis: A broad spectrum approach was used for sample analysis to analyse both acid and neutral herbicides (e.g., 2,4-D and atrazine). PUFs were extracted using Soxhlet with DCM for 4 h. Deuterated (d_5) atrazine was used as an internal recovery standard which was added at the extraction step.

Creek water and precipitation samples were extracted in the sampling containers using DCM. Creek water samples were filtered (1 μm glass fibre) prior to extraction. Water was adjusted to pH 2 to extract acid herbicides and phenolics and then taken to pH 10 to recover hydrophobic organics. Extracts were evaporated to small volumes and methylated using freshly prepared diazomethane. Cleanup was performed using 5% deactivated Florisil, eluting with 20 mL hexane to remove polychlorinated biphenyls, followed by 85 mL, 18% ethyl acetate (EtAc) in hexane. Using this procedure, the acid and phenolic herbicides were analysed as methyl ester derivatives and results will be reported elsewhere. Atrazine was quantified using gas chromatography-mass spectrometry (Hewlett Packard 5971 MSD) using a 60 m x 0.25 μm DB-5 column with helium as the carrier gas. Detection criteria were the correct ratios of two characteristic ions and retention times. Herbicides were quantified using external standard solutions and corrected for volume changes using PCB-104.

Quality assurance: Blank PUFs, filters and blanks of "Super Q" water were analysed with each batch of samples. PUF and air filter samples were corrected for internal standard recoveries. Average internal standard recovery for precipitation and creek water was 107%. Water and precipitation results were not corrected for d_5-atrazine standard recoveries. Desethylatrazine (DEA) was detected in water and precipitation, although recovery from water may be inefficient using liquid-liquid partitioning, resulting in lower observed concentrations in the present study (Thurman, E.M., U.S. Geological Survey, Lawrence, KA. personal communication, 1996), therefore, DEA concentrations are not reported for precipitation or creek water samples.

Results and Discussion:

Air: Temporal patterns of atrazine in ambient air were similar between the South Tobacco Creek Watershed (STCW) and the Experimental Lakes Area (ELA) sites (Figure 3). Atrazine concentrations at the ELA location were generally lower than those

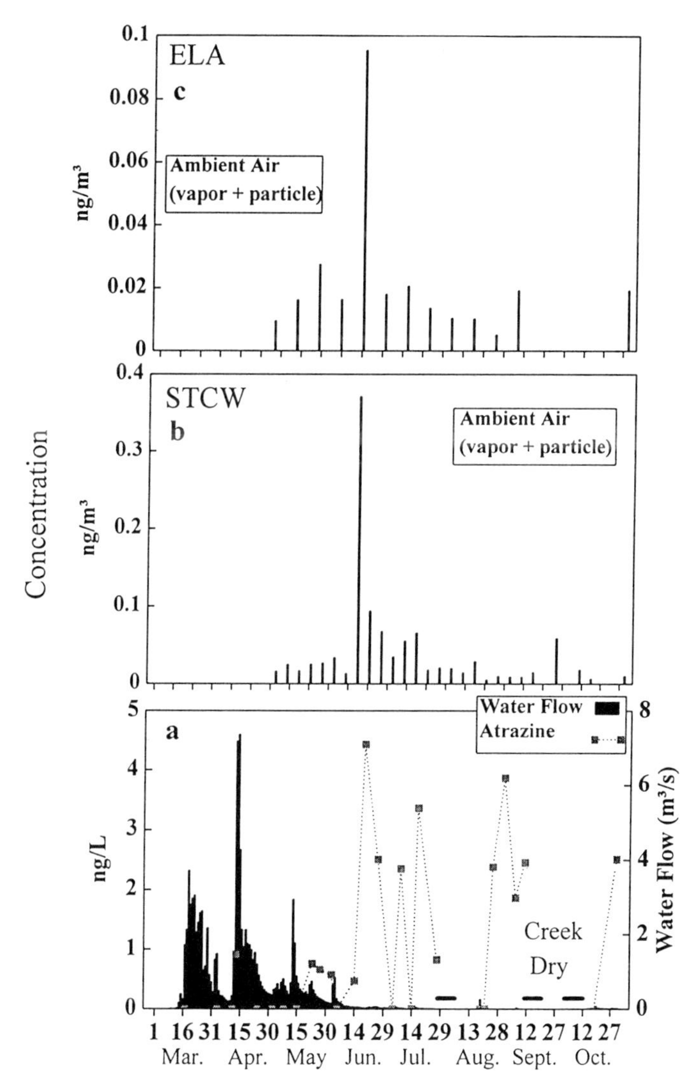

Figure 3. Atrazine concentrations - 1995 a) in creek water (ng/L) in South Tobacco Creek, b) in ambient air (ng/m^3) in the South Tobacco Creek Watershed and c) in ambient air (ng/m^3) at the Experimental Lakes Area.

measured in the STCW. Maximum concentrations were detected at both sites during mid-June sampling times (0.3 ng/m^3 STCW; 0.09 ng/m^3 ELA). Atrazine presence in ambient air reflected its probable use pattern, with maximum values obtained during the period between May and July when high herbicide use would be expected (Figure 3). Hoff and coworkers (*29*) found a similar pattern for trifluralin levels in air samples collected in southern Ontario.

Water: Atrazine concentrations during the pre-application period were at or near detection limits (0.04 ng/L) in South Tobacco Creek water and remained at low levels until early June sampling times. Elevated concentrations were detected in the water column at most sampling times through June and July (Figure 3a). Temporal patterns of atrazine in the South Tobacco Creek follow similar trends to those determined in Roberts Creek (Iowa) which also drains a small (16 km^2) watershed, where maximum concentrations were found in June (*30*). Maximum concentrations (8.9 μg/L) were much higher in Roberts Creek than observed in the southern Manitoba watershed, (4.4 ng/L) where no atrazine use occurred.

Maximum seasonal atrazine concentrations in South Tobacco Creek water coincided with seasonal maximum levels observed in ambient air and elevated levels in precipitation (Figure 4a,b). Low concentrations of atrazine were detected in creek water samples collected during spring runoff events. Similarly, 1994 results indicate low atrazine levels in South Tobacco Creek following an extended (July-early September) dry period. Maximum atrazine concentrations observed during the 1994 field season were found during the May sampling times, which corresponded to maximum concentrations in both precipitation and ambient air samples.

Precipitation: Seasonal patterns of atrazine in precipitation within the STCW varied between the 1994 and 1995 field seasons (Figure 4a,b). The 1994 pattern is in better agreement with the majority of other research. High atrazine levels in precipitation, during the period of application in areas of use, have been found in Germany (*13*), Switzerland (*12*) and in the US (*11*; *2*). Glotfelty and coworkers (*4*) found highest atrazine concentrations in rain samples prior to local application, which they attributed to atmospheric transport from more southern locations, where atrazine use began earlier in the year than in the Chesapeake Bay region. In this study, maximum observed atrazine concentrations were much lower during 1995 (10 ng/L) than 1994 value (57 ng/L). The 1995 pattern of atrazine in precipitation is similar to the observed pattern found for the first year of sampling by Hall and coworkers (*21*), where maximum concentrations were found later in the field season (June), rather than during local application times at three sampling sites.

The trend in atrazine levels observed in precipitation at the remote ELA site and the STCW suggest different input sources. Atrazine concentrations at ELA correspond to sources from spring application (Figure 4c), similar to the 1994 determinations in the STCW. Temporal relationships within a precipitation event were not possible because rainfall was collected on a continuous basis for 2 to 4 weeks to obtain bulk rainfall samples. Precipitation weighted concentrations observed in this study for both sites, where atrazine is not used, were three to four orders of magnitude below values obtained in the high herbicide use areas in the US (Table IV).

Sources: Air-mass back trajectories were performed by the Canadian Meteorological Centre, (Environment Canada) for 5 day periods at 6 hour intervals at the 900 mbar

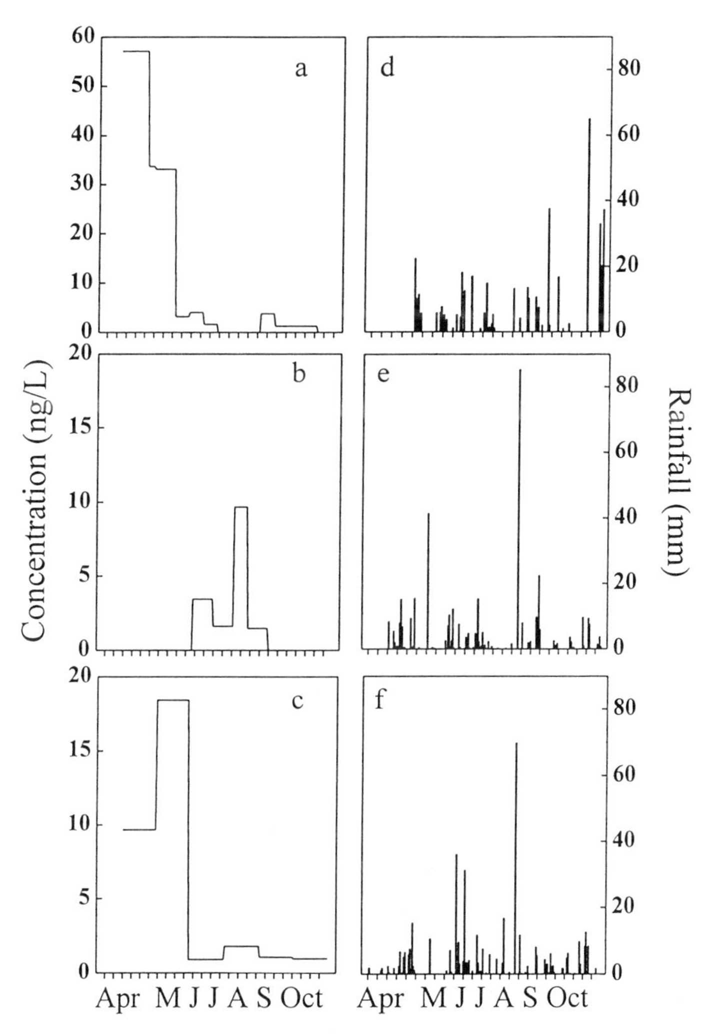

Figure 4. Atrazine concentrations (ng/L) in precipitation samples collected in the South Tobacco Creek Watershed: a) 1994; b) 1995; c) Experimental Lakes Area 1995; corresponding rainfall (mm) indicated in d) e) and f), respectively.

Table IV: Precipitation weighted atrazine concentrations.

Location/Year/Reference	Atrazine Concentration (ng/L)	Weighted Concentration[1] (ng/L)
1993 STCW[2]	0.42 - 12.55	5.80
1994 STCW	<0.04 - 12.55	11.59
1995 STCW	<0.04 - 9.68	2.85
1995 ELA	0.9 - 18.41	5.64
1990 - 1992 ELA (*1*)	<0.03 - 51	27.29
Midwest US (*2*)	-	200 - 400
North Dakota 1990 (*3*)	<50 - 420	56.76[3]
Minnesota 1991 (*3*)	<50 - 440	43.89[3]
Iowa, Illinois and Indiana (*2*)	-	600 - 1,000

[1]Pptn weighted concentration = $\Sigma(\text{Precipitation}_{\text{amount}} \cdot [\text{atrazine}])/\Sigma\text{Precipitation}_{\text{amount}}$
[2] Based on June-October sampling
[3] Estimated by taking atrazine levels reported to be below detection limits (50 ng/L) as 0.5 x detection limit concentrations (25 ng/L)

pressure level. Results from the 1994 field year indicated that elevated concentrations of atrazine were associated with air movement from the south. During the period of maximum atrazine in air samples in 1995, air masses had moved in from the south, at both sampling sites (Figure 5a,b). Low concentrations of atrazine were generally observed in samples collected over periods when air masses entered from non-source regions.

Atrazine concentrations in the gas and particle phase were used to calculate the fraction of atrazine associated with the particle phase in order to estimate dry deposition. Previously Eisenreich and Strachan (*18*) had estimated this fraction ($\varnothing$) to be 0.75 during summer months and 0.85 during winter periods. The value obtained in the present study was much lower (0.40), based on 85 sampling observations producing detectable levels of atrazine in both vapor and particle phases, from both ELA and STCW (1993-1995). A wide range of atrazine in the gas/particle distribution was observed in ambient air samples (Figure 6a). Atrazine presence in the vapor phase was higher during periods of application, relative to particle phase levels (Goolsby, D.A., U.S. Geological Survey Lakewood, CO. personal communication, 1996). Problems with stripping of SOCs from filter surfaces during extended sampling times have been reported for highly volatile compounds, such as trifluralin (*29*), however the vapor pressure of atrazine is sufficiently low (0.0399 mPa) and the total air volumes of ~350 m^3 were small enough to suggest that this was not an important source of error.

Hoff and coworkers (*31*) plotted log concentrations of several SOCs against inverse temperature in Antoine type plots and showed a clear relationship between ambient air concentration and temperature for several organochlorine insecticides. A similar Antoine plot of vapor phase atrazine concentrations measured in this study with temperature, indicated a poor relationship between these parameters (correlation coefficient = 0.02) (Figure 6b) and suggests that atrazine sources are not temperature

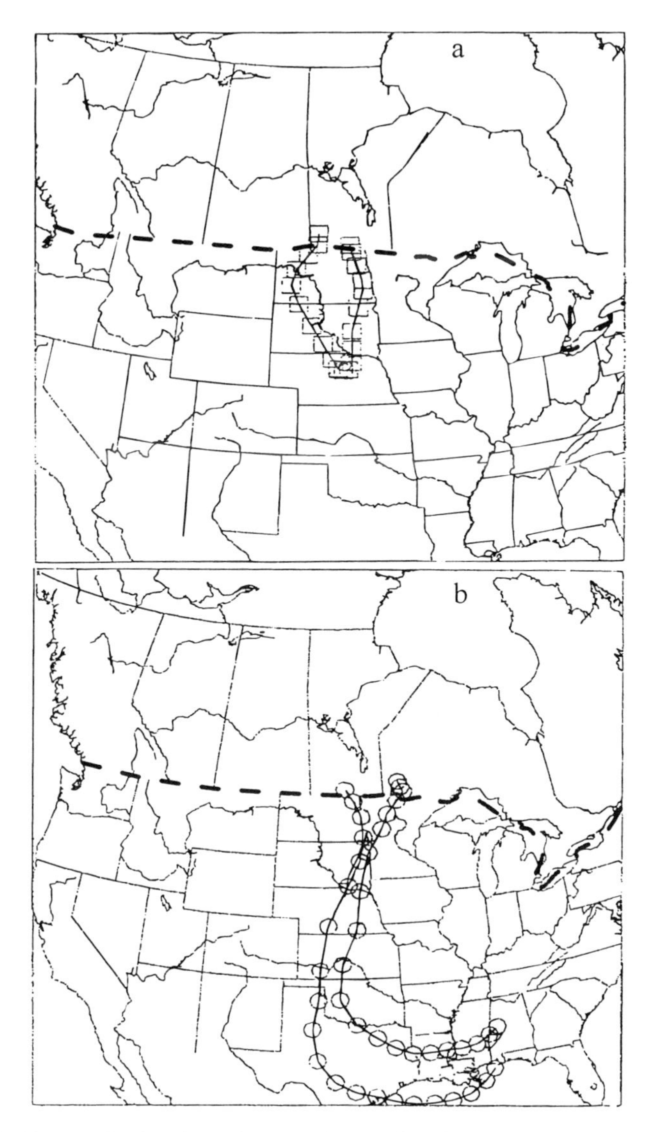

Figure 5. Air mass back trajectories a) to the South Tobacco Creek Watershed: June 14, 1995 (981 mbar); b) to the South Tobacco Creek Watershed and Experimental Lakes Area: June 20, 1995 (950 mbar).

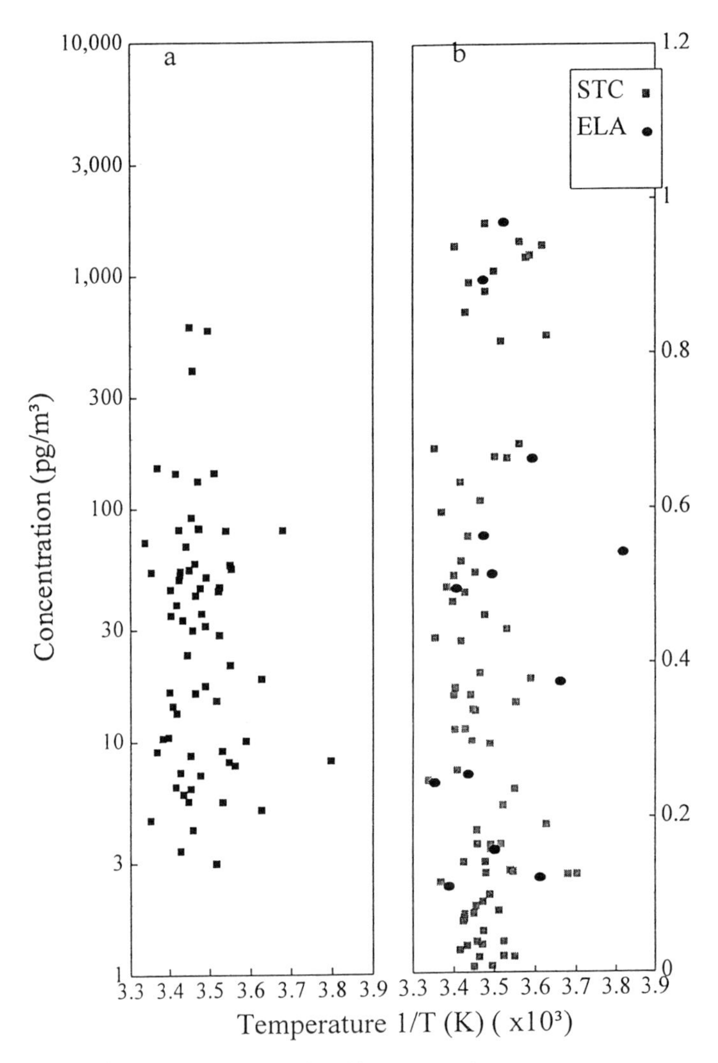

Figure 6. Atrazine temperature relationships: a) atrazine concentration in the vapor phase of ambient air; b) fraction of atrazine on particles. Data from the 1993 through 1995 sampling years at the South Tobacco Creek Watershed and measurements during 1995 at the Experimental Lakes Area site were used for the Antoine plot (a).

related. Sources at a distance removed from sampling sites appear to be controlling local concentration, not sorption/desorption processes.

Flux: Deposition via precipitation (N_{pptn}) to the STCW was calculated for each of the compartments under investigation, for both the 1994 and 1995 sampling seasons and for the ELA site for 1995, by relating atrazine concentration to the volume of precipitation collected, corrected for surface area and duration over which the sample was collected:

$$N_{pptn} = \frac{C_{TP} \cdot \text{Sample Volume (L)}}{\text{S.A.} \cdot \text{\# days sample was collected}} \ \text{ng/m}^2\cdot\text{d} \quad (1)$$

where C_{TP} = atrazine concentration in precipitation (sum of particle and dissolved phases) and S.A. = collection surface area.

Deposition of atrazine from dry particles (N_{dry}) was estimated using the relationship (*18*):

$$N_{dry} = C_{TA} \cdot \varnothing \cdot V_d \cdot \text{S.A.} \ \text{ng/m}^2\cdot\text{d} \quad (2)$$

where the fraction of atrazine associated with the particle phase, $\varnothing$, was taken to be 0.4, based on measured results in this study. Total atrazine concentrations (C_{TA}) in air (vapor + particle) were required for the dry deposition estimates. Particle size measurements were not performed in this study and, therefore, the deposition velocity (V_d) was taken to be 0.2 cm s^{-1}, which is more representative of small particles and used by Eisenreich and Strachan for deposition estimates in the Great Lakes (*18*). The coarse particle fraction which has greater deposition velocity, (*32*) was not considered. Dry deposition estimates were calculated over a 1 m^2 surface area (S.A.).

Particle size distribution ratios and deposition velocities have been determined and used in flux estimates of metals (*33*; *34*), however, similar results for organic contaminants are not available. Meteorological factors such as wind speed and direction, ground condition and atmospheric stability play an important role in the deposition of particles (*32*).

Gas exchange of organic contaminants between a water surface and the air above the water body can result in deposition of contaminants to the water system or export of the chemical from the water to the atmosphere. Gas exchange can be calculated for lake environments using the two-film model (*35*). For small streams, where water depths are low and where bottom topography may affect the air-water interface, the surface renewal model can be used (*35*). The total mass transfer velocity (v_{tot}) is calculated as two separate factors, air and water (v_a and v_w, respectively).

Air transfer velocities (v_a) were calculated using the relationship between water and wind speed at 10 m above the water surface (u_{10}) : $v_a(H_2O) \approx 0.2\, u_{10}(m\cdot s^{-1}) + 0.3\, (cm\cdot s^{-1})$ (*35*) and extrapolated to the chemical of interest by the relationship:

$$v_a(\text{atrazine}) = v_a(H_2O)\, [D_a(\text{atrazine})/D_a(H_2O)]^{0.67}. \quad (3)$$

Atrazine diffusivity in air was estimated using a molecular weight ratio (*35*):

$$D_a(\text{atrazine})/D_a(H_2O) \approx [MW(H_2O)/MW(\text{atrazine})]^{0.5}. \quad (4)$$

The calculation of the transfer velocity in water (v_w) takes into account factors such as water currents and water depths which affect v_w in a small stream. Molecular diffusivities in water can be calculated, using oxygen as the reference compound because gas exchange of O_2 is only dependant on v_w and not v_a (*35*). The transfer velocity of a shallow stream can be calculated using a relationship that includes water velocity and depth of the water (*35*):

$$v_w \approx [(D_w(\text{atrazine})\ (cm^2 \cdot s^{-1}) \cdot \text{water velocity}\ (cm \cdot s^{-1})/\text{water depth}\ (cm)]^{0.5}. \quad (5)$$

For a lake situation, the transfer velocity is calculated using wind speed at the 10 m height (u_{10}) (*35*):

$$v_w = (Dw/Dw\ (O_2)) \cdot (4 \times 10^{-5}\ u^2{}_{10}\ (m \cdot s^{-1}) + 0.0004). \quad (6)$$

Once the two transfer velocities (v_a and v_w) are known for a given compound, the overall transfer velocity (v_{tot}) $1/v_{tot} = 1/v_w + 1/v'_a$ [$v'_a = v_a\ (H/RT)$] can be calculated, where: H = Henry's Law Constant, R is the gas constant and T = temperature (K). This overall transfer velocity is then used to calculate the flux (N) by:

$$N = v_{tot}\ (C_w - C_a/K'_H) \approx N = C_a/K'_H \quad [\text{where:} K'_H = H/RT]. \quad (7)$$

Atrazine gas exchange across the air-water interface was calculated for South Tobacco Creek with the surface renewal model, using water flow and depths in combination with molecular diffusivities, as discussed above. Calculations of gas exchange were performed using air and water samples collected at approximately the same time. Gas exchange calculations required the estimation of the H for the temperatures over which samples were collected. H values were estimated from a linear regression, using the averages of several pesticide temperature:H relationships (*36*). Gas exchange with lake water at the ELA site was calculated using 1992 lake water data (*1*), in combination with the air data (Table V) for atrazine taken from 1995, using the two film model, corrected for wind speed.

Seasonal fluxes from each of the precipitation, dry deposition and gas exchange pathways are compared in Table VI to measurements within the STCW during the 1993 and 1994 seasons. The 1995 precipitation flux in the South Tobacco Creek Watershed did not fit the trends observed in 1993 and 1994, flux values were much lower and the greatest flux via precipitation occurred during August. Flux values at ELA were greatest for precipitation samples collected in May and June. Maximum flux at the ELA site was an order of magnitude below maximum values estimated in the STCW during the 1994 field season. The results show that precipitation has the greatest relative contributions to atrazine flux at each of these locations. Gas exchange of atrazine is most important during the June-July period. Dry deposition is also greatest during June and July, when air concentrations of atrazine are at maximum values, however, dryfall is not predicted to be as important a route of entry as either the precipitation or gas exchange pathways.

Precipitation and gas exchange fluxes calculated for ELA were in the same order of magnitude as those calculated by Muir and Grift (*1*) for ELA during 1992. Atrazine flux results in these regions, where there is zero atrazine use, indicate there can be

Table V: Average monthly conditions and transfer velocities during sample times in the South Tobacco Creek Watershed (STCW) (1993-95) and the Experimental Lakes Area (ELA) (1995).

Month	STCW (Stream)[1,2]				ELA (Lake)		
	v_{tot} (x 10^{-5}) $m \cdot d^{-1}$	Wind Speed $m \cdot s^{-1}$	Water Temperature °C	Water Flow $m^3 \cdot s^{-1}$	v_{tot} (x 10^{-5})	Wind Speed $m \cdot s^{-1}$	Water Temperature °C
May	1.66	4.5	10	0.60	1.78	3.1	10
June	2.24	2.9	22	0.08	2.54	3.2	16
July	2.12	3.3	20	0.02	2.56	2.9	18
Aug.	2.19	3.4	17	0.09	2.02	1.9	19
Sept.	2.75	6.7	17	0.01	1.91	1.6	14
Oct.	1.76	2.8	3	0.02	1.38	1.9	9

[1]assumes average stream width of 1.4 m
[2]monthly average depth range 0.02 - 0.29 m

Table VI: Seasonal Flux estimates for the South Tobacco Creek Watershed (STCW) 1994-1995 with Experimental Lakes Area (ELA) results.

Location	Flux $ng \cdot m^{-2}$			
	Dry Deposition	Gas Exchange	**Total Dry Process**	Precipitation
1995 ELA	235	421	656	2006
1995 STCW	452	1200	1652	1269
1994 STCW	670	1151	1821	17245
1993 STCW[1]	506	812	1318	3101

[1]Does not include May - June period, when potentially high precipitation concentrations would be observed.

deposition to the watershed and lake surface during all sampling times. Our atrazine gas exchange flux values indicated deposition to surface waters throughout the field season. McConnell and coworkers (1993) (*37*) reported both deposition and export fluxes for α- and γ- HCH in the Great Lakes. Water temperature and degree of mixing of the water affect the direction of flux of SOCs (*37*), however, atrazine flux is likely to favour deposition rather than volatilization because of low H values.

Flux values for atrazine from other studies are not readily available, however, estimates based on atrazine measurements in precipitation can be made (Table VII). Precipitation inputs into Northern Germany are very close to the measured flux values of the present study. Atrazine use was banned in Germany during the study period (*14*), therefore, long range transport and deposition pathways may be similar to those in the present study. Flux values via wet deposition in Switzerland were higher than maximum annual values estimated for the sites where atrazine was not used. Higher flux values were estimated for the Great Lakes region (*18*). This area has significant local atrazine use and may receive inputs via atmospheric deposition and transport from the Midwest (*2*).

Table VII: Comparison with other predicted or observed fluxes for atrazine.

Location	**Flux ($\mu g \cdot m^{-2} yr^{-1}$)**		**Reference**
	Wet	**Dry**	
Great Lakes	100	35	*23*
N. Germany	1-5	not determined	*15*
Switzerland	20	not determined	*12*
Minnesota	12	not determined	*7*
North Dakota	16	not determined	*7*
South Tobacco Creek	1.3-13.7	0.4-0.6	Present study
Experimental Lakes Area	2.7	0.2	Present study

Numerous uncertainties in the dry deposition and gas exchange estimates exist, due to the wide range of observed results (e.g., gas/particle partitioning) or lack of knowledge of the physical properties (e.g., subcooled liquid vapor pressure and variation of H with temperature). Dry deposition estimates are calculated by using estimated $\varnothing$ values, which can range between 0.1 and 0.9. Altering this value between the extremes would result in a 3-fold decrease or a 2-fold increase in dryfall estimates, from those obtained using the measured $\varnothing$ value (0.4) in this study. V_d of small particles sampled by high volume air samplers range between 0.1 and 0.8 cm/s (*38*). Because each of these factors ($\varnothing$ and V_d) are linearly related to dryfall flux, their effect on uncertainty is additive (equation 2). The critical factor in dryfall estimations, is the $\varnothing$ value, because it has a greater range than V_d. Minor differences in gas exchange results were observed for South Tobacco Creek, when calculations were performed using the surface renewal model, which most accurately represents a moving stream (*35*) or using the two-film model for this creek, indicating the air mass transfer coefficient (v_a) of these calculations drives gas exchange. Flux is temperature dependent (equation 7) through effects of temperature on H. However, sensitivity analysis showed that wind speed was a more critical parameter because of its effect on $v_a(H_2O)$. Therefore, the wind speed is the critical value in gas exchange flux calculations.

Conclusions:

Although these results have been obtained for a region of low atrazine use, we believe they have relevance for estimating atrazine deposition in use areas. Atrazine was found in air, both gas and particle phase, precipitation and surface water throughout the field season in the present study. Temporal relationships correlated to atrazine use times in the US Midwest, although atrazine is not used in either of these sampling locations.

Atrazine concentrations in air and its partitioning to particles were not related to air temperature. This indicates that distant sources, rather than local sorption/desorption processes control the atrazine air concentration at these sites. Particle fraction estimates previously used for flux calculations may be high which could lead to overestimating dry deposition of atrazine at least for areas remote from major use.

Precipitation was found to be the most important route of entry of atrazine into a small lake in northwestern Ontario and into southern Manitoba surface waters. Results also indicate that dry processes, gas exchange and dry deposition contribute most significantly during the early part of the field season. The low concentrations of atrazine observed in each of these areas is indicative of regional or long range transport and deposition. Atrazine deposition via precipitation in southern Manitoba and southern Ontario is generally lower than found in the US, where atrazine use occurs. Dry deposition values estimated for the South Tobacco Creek Watershed and the Experimental Lakes Area are also lower than the values determined for the Great Lakes region (*16*). There is considerable uncertainty in the dry deposition and gas exchange estimates. Nevertheless extrapolating to high use areas, by assuming concentrations in air are similar to those measured by Glotfelty *et al.* (*4*), an additional 120% deposition of atrazine could be accounted for by gas exchange to water during the intensive use of the herbicide. We have not considered gas exchange to other surfaces, such as plants and soil nor the deposition of atrazine degradation products.

Acknowledgments:
We thank K. Beaty and M. Lyng (Freshwater Institute, ELA) for air sample collection and Bevan Lawson, (Atmospheric Environment Service, Environment Canada, Winnipeg, Manitoba) for providing the air mass trajectories. The Deerwood Soil and Water Management Association provided detailed land use information and field assistance. We thank the Canada - Manitoba Agreement on Agricultural Sustainability (CMAAS) and Environment Canada (Monitoring and Systems Branch) for funding. We thank R. Hoff and D. MacKay for their helpful reviews of the manuscript.

Literature Cited:

(1) Muir, D.C.G.; Grift, N.P. In Eighth International Congress of Pesticide Chemistry: Options 2000. Ragsdale, N.N.; Kearney; P.C.; Plimmer, J.R.. Eds. American Chemical Society, Washington, D.C. **1995**, 141-156.
(2) Goolsby, D.A.; Thurman, E.M.; Pomes, M.L.; Battaglin,W.A. In New Directions in Pesticide Research, Development, Management and Policy, Proc. of the Fourth National Pesticide Conference, Weigmann, D.L. Ed. **1994**, 696-710. Richmond VA. 1993.
(3) Goolsby, D.A.; Scribner, E.A.; Thurman, E.M.; Pomes, M.L.; Meyer, M.T. U.S. Geological Survey Open-File Report 95-469. **1995**, Lawrence, KA. 341 pp.
(4) Glotfelty, D.E.; Williams, G.H.; Freeman, H.P.; Leech,M.M. In Long Range Transport of Pesticides. Kurtz, D.A. Ed. **1990**, Lewis Publishers, Chelsea, MI. 199-221.
(5) Clendening, L.D.; Jury, W.A.; Ernst, F.F. In Long Range Transport of Pesticides. Kurtz, D.A. Ed. **1990**, Lewis Publishers, Chelsea, MI. 47-60.
(6) Glotfelty, D.E.; Leech, M.M.; Jersey, J.; Taylor, A.W. *J. Agric. Food Chem.* **1989**, *37*, 546.
(7) Wauchope, R.D. *J. Environ. Qual.* **1978**, *7*, 459.

(8) Pereira, W.E; Rostad, C.E.; Leiker T.J. *J. Contam Hydrol.* **1992**, *9*, 175.
(9) Shiu, W.Y.; Ma, K.C.; Mackay, D.; Seiber, J.N.;Wauchope, R.D. *Rev. Environ. Contam. Toxicol.* **1990**, *116*. 1.
(10) Wauchope, R.D.; Butler, T.M.; Hornsby, A.G.; Augustijn-Beckers, P.W.M.; Burt, J.P. *Rev. Environ. Contam. Toxicol.* **1992**, *123*. 1.
(11) Richards, R.P.; Kramer, J.W.; Baker,D.B.; Krieger, K.A. *Nature*. **1987**, *327*, 129.
(12) Buser, H.R. *Environ. Sci. Technol.* **1990,** *24*, 1049.
(13) Seibers, J.; Gottschild, D.; Nolting, H. -G. *Chemosphere*. **1994**, *28*, 1559.
(14) Bester, K.; Hühnerfuss, H.; Neudorf, B.; Thiemann, W. *Chemosphere*. **1995**, *30*, 1639.
(15) Trevisan, M.; Montepiani, C.; Ragozza, L.; Bartoletti, C.; Ioannilli, E.; Del Re, A.A.M. *Environ. Poll.* **1993**, *80*, 31.
(16) Chevreuil, M.; Garmouma, M. *Chemosphere*. **1993**, *27*, 1605.
(17) Lode, O.;Eklo, O.M.; Holen, B.; Svensen, A.; Johnsen, A.M. *Sci. Total Environ.* **1995**, *160/161*, 421.
(18) Eisenreich, S.J.; Strachan, W.M.J. Estimating Atmospheric Deposition of Toxic Substances to the Great Lakes. An Update. **1992**, National Water Research Institute. Environment Canada. Burlington,. ON. 59 pp.
(19) Majewski, MJ.; Capel, P.D. Pesticides in the Atmosphere. Gilliom, R.J. Ed. Pesticides in the Hydrologic System, Vol. 1; Ann Arbor Press: Chelsea, MI. **1995.**
(20) Bidleman, T.F.; Foreman, W.T. In Sources and Fates of Aquatic Pollutants. Hites, R.A. and Eisenreich, S.J. Eds. **1987**, American Chemical Society, Washington, D.C. 27-56.
(21) Hall, J.C.; Van Deynze; T.D., Struger, J.; Chan, C.H. *J. Environ. Sci. Health.* **1993**, *B28*, 577.
(22) Nations, B.K.; Hallberg, G.R. *J. Environ. Qual.* **1992**, *21*, 486.
(23) Capel, P.D. In: Proceedings of the Technical Meeting, US Geological Survey Toxic Substances Hydrology Program. **1991**, 334.
(24) Koester, C.J.; Hites, R.A. *Environ. Sci. Technol.* **1992**, *26*, 1375.
(25) Noll, K.E.; Fang.; K.Y.P.; Watkins, L.A. *Atmos. Environ.* **1988**, *22*, 1461.
(26) Bidleman, T.F. *Environ. Sci. Technol.* **1988**, *22*, 361.
(27) Pankow, J.F. *Atmos. Environ.* **1987**, *21*, 2275.
(28) Haraguchi, K.;Kitamura, E.; Yamashita, T.; Kido., A. *Atmos. Environ.* **1994**, *28*, 1319.
(29) Hoff, R.M.; Muir, D.C.G.; Grift., N.P. *Environ. Sci. Technol.* **1992**, *26*, 266.
(30) Kolpin, D.A.; Kalkhoff.,S.J. *Environ. Sci. Technol.* **1993**, *27*, 134.
(31) Hoff, R.M.; Muir, D.C.G.; Grift, N.P. *Environ. Sci. Technol.* **1992**, *26*, 276.
(32) Noll, K.E.; Pontius, A.; Frey, R.; Gould, M. *Atmos. Environ.* **1985**, *19*, 1931.
(33) Holsen, T.M.; Noll, K.E. *Environ. Sci. Technol.* **1992**, *26*, 1807.
(34) Noll, K.E.; Yuen, P.F.; Fang, K.Y.P. *Atmos. Environ.* **1990**, *24A*, 903.
(35) Schwarzenbach, R.P.; Gschwend, P.M.; Imboden, D.M. Environmental Organic Chemistry. **1993**, John Wiley and Sons. New York, NY 681 pp.
(36) Chernyak, S.; McConnell, L.L.; Rice, C. *Mar. Poll. Bull.* **1996**, In press.
(37) McConnell, L.L.;Cotham, W.E.; Bidleman, T.F. *Environ. Sci. Technol.* **1993**, *27*, 1304.
(38) Holsen, T.M.;Noll, K.E. Environ. Sci. Technol. **1992**, *26*, 1807.

Chapter 16

Integrated Chemical and Biological Remediation of Atrazine-Contaminated Aqueous Wastes

S. M. Arnold[1], W. J. Hickey[1], R. F. Harris[1], and R. E. Talaat[2]

[1]Soil Science Department, University of Wisconsin—Madison, Madison, WI 53706
[2]Corning Hazleton, Inc., Madison, WI 53707

Treatment of [2,4,6-^{14}C]atrazine with Fenton's reagent (FR) produced seven major dechlorinated, dealkylated, and/or partially oxidized products identified by high-performance liquid chromatography megaflow electrospray tandem mass spectrometry. The best FR mixture, 2.69 mM (1:1) $FeSO_4$:H_2O_2, completely degraded atrazine (140 μM) in ≤30s, primarily to two dealkylated, chlorinated products: 23% diaminochloro-*s*-triazine and 28% deisopropylatrazine amide. About 55% chloride release indicated dehalogenated *s*-triazines accounted for the balance or products. *Rhodococcus corallinus* degraded these chlorinated products in ≤10 minutes and converted 47% of [2,4,6-^{14}C]atrazine to $^{14}CO_2$ in 7 d. *R. corallinus* combined with *Pseudomonas* sp. strain D increased $^{14}CO_2$ production to 73%. When applied to a pesticide rinse water containing atrazine, cyanazine, alachlor, metolachlor, and EPTC, ≥99% of the pesticides were degraded with 12.2 mM FR. Subsequent treatment with *R. corallinus* and *Pseudomonas* sp. strain D degraded all chlorinated *s*-triazine intermediates and released 70% $^{14}CO_2$ from an [2,4,6-^{14}C]atrazine tracer in 10 d. Collectively, these studies have demonstrated that the integrated approach has potential as an on-site treatment for pesticide rinse water.

In this study, Fenton's reagent (FR) in combination with *Rhodococcus corallinus* and/or *Pseudomonas* sp. strain D were used to treat atrazine alone and in mixed pesticide wastes. Fenton's reagent (Fe^{2+} and H_2O_2) generates hydroxyl radical ($HO^{\bullet}$), a powerful non-specific oxidant (*1*).

$$Fe^{2} + H_2O_2 \rightarrow FE^{3+} + HO^{\bullet} + HO \qquad (1)$$

Previous research has focused on degrading *s*-triazine herbicides using $HO^\bullet$ generated by ozone (*2-6*), TiO_2/UV light (*7-9*), H_2O_2/UV light (*10*), photodecomposition of $Fe(OH)^{2+}$ (*11*), and Fenton's reagent (FR) (12-13). The main advantage of Fenton's reagent (FR) over other $HO^\bullet$ systems is its simplicity: the chemicals are commonly available and inexpensive, and there is no need for special equipment like UV lamps, complex reaction vessels, TiO_2 particles, or ozone generators. Because of its simplicity, FR has the potential for widespread use in treating atrazine wastes. The drawback of treating *s*-triazines with $HO^\bullet$-generating systems is that one or more stable chlorinated products commonly accumulate. Thus, treatment is not complete in a remediation context because the toxicity of chlorinated *s*-triazine products may be as great as that of the parent compound (*14*).

Microbial processes are often well-suited for degradation of single pesticides, but may be limited in treating pesticide mixtures because of the limitations of most enzymes in attacking structurally diverse chemicals. Moreover, wastes such as pesticide rinse water generated during field applications may contain a variety of pesticides, formulating agents, surfactants, emulsifiers and fertilizers, which may inhibit microbial growth (*15*). Thus, chemical pretreatment can eliminate microbial inhibitors and breakdown target pesticides into common substrates for bacterial degradation. We hypothesized that the collective catabolic activities of *R. corallinus* (contains an inducible hydrolase capable of dechlorination and deamination of partially dealkylated *s*-triazines) and *Pseudomonas* sp. strain D (metabolizes a broader spectrum of dechlorinated, partially dealkylated *s*-triazines degradation products) could be used to efficiently degrade the stable chlorinated end-products generated by FR treatment of atrazine.

Materials and Methods

Chemicals. The Glossary of this volume lists the common names and chemical names of *s*-triazines and pesticides described in this study. Atrazine (99% pure), cyanazine (99%), alachlor (99%), and EPTC (98%) were purchased from Chem Service (West Chester, PA). [2,4,6-^{14}C]Atrazine (19.4 µCi/mg, 97%), desethylatrazine (99%), desisopropylatrazine (98%), diaminochloro-*s*-triazine (90%), desethylhydroxyatrazine (98%), ammeline (95%), desisopropylhydroxyatrazine (97%), cyanazine (96%), and metolachlor (98%) were provided by Ciba-Geigy Corporation (Greensboro, NC). Pesticide rinse water, collected from a commercial pesticide application facility in Dane County, WI, was a yellowish liquid (pH 8.1) in which the following pesticides were detected: atrazine (131 µM), cyanazine (132 µM), EPTC (159 µM), metolachlor (209 µM), and alachlor (98 µM). The pesticide rinse water contained 235 mg/L total organic carbon; Fe, Mg, Mn, P, S, Zn were detected at 0.01, 0.55, 0.06, 44, 193, and 0.63 mg/L, respectively. Other heavy metals were not detected (*16*). Suspended solids were removed by centrifugation (10,000 rpm; 14,336 x g) with a Sorvall Szent-Gyorgyi and Blum continuous flow-through system.

Fenton's Reagent Treatments. For treatment of atrazine, ratios ($FeSO_4$:H_2O_2) of 1:11, 1:100, and 2:1 were examined at concentrations from 0.1 to 25 mM. Aliquots of 0.05 ml to 1.5 ml of 50 mM $FeSO_4$ were mixed with 25 ml of 135 μM atrazine in 50-ml Erlenmeyer flasks; $HO^{\bullet}$production was initiated by adding 0.05 ml to 1.5 ml of 50 mM H_2O_2. For pesticide rinse water treatment, 0.15-2.0 ml of 50 mM $FeSO_4$ and 50 mM H_2O_2 were mixed with 10 ml pesticide rinse water. The aluminum foil-wrapped flasks were incubated for 24 h on a rotary shaker at 200 rpm (25 ± 1°C). Treated atrazine and pesticide rinse water samples (0.5 ml) were mixed with methanol (0.5 ml) to quench the reaction, centrifuged (10 min; 3,200 rpm), and the supernatants analyzed by high-performance liquid chromatography (HPLC). Treated pesticide rinse water samples were also extracted and analyzed by gas chromatography (GC) as described later.

Bacterial Degradation Experiments. *Rhodococcus corallinus* (NRRL B-15444R) was a gift from Ciba-Geigy. *Pseudomonas* sp. strain D (NRRL B-12228) was obtained from the USDA, National Center for Agricultural Utilization Research, Peoria, IL. Cells were grown on cyanuric acid-glucose (CAG) medium that consisted of potassium phosphate buffer (10 mM, pH 7.3), $MgSO_4$ (1 mM), glucose (10.0 mM), cyanuric acid (1.7 mM), and a trace element mixture (*17*). Cultures were harvested during their early stationary phase of growth 1 L CAG, washed four times with potassium phosphate buffer (130 mM, pH 7.4), and resuspended to an optical density (measured at 600 nm) of 2.5 in 80 ml potassium phosphate buffer. Duplicate incubations were established by adding 10 ml cell suspension to 10 ml FR-treated atrazine or pesticide rinse water. Duplicate controls had 10 ml of FR-treated solutions added to 10 ml of 130 mM potassium phosphate buffer. Reactions were terminated at selected intervals by immersing 1-ml aliquots in dry ice baths. The samples were thawed by boiling for 10 min, clarified by centrifugation (10 min, 3200 rpm), and the supernatant analyzed by HPLC.

Mineralization Studies of Atrazine and its FR Products. FR-treated atrazine and/or pesticide rinse water spiked with [2,4,6-^{14}C]atrazine (0.026 to 0.046 μCi/ml in 2.0 ml) was added to a 25-ml serum bottle and crimp sealed with a Teflon-lined septum. Next, 2.0 ml of either *R. corallinus* or *Pseudomonas* sp. strain D, 1.0 ml of both cultures, or 2.0 ml of 130 mM buffer, was injected and the samples incubated on a rotary shaker (150 rpm) for 7-10 days. All treatments were incubated in duplicate. At selected times the flask headspaces were flushed for 1 hour to trap $^{14}CO_2$ in carbon 14 cocktail (R. J. Harvey Instrument Co., Hillsdale, NJ) and volatile organic compounds in ethylene glycol monoethyl ether. After incubation, 1 ml was acidified with 0.5 ml 6 N H_2SO_4 and flushed for 1 hour to release dissolved $^{14}CO_2$. Radioactivity was quantified using a Rackbeta model 1209 liquid scintillation counter (LKB-Wallac, San Francisco, CA).

HPLC/UV Analysis for Atrazine and s-Triazine Transformation Products. Analysis for atrazine and *s*-triazine transformation products was done with a Hewlett-Packard (Palo Alto, CA, USA) Model 1050 HPLC equipped with a variable-wavelength UV detector at 220 nm. Analytes were separated using a Hewlett-Packard ODS Hypersil C-18 reversed-phase column (200 mm x 4.6 mm i.d.; 5 μm mean particle diameter) at a flow-rate of 1 ml/min. The mobile phase, potassium phosphate (5 mM, pH 4.6)-acetonitrile, was run in a gradient from 5 to 89% acetonitrile in 18 min. The column was re-equilibrated at the starting conditions in a 10 min post-run. Five-point calibration curves were run for atrazine, desethylatrazine, desisopropylatrazine, diaminochloro-*s*-triazine, desethylhydroxyatrazine, desisopropylhydroxyatrazine, and ammeline, which had detection limits of 0.24, 0.24, 0.14, 0.29, 0.63, 0.39, and 0.77 μM, respectively. Atrazine amide, simazine amide, N-isopropylammelide, and hydroxyatrazine were quantified using a response factor, which is the ratio of the UV (220 nm) response to ^{14}C activity in fraction-collected peaks. The *s*-triazine transformation products were identified by HPLC/Electrospray (ES)-MS/MS equipped with a radioactivity detector, and confirmatory analysis was done using high resolution-electron impact-mass spectrometry (HR-EI-MS) and GC/MS as described previously (*18*). Chloride ion was determined using an ion-specific electrode (*12, 19*).

GC Analysis for Pesticides in Pesticide Rinse Water. Atrazine, cyanazine, metolachlor, alachlor, and EPTC were extracted from FR-treated pesticide rinse water by shaking for 2 min with 2 x 25 ml dichloromethane and 1.25 g of NaCl. The organic phase was separated and then dried by filtering through 50 g anhydrous Na_2SO_4, and concentrated to 5 ml by rotoevaporation. The aqueous phase was extracted with ethyl acetate and the organic phases were combined, rotoevaporated to approximately 2 ml, and dried with nitrogen. The residue was resuspended in 5 ml ethyl acetate, filtered with a 0.22-μm nylon syringe filter, and analyzed by GC. Percent recoveries (all ± <1%) were 109, 94, 116, 104, and 91 for atrazine, cyanazine, metolachlor, alachlor, and EPTC, respectively. Pesticides and transformation products were analyzed using a Hewlett-Packard Model 5890 Series II gas chromatograph equipped with a Hewlett-Packard Model 7673 auto injector and nitrogen-phosphorus detector. Samples (1 μl) were separated with an AT-1701 (Alltech and Associates, Deerfield, IL) capillary column (15 m x 0.25 mm) with the following oven temperature program: 70°C (1 min), 70 to 140°C (50°C/min), 140 to 190°C (2°C/min), 190 to 220°C (30°C/min), 220°C (12 min). Injector and detector temperatures were 240°C.

Results and Discussion

Identification of FR-Generated Products. FR reactions generated a complex mixture containing residual [2,4,6-^{14}C]atrazine and 15 oxidation products detectable by HPLC/UV. In the 0.73 mM 1:1 FR mixture, seven major products were identified by HPLC/ES-MS/MS: atrazine amide, desethylatrazine, simazine amide, desisopropylatrazine, N-isopropylammelide, hydroxyatrazine, and diaminochloro-*s*-triazine (*18*). Structural assignments for all eight peaks based on HPLC/ES-MS/MS, HR-EI-MS and GC/MS spectra and/or co-elution with authentic standards are given in Table I.

Table I. Atrazine degradation products produced by Fenton's reagent

structure	name	Identification method			
		HPLC/ES-MS/MS	HPLC[a]	HR-EI-MS	GC/MS
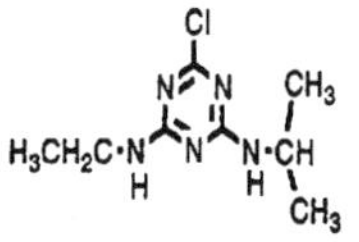 Cl N N CH_3 H_3CH_2C·N N N·CH H H CH_3	2-Chloro-4-(ethylamino)-6-(isopropylamino)-*s*-triazine (atrazine)	X[b]	X		
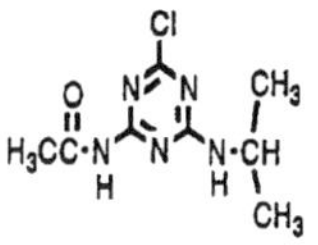 Cl O N N CH_3 H_3CC·N N N·CH H H CH_3	2-Acetamido-4-chloro-6-(isopropylamino)-*s*-triazine	X		X	
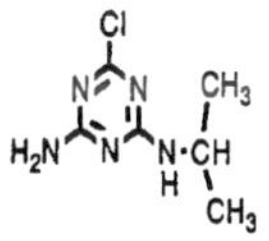 Cl N N CH_3 H_2N N N·CH H CH_3	2-Amino-4-chloro-6-(isopropylamino)-*s*-triazine (deethylatrazine)	X[b]	X		
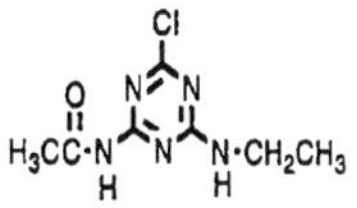 Cl O N N H_3CC·N N N·CH_2CH_3 H H	2-Acetamido-4-chloro-6-(ethylamino)-*s*-triazine	X		X	
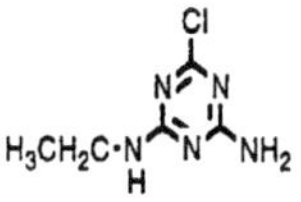 Cl N N H_3CH_2C·N N NH_2 H	2-Amino-4-chloro-6-(ethylamino)-*s*-triazine (deisopropylatrazine)	X[b]	X		
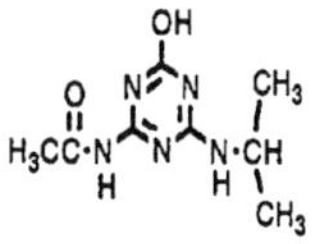 OH O N N CH_3 H_3CC·N N N·CH H H CH_3	2-Acetamido-4-hydroxy-6-(isopropylamino)-*s*-triazine	X			
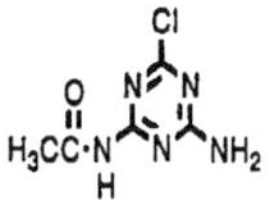 Cl O N N H_3CC·N N NH_2 H	2-Acetamido-4-amino-6-chloro-*s*-triazine	X			
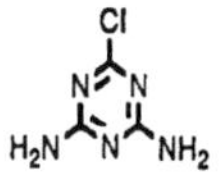 Cl N N H_2N N NH_2	2-Chloro-4,6-diamino-*s*-triazine (chlorodiamino-*s*-triazine)	X	X		X[b]

[a]Co-elution with authentic standard.
[b]Spectra matched authentic standard.

Effects of $FeSO_4$,:H_2O_2 Ratio an Concentration on Atrazine Degradation. From a remediation standpoint (i.e. chlorinated product depletion), the 1:1 $FeSO_4$:H_2O_2 was the most efficient ratio of FR used (*12*). With 1.42 mM 1:1 FR, atrazine was completely transformed in ≤30 seconds to a mixture of oxidized and dealkylated products including atrazine amide, desethylatrazine, desisopropylatrazine, N-isopropylammelide, hydroxyatrazine, diaminochloro-*s*-triazine, and five unknowns. Reaction mixtures were totally depleted of atrazine amide, desethylatrazine, desisopropylatrazine, and N-isopropylammelide by increasing FR concentrations to 2.69 mM (Figure 1). But hydroxyatrazine, diaminochloro-*s*-triazine, and six minor unidentified atrazine derivatives persisted at FR concentrations up to 25 mM (data not shown). Chloride ion release in the 2.69 mM FR treatment was 55 ± 9%, which was consistent with HPLC/UV and ^{14}C mass balance data (*12*) indicating that hydroxyatrazine and diaminochloro-*s*-triazine collectively accounted for 100% of the residual chloro-*s*-triazines. Dechlorinated compounds were detected that could account for the Cl-release: N-isopropylammelide was detected by HPLC/UV and HPLC/ES-MS/MS, and ammelide was tentatively identified by single ion monitoring using HPLC/ES-MS/MS (*18*). Other dechlorinated products, like ammelide and cyanuric acid could have been produced, but were not detectable by our analytical methods. Further FR treatment of these samples was ineffective in degrading hydroxyatrazine and diaminochloro-*s*-triazine because of the low reactivity of these oxidized products toward $HO^\bullet$. In sequential batch treatments of atrazine using 2.69 mM FR, the sum of hydroxyatrazine and diaminochloro-*s*-triazine decreased by only 5.8%. Thus, the mixture containing hydroxyatrazine, diaminochloro-*s*-triazine, and minor unknowns represented the terminal end products resulting from "complete" FR treatment of atrazine.

Using 1:100 $FeSO_4$:H_2O_2 ratios, a "complete" treatment was achieved with lower Fe^{2+} concentrations as compared to the 2.69 mM (1:1) treatments, but increasing the H_2O_2 concentration 100-fold lowered the reaction's efficiency (from a remediation standpoint) since larger amounts of the chlorinated products (hydroxyatrazine and diaminochloro-*s*-triazine) remained compared to the 1:1 FR treatments (cf. Figure 1A,B). Using excess H_2O_2 may have favored dealkylation, thus decreasing dechlorination. Increasing Fe^{2+} levels lowered reaction efficiency, presumably because excess Fe^{2+} reacted with $HO^\bullet$; at an $FeSO_4$: H_2O_2 concentration ratio of 2:1, atrazine and the alkylated products remained at FR concentrations up to 5.38 mM $FeSO_4$ and 2.69 mM H_2O_2, (Figure 1C).

Results from our FR study indicate that dechlorination and dealkylation occur simultaneously, and sequential batch treatments showed dechlorination occurs more readily with the alkylated *s*-triazines. Chlorinated products accounted for a large part of the *s*-triazines present at FR's end-point, thus it is important to determine dechlorination during atrazine treatment because of the potential toxicity of the chlorinated products (*14*).

R. corallinus **and** ***Pseudomonas*** **sp. strain D Metabolism of Atrazine Degradation Products.** *R. corallinus* completely degraded the remaining chlorinated end-products, diaminochloro-*s*-triazine and hydroxyatrazine, from the

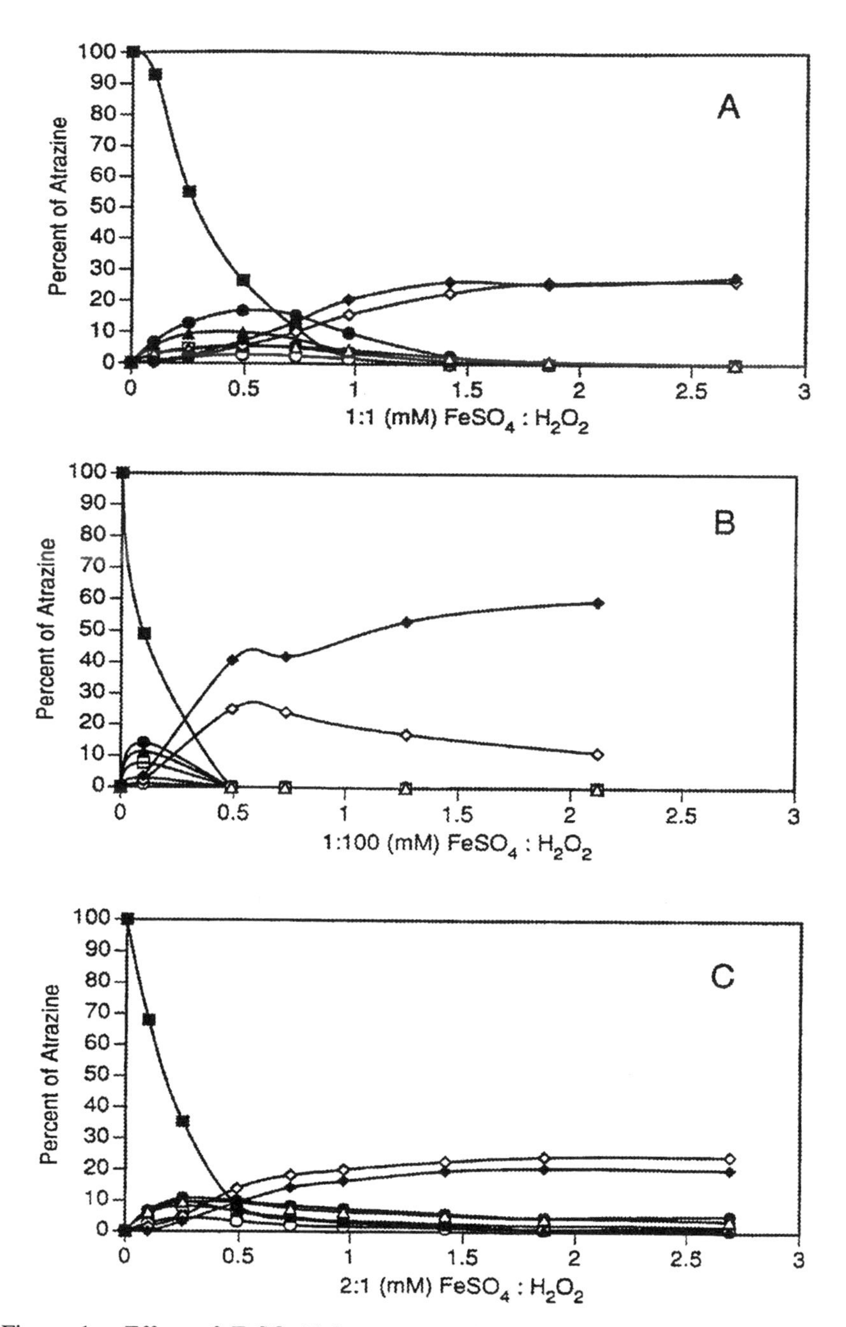

Figure 1. Effect of $FeSO_4$:H_2O_2 concentration ratios on atrazine (135 μM) degradation and formation of selected products after 24-hour incubation. Panel A, 1:1 (mM); Panel B, 1:100 (mM); Panel C, 2:1 (mM). Key to symbols: atrazine (■), atrazine amide (❒), desethylatrazine (●), simazine amide (❍), desisopropylatrazine (▲), N-isopropylammelide (Δ), hydroxyatrazine (✧), diaminochloro-*s*-triazine (✦).

most efficient FR treatment (2.69 mM 1:1 FR) in ≤10 min. A previous study reported that the *R. corallinus* inducible dehalogenase/deaminase effected deamination but not dechlorination of diaminochloro-*s*-triazine (*20*). In our study, *R. corallinus* completely degraded diaminochloro-*s*-triazine (381 mM) as an isolated substrate in ≤2 minutes as assayed by HPLC. Chloride release in this period was only 50 ± 5%, but increased to 87 ± 8% in 4 days. We interpreted this result to indicate that dechlorination of diaminochloro-*s*-triazine occurred, but at a slower rate than deamination. In tests with [2,4,6-^{14}C]diaminochloro-*s*-triazine (452 mM, 0.038 mCi/ml), 38% of the radioactivity was recovered as $^{14}CO_2$ in 7 days, showing that *R. corallinus* could completely degrade this compound. diaminochloro-*s*-triazine mineralization was further improved by using cell suspensions containing both *Pseudomonas* sp. strain D and *R. corallinus*: in 7 days 65% of the ^{14}C was released from [2,4,6-^{14}C]diaminochloro-*s*-triazine as $^{14}CO_2$.

When *R. corallinus* and *Pseudomonas* sp. strain D were incubated separately for 7 days with 2.69 mM 1:1 FR-generated [2,4,6-^{14}C]atrazine degradation products, $^{14}CO_2$ evolution was 50% and 29%, respectively. Combining the cultures increased $^{14}CO_2$ evolution to 73% (Figure 2). No radioactivity was detected as volatile organic compounds in any treatment (*19*). In the dual culture incubation, *Pseudomonas* sp. strain D was dependent on *R. corallinus* to dechlorinate the *s*-triazines and thereby generate substrates the former could metabolize. Since *Pseudomonas* sp. strain D was unable to degrade diaminochloro-*s*-triazine and hydroxyatrazine, degradation of dechlorinated dealkylated *s*-triazines (ammelide, cyanuric acid, and hydroxyatrazine) or dechlorinated partially alkylated compounds not detected by HPLC probably accounted for the evolved $^{14}CO_2$ in the treatments not containing *R. corallinus*. The ^{14}C and chloride data presented here, combined with prior information on *s*-triazine degradation by *R. corallinus* (*17*), indicate this organism dechlorinated diaminochloro-*s*-triazine and hydroxyatrazine forming cyanuric acid, hydroxyatrazine, biuret, urea and ultimately NH_4^+ and CO_2.

FR Treatment of Pesticide Rinse Water. The 1:1 FR ratio was then applied to pesticide rinse water. Over 98% of EPTC (159 μM), metolachlor (209 μM), and alachlor (98 μM) were degraded with 5.3 mM FR (Figure 3). Atrazine (131 μM) and cyanazine (132 μM) were more recalcitrant with 11 and 22% of the initial compounds remaining, respectively. The relative susceptibility of the former compounds to oxidation by $HO^•$was consistent with their chemical structures. EPTC has three saturated alkyl chains that are easily attacked by $HO^•$. Metolachlor and alachlor each contain several saturated alkyl chains and, unlike the *s*-triazine ring, the benzene ring is easily attacked and cleaved by $HO^•$. The nitrilo alkyl group in cyanazine is more resistant to oxidation than atrazine's alkyl groups (*15*). When FR was increased to 12.2 mM, ≤1% of the initial parent compounds remained after 24 hours. This amount of FR was 4.5-fold greater on a molar basis than that needed to degrade aqueous solutions of atrazine (*12*) and reflected decreased reaction efficiency likely caused by $HO^•$ scavengers present in pesticide rinse water. A FR concentration of 14.4 mM increased chlorinated *s*-triazine

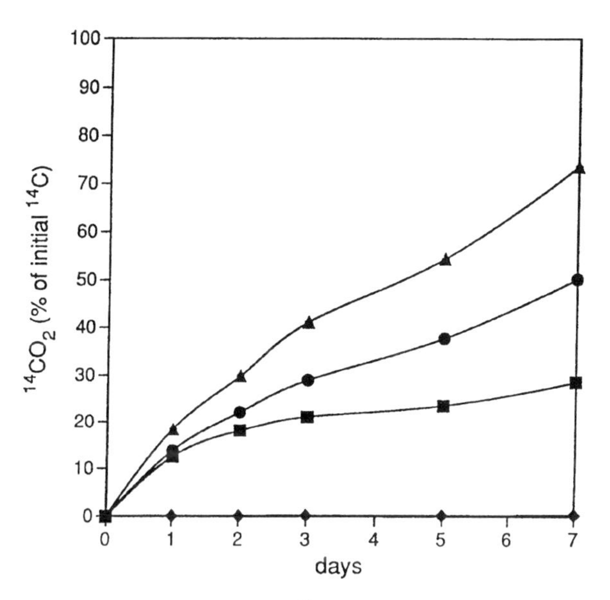

Figure 2. Mineralization of [2,4,6-^{14}C]atrazine (175 μM, 0.042 μCi/ml) degradation products generated by 2.69 mM $FeSO_4$:H_2O_2. Key to symbols: control (◆), *Pseudomonas* sp. strain D (■), *R. corallinus* (●), *Pseudomonas* sp. strain D and *R. corallinus* (▲). The standard deviation of the mean of duplicate samples was less than the size of the symbols.

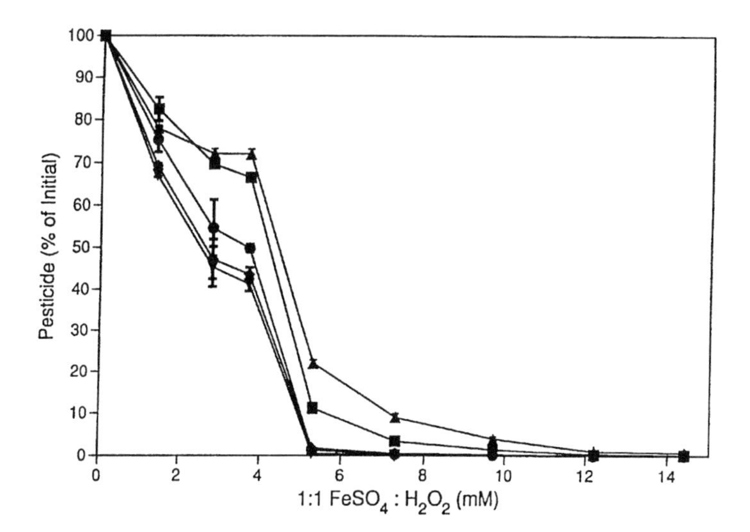

Figure 3. Effect of $FeSO_4$:H_2O_2 concentration on pesticide degradation in pesticide rinse water after 24-hour incubation. Key to symbols: atrazine (■), cyanazine (▲), EPTC (●), metolachlor (◆), alachlor (▼). Error bars represent standard deviation of the mean of duplicate samples.

elimination, and atrazine and cyanazine were transformed to diaminochloro-*s*-triazine (53 μM), hydroxyatrazine (20 μM), desisopropylatrazine (0.6 μM), desethylatrazine (2 μM), and two cyanazine products: deethylcyanazine and cyanazine amide.

Metabolism of FR-Generated Pesticide Rinse Water Products by *R. corallinus* and *Pseudomonas* sp. strain D. To complete the pesticide rinse water remediation process, *R. corallinus* was incubated with the products resulting from pesticide rinse water pretreatment with 14.4 mM 1:1 FR. Diaminochloro-*s*-triazine was degraded 95% in 12 hours, hydroxyatrazine, desethylatrazine, and desisopropylatrazine were completely degraded in ≤3 hours, 30 minutes, and 1 minute, respectively. The decreased rate of degradation of chlorinated *s*-triazines by *R. corallinus*, compared to incubations with the 2.69 mM 1:1 FR pretreatment of atrazine, could have reflected competitive inhibition of the dechlorination/deamination hydrolase by unknown compound(s) in the mixture. Since analytical standards for the cyanazine products identified by HPLC/ES-MS/MS were not available, peak areas were used to measure relative degradation amounts. Desethylcyanazine and cyanazine amide decreased 31 and 81% in 12 hours, respectively.

[2,4,6-^{14}C]Atrazine was used as a tracer to monitor *s*-triazine degradation in the pesticide rinse water. Individually, the mineralization levels effected by the cultures in a 10-day incubation were 61% for *R. corallinus* and 19% for *Pseudomonas* sp. stain D (Figure 4). Combining the cultures increased $^{14}CO_2$ evolution from atrazine in the pesticide rinse water mixture to 71%. These results are similar to those with atrazine in pure solution co-culture treatments (see Figure 1). However, $^{14}CO_2$ evolution was lower for *Pseudomonas* sp. strain D in the pesticide rinse water treatments. Possibly, the 14.4 mM FR treatment of pesticide rinse water resulted in less dechlorination of atrazine and its degradation products because of excess $HO^{\bullet}$quenchers in pesticide rinse water. After 10-day incubation, HPLC analysis showed that diaminochloro-*s*-triazine, hydroxyatrazine, desisopropylatrazine, desethylatrazine, and desethylcyanazine were completely degraded by *R. corallinus* and only 3% of cyanazine amide remained.

Conclusions

Combining chemical and microbial treatments completely degraded FR-generated chlorinated products produced from atrazine in aqueous solutions. Treatment effectiveness was not diminished in an agricultural pesticide waste containing atrazine as well as cyanazine and several other herbicides. For use as a remediation technology, the effectiveness of the chemical-biological treatment to effect degradation of atrazine and its chlorinated intermediates was important as the toxicity of latter compounds may be as great as that of the parent compound.

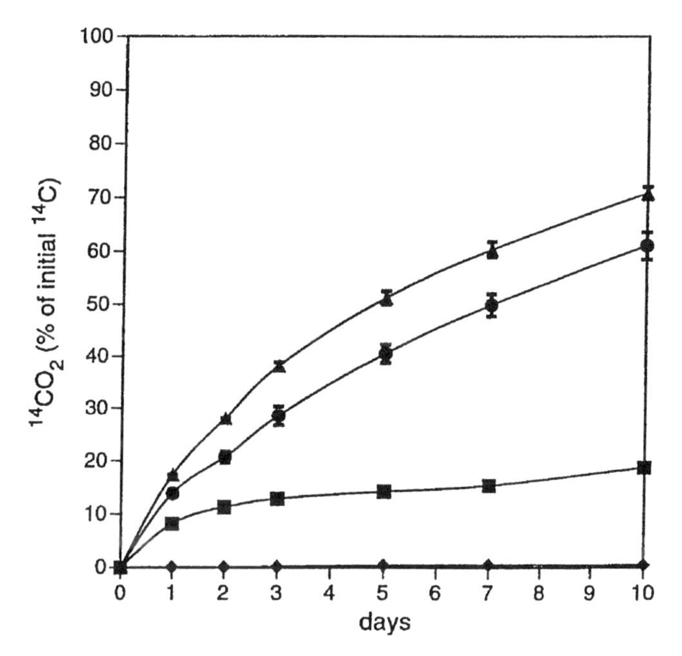

Figure 4. Mineralization of [2,4,6-^{14}C]atrazine (0.026 μCi/ml) degradation products in pesticide rinse water generated by 14.4 mM $FeSO_4$:H_2O_2. Key to symbols: control (◆), *Pseudomonas* sp. strain D (■), *R. corallinus* (●), *Pseudomonas* sp. strain D and *R. corallinus* (▲). Error bars represent standard deviation of the mean of duplicate samples.

Acknowledgments

HR-EI-MS and GC/MS data were provided by Cornelius Hop (Department of Chemistry, University of Wisconsin-Madison). The authors thank Michael Pelech for skillful GC analysis and Bruce Thede of Ciba-Geigy Corporation for providing *R. corallinus* and the cyanazine standard. This work was funded by the University of Wisconsin-Madison College of Agricultural and Life Sciences Hatch project number 3581 (to R.F.H).

Literature Cited

1. Fenton, H.J.H. *J Chem. Soc.* **1894**, *65*, pp. 899-910.
2. Beltrán, F.J.; García-Araya, J.F.; Acedo, B. *Water Res.* **1994**, *28*, pp. 2153-2164.
3. Beltrán, F.J.; García-Araya, J.F.; Acedo, B. *Water Res.* **1994**, *28*, pp. 2165-2174.
4. Hapeman-Somich, C.J.; Zong, G-M; Lusby, W.R.; Muldoon, M.T.; Waters, R. *J. Agric. Food Chem.* **1992**, *40*, pp. 2294-2298.

5. Adams, C.D.; Randtke, S.J. *Environ. Sci. Technol.* **1992**, *26*, pp. 2218-2227.
6. Legube, B.; Guyon, S.; Dore, M. *Ozone Sci. Eng.* **1987**, *9*, pp. 233-246.
7. Pelizzetti, E.; Maurino, V.; Minero, C.; Carlin, V.; Pramauro, E.; Zerbinati, O.; Tosato, M.L. *Environ. Sci. Technol.* **1990**, *24*, pp. 1559-1565.
8. Pelizzetti, E.; Minero, C.; Carlin, V.; Vincenti, M.; Pramauro, E.; Dolci, M. *Chemosphere.* **1992**, *24*, pp. 891-910.
9. Hustert, K.; Moza, P.N. *Toxicol. Environ. Chem.* **1991**, *31-32*, pp. 97-102.
10. Beltrán, F.J.; Ovejero, G.; Acedo, B. *Water Res.* **1993**, *27*, pp. 1013-1021.
11. Larson, R.A.; Schlauch, M.B.; Marley, K.A. *J. Agric. Food Chem.* **1991**, *39*, pp. 2057-2062.
12. Arnold, S.M.; Hickey, W.J.; Harris, R.F. *Environ. Sci. Technol.* **1995**, *29*, pp. 2083-2089.
13. Plimmer, J.R.; Kearney, P.C.; Klingebiel, U.I. *J. Agric. Food Chem.* **1971**, *19*, pp. 572-573.
14. U.S. Environmental Protection Agency. Atrazine, simazine and cyanazine; notice of initiation of special review. OPP-30000-60; FRL4919-5. **1994**. Office of Pesticide Programs, Washington, DC.
15. Somich, C.J.; Muldoon, M.T.; Kearney, P.C. *Environ. Sci. Technol.* **1990**, *24*, pp. 745-749.
16. Chesters, G.; Read, H.W.; Harkin, J.M.; Chen, C-P. Safe on-farm disposal of dilute pesticide wastes. Final Report. Project WIS-9000469. **1993**. U.S. Department of Agriculture, Washington, DC.
17. Cook, A.M.; Hütter R. *J. Agric. Food Chem.* **1984**, *32*, pp. 581-585.
18. Arnold, S.M.; Talaat, R.E.; Hickey, W.J.; Harris, R.F. *J. Mass Spectrom.* **1995**, *30*, pp. 452-460.
19. Arnold, S.M.; Hickey, W.J.; Harris, R.F.; Talaat, R.E. *Environ. Toxicol. Chem.* **1996**, in press.
20. Mulbry, W.W. *Appl. Environ. Microbiol.* **1994**, *60*, pp. 613-618.

Chapter 17

Source and Transport of Desethylatrazine and Desisopropylatrazine to Groundwater of the Midwestern United States

E. M. Thurman[1], D. W. Kolpin[2], D. A. Goolsby[3], and M. T. Meyer[4]

[1]U. S. Geological Survey, 4821 Quail Crest Place, Lawrence, KS 66049
[2]U. S. Geological Survey, 400 South Clinton St., Iowa City, IA 52244
[3]U. S. Geological Survey, Box 25046, DGC, MS 406, Denver, CO 80225
[4]U. S. Geological Survey, 4308 Mill Village Road, Raleigh, NC 27612

Based on usage of the parent compounds and studies of their dissipation in corn fields, atrazine (6-chloro-N-ethyl-N'-(1-methylethyl-1,3,5-triazine-2,4-diamine)), cyanazine (2-[[4-chloro-6-(ethylamino)-1,3,5-triazin-2-yl]amino]-2-methylpropionitrile), and simazine (6-chloro-N,N'diethyl-1,3,5-triazine-2,4-diamine) are thought to be the important contributors of desethylatrazine (6-chloro-N-(1-methylethyl)-1,3,5-triazine-2,4-diamine) or DEA and desisopropylatrazine (6-chloro-N-ethyl)-1,3,5-triazine-2,4-diamine) or DIA to ground water. Atrazine degrades to both DEA and DIA by dealkylation. DEA is transported through the unsaturated zone more readily than DIA because of the more rapid degradation of DIA in the shallow unsaturated zone. Both cyanazine and simazine degrade to DIA by dealkylation, and DIA may be an indicator of leaching and subsequent degradation from both parent compounds. Because cyanazine has an acid intermediate that is mobile in the unsaturated zone, it may be an important source for DIA transport to ground water. Based on a regional survey of ground water of the Midwest during the Spring and Summer of 1991, DEA occurs most frequently with detections of DEA (15.4%) > atrazine (14.7%) > DIA (4%) > simazine (0.7%) > cyanazine (0.3%) > propazine (no detections). The DEA to atrazine ratio (DAR) is an indicator of atrazine source and transport, with the lowest ratios indicating most rapid transport.

Agricultural practices may cause widespread degradation of water quality in the Midwestern United States (1-4). Approximately three-fourths of all pre-emergent herbicides are applied to row crops in a 9-State area, called the "Corn Belt" (5). Because many herbicides are water soluble, they may leach into ground water (6-7), as well as be transported in surface runoff (8-9). Monitoring studies in the Midwest have shown widespread detection of herbicides, such as atrazine and its metabolites in surface water (2-3, 10) and in ground water (6-7). Studies of the Mississippi River and its tributaries (3) show that two metabolites of atrazine, DEA, and DIA, are also widely detected.

It has been demonstrated that ground water is a source of atrazine to surface water (10-13). Furthermore, the frequency of detection of the dealkylated degradation products of atrazine in surface and ground water has prompted research demonstrating that atrazine's degradation products may be indicators of surface- and ground-water interaction (10). Adams and Thurman (14) studied the movement of DEA through the unsaturated zone and found that the DEA-to-atrazine ratio (called the DAR) may indicate the rate of transport of atrazine to ground water, thus differentiating non-point source from point-source contamination. The concept of the DAR was applied to surface-water effects on ground water in alluvial aquifers by Thurman et al. (10), who suggested that the DAR values found in surface water of the Midcontinent indicate ground water discharging to streams in early spring and late fall and surface water recharging alluvial aquifers during late spring.

Recently, Thurman et al. (15) have shown that DEA and DIA occur in surface water of the Midwestern United States and that the source of these two metabolites is chiefly a combination of atrazine and cyanazine. Furthermore, there have been regional studies of herbicides in ground water, including a recent study by Kolpin et al. (16) that show that the most commonly detected compound was the dealkylated triazine metabolite, DEA. Less commonly detected was DIA.

This paper examines the source of these two dealkylated metabolites in light of previous field-dissipation studies for atrazine (14-15), cyanazine (15,17), propazine, and simazine (18). Data are collated from these studies and from a regional ground-water survey in order to document the sources and transport of DEA and DIA in ground water of the Midwestern United States. The major objectives of this paper are: (1) to present evidence for the source of DEA and DIA in ground water from four parent triazine compounds, (2) to show relative transport through the unsaturated zone of the four triazine compounds and their two dealkylated metabolites, and (3) to interpret the detection of DEA and DIA in ground water relative to the use of the four triazine compounds in the Corn Belt of the Midwestern United States.

Experimental Methods

Field-Dissipation Studies. Four field-dissipation studies of atrazine, cyanazine, propazine, and simazine were conducted at the same site, Kansas River Valley Experimental Field near Topeka, Kansas, from 1989 through 1992. Results of one of the atrazine studies conducted in 1989 were published by Adams and Thurman (14) and dealt with atrazine and DEA transport in the vadose zone. The second atrazine study of 1989 was published with the cyanazine field-dissipation study of 1992 (15, 17). The propazine and simazine study of 1990 was published by Mills and Thurman (18) and dealt with metabolite movement in both surface water and ground water. The experimental conditions from these four studies are presented in general format here, and further details are given in the cited papers. The field-dissipation studies of DEA and DIA are summarized for the first time in this paper with emphasis on the movement of DEA and DIA in the unsaturated zone to ground water.

Two studies on atrazine were conducted between May and November 1989 and 1992. Eudora silt-loam plots (65-m^2) were planted with corn (*Zea mays L.*), and AAtrex Nine-O (Ciba-Geigy Corp.) was applied as a pre-emergent herbicide at a rate of 2 kg/ha for the atrazine study in 1989 (14) and as AAtrex 4L in the 1992 study (15). In the 1992 study, replicate Eudora silt-loam plots were planted with corn, and cyanazine (Bladex, Dupont Corp.) was applied to both at 2 kg/ha. Propazine (Milogard, Ciba Geigy) and simazine (Princep, Ciba Geigy) were applied to different Eudora silt-loam plots in 1990 at 2 kg/ha.

In all four studies, both sprinkler-applied irrigation and natural precipitation maintained crop growth. All herbicides were applied prior to corn planting and incorporated into the soil, which was a silt loam with a particle size distribution of 44-63% silt, 26-50% sand, and 5-21% clay. The soil pH was in the range of 6.8 to 7.8 and the organic carbon content was 0.99, 0.69, and 0.22% at 15, 30, and 45 cm, respectively. Each plot had a 1% slope and was isolated from the other plots by perimeter berms. In all studies, suction-cup lysimeters (5-cm diameter; Soil-Moisture Equipment, Santa Barbara, CA) were used to sample soil pore water, were placed at depths of 30, 60, 90, and 120 cm, and were sampled following each rain or irrigation event and monthly during the dry harvest season.

Soil cores were collected from one location near the center of each plot using a split-tube sampler (CME Co., St. Louis, MO) before application, after application, and then monthly. Soil cores were separated into 15-cm intervals, placed in polypropylene bags, and frozen until analyzed. Samples of surface runoff drained into a bucket that was level with the field surface at the downslope end of each plot and pumped continuously into a 380-L storage tank using a sump pump. The volume of runoff was recorded and a 4-L sample of water and suspended sediment was collected after each event for measurement of surface-runoff losses from the plot.

Ground-Water Survey. The ground-water reconnaissance network consisted of 303 wells distributed across 12 States (Figure 1), with the entire network being sampled twice during 1991 (299 wells March-April and 290 wells July-August) (16). Additional water samples were collected from 100 randomly selected wells from the network during July-August 1992, from 110 wells collected in unconsolidated aquifers during September-October 1993, and from 38 wells collected in unconsolidated aquifers during July-August 1994. See recent publication by Kolpin et al. for complete details (19).

All samples were collected by USGS personnel using equipment constructed of materials, such as glass and stainless steel that would not leach or sorb pesticides (19). Decontamination procedures, which included the thorough rinsing and cleaning of all equipment, were implemented to prevent cross contamination between wells and samples. Wells were purged before sampling until pH, water temperature, and specific conductance stabilized. The pumping time to reach chemical stability for each well varied but required a minimum of 15 minutes. All water samples were stored in amber, baked-glass bottles and chilled upon collection. No other preservation techniques were required. A quality-control program using a series of field blanks, field duplicates, and spikes verified effectiveness of the sampling protocol and the analytical procedures.

Analysis. Methanol (Burdick and Jackson, Muskegon, MI), ethyl acetate, and isooctane (Fisher Scientific, Springfield, NJ) were pesticide-grade solvents. Deionized water was charcoal filtered and glass distilled prior to use. Atrazine, propazine, and simazine were obtained from Supelco (Bellefonte, PA); terbuthylazine standards were

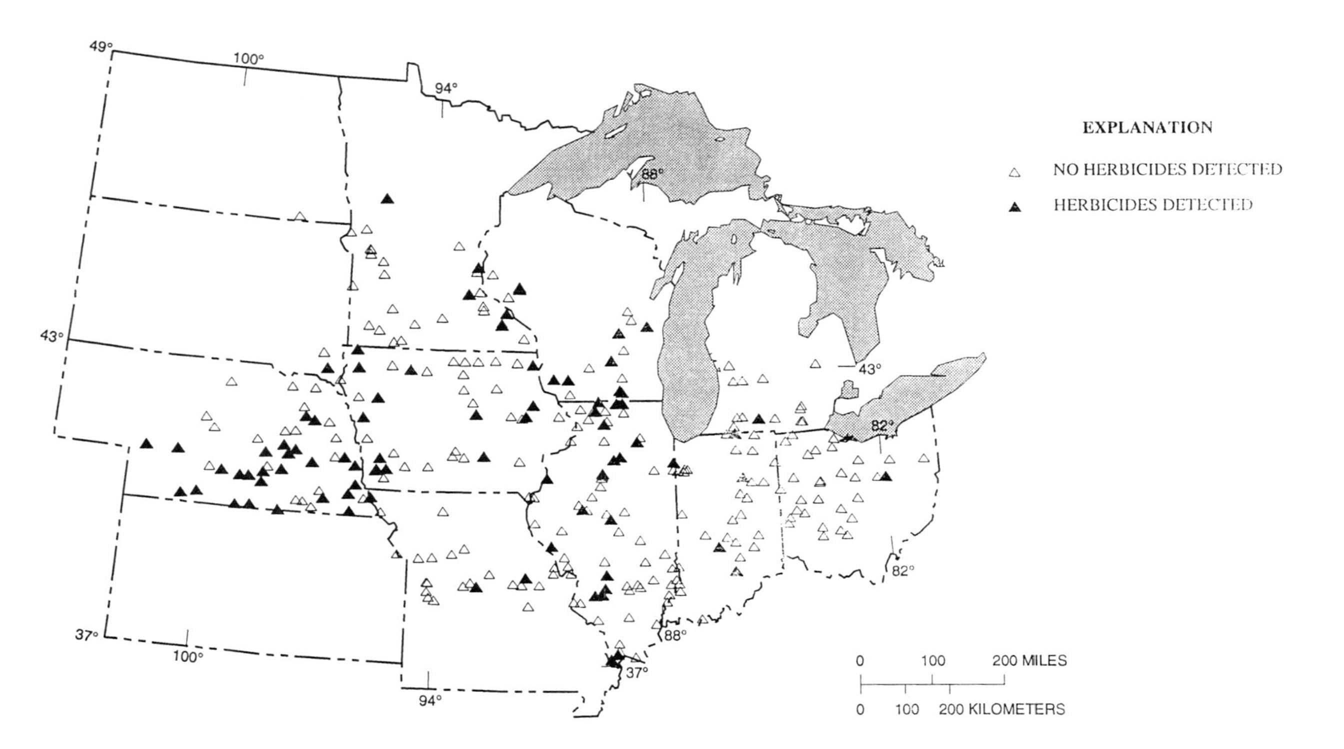

Figure 1. Location of sampled wells and herbicide detections in ground water from a 12-state area in the Corn Belt of the Midwestern United States, 1991 (16).

obtained from the U.S. EPA (U.S. Environmental Protection Agency) Pesticide Chemical Repository (Research Triangle Park, NC); and the triazine metabolites, DEA and DIA, were obtained from Ciba Geigy (Greensboro, NC). The C_{18} cartridges (SEP-PAK from Waters, Milford, MA) contained 360 mg of 40-mm C_{18} bonded silica. Standard solutions were prepared in methanol, and d_{10} phenanthrene (U.S. EPA, Cincinnati, OH) was used as an internal GC/MS quantitation standard.

The method of Thurman et al. (20) was used for herbicide analysis and consists of using a Waters Millilab Workstation (Milford, MA) for solid-phase extraction with C_{18} cartridges. Each 123-mL water sample was spiked with a surrogate standard, terbuthylazine (2.4 ng/mL, 100 mL), and pumped through the cartridge at a rate of 20 mL/min by the robotic probe. Analytes were eluted with ethyl acetate and spiked automatically with d_{10} phenanthrene. The extract was evaporated automatically by a Turbovap (Zymark, Palo Alto, CA) at 45 °C under a nitrogen stream to 100 µL.

Automated GC/MS analyses were performed on a Hewlett Packard model 5890 GC (Palo Alto, CA) and a 5970A mass selective detector (MSD). Operating conditions were: ionization voltage, 70 electron volts; ion source temperature, 250 °C; electron multiplier, 2,200 volts; direct capillary interface at 280 °C, tuned daily with perfluorotributylamine; dwell time, 50 milliseconds. Separation of the herbicides was carried out using a 12 meter fused-silica capillary column, 0.2 millimeter in diameter with a methyl silicone stationary phase, 0.33 µm thick. Helium was used as the carrier gas at a flow rate of 1 mL/min and a head pressure of 35 kilopascals. The column temperature was held at 50 °C for 1 minute, then increased at 6 °C per minute to 250 °C where it was held for 10 minutes. Injector temperature was 210 °C. Quantification of the base peak of each compound was based on the response of the 188 ion of the internal standard, phenanthrene-d_{10}. Confirmation of the compound was based on the presence of the molecular ion and one to two confirming ions with a retention-time match of $\pm$ 0.2 percent relative to d_{10}-phenanthrene (20).

Results and Discussion

Sources of Parent Triazines. Atrazine, cyanazine, propazine, and simazine degrade in soil to DEA and DIA (Figure 2). Figure 2 was summarized from published studies (15), including recent work by the authors on all four compounds using field disappearance studies (14, 15, 17-18). The fact that cyanazine forms DIA has been reported in earlier studies (21-23) and was recently confirmed in field studies (15, 17). Furthermore, it was found that cyanazine may form as much as 25% of the DIA found in surface waters. Thus, all four of the parent triazines are important sources for DIA and DEA in surface water.

The use of these four compounds in the Midwestern United States in 1995 varied from 20 million kilograms of atrazine applied in the 12 state area (Figure 1), followed by cyanazine at 11 million kilograms, followed by simazine at 0.3 million kilograms, and propazine has not been sold since 1990. In spite of the lower use figures for both simazine and propazine, both compounds occur regularly in surface water of the Midwestern United States as reported by Thurman et al. (10) in a large water quality survey of the Midwest. Simazine was detected in 55% of the post planting samples and propazine in 40% with median concentrations of 0.07 µg/L for simazine and <0.05 µg/L for propazine (10). Atrazine and cyanazine are used chiefly on corn in the

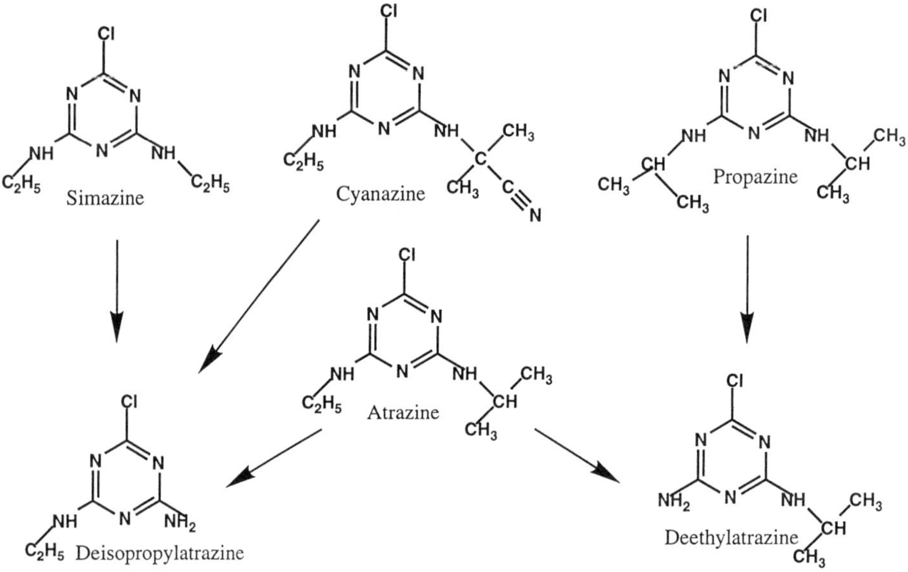

C = Carbon
H = Hydrogen
N = Nitrogen
Cl = Chlorine
C_2H_5 = Ethyl
CH_3 = Methyl

Figure 2. Degradation pathway for formation of DEA and DIA from four major triazine herbicides used or found in the Corn Belt.

Midwest. Simazine is used on corn, citrus, alfalfa, and miscellaneous orchard crops. The majority of the application in the 12-state area is for corn. Propazine on the other hand has not been sold in the United States since 1990. Thus, why is there 40% detection in the surface water study reported by Thurman et al. (10) and is there an unknown source of propazine to ground water?

One possible explanation is shown in the correlation of the propazine concentrations to atrazine concentrations from the 1992 study (Figure 3), which shows that the correlation coefficient is 0.90 (significant at 0.01) with a slope of 0.012. Or the concentration of propazine is approximately 1% the concentration of atrazine. These data suggest that atrazine is contaminated with propazine at the level of ~1%. Analysis of the AAtrex used at the Kansas River Valley Experimental Field plot study showed that both propazine and simazine are present in the starting material at approximately 1% for propazine and 0.5% for simazine. The positive correlation between propazine and atrazine in surface water suggests that the degradation half lives of both compounds are similar. However, propazine has a longer half life than atrazine (18) because of the isopropyl-isopropyl substitution pattern. One explanation for the good correlation in spite of the differences in half life is that the propazine-atrazine correlation was obtained early in the season when degradation has not occurred to a significant degree in soil. Furthermore, the degradation process is much slower in the aqueous phase in runoff waters that may be stored in alluvial aquifers and discharged later in the growing season to the streams. Thus, a major source of propazine for alluvial ground water is from trace contamination of atrazine that is applied to fields, at approximately 1%.

Correlation of simazine to atrazine concentrations from the same 1992 data set showed no significant correlation. This is probably caused by two factors. First, simazine is used on orchards and atrazine is not. Thus, the amount of atrazine and simazine should not necessarily be correlated. Second, simazine has a considerably shorter half life than atrazine (23). Thus, simazine has the opportunity to degrade considerably more in the soil than atrazine does, which would affect the concentrations in surface water. Thus, in spite of the recent use of only three triazines, there are sources for all four triazines into ground water of the Midwestern United States, including propazine.

DEA and DIA from Atrazine. Adams and Thurman (14) and Mills and Thurman (18) have shown in field dissipation studies that the rate of degradation of atrazine in soil interstitial water is rapid for the production of DEA and slower for the production of DIA. This suggests that the removal of the ethyl group is apparently quicker than the loss of the isopropyl group (Figure 2), probably because of steric hindrance of the isopropyl group. Furthermore, laboratory studies by Kruger et al. (24-25) showed that the production of DEA exceeds DIA at a ratio of 2:1 to 3:1 using ring labeled ^{14}C-atrazine. They found that 7% of the ^{14}C-atrazine degraded to DEA and 3% to DIA in studies of an Iowa silt loam. Thus, these four field-dissipation studies of DEA and DIA occurrence in the unsaturated-zone soils and soil water are consistent with laboratory studies of ^{14}C-atrazine.

For example, Figure 4 shows the disappearance of atrazine and the appearance of DEA in soil water from different lysimeters at the Kansas River Valley Experimental Field plot. DIA occurred much less frequently in lysimeters and cores, about 4 to 5 times less than DEA. Thus, DIA is removed by degradation from the soil waters and only trace amounts are transported to deeper lysimeters. Whereas, Figure 4 shows that DEA is transported to a depth of 460 centimeters with concentrations as large as atrazine and apparently a slower degradation rate than DIA. DIA, on the other hand, is present

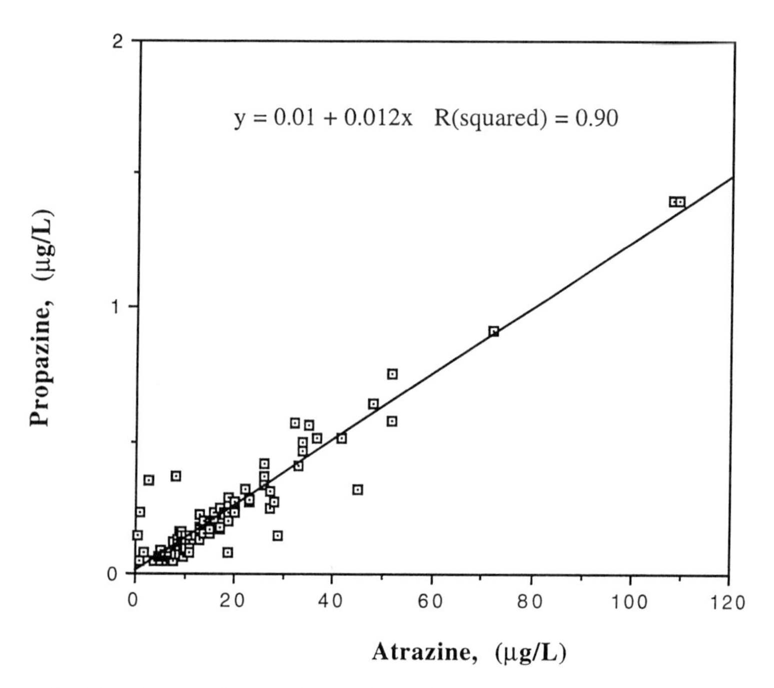

Figure 3. Correlation of propazine and atrazine concentrations from 145 surface-water samples from the Midwestern United States, using the data base published by Scribner et al. (27).

in only trace amounts at depth, which may be accounted for as DIA being formed at depth and only trace amounts being transported from the surface soil.

If one assumes that the production ratios of atrazine degradation products in the unsaturated zone are similar to that in the shallow soil (24-25), then approximately 12% of the atrazine degrades to dealkylated products (9% DEA and 3% DIA), which are the results reported by Mills and Thurman (18). Thus, the concentrations of atrazine shown in Figure 4 at a depth of 370 centimeters (0.50 μg/L) would indicate that approximately 0.05 μg/L of DEA and 0.02 μg/L of DIA would be generated by decomposition. However, the concentration of DEA at 370 centimeters is 0.50 μg/L, which indicates that the majority of the DEA is being transported down the soil column from overlying layers that contain high concentrations of atrazine and DEA (>1.0 μg/L). Soil-core analysis from this same site confirmed that only trace levels of DIA were present (generally 5-10 times lower than DEA), which presumably came from the degradation of atrazine at that depth.

The DEA-to-atrazine ratio (DAR) increased with depth and with time in the soil water. For example, Figure 4 shows that the concentrations of atrazine and DEA in soil water and that the DAR increased from 0.14 at the beginning of the study (day 1), which represents the DEA and atrazine concentrations from application, to 1.0 on August 6th, which is 140 days after application (at 0.6m depth). The large increase in the DAR indicates that DEA was preferentially transported relative to atrazine from the upper soil horizon. This preferential transport is called a chromatographic effect (18). The chromatographic effect consists of the sorption of atrazine (organic carbon partition coefficient, K_{OC}, of 160) relative to DEA (estimated from solubility data, K_{OC}=16) and DIA (estimated from solubility data, K_{OC}=5) followed by the loss of DIA by degradation. This combined effect causes the rapid movement of DEA relative to atrazine and DIA. The values for the DAR for the two soil cores were considerably lower than the DAR values for the soil-water samples. This fact is caused by the preferential release of DEA to soil water over atrazine; atrazine is much less soluble and is more tightly sorbed to the soil. There may also be conversion of atrazine to DEA at depth, but this is considered to be a small value relative to the transport of DEA from above because of the low concentrations of DIA that are found at depth. In the shallow soil where the microbial activity is high, there is a 2 to 1 ratio of DEA to DIA, but at depth this ratio increases to 5 to 1 to 10 to 1, which indicates that conversion of atrazine to DEA and DIA is not a major source.

To summarize, unsaturated-zone studies of atrazine, DEA, and DIA show that atrazine is sorbed to soil organic matter in the shallow soil and undergoes dealkylation to DEA and DIA. Because of the chromatographic effect, DEA is preferentially carried to the deeper unsaturated zone, causing the DAR to increase to values greater than 1.0. Deisopropylatrazine also is transported downward slightly but it rapidly degrades to less than trace levels, creating a ratio of DIA to DEA of 0.2 or less. Thus, there is 5 to 10 times more DEA present in the unsaturated zone at depth, which is a result that has been shown in several unsaturated-zone studies. These results will be correlated to the ground-water survey later in this paper.

DIA from Cyanazine. Figure 5 shows the concentration of cyanazine, cyanazine amide, and DIA in lysimeters on the Kansas River Valley Experimental Field. DIA did not occur during the first 2 months of the study at detectable concentrations, rather it occurred later in the year at both of the lysimeter depths, 0.9 and 1.2 meters. The DIA concentrations at these depths are considerably greater than the concentrations of DIA

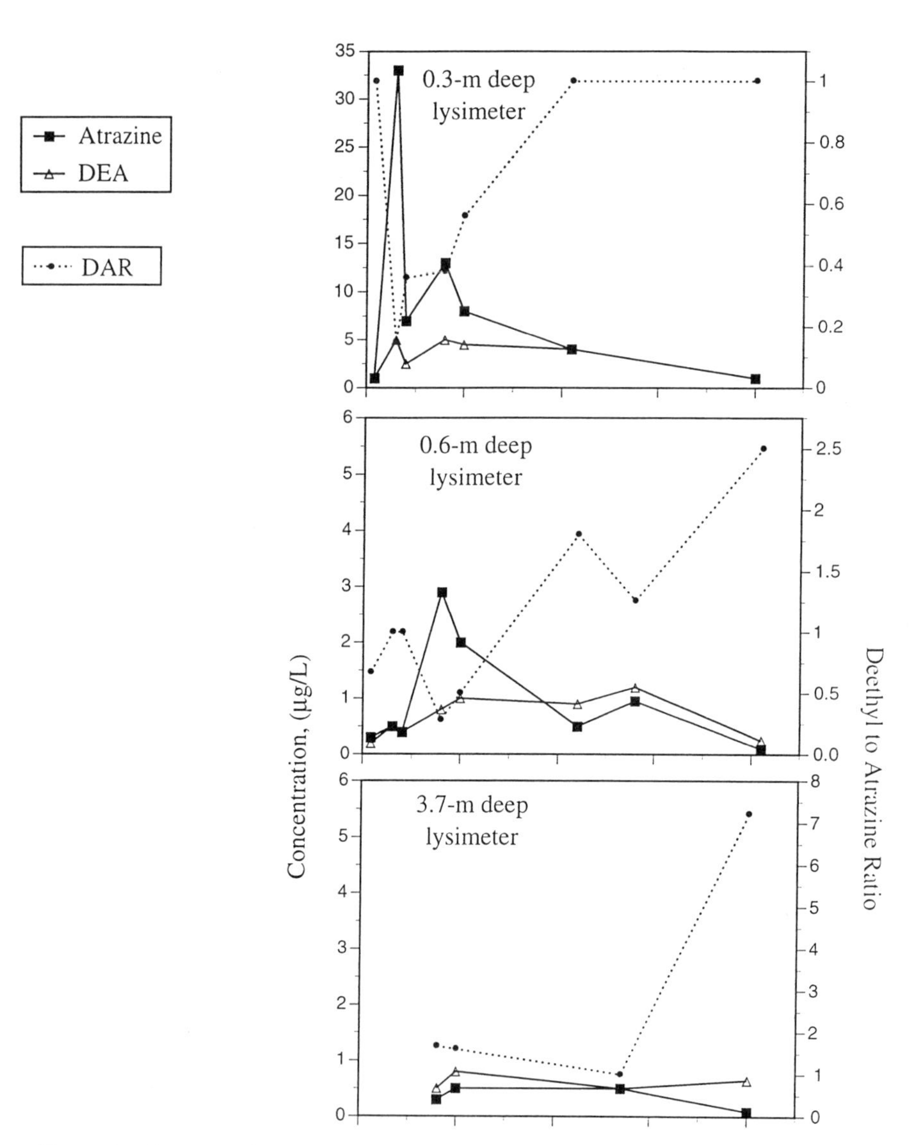

Figure 4. Concentrations of atrazine and DEA in soil-water lysimeters at Kansas River Valley Experimental Field, Topeka, Kansas, data from Adams and Thurman (14).

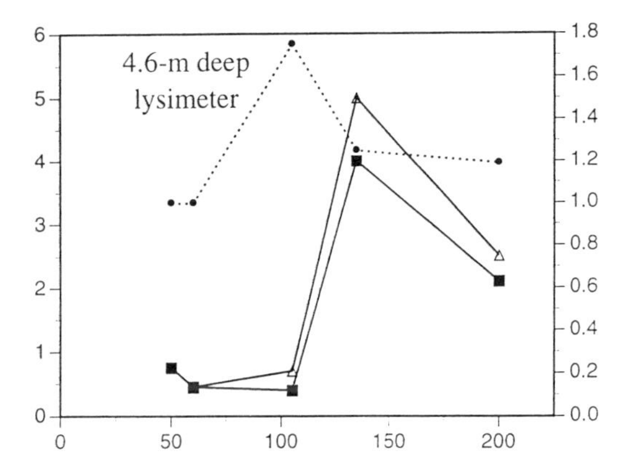

Figure 4. *Continued.*

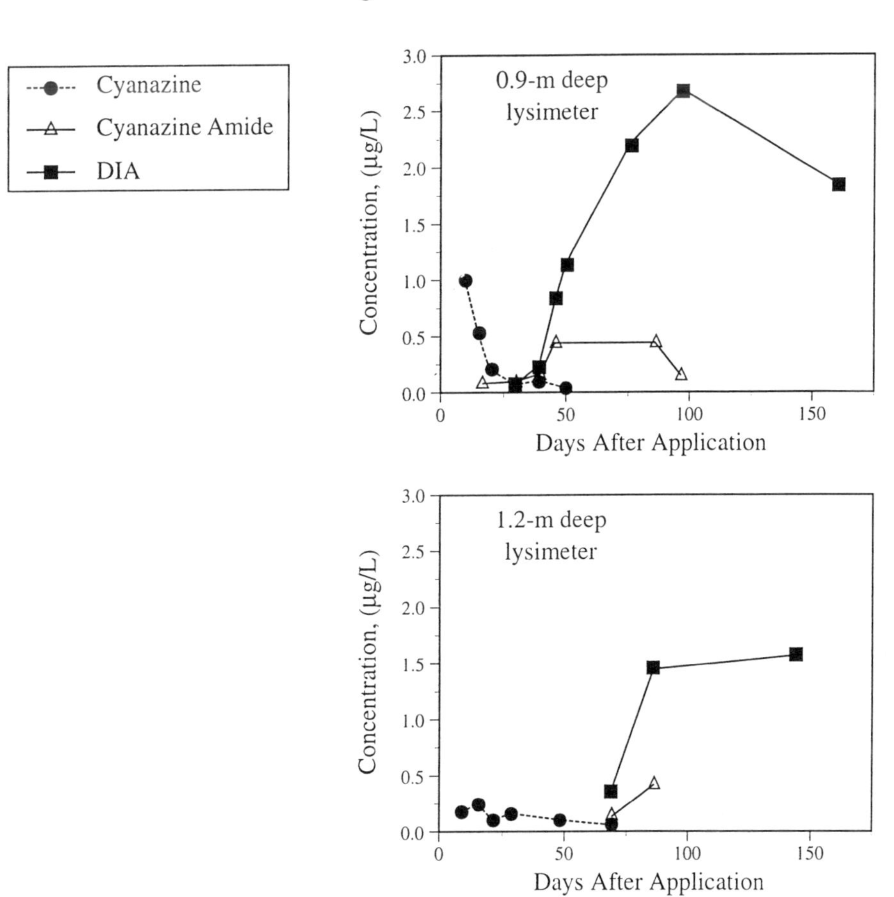

Figure 5. Concentrations of cyanazine, cyanazine amide, and DIA in soil-water lysimeters at Kansas River Valley Experimental Field, Topeka, Kansas, data from Meyer (17).

that were found in the atrazine field plot studies. The current hypothesis from Meyer (17) is that DIA forms from the pathway shown in Figure 6, which includes the formation of the cyanazine amide, followed by the formation of cyanazine acid, and finally the formation of the DIA from cyanazine acid.

The cyanazine acid has the potential to leach downward in the soil profile because its pKa is ~4.6, which means that the cyanazine acid will be anionic at the pH of most soil water (pH 6-7). Thus, the cyanazine acid will have a low affinity for sorption to the soil and will be much more mobile than either the parent compound or its metabolites. The rapid transport of the cyanazine acid would explain why DIA is found at depth when cyanazine is applied to soil, but DIA is found at much lower concentrations when atrazine is applied to soil. Thus, cyanazine is a potential source of DIA to ground water through the transport of cyanazine acid to the unsaturated zone and its subsequent degradation to DIA.

DEA and DIA from Propazine and Simazine. Figure 7 shows the changes in concentration of propazine, simazine, DEA, and DIA in soil cores from three depths (15, 30, and 45 cm) throughout the season at the Kansas River Valley Experimental Field plots. Maximum concentrations of all compounds were observed early in the growing season in the 15-cm cores, in the order propazine ≥ simazine > DIA ≥ DEA. Early in the growing season, DIA concentrations were equal to or exceeded DEA because of the more-rapid production of DIA from simazine. DIA continued to be greater in concentration then DEA throughout the growing season at the shallowest depth. DIA dissipated to trace levels by late season at the deepest soil cores (45 cm) and DEA became the major dealkylated metabolite in the propazine-simazine field plot at depth (45 cm).

These data show that propazine and DEA are the most resistant to degradation, both containing only isopropyl moieties. Simazine and DIA degrade more rapidly, both containing only ethyl moieties. This preferential dealkylation is most evident from the changing concentrations of metabolites. As a useful comparison of the relative concentrations of metabolites, the ratio of DIA to DEA is used, called the D^2R (DIA to DEA ratio). A D^2R of less than 1 indicates that DEA concentrations exceed DIA concentrations.

Figure 8 summarizes how the D^2R changes through time in the unsaturated zone on both the atrazine and the propazine-simazine plots. The D^2R remained constant in shallow soil cores from both plots throughout the growing season (~0.5 atrazine plot, ~3.0 propazine-simazine plot). These values match the reported degradation ratios of DEA and DIA for atrazine in soil by Kruger et al. (22-23) with values of 2-3 to 1 for DEA to DIA. In shallow soil, the parent source term is large and metabolite production is large, overshadowing continued degradation of the metabolites. In all soil cores taken below 15 cm, however, the D^2R declines to less than 1 through time regardless of the initial size of the ratio. At greater depths, the metabolites have moved chromatographically ahead of the parent triazine due to a large increase in aqueous solubility of the metabolites (17). At this point degradation becomes the dominant reaction, and the ratio declines as DIA is preferentially decomposed over the more resistant DEA.

The relative importance of deethylation versus deisopropylation in the atrazine plot appears comparable to the propazine-simazine plot. The ratio of DIA:DEA is equal to 0.47, and the ratio of deisopropylpropazine:deethylsimazine is equal to 0.3. Thus, the deethylation rates of atrazine and simazine appear similar and approximately two to three times more rapid than the deisopropylation rates of atrazine and propazine. Again,

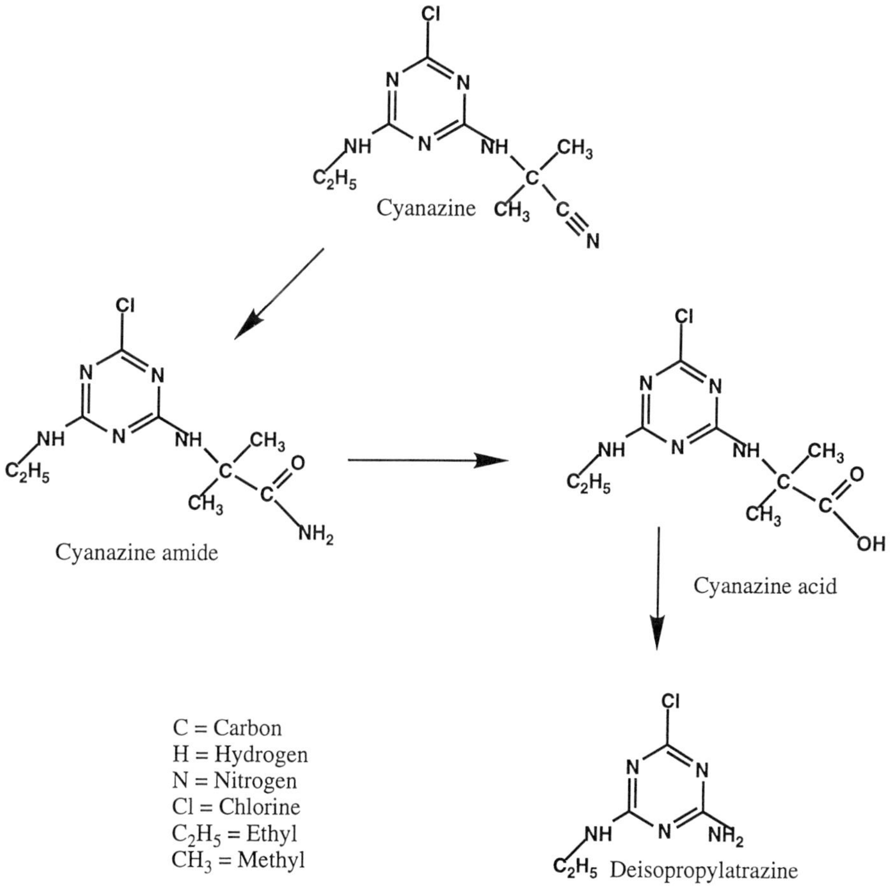

Cyanazine

Cyanazine amide

Cyanazine acid

Deisopropylatrazine

C = Carbon
H = Hydrogen
N = Nitrogen
Cl = Chlorine
C_2H_5 = Ethyl
CH_3 = Methyl

Figure 6. Degradation pathway for dealkylation reactions of cyanazine to DIA.

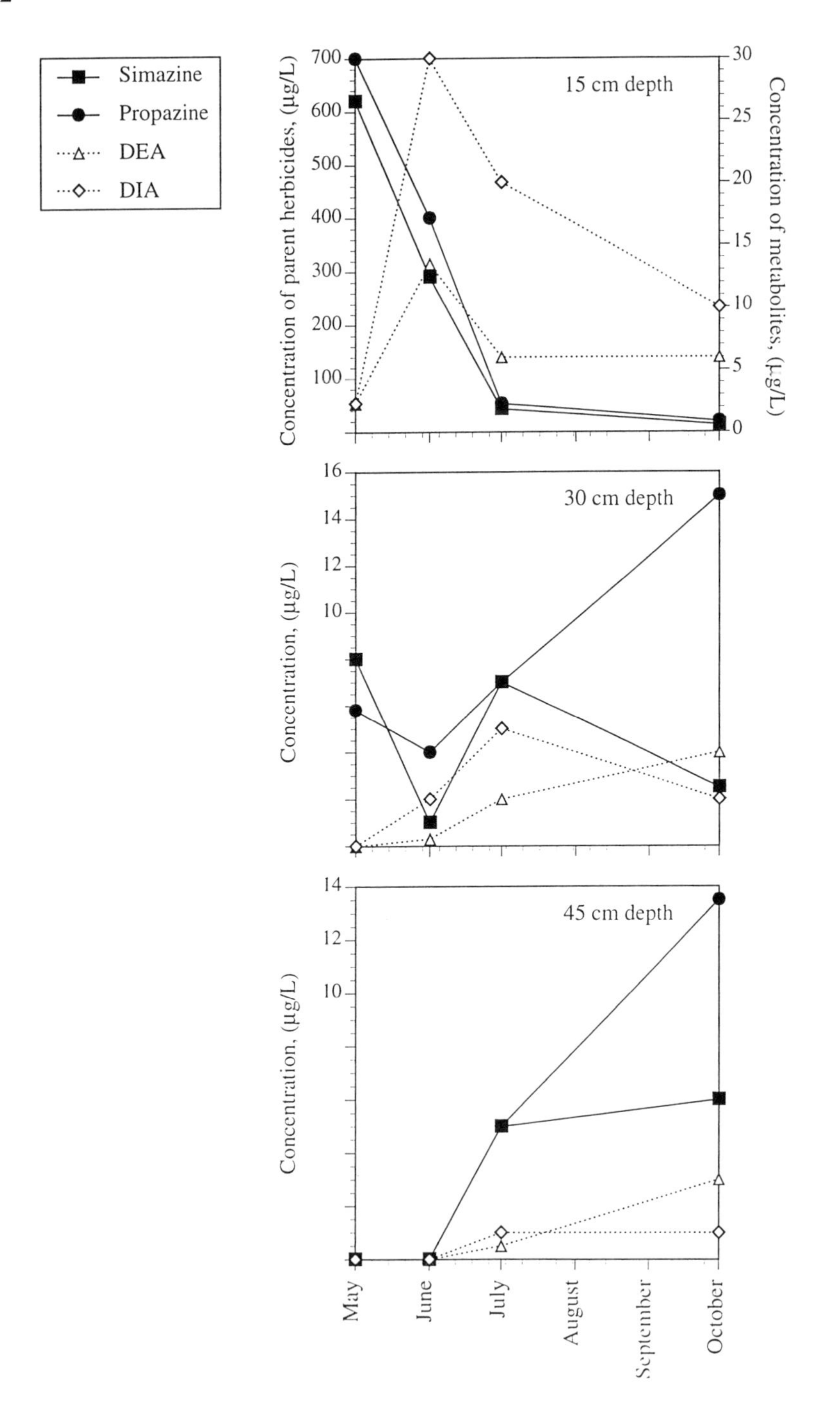

Figure 7. Concentrations of propazine, simazine, DEA, and DIA in soil cores from 15-, 30-, and 45-cm at Kansas River Valley Experimental Field, Topeka, Kansas, data from Mills and Thurman (18).

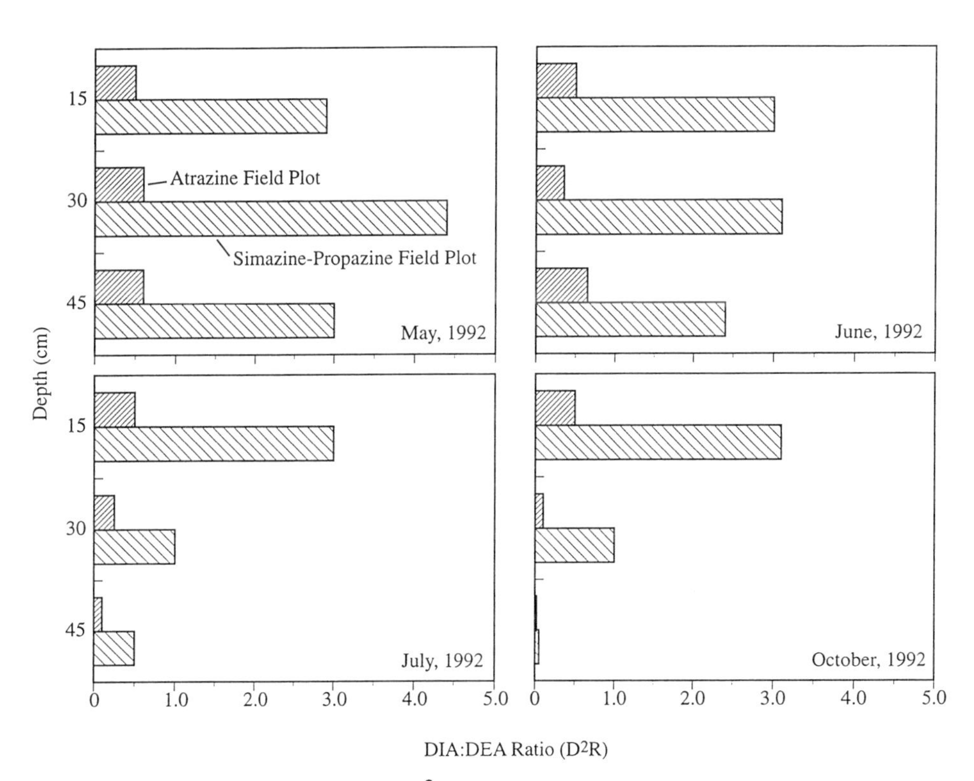

Figure 8. Ratio of DIA to DEA (D^2R) in soil cores from 15-, 30- and 45-cm on the atrazine and propazine-simazine plots at Kansas River Valley Experimental Field, Topeka, Kansas, data from Mills and Thurman (18).

comparable to values reported by Kruger et al. (23-25). Furthermore, this indicates that removal of an ethyl side chain is preferential over an isopropyl side chain regardless of parent triazine.

Finally, the concentrations of parent triazines and metabolites were followed in soil pore water collected from suction lysimeters installed at 30-cm intervals from 30 to 180 cm in the unsaturated zone. Parent herbicides and metabolites were not detected in the 60-cm lysimeters, correlating to the small concentrations present in soil cores from the same depth. This lack of transport presumably is due to sorption of the herbicides on organic matter and the clay fraction of the Eudora silt loam.

Ground-Water Studies. Herbicides and their metabolites were surveyed during the spring and summer of 1991 in water from 303 wells screened in near surface unconsolidated and bedrock aquifers. The wells were located in 12 States in the Midcontinent, as shown in Figure 1 (16, 26). The first sampling was before planting of herbicides (March and April) and 299 samples were collected. After planting (July and August) 280 samples were collected for a total of 579 samples.

Triazines or triazine metabolites were detected in 19.3% of the 303 wells collected. Deethylatrazine was detected most frequently at 15.4%, followed by atrazine at 14.7%, then DIA at 4.0%, simazine 0.7 %, and cyanazine 0.3% (Table I). Propazine was not detected in samples from any of the wells. The detection limit for all analyses was 0.05 μg/L. Figure 1 shows the distribution of the detections of triazine herbicides in the Midcontinent.

Table I. Summary of percent detections of herbicides in ground water of the Midcontinent from work of Kolpin et al. (16, 19, 26).

Herbicide or Metabolite concentration	Detections (%)	Maximum (μg/L)
Deethylatrazine (DEA)	15.4	2.3
Atrazine	14.7	2.1
Deisopropylatrazine (DIA)	4.0	0.60
Simazine	0.7	0.27
Cyanazine	0.3	0.68
Propazine	0.0	--------

The maximum concentrations found were highest for the metabolite, DEA at 2.3 μg/L, followed by atrazine at 2.1 μg/L, cyanazine at 0.68 μg/L, DIA at 0.60 μg/L, and simazine at 0.27 µg/L. Deethylatrazine was the major triazine metabolite found in ground water, with the highest concentrations and the greatest number of detections. In fact in 48% of the ground-water samples, the DEA concentration exceeded the atrazine concentration; however, in surface water, DEA only exceeded atrazine concentrations in 1.5% of the 450 samples collected in the 1989 reconnaissance of the Midwest (16). This result indicates that DEA is readily leached from soil and is at least as stable as atrazine in the ground water environment. Furthermore, it suggests that DEA may be forming from the degradation of atrazine in situ in ground water. However, the fact that DIA is present in ground water at ratios much less than in shallow soil suggests that the degradation of atrazine is either not a major source term or that the subsequent degradation of DIA is much more rapid than DEA; thus, DEA would accumulate relative to DIA.

If the ratio of DEA to atrazine (DAR) is ranked against the histogram of concentration ranges found in the ground-water samples (Table II), then one sees a trend of decreasing DAR values with increasing concentration of atrazine. The median DAR was 1.8 for atrazine concentrations ranging from 0 to 0.10 µg/L for 61 samples. The median DAR decreased to 0.58-0.60 in the two ranges from 0.10 to 0.20 µg/L and from 0.20 to 0.40 µg/L. The median DAR decreased to 0.23 again in the concentration ranges from 0.40 to 0.80 µg/L and from 0.80 to 2.1 µg/L. These low DAR values suggest that when the concentration of atrazine is greater than 0.40 µg/L and the DAR values are 0.2-0.3, that point source contamination or rapid transport of atrazine may be occurring. On the other hand, when the DAR is greater than 1 and the concentrations are less than 0.20 µg/L, then non point source contamination of ground water may be responsible for these concentrations. The DAR values and concentration ranges between these two extremes are not so easily interpreted and may indicate degradation of atrazine to DEA along the flow path or in the ground water.

Table II. Range of atrazine concentrations found in ground-water survey compared to median DAR values.

Range of Atrazine Concentration	Median DAR	Number of Samples
0.05 to 0.10	1.8	61
0.10 to 0.20	0.58	14
0.20 to 0.40	0.60	14
0.40 to 0.80	0.23	12
0.80 to 2.1	0.23	9

Deethylatrazine occurred more frequently than DIA (15.4% versus 4.0%), and DIA only exceeded DEA concentration in 5 of the 91 detects of both metabolites in ground water. In three of these cases, cyanazine and simazine were also detected, which suggests that these two parent triazine herbicides are contributing to the pool of DIA in ground water but on a more limited basis than atrazine. Thus, in conclusion, the significant degradation product of the triazine compounds occurring in ground water is DEA, which is degrading from atrazine in the soil, unsaturated zone, and in ground water.

Furthermore, the slow transport atrazine through the unsaturated zone commonly results in concentrations of DEA that are greater than atrazine. Thus, a large ratio of DEA to atrazine (DAR>1.0) is a good indicator of nonpoint-source contamination of ground water. On the other hand, low DAR values and high concentrations of atrazine indicate point source contamination of ground water. Finally, the detections of DIA in ground water suggest that both cyanazine and simazine are contributing to the concentrations of this metabolite in ground water. Cyanazine, because of its usage, is most likely an additional source of DIA to ground water in the Corn Belt.

Acknowledgments

We thank David Morganwalp and the Surface Water and Ground Water Toxics Program of the U.S. Geological Survey for support of this research and Dr. Homer LeBaron and Dr. K. Balu of Ciba Geigy for their support of student staff from the University of Kansas for work on metabolites of atrazine. The use of brand, trade, or firm names in this paper is for identification purposes only and does not constitute endorsement by the U.S. Geological Survey.

Literature Cited

(1) Humenik, F.J.; Smolen, M.D.; Dressing, S.A., *Environ. Sci. Technol.*, **1987**, *21*, 737-742.
(2) Thurman, E.M.; Goolsby, D.A.; Meyer, M.T.; Kolpin, D.W., *Environ. Sci. Technol.*, **1991**, *25*, 1794-1796.
(3) Pereira, W.E.; Rostad, C.E., *Environ. Sci. Technol.*, **1990**, *24*, 1400-1406.
(4) Kolpin, D.W.; Thurman, E.M.; Goolsby, D.A., *Environ. Sci. Technol.*, **1996**, *30,* 1413-1417.
(5) Gianessi, L.P.; Puffer, C.M., *Use of selected pesticides for agricultural crop production in the United States,* **1982-1985**. NTIS: Springfield, VA.
(6) Hallberg, G.R., *Agric. Ecosys. Envir.*, **1989**, *26,* 299-367.
(7) Burkart, M.R.; Kolpin, D.W., *J. Environ. Qual.*, **1993**, *22*, 646-656.
(8) Glotfelty, D.E.; Taylor, A.W.; Zoller, W.H., *Science,* **1983**, *219*, 843-845.
(9) Leonard, R.A., In *Environmental Chemistry of Herbicides;* (Grover, R. Ed., CRC: Boca Raton FL, **1988**), pp. 45-88.
(10) Thurman, E.M., Goolsby, D.A.; Meyer, M.T.; Mills, M.S.; Pomes, M.L; Kolpin, D.W.; *Environ. Sci. Technol.*, **1992**, *26*, 2440-2447.
(11) Squillace, P.J.; Thurman, E.M., *Environ. Sci. Technol.*, **1992**, *26*, 538-545.
(12) Squillace, P.J.; Thurman, E.M., Furlong, E., *Water Resour. Res.,* **1992**, *29*, 1719-1729.
(13) Squillace, P.J., *Ground Water*, **1996**, *34*, 121-134.
(14) Adams, C.D.; Thurman, E.M., *J. Environ. Qual.*, **1991**, *20*, 540-547.
(15). Thurman, E.M.; Meyer, M.T.; Mills, M.S.; Zimmerman, L.R.; Perry, C.A., *Environ. Sci. Technol.,* **1994**, *28*, 2267-2277.

(16) Kolpin, D.W.; Burkart, M.R.; Thurman, E.M., *U.S. Geol. Survey, Water-Supply Paper 2413*, **1994**.
(17) Meyer, M.T., In *Geochemistry of cyanazine and its Metabolites: Indicators of contaminant transport in surface water of the midwestern United States,* Ph.D. Thesis, Geology Department, University of Kansas, 1994, pp 1-362.
(18) Mills, M.S.; Thurman, E.M., *Environ. Sci. Technol.,* **1994**, *28*, 600-605.
(19) Kolpin, D.W.; Goolsby, D.A.; Thurman, E.M., *J. Environ. Qual.*, **1995**, *24*, 1125-1132.
(20) Thurman, E.M.; Meyer, M.T.; Pomes, M.L., Perry, C.A.; Schwab, A.P., *Anal. Chem.,* **1990**, *62*, 2043-2048.
(21) Benyon, K.I.; Stoydin, G.; Wright, A.N., *Pestic. Sci.*, **1972***3*, 293-297.
(22) Benyon, K.I.; Stoydin, G.; Wright, A.N., *Pestic. Biochem. Phys.,* **1972**, *2*, 153-157.
(23) Weed Science Society of America, **1992**, *Herbicide Handbook*, 7th Edition, Weed Science Society of American, Champaign, Illinois, 301p.
(24) Kruger, E.L.; Somasundaram, L.; Kanwar, R.S.; Coat, J.R., *Environ. Toxicol. Chem.,* **1993**, *12*, 1959-1967.
(25) Kruger, E.L.; Somasundaram, L.; Kanwar, R.S.; Coat, J.R., *Environ. Toxicol. Chem.,* **1993**, *12*, 1969-1975.
(26) Kolpin, D.W.; Burkart, M.R.; Thurman, E.M., *U.S. Geol. Survey, Open-File Report 93-114,* **1993**.
(27) Scribner, E.A.; Thurman, E.M., Goolsby, D.A., Meyer, M.T., Mills, M.S., Pomes, M.L., *U.S. Geological Survey, Open-File Report 93-457,* **1993**.

Chapter 18

Temporal and Spatial Trends of Atrazine, Desethylatrazine, and Desisopropylatrazine in the Great Lakes

S. P. Schottler[1,4], S. J. Eisenreich[2], N. A. Hines[2], and G. Warren[3]

[1]Gray Freshwater Biological Institute, Navarre, MN 55455
[2]Department of Environmental Sciences, Cook Colllege, Rutgers University, New Brunswick, NJ 00123
[3]U. S. Environmental Protection Agency, Great Lakes National Program Office, Chicago, IL 60604

Temporal and spatial trends of atrazine, desethylatrazine (DEA), and desisopropylatrazine (DIA) concentrations were determined in the Great Lakes from 1991 to 1994. Concentrations of atrazine ranged from 3 ng/l in Lake Superior to > 100 ng/l in Lake Erie. Lakes Michigan, Huron and Superior exhibited no vertical, lateral, or temporal trends in atrazine concentrations suggesting a slow water column transformation rate. Concentrations in Lakes Erie and Ontario varied spatially and showed an overall decrease from 1991 to 1994. DEA and DIA concentrations exhibited trends similar to those of atrazine, and DEA to atrazine ratios (DAR) were consistently about 0.5-0.6. The magnitude of the DAR is greater than what would be expected from annual atrazine and DEA loadings to the lakes. This suggests that an estimated 1 to 5% of the atrazine inventory in each lake is converted *in situ* to DEA annually.

High use agricultural herbicides are delivered to the Great Lakes by non-point source runoff from agricultural regions, and atmospheric deposition [*1-7*]. Schottler and Eisenreich [*8*] have detected atrazine, DEA, alachlor, and metolachlor in the Great Lakes at concentrations ranging from 1-100 ng/l. The Laurentian Great Lakes represent 20% of the earth's freshwater resource [*9*] and are potentially threatened by loadings of these herbicides. Thus, an understanding of the environmental behavior of these compounds within the Great Lakes systems is imperative. Previously, few data were published on the temporal and spatial distributions of herbicide concentrations within the lakes over a long period of time. A careful study of the distributions of these herbicides will yield insight into their transport and sources to the Great Lakes, and in-lake elimination processes.

[4]Current address: Department of Civil Engineering, 500 Pillsbury Drive, University of Minnesota, Minneapolis, MN 55455.

Because the Great Lakes have such long water residence times, they may act as long-term integrators of environmental processes and provide a unique opportunity to examine the fate of herbicides in large aquatic systems. Approximately 3000 metric tonnes of atrazine are applied annually in the Great Lakes basin during the months of April through June [*10-12*]. Surface runoff and atmospheric washout have been shown to remove ~0.5 - 1.5% of annual atrazine applications to watersheds with the greatest fluxes in the 60 days following application [*6, 13-18*]. This means that 15 to 45 tonnes of atrazine could potentially enter the Great Lakes annually.

An intensive four year study was conducted from 1991 through 1994 to quantify the environmental behavior of selected high use herbicides in the Great Lakes. During this period 590 samples were collected throughout the major basins of the five Great Lakes (Figure 1), and analyzed for atrazine, DEA, DIA, cyanazine, alachlor, and metolachlor. Water column concentration profiles were constructed for each site permitting detailed examination of vertical, lateral, and temporal trends in concentration. The overall objective of this research was to use the temporal and spatial distributions of herbicide concentrations as the framework for determining the mechanisms and magnitudes of inputs to the lakes, as well as the in-lake elimination rates. This paper will describe the spatial distributions of observed concentrations over a four-year period, and provide qualitative interpretations of transport and *in situ* elimination mechanisms. The ratios of DEA and DIA concentrations to atrazine concentrations will also be examined, and evidence of how these ratios reflect *in-situ* degradation of atrazine will be presented. In addition, the concentration data in this paper were also used by Schottler and Eisenreich to construct a mass balance model for atrazine in the Great Lakes [*20*]. Results from this model were integral in the discussion of concentration variations and transformation product ratios presented in this paper.

Sampling and Analytical Procedures

Synoptic vertical profiles of atrazine, DEA, DIA, cyanazine, alachlor, and metolachlor were constructed for 27 sites throughout the Great Lakes between 1990 and 1994 (Figure 1). Water column profiles of herbicide concentrations representing 4 to 10 depths per site were constructed for 10 sites in Lake Michigan, 3 sites in Lake Huron, 5 sites in Lake Erie, and 7 sites in Lake Ontario. A combined total of 592 samples were collected from the five lakes from September 1991 through October 1994. Lakes Michigan, Erie, and Ontario were intensively sampled in August/September of 1991, 1992 and 1994. Lake Michigan was also intensively sampled in April, June, and October of 1994. Three sites in Lake Huron were sampled in August/September of 1991 and 1992, and the northern most station was sampled again in April of 1994. Lake Superior was sampled in August 1990 (stations DTL and 12), in July of 1993 (station 17) and in May of 1994 (stations 1, 2, 8, 12, 17).

Water samples (2 liters) were collected using a General Oceanics (Model 1015) rosette sampler on board the U.S. EPA Research Vessel *Lake Guardian*. Solid Phase Extraction (SPE) cartridges were used to isolate herbicides from water samples. Samples collected from 1991 and 1992 were processed using 5-

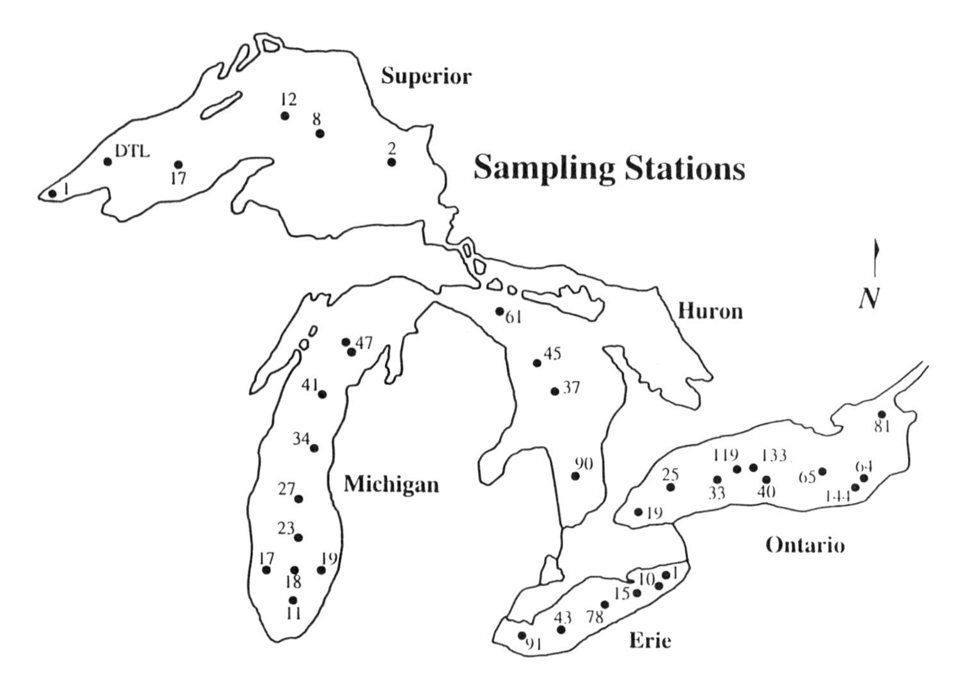

Figure 1. Great Lakes Sampling Stations

gram C-18 SPE cartridges (Varian Associates). To improve recoveries and detection limits, samples collected in 1993 and 1994 were processed using 250 mg Envi-Carb graphitized carbon black (GCB) SPE cartridges (Supelco). Prior to isolation on the SPE cartridge, samples were spiked with 500 ng of terbuthylazine and butachlor as surrogates for the triazines and acetamide herbicides. For two liter sample volumes, DEA and DIA exhibit about 40% and 90% breakthrough from the C-18 extraction cartridges, respectively. To compensate for this duplicate samples of 0.35 liters were collected in 1991-1992 to achieve greater than 85% retention of the metabolites. The GCB SPE cartridges demonstrated > 95% recovery of DEA and DIA from a 2 liter sample and were thus used for all 1993 - 1994 samples.

Switching from C-18 to GCB could have potentially introduced a systematic bias to 1994 results. To test this possibility, 9 samples from Lake Michigan and 5 samples from Lake Erie and Ontario were collected in triplicate and processed using both C-18 and GCB. GCB and C-18 results were compared using a standard paired t-test. Concentrations of atrazine and DEA showed no statistical difference between GCB and C-18 methods at the 95% confidence interval. Detection limits for DIA in a 0.35 liter sample are close to observed concentrations in the Great Lakes, thus making precise quantification difficult. Therefore, the 1994 method employing a 2 liter sample and GCB SPE cartridges produces superior results for DIA as compared to the method using C-18 .

GCB SPE cartridges were extracted with ~ 6 ml of DCM:methanol (85:15, v/v) and 2 ml of methanol; C-18 SPE cartridges were extracted with 10 ml of diethyl ether. Concentrated extracts were spiked with d_{10}-anthracene and 4,4'-dibromobiphenyl as internal standards and analyzed by gas chromatography (GC; HP 5890) coupled to a mass spectrometer (MS; HP 5971A). Details of analytical procedures are presented in Schottler et al. [*8, 15*]. Surrogate recoveries ranged from 50-135% and all sample results were corrected for surrogate recovery.

The detection limit for atrazine, alachlor, cyanazine, and metolachlor was ~5 ng/l based on a two liter sample. Detection limits for atrazine metabolites required a sample concentration of ~ 10 ng/l. For Lake Superior samples, four liters of water were passed through the extraction cartridge, and extracts were concentrated to less than 75 ul. Using this technique with the GCB SPE cartridges, the detection limit for atrazine, DEA, DIA, alachlor, and metolachlor in Lake Superior samples was lowered to ~1 ng/l. However, using these large volumes of water results in greater than 80% breakthrough of DEA and DIA from the C-18 SPE cartridge, effectively increasing the detection limit to greater than 100 ng/l. Thus, for Lake Superior, DEA and DIA were only detected in samples employing the GCB SPE cartridges (i.e. 1993 and 1994).

The relative error in any particular concentration is an important measurement when examining trends in the data. The precision of measured concentrations was evaluated though the use of duplicate samples. Three to nine sets of duplicate samples were collected from each lake, and a total of 40 sets of duplicates were collected from 1991-1994. Duplicate samples were compared as relative percent difference, (RPD):

$$RPD = (S1 - S2 * 100) / ((S1 + S2)/2) \tag{1}$$

where S1 is one sample and S2 is a duplicate of that sample. Overall RPD values for atrazine ranged from 0% to 22% with average RPD of duplicates for each lake from 1991-1994 as follows: Lake Michigan 5%; Lake Huron 1%; Lake Erie 6%; Lake Ontario 11%; Lake Superior 34%. The RPD between duplicates can be used to evaluate the precision of measured concentrations at a particular site. For example the average atrazine concentration observed for Lake Michigan is ~36 ng/l and the average RPD is 5%. Thus, for Lake Michigan, the precision of any particular concentration value is approximately $\pm$ 2 ng/l; (36 x 0.05).

Results and Discussion

Herbicide Concentrations. Atrazine was detected in 100% of the samples collected in the Great Lakes from 1991-1994. Average concentrations of atrazine ranged from about 20-35 ng/l in Lakes Huron and Michigan to 60-120 ng/l in Lakes Erie and Ontario (Table I). The highest concentrations were observed in Lake Erie at about 120 ng/l and are well below the established maximum contaminant level (MCL) for atrazine of 3000 ng/l. The lowest observed concentrations of atrazine were in Lake Superior at just above the detection limit of 1-3 ng/l. Cromwell and Thurman have detected similar atrazine concentrations in lakes on Isle Royale in Lake Superior [*19*]. While the concentrations of atrazine in Lake Superior are very low, the presence of atrazine in a non-agricultural watershed is significant because it suggests long range transport and slow water column transformation rates.

DEA was detected in all samples collected from Lakes Michigan, Huron, Erie, and Ontario, and from 1994 samples of Lake Superior (Table I). DIA was detected at trace levels (< 30 ng/l) in samples collected from 1991 to 1993. In 1994, because of the use of GCB SPE cartridges, DIA was detected in > 80% of the samples at concentrations of 5 - 35 ng/l (Table I). Metolachlor was detected in > 95% of the samples collected. Metolachlor concentrations ranged from ~ 5 ng/l in Lake Michigan to 20 ng/l in Lake Erie. Metolachlor was not detected in samples from Lake Superior. Alachlor was detected at trace levels (< 5 ng/l) in less than 50 % of samples collected in 1991-1993. In 1994, alachlor was detected in all samples (except Lake Superior) at trace concentrations (1-7 ng/l). Measured values for alachlor were approximately the same for all lakes. Alachlor concentrations might be expected to be higher in Lake Erie since this is a region of intense row crop agriculture; however, alachlor is not applied in the Ontario Province (Canada) watersheds draining to Lake Erie. Cyanazine was confirmed in <10% of the samples collected from 1991-1993. Chromatography of cyanazine in these samples was poor, making quantification difficult. Incorporating GCB solid phase extraction in 1994, the chromatography of cyanazine analyses improved. Subsequently, cyanazine was confirmed in >85% of the 1994 samples; although, analytical precision of these samples still varied greatly. Cyanazine concentrations in the Great Lakes for 1994 range from 5-40 ng/l, with concentration magnitudes in order of Erie > Ontario > Michigan > Huron (Table I). Cyanazine was not detected in Lake Superior at a detection limit of 5 ng/l.

Table I. Summary of Herbicide Concentrations in the Great Lakes 1991-1994.

	Concentration (ng/l)[a]				
	Superior	Michigan	Huron	Erie[b]	Ontario
Atrazine	**3** (1.5)	**36** (2.6)	**21** (1.6)	**60 - 120**	**75 - 100**
DEA	**2** (1.5)	**21** (2.4)	**19** (2.0)	**44 - 65**	**40 - 55**
DIA[c]	**1** (2.0)	**11** (2.5)	**11** (2.1)	**17 - 33**	**20 - 27**
Cyanazine[c,d]	**ND**	**9** (4.0)	**5** (1.5)	**20 - 55**	**15 - 35**
Metolachlor	**ND**	**5** (1.5)	**< 3** (0.9)	**10 - 35**	**10 - 30**
Alachlor[e]	**ND**	**< 5**	**< 5**	**< 7**	**< 5**

[a] Concentrations are lakewide averages of all years. Numbers in parentheses are one standard deviation of annual averages, Lakes Erie and Ontario had large temporal variations, thus a range of concentrations are given. Standard deviations Of average concentrations at individual sites are shown in Figure 2.
[b] Concentrations given for Lake Erie represent the central and eastern basins.
[c] Only 1994 values were used for DIA and cyanazine.
[d] Precision of cyanazine quantification was low creating large variations in concentrations.
[e] Quantified values of alachlor were 1-7 ng/l which is at the detection limit.

Vertical Concentration Trends. Water column concentration profiles of atrazine and DEA were constructed for all sites in 1991, 1992 and 1994; profiles of DIA were constructed for 1994 sample sites only. Vertical profiles of atrazine, DEA and DIA concentrations were statistically constant throughout the water column with much of the variation falling within the range of the RPD between duplicates. While concentrations do show some variation with depth, no trends are readily apparent.

Vertical differences in atrazine and DEA concentrations were evaluated statistically at each station and on a lakewide basis. At stations with at least six samples, epilimnetic concentrations were grouped against hypolimnetic concentrations, and evaluated using a single factor ANOVA test. None of the sites evaluated demonstrated any statistical differences between epilimnetic and hypolimnetic atrazine concentrations (For all vertical profiles evaluated, $p = 0.1$ to 0.64.) Lake-wide evaluations of vertical differences were tested using paired two tailed t-tests. Several types of pairing schemes were used: 1) average epilimnetic concentrations at each site were paired against average hypolimnetic concentration for that site; 2) average concentration of the metalimnion was paired against average epilimnetic and hypolimnetic concentrations for each site; 3) average concentrations at depths of less than 10 m for each site were paired against average concentrations of depths greater than 10 m. Using these paring criteria, sites from each lake were grouped together and evaluated each year for differences. No statistical difference between concentrations at different depths for each lake could be identified with this method (p levels for t-tests; $p = 0.22$ to 0.75 with degrees of freedom; $df = 5$ to 11.).

The lack of variation with depth suggests a water column residence time long enough to allow for thorough vertical mixing. While most vertical differences fall within the average error of measurement, some of the variation in concentration must be real. These differences cannot be predicted or correlated with temperature, depth or season and may simply represent pockets of water originating from periods of higher or lower inputs, slow mixing with tributary plumes, or depths with different loss rates.

Lateral and Temporal Variations. Since atrazine concentrations did not vary significantly with depth, average concentrations for each site were calculated. Figure 2 shows average atrazine concentrations ($\pm$1 S.D.) at each site from 1991-1994, with the data arranged in progressive order from southern Lake Michigan through eastern Lake Ontario. Because concentrations for Lake Superior are near the detection limit, and concentrations did not vary spatially, all sites were combined into one average value. (Spatial differences in Lake Superior concentrations were within the RPD for Lake Superior duplicates.)

Lakes Superior, Michigan, and Huron. Lakes Michigan and Huron show no apparent temporal or lateral trends in atrazine concentrations. Given that the use of atrazine is confined to the southern portions of both the Michigan and Huron watershed [*10, 12*], the lack of north - south variation in concentration is remarkable (Figure 2). Lateral differences in concentrations were quantified using a single factor ANOVA test, which compared average concentrations of northern stations (stations 34 to 47) against those of southern stations (stations 11 to 27) on an annual basis for 1991, 1992 and 1994. No statistical differences between southern and northern Lake Michigan could be determined (p levels for ANOVA tests = 0.34 to 0.65).

Using a volume-weighted mean of connecting channel inputs from Lake Michigan and Lake Superior, Schottler and Eisenreich [*8*] predict a steady state concentration of 20 ng/l for northern Lake Huron, with concentrations increasing in the south as the use of atrazine intensifies. Measurements made in 1991 and 1992 support this scenario. Average concentrations in northern Lake Huron are 20 ng/l and increase slightly to 22 ng/l in southern Lake Huron. However, it is difficult to determine if this lateral trend is real since the difference is within the range of the RPD between Lake Huron duplicates.

Concentrations were measured over a four-year period to examine any temporal trends, and to determine if atrazine may be accumulating in the lakes. Average concentrations from any two years were paired on a site-by-site basis for Lake Michigan and Huron and evaluated using a two-tailed paired t-test for means. Concentrations of atrazine showed no significant changes from 1991 through 1994 (p levels of t-tests = 0.06 to 0.2, df = 3 to 7). Average concentrations in 1992 were slightly higher than those of 1991 (p = 0.06), but the difference is small and the trend did not continue through 1994. The unchanging concentrations from 1991-1994 suggest that Lakes Michigan and Huron (and probably Superior) are at steady state with respect to atrazine.

Use of atrazine [*10-12*] in the Great Lakes basin, and inputs of atrazine [*1*] to the lakes varies annually depending on environmental factors, marketing, and

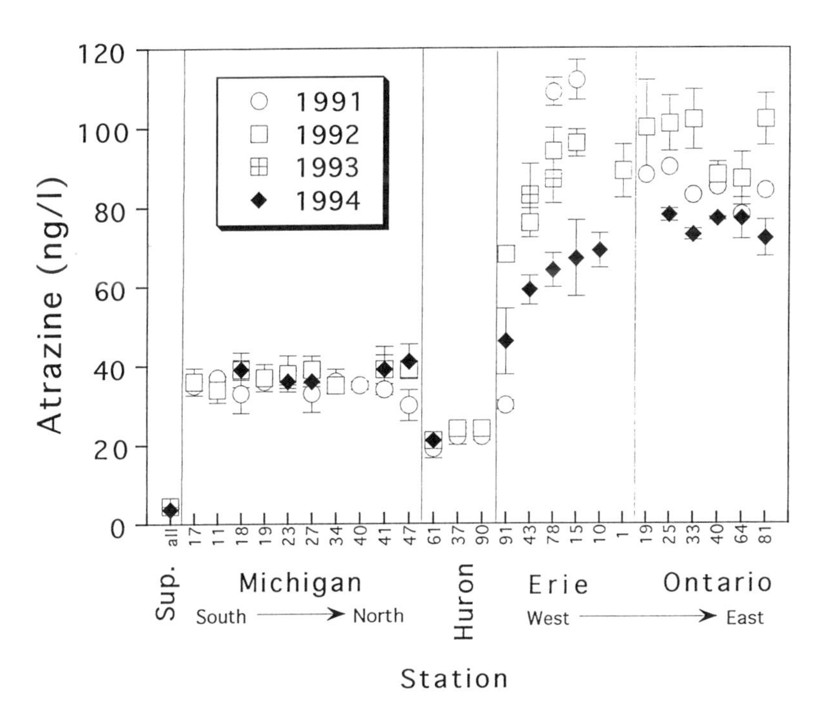

Figure 2. Temporal and spatial distributions of average atrazine concentrations. Concentrations were near detection limits throughout Lake Superior, thus all stations were averaged into one value. For 1994, in Lake Ontario, stations 119, 133, and 144 were substituted for stations 33, 40, and 64 respectively.

agricultural trends. Loadings to Lakes Huron and Michigan also vary spatially with the greatest inputs from the agricultural regions in the southern portion of the watersheds. The lack of vertical, lateral, or temporal variation in concentrations indicates that annual and spatial variations in loadings are not large enough to be reflected in the distribution of open water concentrations. To achieve the observed uniform concentrations in Lakes Michigan and Huron, the annual loadings of atrazine must be small with respect to the mass accumulated in the lake, and the *in situ* transformation rate must be slow. This hypothesis is supported by a mass balance model constructed by Schottler and Eisenreich [*20*] which demonstrates that annual loadings to Lake Michigan and Huron are less than 15% of the lakewide inventory, and that internal transformation half-lives are >5 years.

Inputs of atrazine to the Great Lakes are highly seasonal. Numerous studies [*1, 4, 15, 21-23*] have shown a spring flush phenomenon, where ~80% of the annual riverine flux of herbicide occurs in the 60 days following application. Loadings of atrazine from Midwestern precipitation have similar seasonal trends [*3, 5, 7*]. It is hypothesized that these seasonal loadings could be reflected in the open water concentrations of Lake Michigan. To test this, samples were collected from 3 to 5 stations in Lake Michigan in late April, June, late August and October of 1994. Average concentrations for each station, during each season, were generated and compared using a single factor ANOVA. No seasonal variations could be identified for Lake Michigan ($p > 0.4$). While loadings to Lake Michigan certainly vary seasonally, the mass of herbicide delivered to the lake is small in comparison to the lakewide inventory. The total annual load of atrazine to Lake Michigan is estimated at ~12 tonnes [*20*], as compared to a Lake Michigan atrazine inventory of ~175 tonnes [*20*]. Thus, seasonal variations are not large enough to be detected, and any concentration differences due to seasonal or annual variations in loading are probably within the precision of the analytical measurement.

Lakes Erie and Ontario. Lakes Erie and Ontario exhibited much greater temporal and spatial variations in atrazine concentrations (Figure 2). In each of the years sampled, Lake Erie shows a strong west to east gradient of increasing concentration. The largest inputs of atrazine occur in the western basin (station 91) as a result of tributaries draining intensively farmed land in the U.S. and Canada. However the western basin is also shallow and inputs from the Detroit River and other tributaries are poorly mixed [*24*]. The western basin station that was sampled in this study probably reflects Detroit River inputs more so than other tributary inputs. Detroit River inputs are a combination of Lake Huron and Lake St. Claire inputs, and should have lower concentrations than inputs from other tributaries. Western basin concentrations may also be lower because the residence time of water in the western basin is short (0.13 yr) [*24*] and inputs to the basin in August/September (samples were collected in early-mid September) are relatively low. Lake Erie concentrations increase to the east as tributary and Detroit river inputs become mixed in the central basin (stations 43, 78). The residence time of the central and eastern basins is also long enough (~2.5 yr) [*24*] to allow seasonal inputs to accumulate.

Concentrations in the central and eastern basins of Lake Erie show a consistent decrease from 1991 through 1994 (Figure 2). Total annual loadings to Lake Erie are about 35% of the whole lake inventory [*20*]. These large loadings (relative to inventory), coupled with a short water residence time allow Lake Erie to reflect annual and spatial variations of inputs. Decreases in tributary loadings and atrazine use contribute to the decreasing trend in atrazine concentration. Estimates of atrazine use in the Lake Erie basin show a ~25% decrease between 1990 and 1994 [*10, 11, 25*]. In addition, combined loadings calculated by Richards et al. [*25*] for five U.S. tributaries to Lake Erie during 1992 and 1993 are ~50% lower than loadings during 1990 and 1991. Thus, trends of both atrazine use and tributary loading estimates are generally consistent with the observed decreasing Lake Erie concentration trend.

Concentrations in Lake Ontario vary temporally, but do not show any consistent trends or gradients (Figure 2). In general, western Lake Ontario concentrations tend to reflect the previous year concentrations of eastern Lake Erie. That is, Lake Ontario concentrations in 1992 and 1994 are approximately equal to Lake Erie concentrations in 1991 and 1993, respectively. Lake Ontario receives ~ 85% of its annual water input from Lake Erie via the Niagara River and Welland Canal. In addition, ~65% less atrazine is used in the Lake Ontario basin than the in the Lake Erie basin [*10-12, 20*]. This information and the observed concentration patterns suggest that Lake Ontario concentrations are influenced primarily by inputs from Lake Erie. To produce the observed variation in Lake Ontario concentrations two factors must apply: 1) Annual loadings of atrazine to the lake must be a significant percentage of the mass in the lake; and 2) The water retention time of the lake and atrazine transformation rate must be long enough to allow atrazine accumulation to reflect trends, but not so long that the accumulated mass outweighs the importance of inputs. This relationship between loadings, hydrology, and accumulation rates on Lake Ontario concentrations is supported by Schottler and Eisenreich's mass balance results [*20*], in which the water column half-life of atrazine is estimated to be about 5 years, and the annual inputs to be about 15 % of the lake-wide inventory.

DEA and DIA Concentrations and Ratios

Desethylatrazine (DEA) was detected in 100% of the samples from 1991-1994, with concentrations ranging from near detection limits in Lake Superior (~2 ng/l) to nearly 70 ng/l in Lakes Erie and Ontario in 1992-1993. Desisopropylatrazine (DIA) was detected in all samples analyzed using GCB SPE cartridges. DIA concentrations were ~10 ng/l in Lakes Michigan and Huron, and ~20 ng/l in Lakes Erie and Ontario. DEA and DIA concentrations in Lake Superior were 2 to 4 ng/l, which is very close to the detection limit of ~ 1 ng/l. Temporal, spatial, and seasonal trends in DEA and DIA concentrations were similar to those observed for atrazine.

To compare Great Lakes atrazine and transformation product concentrations to other systems it is useful to use ratios of observed concentrations. Two important ratios are the DEA to atrazine ratio, DAR, and the DIA to DEA ratio,

D^2R, as defined by Thurman et al. [*2, 26*]. The temporal and spatial similarity between atrazine and transformation product concentrations produces a remarkably constant DAR and D^2R throughout the lakes (Figures 3 & 4). The DAR is consistently about 0.5 to 0.6 in the Great Lakes from 1991 to 1994 (Figure 3A), and the D^2R is about 0.45 to 0.55 (Figure 3B). Lake Huron is an exception, and will be discussed later. The consistency in the proportionality between atrazine and DEA is demonstrated by the box-plots in Figure 4. The uniformity of the DAR between the lakes and within the lakes suggests that: 1) atrazine and DEA have similar sources and environmental processing in the Great Lakes system, and/or 2) atrazine concentrations are driving the DEA concentrations.

The environmental processing, and source functions for DEA and atrazine have been shown to be similar [*1, 3-5, 15, 17, 21, 26-29*]. Thus, the DAR values of 0.6 observed throughout the Great Lakes could be used to suggest that the sources to each of the lakes are similar and indicative of a source with DAR values of about 0.6. While DAR values of ~0.1 to 0.8 have been measured in precipitation, groundwaters and rivers [*3, 4, 14, 15, 19, 21, 26, 30, 31*], a comparison of the DAR within the lakes to the DAR expected from annual tributary and precipitation inputs reveals that these sources alone cannot produce a DAR of 0.6 in the Great Lakes. Since the observed DAR is greater than the DAR predicted from inputs, it is hypothesized that *in situ* transformation of atrazine to DEA is supplying the additional DEA needed to maintain a DAR of ~0.6. A discussion of the DAR expected from source inputs and the relationship to the observed DAR follows.

Tributaries: DAR values of >0.4 are common in Midwestern tributaries during the late summer and fall [*4, 15, 21, 26*], while DAR values of <0.1 are observed in early post application flow periods. Water residence times of the Great Lakes are long, and integrate the annual overall DAR of tributary inputs. Thus, the annual mass loading of DEA divided by the annual mass loading of atrazine from tributaries would be the overall representative DAR of tributary inputs, equation 2.

$$\text{Annual DAR}_{\text{tributary}} = \Sigma\ \text{DEA load}_{\text{tributary}} \Big/ \Sigma\ \text{Atrazine load}_{\text{tributary}} \qquad (2)$$

Annual DAR values of this type can be calculated from several studies. Annual mass loadings of DEA and atrazine were determined by Goolsby et.al. [*4*] at four locations along the Mississippi river, and four major tributaries to the Mississippi River in 1991. Annual overall DAR values calculated from their study were 0.09 to 0.21. Schottler et al. [*15*] calculated annual DAR values for the Minnesota River of 0.22 and 0.24 in 1990 and 1991, respectively. Bodo et al. [*6*] have calculated annual atrazine and DEA loads for three Canadian tributaries to Lake Erie from 1981-1988. Annual DAR values for these tributaries range from 0.19 to 0.5, with an average annual DAR of ~0.35. Results from these studies suggest that if tributaries were the major source of both atrazine and DEA to the Great Lakes, DAR values for the lakes should be 0.2 to 0.3. Since tributaries account for > 70% of atrazine inputs to all the Great Lakes except

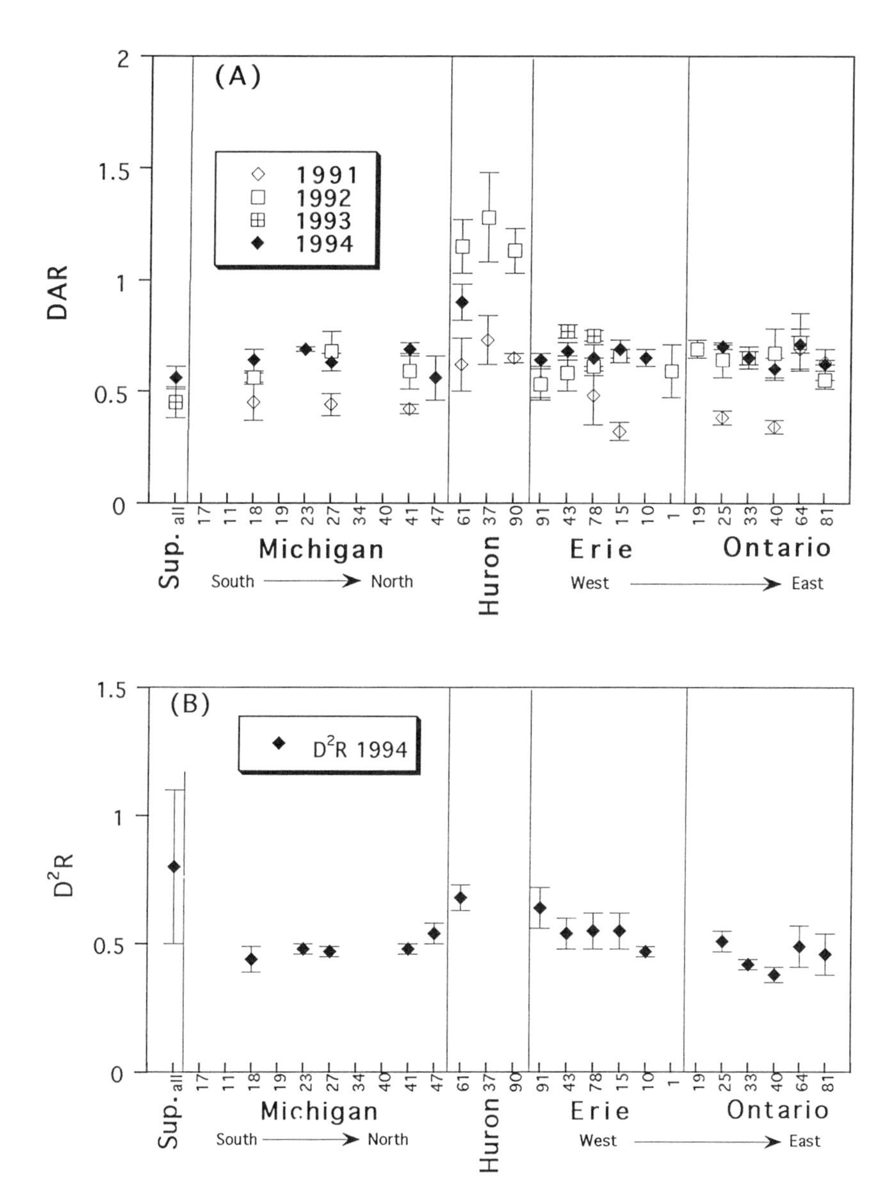

Figure 3. (**A**) Temporal and spatial distribution of the DEA to atrazine ratio, DAR. (**B**) Spatial distribution of the DIA to DEA ratio, D^2R.

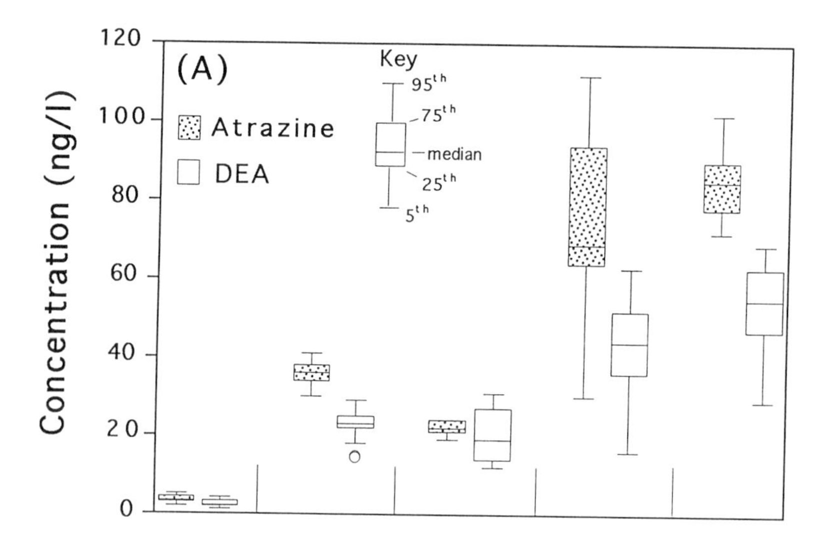

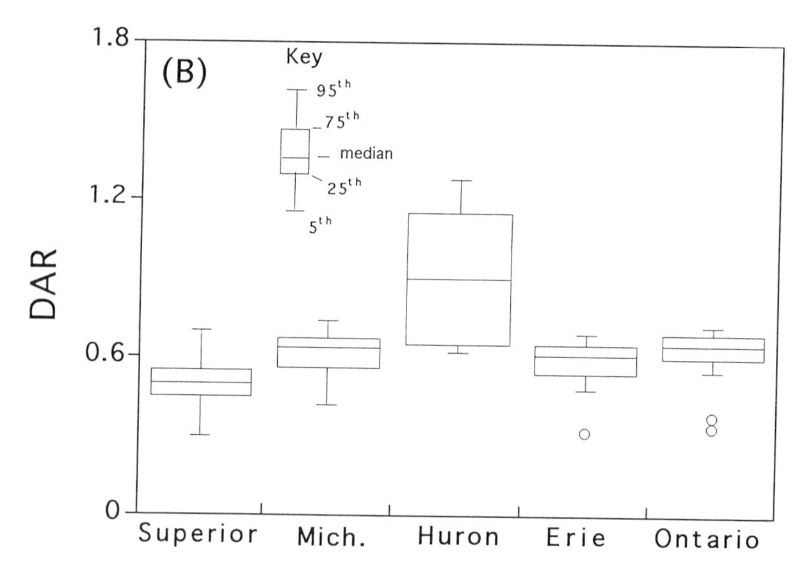

Figure 4. **(A)** Box plot of atrazine and DEA values for each lake. The box encompasses the 75th and 25th percentile of data, the bars define the 95th and 5th percentile, the line inside the box is the median concentration, and the open circles are outliers. **(B)** Box plot of DAR values for each lake; demonstrating the consistency in proportionality of DEA to atrazine concentrations.

Superior [*20*], there must either be an additional source of DEA to the Lakes or a loss mechanism of atrazine to produce the observed DAR values of ~0.6.

Groundwaters: DAR values of 0.4 or greater are typical of groundwaters in the Midwest [*15, 32*]. However, groundwaters generally account for <15% of the total water budget to the Great Lakes [*33, 34*]. Groundwaters throughout the Midwest have shown that atrazine and DEA are detected much less frequently in groundwaters compared to surface waters, and generally in lower concentrations [*30, 32, 35*]. Thus, groundwater contributions to the Great Lakes should be small compared to tributary and precipitation loadings, and have little impact on the observed DAR values.

Precipitation: DAR values measured by Goolsby et al. [*31*] for Midwestern precipitation range from ~0.1 to > 1. Cromwell and Thurman [*19*] calculated annual precipitation fluxes of atrazine and DEA to lakes on Isle Royale in Lake Superior. Annual overall DAR values based on these fluxes are 0.28, 0.47 and 0.41 for 1992, 1993, and 1994, respectively. Atmospheric inputs account for > 95% of the annual loading of atrazine to Lake Superior [*19*]. Thus the DAR value of ~0.5 observed in the open waters of Lake Superior is likely representative of atmospheric deposition. Mass loadings from precipitation to the other lakes however, are not large enough to produce a DAR value of > 0.5. This can be demonstrated using Lake Michigan as an example. Based on the mass balance model of atrazine in the Great Lakes [*20*], tributaries add about 9 tonnes per year of atrazine to Lake Michigan, and atmospheric inputs an additional 2.8 tonnes. If an annual DAR of 0.25 is applied to the tributaries, then about 2.3 tonnes per year of DEA would enter the lake from tributaries. Using a high estimate of 0.8 for the DAR in precipitation would yield an annual precipitation loading for DEA of 2.2 tonnes. Assuming atrazine and DEA are lost from the lake at equal rates, the combination of tributary and precipitation inputs would produce a DAR less than 0.4. Thus, atmospheric inputs may increase the DAR, but they are not large enough to increase the DAR to the observed value of ~0.6.

The above discussion demonstrates that inputs from tributaries, groundwater, and precipitation probably cannot account for the DAR values of >0.5 in the Great Lakes. Therefore it is necessary to invoke either a preferential loss of atrazine or an additional source of DEA to account for the high DAR values. It is difficult to identify a preferential external source of DEA other than those previously discussed and the only loss processes identified for atrazine in the Great Lakes are outflow, sedimentation and internal transformation. Outflow would affect DEA and atrazine equally. Removal by sedimentation has been shown to account for << 1% of all annual outputs [*16, 20*]. Even increasing values for sedimentation by a factor of 10 does not significantly change the importance of sedimentation as a removal process. This leaves internal transformation as the source of the high DAR values.

There are three ways in which internal transformation could produce elevated DAR values: 1) The rate of atrazine transformation is faster than the rate of DEA transformation; 2) additional DEA is produced *in situ* through transformation of atrazine; or 3) a combination of 1 and 2. It is difficult to determine the relative rates of transformation of atrazine and DEA.

Transformation rates for atrazine in the Great Lakes have been determined [*20*], but there were insufficient data to determine rates for DEA transformation. Evoking a faster degradation rate for atrazine over DEA in the open waters would produce an increased DAR; however, there is no apparent reason for this to be true since, in soils, the rate of DEA transformation is faster than the rate of atrazine transformation [*36*].

The most likely scenario to produce the high DAR values involves *in situ* transformation atrazine to DEA within the lakes. Because Lake Michigan has no connecting channel inputs, it is the simplest lake to illustrate the potential for atrazine conversion to DEA. Inputs of atrazine to the lake are: 9 tonnes from tributaries, and 2.8 tonnes from total atmospheric inputs [*20*]. Using a DAR of 0.25 for tributaries and 0.8 for atmospheric inputs would result in annual loadings of about 4.5 tonnes of DEA. From these inputs alone, the lake would have a DAR of ~0.4 {4.5/(9+2.8)}, and an additional 2.5 tonnes of DEA is required to produce an annual DAR of 0.6. Schottler and Eisenreich [*20*] estimate that about 8.9 tonnes of atrazine in Lake Michigan are lost each year through *in situ* transformation. Hence, 28 % (2.5/8.9) of atrazine lost annually through transformation must be converted to DEA. Similarly, since Lake Michigan has an atrazine inventory of ~175 tonnes [*8, 20*], and assuming that DEA and atrazine degrade at the same rate, then 1.4% (2.5/175) of the atrazine inventory must degrade annually to DEA to maintain a DAR of 0.6. Similar calculations were done for the other Great Lakes to estimate the necessary percentage of atrazine inventory converted *in situ* to DEA required to maintain a DAR of 0.6. The percent of atrazine inventory that must be transformed to DEA annually in Lakes Ontario, Erie and Huron was 0.9%, 4% and 5%, respectively.

The above calculations are estimates for the amount of atrazine converted to DEA in the Great Lakes. The amount of atrazine transforming to DEA depends on the transformation rate of DEA. If DEA degrades faster than atrazine then a greater percentage of the atrazine lost annually must transform to DEA. This transformation scenario also assumes that atrazine and DEA concentrations are at steady state (a reasonable assumption based on the temporal trends). If the percent of atrazine inventory degrading to DEA in the above example for Lake Michigan were slightly greater than 1.6% then the DAR should slowly increase over time. The above calculations also assume a DAR of 0.8 for atmospheric inputs, which may be a high estimate based on the atmospherically driven DAR in Lake Superior of 0.5. A high estimate for the DAR of atmospheric inputs was deliberately chosen to illustrate the point that external sources alone could not produce the DAR values observed in the lakes. If a DAR value of 0.5 is used instead of 0.8 for atmospheric inputs in the above calculations, the estimates for the percentage of atrazine inventory converted annually to DEA would only increase by a factor of <1.3.

A comparison of DAR values shows that the DAR in Lake Huron is conspicuously higher than the other lakes (Figure 4). DAR values near unity were measured in Huron for 1992 and 1994. It is difficult to identify a source of DEA or loss process of atrazine that is unique or magnified in Lake Huron. Connecting channel inputs of DEA from Lakes Michigan and Superior certainly contribute to the high DAR value. However, ~ 50% of annual atrazine inputs (and probably DEA) to Lake Huron are from tributaries. Thus a DAR closer to 0.4 would be expected based on tributary and connecting channel inputs. Adding in precipitation inputs of DEA, the DAR would still only approach 0.6.

The most likely source of the high DAR values is an accelerated rate of *in situ* transformation of atrazine to DEA. Either a greater percentage of the atrazine inventory is converted to DEA annually in Lake Huron or the subsequent degradation of DEA is slower than in the other Great Lakes. Performing calculations similar to those above predicts that about 5% of the atrazine inventory in Lake Huron must be converted to DEA annually to maintain a DAR of 0.8 (assuming DEA and atrazine have similar transformation rates within the lake). This is not an unreasonable percentage, since about 7% of the atrazine inventory in Lake Huron is estimated [*20*] to be lost annually through internal transformation each year. It is also possible that the relative degradation rate of DEA in Lake Huron is slower than the rate in the other lakes. Without a better accounting of the DEA inputs to Lake Huron it is impossible to determine the exact cause of the high DAR. However, conversion of atrazine to DEA at some rate must be occurring, and the internal processing must be such that the subsequent accumulation of DEA in Lake Huron is greater than in the other Great Lakes.

The ratio of DIA to DEA (D^2R) in the Great Lakes ranges from ~0.45 to 0.6 (Figure 3B), which is similar to ratios measured in agricultural runoff. D^2R values of ~0.45 have been measured in surface and shallow unsaturated zone runoff from field plots by Mills and Thurman, [*37*]. On a regional scale, Thurman et al.[*26*] have identified a strong correlation between DIA and DEA in Midwestern rivers. From samples collected at more than 55 sites in 1989 and 1990, Thurman et al. [*26*] estimate a D^2R in runoff of 0.6 ± 0.1. Previous discussion of Great Lakes DAR values demonstrated that DEA is generated *in situ* from the transformation of atrazine. If this is true, and inputs to the lake have a D^2R of 0.6, then DIA must also be being generated *in situ* to maintain the observed D^2R of ~0.6. If DIA were not being formed *in situ*, the internal production of DEA would reduce the D^2R to less than the 0.6 value expected from external loadings.

Discussion

The presence of atrazine, DEA and DIA throughout the Great Lakes at ng/l concentrations supports the notion of the Great Lakes as long-term integrators of the environmental processing of herbicides. Temporal and spatial trends indicate a long water column residence time for atrazine, and suggest that atrazine has accumulated in the lakes such that current annual loadings are a small fraction of the overall lake-wide inventories. Ratios of transformation products to parent atrazine suggest that atrazine is being converted *in situ* to DEA and DIA. A comparison of the observed DAR values against the DAR value expected from tributary and precipitation loadings, predicts that an estimated 1-5% of the atrazine inventory in each lake is converted *in situ* to DEA annually.

Dealkylation of atrazine is a biotic process [*36, 38-42*]. DIA and DEA observed in rivers is generally thought to be the result of microbial processes occurring in the soil, with limited degradation in alluvial aquifers and the rivers. Thus, the *in situ* conversion of atrazine to DEA and DIA, suggests that at least a portion of the atrazine transformation in the Great Lakes is also biological. More information is needed on fluxes of DEA and DIA to permit calculations of DEA and DIA transformation rates. Once these values have been established, it will be possible to more accurately quantify the portion of atrazine that is converted to DEA and DIA *in situ*, and further define the role of biodegradation in large aquatic systems.

Acknowledgments

This research was supported by a grant from the US Environmental Protection Agency, Great Lake National Program Office (EPA/X995248 and EPA/GL995592-01-1); Project Officer Glenn Warren, Principle Investigator Steve Eisenreich. We wish to express our appreciation to the Captain and crew of the EPA RV Lake Guardian, the laboratory assistance of Joe Hallgren, and Mari-Pat Von Feldt, and to our colleagues, especially Paul Capel and Steve Larson, for many helpful comments. We also sincerely thank Captain R. Ingram and the crew of the EPA RV Lake Guardian for their cooperation and assistance. We especially note the assistance given by Kelley Jenkins.

Literature Cited

1. Richards, R.P.; Baker, D.B.; *Environ.Tox. Chem.* **1993**, *12*, pp. 13-26.
2. Thurman, E.M.; Goolsby, D.A.; Meyer, M.T.; Kolpin, D.W. *Environ Sci. Technol.* **1991**, *25*, pp. 1794-1796.
3. Goolsby, D.A.; Thurman, E.M.; Pomes, M.L.; Meyer, M.; Battaglin, W.A. In *Selected Papers on Agricultural Chemicals in Water Resources of the Midcontinental United States - U.S. Geological Survey,* Open-File Report 93-418; D.A. Goolsby, L.L. Bolyer, and G.E. Mallard Eds.; U.S. Department of the Interior: Denver, CO, 1993, pp. 75-80.
4. Goolsby, D.A; Battaglin, W.A. In Selected Papers on Agricultural Chemicals Water Resources of the Midcontinental United States - U. S. Geological Survey - Open-File Report 93-418; D.A. Goolsby, L.L. Boyer, and G.E. Mallard Eds.; U.S. Department of the Interior: Denver, CO, 1993, pp. 2-22.
5. Glotfelty, D.E.; Williams, G.H.; Freeman, H.P.; Leech, M.M. In *Long Range Transport of Pesticides,* D.A. Ed.; Lewis Publishers Inc.: Chelsea, MI, 1990, p. 199-222.
6. Bodo, B.A.; *Environ.Tox. Chem.* **1991**, *10*, pp. 1105-1121.
7. Richards, R.P.; Kramer, J.W.; Baker, D.B.; Krieger, K.A. *Nature.* **1987**, *327*, pp. 129-131.
8. Schottler, S. P.; Eisenreich, S.J. *Environ. Sci. and Technol..* **1994**, *28*, pp. 2228-2232.

9. General Accounting Office. *Issues Concerning Pesticides Used in the Great Lakes Watershed*, GAO/RCD-93-128; Resources, Community and Economic Development Division, Washington D.C., June 14, 1993, pp. 1-20
10. Gianessi, L.P. *U.S Pesticide Use Trends: 1966-1989;* Report for U.S. Environ. Protection Agency, Office of Policy Analysis, Jan.,1991, pp. 1-12.
11. Gianessi, L.P.; Anderson, J.E. *Pesticide Use in U.S. Crop Production*, National Center for Food and Agricultural Policy, Washington, D.C. Feb. 1995
12. Kirschner, B.; In*1993-1995 Priorities and Progress Under the Great Lakes Water Quality Agreement*; International Joint Commission, Windsor, Canada: 1995.
13. Goolsby, D.A., Coupe, R.C., and Markovchick, D.J. *Distribution of selected herbicides and nitrogen in the Mississippi river and its major tributaries, April through June 1991*; Water Resources Investigation Report 91-4163; U. S. Geological Survey: Denver, CO, 1991, pp. 35
14. Muir, D.C.G., Yoo, J.Y., and Baker, B.E.; *Arch. Environ. Contam. Tox.* **1978**, *7*, pp. 221-235.
15. Schottler, S.P., Eisenreich, S.J., and Capel, P.D.; *Envriron.Sci.Technol.* **1994**, *28*, pp. 1079-1089.
16. Pereira, W.E.; Rostad, C.E.; *Environ. Sci. Technol.* **1990**, *24*, pp. 1400-1406.
17. Squillace, P.J.; Thurman, E.M.; *Environ. Sci. and Technol* . **1992**,*26*, pp. 538-545.
18. Larson, S.J.; Capel, P.D.; Goolsby, D.A.; Zaugg, S.D.; Sandstrom, M.W.; *Chemosphere.* **1995**, *31*, pp. 3305 -3321.
19. Cromwell, A.E.; Thurman, E.M. In *Toxic Substances Hydrology Program: Proceedings of the Sept. 20-24, 1993 Meeting, Colorado Springs, CO.*; Morganwalp D.W., Aronson, D.A., Ed.; U.S.G.S WRI 94-4105, Tallahassee, FL, 1996, Vol. 1, pp 423-428.
20. Schottler, S.P. and Eisenreich, S.J.; A mass balance model to quantify atrazine sources, transformation rates and trends in the Great Lakes, *Environ. Sci. Technol.* 1997, In Press.
21. Thurman, E.M.; Goolsby, D.A.; Meyer, M.T.; Mills, M.S.; Pomes, M.L.; Kolpin, D.W. *Environ. Sci. Technol* .**1992**, *26*, pp. 2440-2447.
22. Baker, D.B.; Richards, R.P. In *Long-Range Transport of Pesticides*, D.A. Kurtz Eds.; Lewis Publishers: Chelsea, MI, 1990, pp. 241-271.
23. Spalding, R.F.; Snow, D.D. *Chemosphere.* **1989**, *19*, pp. 1129-1140.
24. Bolsenga, S.J. and Herdendorf, C.E., Eds.; *Lake Erie and St. Clair Handbook*: Wayne State University Press: Detroit, MI, 1993, pp. 467.
25. Richards, P.R..; Baker, D.B.; Kramer, J.W.: Ewing, E.D. *J. Great Lakes Research*,**1996**, 22, pp. 41-428
26. Thurman, E.M.; Meyer, M.T.; Mills, M.S.; Zimmerman, L.R.; Perry, C.A. *Environ. Sci. and Technol.* **1994**, *28*, pp. 2267-2277.
27. Muir, D.C.G.; Grift, N.P.; *J. Environ. Sci. Health.* **1987**, *22,* pp. 259-284.
28. Muir, D.C., In *Environmental Science of Herbicides*, Grover,R.; Cessna, A.J. Eds.; CRC Press Inc.: Boca Raton, FL, 1991, p. 1-88.
29. Adams, C.D.;Thurman, E.M.; *J. Environ. Qual..* **1991**, *20*, pp. 540-547.

30. Squillace, P.J., Thurman, E.M.; Furlong, E.T.*Water Resources Research.* **1993**, *29*, pp. 1719-1729.
31. Goolsby, D.A.; Schribner, E.A.; Thurman, E.M.; Pommes, M.L.; Meyer, M.T. In. *Data on selected herbicides and two triazine metabolites in precipitation of the Midwest and Northeastern U.S. 1990-1991*, U.S.G.S. Open File Report 95-469, U.S. Department of the Interior: Lawrence, KA, 1995 pp. 1-341
32. Kolpin, D.W.; Goolsby, D.A.; Aga, D.S.; Iverson, J.L.; Thurman, E.M., In *Selected Papers on Agricultural Chemicals in water Resources of the Midcontinental United States* , U.S.G.S. Open-File Report 93-418, Goolsby, D.A.; Boyer, L.L.; Mallard, G.E. Eds.; U.S. Department of the Interior: Denver, CO, 1993, p. 64-71.
33. Cartwright, K.; *J. Hydrology.* **1979**, *43*, pp. 67-78.
34. Great Lakes Basin Commission-U.S.G.S.; *Great Lakes Basin Framework Study, Appendix 3: Geology and Groundwater,* U.S. Dept. of Interior Public Information Office: Ann Arbor, MI. **1975**, pp. 1-11.
35. Cohen, S.Z.; Eiden, C.; Lorber, M.N. In *Evaluation of pesticides in groundwater*; Garner, W.Y., Honeycutt, R.C., Nigg, H.N. Eds.; ACS Symposium Series 315; American Chemical Society: Washington DC, **1984**.
36. Giardi, M.T.; Giardina, M.C.; Filacchioni, G. *Agric. Biol. Chem..* **1985**, *49*, pp.1551-1558.
37. Mills, M.S.; Thurman, E.M. *Environ. Sci. Technol..* **1994**, *28*, pp. 600-605
38. Khan, S.U.; Marriage, P.B. *J. Agric. Food Chem.* **1977**, *25*, pp. 1408-1413.
39. Giardina, M.C.; Giardi, M.T.; Filacchoini, G.; *Agric. Biol. Chem.* **1980**, *44*, pp. 2067-2072.
40. Giadina, M.C.; Giardi, M.T.; Filacchoini, G. *Agric. Biol. Chem.* **1982**, *46*, pp.1439-1445.
41. Winkelmann, D.A.; Klaine, S.J. *Environ. Sci. Technol.* **1991**, *10*, pp. 335-345.
42. Mandelbaum, R.T.; Wackett, L.P.; Allan, D.L. *Environ. Sci. Technol.* **1993**, *27*, pp. 1943-1946.

Chapter 19

Summary of Ciba Crop Protection Groundwater Monitoring Study for Atrazine and Its Degradation Products in the United States

K. Balu[1], P. W. Holden[2], L. C. Johnson[2], and M. W. Cheung[3]

[1]Waterborne Environmental Inc., 7031 Albert Pick Road, Suite 100, Greensboro, NC 27409
[2]Waterborne Environmental Inc., 897-B Harrison Street, SE, Leesburg, VA 20175
[3]Novartis Crop Protection, Inc., P.O. Box 18300, Greensboro, NC 27419

Ciba Crop Protection (Ciba) has completed a private well monitoring program in cooperation with nineteen states to determine the levels of atrazine and its degradation products in groundwater in vulnerable regions of major use areas within the United States. The nineteen states were selected based on high atrazine use. In each state, between 30 - 200 wells were selected for monitoring based on high atrazine use, groundwater vulnerability and previous atrazine detections. Along with atrazine, the following major degradation products were monitored: desethylatrazine, deisopropylatrazine, diaminochlorotriazine, hydroxyatrazine, desethylhydroxyatrazine, deisopropylhydroxy-atrazine and ammeline.

A total of 1,505 wells were sampled and analyzed for chlorotriazines by GC/MS and hydroxytriazines by LC/MS/MS at the Limit of Quantitation/Limit of Detection (LOQ/LOD) of 0.10 ppb for each analyte. Of the 1,505 wells analyzed, 76.1% showed no detections of atrazine and 0.5% had atrazine concentrations exceeding the EPA Maximum Contaminant Level (MCL) of 3 ppb. Frequencies of detections of desethylatrazine and diaminochlorotriazine were similar to atrazine (28.8% and 24.1%), respectively. Deisopropylatrazine was detected in 14.9% of the wells sampled. The hydroxytriazine degradation products, hydroxyatrazine, desethylhydroxyatrazine, deisopropylhydroxy-atrazine and ammeline, were detected in 4.5%, 2.8%, 0.3% and 0.5% of the wells, respectively.

The wells selected for this study were biased for positive detections of atrazine and its degradation products since they were located in areas with high groundwater vulnerability or they had

> previous detections of atrazine. Hence, these monitoring data cannot be used to extrapolate exposure estimates accurately for the general population served by rural drinking water wells.

Although an extensive data base is currently available on atrazine levels in groundwater (*1-11*), data on the degradation products of atrazine are relatively scarce. Atrazine is degraded in the soil environment by microbial degradation with the formation of dealkylated chlorotriazines, desethylatrazine, deisopropylatrazine, and diaminochlorotriazine. Desethylatrazine and deisopropylatrazine have been monitored in groundwater in a very limited number of studies (*12-13*). Atrazine is also degraded by abiotic processes to hydroxyatrazine which is quite polar and remains bound in the soil matrix (*14*). Additional degradation products of atrazine that have been found in soil are dealkylated hydroxytriazines (desethylhydroxyatrazine, deisopropylhydroxy atrazine, and ammeline), which are even more polar than hydroxyatrazine. Structures of atrazine and its degradation products are provided in Figure 1. No monitoring data have been reported in the literature for the hydroxytriazine degradation products of atrazine in groundwater.

The objective of this study was to determine the levels of atrazine and its chloro- and hydroxytriazine degradation products in groundwater samples collected from broad geographic regions in the United States in cooperation with various state agencies.

Nineteen states were selected for monitoring which included: Florida, Hawaii, Illinois, Indiana, Iowa, Kansas, Kentucky, Louisiana, Maryland, Michigan, Minnesota, Mississippi, Ohio, Pennsylvania, Texas, Virginia, Washington, West Virginia and Wisconsin. These states were selected based on 1) high atrazine use within the state, 2) presence of areas considered vulnerable to ground-water contamination where atrazine is used, and 3) the need to cover broad geographic areas in the United States. Well selection and sampling were conducted in cooperation with each state's department of agriculture or its affiliates such as land grant universities.

Experimental

Well Selection Criteria: The wells included in this study were selected through discussions with the representatives of the state agencies and its affiliates. The wells selected for this study were not based on a statistically defined random selection and hence, it is not possible to accurately extrapolate the results of this study to the general population of wells in the United States. The sampling design for this study was a targeted process of well selection meeting certain criteria chosen by each state following the general guidelines summarized below:

a) Wells were selected in areas that met hydrogeologic vulnerability criteria for the requirements of a small-scale retrospective groundwater study (e.g., permeable soils, shallow water tables defined as those less than 50 feet from land surface, absence of layers in the vadose zone with low permeability, high product use areas, etc.).

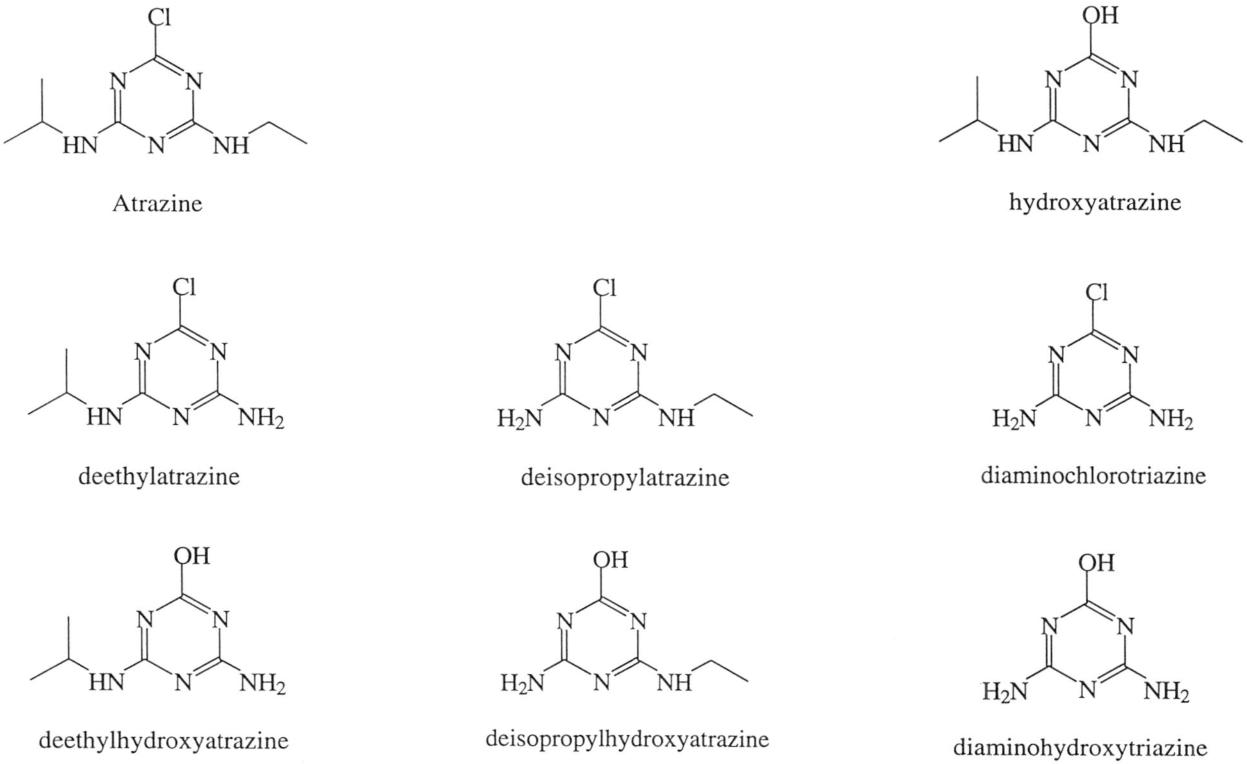

Atrazine

hydroxyatrazine

deethylatrazine

deisopropylatrazine

diaminochlorotriazine

deethylhydroxyatrazine

deisopropylhydroxyatrazine

diaminohydroxytriazine

Figure 1. Structure of Atrazine and Major Degradation Products

b) Priority was given in some states to wells with previous detections of atrazine. This criterion was used in order to obtain sufficient data for the degradation products of atrazine under conditions when atrazine is detected in the wells. Hence, the well selection in this study is biased towards wells with previous detections of atrazine. A number of states used immunoassay analysis as a screening tool to facilitate selection of wells. However, many wells with no previous detections of atrazine were also included in areas of high groundwater vulnerability.
c) Priority was given for selection of rural drinking water wells adjacent to farms. In some states, observation or irrigation wells close to atrazine use areas were also chosen for monitoring.
d) To the extent possible, wells with known point-source contamination due to spills, back-siphoning incidents, or damaged well casings were avoided. However, it is difficult to avoid this problem entirely in a study of this magnitude. A source investigation was conducted separately by Ciba for wells where atrazine concentrations exceeded the life-time MCL value of 3 ppb.
e) Approximately 10% of the wells in many states were resampled to address temporal variability of the concentrations.
f) To the extent possible, wells were sampled in various regions of each sate to obtain a representative cross-section of the state.

Field Phase: The general approach for study initiation in each state was quite similar. A preliminary discussion was held with the key staff to discuss the details of the monitoring program. Following this, each state developed a state-specific protocol. Well sampling was performed by trained staff from the state agency or its designated group, except in Florida, Minnesota and Texas where the sampling was done by independent consultants. A well sampling training session for selected personnel in each state was conducted after completion of the state-specific protocol prior to the initiation of sampling. The training session included an explanation of the purpose and design of the sampling program, an overview of the purpose and need for Good Laboratory Practices (GLP), a discussion of the study's Standard Operating Procedures (SOPs), and an introduction to the field equipment and data collection forms. Every effort was made to ensure that the field phase of the study was conducted in compliance with the GLP requirements of the Federal Insecticide, Fungicide and Rodenticide Act (FIFRA).

Sampling was initiated by having the well owner sign a permission-to-sample form. The wells in each state were identified by a unique well identification number along with the latitude/longitude of the wells. Data collection forms were used to standardize the information collected for all wells sampled. After the well-water system was chosen, the system was purged until the water quality was constant as determined by monitoring three physico chemical parameters at five-minute intervals: pH, electrical conductivity and temperature. Purging was considered complete when all three parameters were stable between two consecutive five-minute sampling intervals. This purging process is identical to procedures used during EPA's National Pesticide Survey (NPS) of drinking water wells (*2*). After purging the well, samples were collected in two 1-liter amber glass bottles. Samples were shipped to Ciba under refrigerated conditions using frozen blue ice packs in specially designed insulated containers.

Analytical Phase: Residues of parent atrazine and the chlorotriazines, desethylatrazine, deisopropylatrazine, and diaminochlorotriazine were determined using a gas chromatograph with a mass selective detector (GC/MSD). In this method, sodium chloride was added to the water sample. The sample was then buffered using a pH 10 buffer solution, and partitioned with ethyl acetate. The organic phase was dried in anhydrous magnesium sulfate and evaporated to dryness. The sample was then reconstituted in acetone and analyzed by gas chromatography-mass spectrometry (GC/MSD) using selected ion monitoring for quantitation.

Water samples were initially analyzed for hydroxytriazine degradates: hydroxyatrazine, desethylhydroxyatrazine, deisopropylhydroxyatrazine and ammeline, by Alta Analytical Laboratories using a thermospray liquid chromatography/mass spectrometry/mass spectrometry (LC/MS/MS) system (Finnigan TSQ-700) and quantified by selected ion monitoring (SIM). In this method, water samples were extracted by passing an acidified sample through SCX solid phase extraction (SPE) column and eluting with a methanol-water-ammonia mixture (75:20:5). The method was subsequently revised to a direct aqueous injection, whereby an aliquot of the sample was evaporated under nitrogen and reconstituted with HPLC-grade water for LC/MS/MS analysis. Because of problems of quantitation of ammeline by direct aqueous injection, this method was further revised using the SCX solid phase extraction cleanup followed by analysis by LC/MS/MS system.

The Limit of Determination (LOD) for atrazine and the chlorotriazine degradation products by GC/MS and the hydroxytriazine degradation products by LC/MS/MS was 0.10 ppb. The method performance for the chloro- and hydroxytriazine degradation products was demonstrated by analysis of laboratory fortification samples in each analytical set. The residue results in the field samples were reported after corrections for the procedural recovery in each analytical set.

RESULTS AND DISCUSSION

National Summary: The frequency distributions of atrazine and the chlorotriazine degradation products, desethylatrazine, deisopropylatrazine and diaminochlorotriazine, for all wells from the nineteen states participating in the atrazine study are shown in Figure 2. These results show that out of 1,505 wells analyzed, 23.9% showed detections of atrazine. Desethylatrazine was detected in more wells than atrazine (28.8%), while diaminochlorotriazine detections were similar to atrazine detections (24.1%). Fewer detections of deisopropylatrazine were observed with 14.9% of the wells having detections. Greater frequency of atrazine and metabolite detections in this study was anticipated because the targeted well selection process, which included sampling in vulnerable groundwater regions and previous detections of atrazine, was biased. In contrast, the NPS by the EPA was conducted as a statistically designed study by random selection of community water systems and rural domestic wells. In the NPS study, atrazine was detected in 0.7% of the rural domestic wells at the detection limit of 0.12 ppb (*2*).

Eight wells (0.5%) out of the 1,505 in the Ciba study had atrazine concentrations exceeding the MCL federal standard of 3.0 ppb. Of these eight wells, three were located in Wisconsin, two were in Kansas and one well each was located in Indiana, West Virginia and Minnesota. The highest atrazine concentration (12 ppb) was found in a well in Wisconsin. Source investigations of these eight wells through site visits were conducted. Follow-up sampling of these wells confirmed that the detections exceeded 3 ppb. The three wells in Wisconsin had shown detections exceeding 3 ppb in the previous sampling program by the Wisconsin Department of Agriculture Trade and Consumer Protection (DATCP) (*15*). High concentrations in one well in Kansas and the well in West Virginia appeared to be caused by point-source contamination associated with a former mixing/loading site. Detection in the well in Indiana was probably caused by a very shallow water table (< 6 feet) and sandy soil. The reason for the high detection of atrazine exceeding 3 ppb in one well in Minnesota is unknown.

The frequency distributions for the hydroxytriazine degradation products (hydroxyatrazine, desethylhydroxyatrazine, deisopropylhydroxyatrazine and ammeline) are shown in Figure 3. These results show that hydroxyatrazine was detected in 68 out of 1,505 wells (4.5%). The maximum concentration of hydroxyatrazine was found in a well in Indiana at 6.5 ppb. This well was also found to contain high levels of atrazine at 9.1 ppb as described above. Desethylhydroxyatrazine was detected in 42 wells (3.8%). Deisopropylhydroxyatrazine and ammeline were detected in only four and six wells in the entire study (0.3 and 0.5%, respectively). The low detections of hydroxytriazine degradation products in groundwater are in agreement with the adsorption/desorption studies which show that these degradation products are tightly bound in the soil substrate and are immobile.

Regional Summary: A summary of atrazine monitoring data in the nineteen states participating in this program is shown in Figure 4. This graph also indicates the total number of wells in each state. The percentage of wells which show detections in each state is shown as % above the LOD. A brief discussion of these results is provided below:

Two hundred private rural wells were sampled in Wisconsin by DATCP. Approximately 135 wells were selected by DATCP to meet the following objectives: a) to study the presence of atrazine and its degradation products in shallow private wells located near irrigated fields in the Central Sands regions of Wisconsin (43 wells); b) to resample wells from the DATCP Rural Well Survey that had previously exceeded Wisconsin's Enforcement Standard of 0.3 ppb for

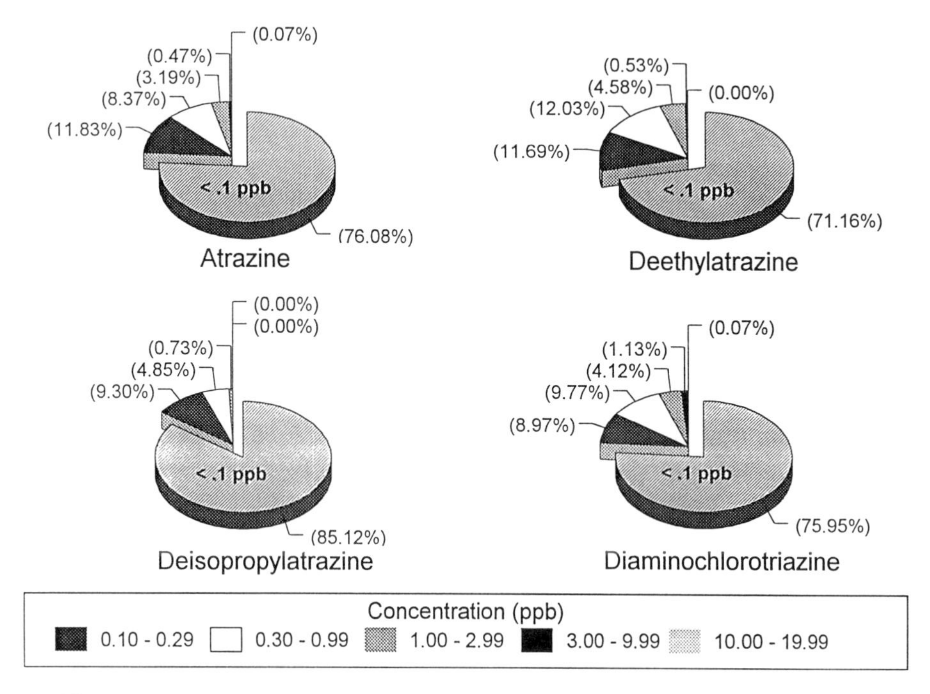

Figure 2. Frequency Distribution of Atrazine and its Chlorotriazine Degradation Products (Total Number of Samples = 1505)

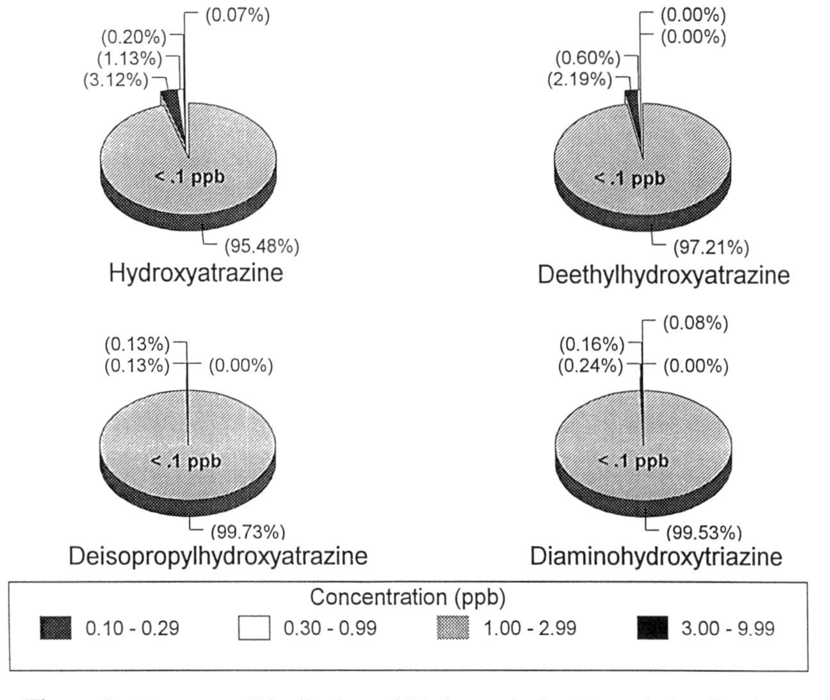

Figure 3. Frequency Distribution of Hydroxytriazine Degradation Products of Atrazine (Total Number of Samples = 1505)

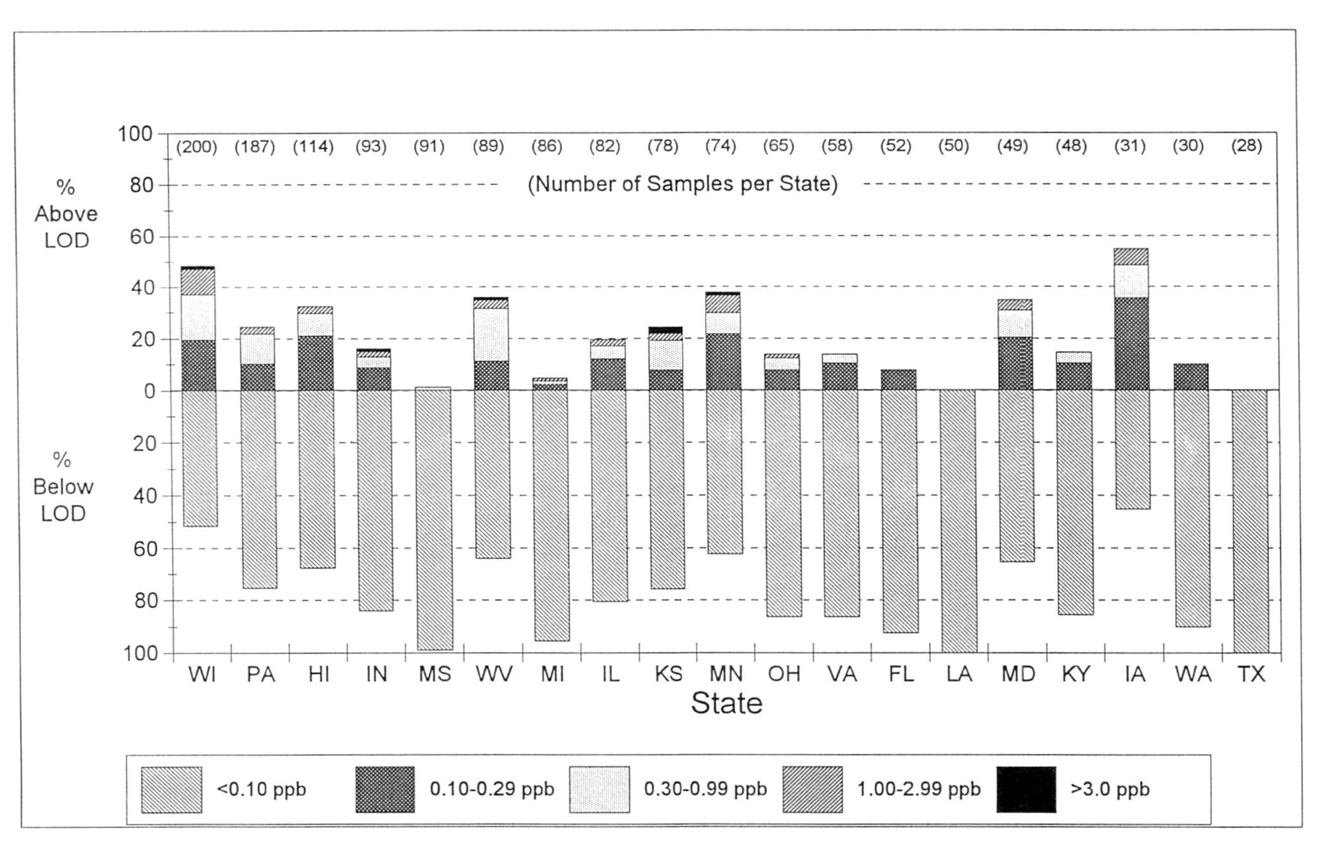

Figure 4. Summary of Atrazine Data for Ciba-Geigy/State Ground Water Monitoring Study

atrazine plus three chlorinated metabolites (44 wells); and c) to evaluate trace level triazine detections detected by immunoassay by splitting the sample with Ciba (48 wells). The remaining 65 wells were sampled for a variety of other reasons, such as responding to the well owner's request. Atrazine was detected in 97 wells (48%). A comparison of the changes in concentrations of atrazine and its chlorotriazine degradation products in 44 wells sampled earlier by DATCP showed significant decline in these levels over a three year period (*15*).

One hundred-fourteen wells were sampled in Hawaii, primarily in the sugar cane producing areas using atrazine. Many of these wells had been sampled earlier for atrazine and some of its degradation products, and were chosen based on previous detections of atrazine by Hawaii Sugar Planters' Association (HSPA). Atrazine was detected in 37 wells (32%) with a maximum concentration of 1.8 ppb. The maximum concentration for desethylatrazine, deisopropylatrazine and diaminochlorotriazine in Hawaii were 1.5, 0.53 and 0.47 ppb, respectively.

Atrazine was detected in 17 of the 31 wells sampled in Iowa (55%). The well selection in Iowa was based on documented atrazine detections in previous monitoring studies conducted by the Iowa Department of Agriculture. From a list of 100 wells with previous detections of atrazine, 31 wells were chosen based on well depth criteria.

Atrazine was detected in 32 out of 89 wells in West Virginia. Significant detections in West Virginia were in corn producing areas of Mason and Jefferson counties where thirty wells were included based on positive detections of atrazine in a previous sampling program by the West Virginia Department of Agriculture. Six of these wells were sampled two to five times each to address temporal variability. The results of these additional samplings were in good agreement with the analysis in the first sampling. Atrazine was detected in 46 of the 187 wells sampled in Pennsylvania. Significant numbers of detections in Pennsylvania were in the karst valleys of the state's Ridge and Valley province and other limestone valleys of southeast Pennsylvania.

Forty-nine wells were sampled in Maryland in high atrazine use areas. Of those, 24 wells were located in the coastal plains and the karst regions of the state and 15 of these wells were USGS observation wells in the eastern shore region. Atrazine was detected in 17 wells (35%). Well selection in Virginia included the Delmarva Peninsula, the southeast portion of the state and several counties in the Blue Ridge Mountain regions. Fifty-eight wells were sampled in Virginia which included five shallow USGS observation wells in Delmarva Peninsula. Atrazine was detected in only eight wells with a maximum concentration of 0.61 ppb.

Thirty wells were sampled in Washington State in atrazine use areas of Christmas tree production. Priority was given to wells with previous detections of atrazine, such as the detections of triazines in the Chehalis River Basin using the immunoassay screens. Atrazine was detected in three out of the 30 wells sampled in Washington State with a maximum concentration of 0.22 ppb. Of the 91 wells sampled in high atrazine use areas in Mississippi, one well showed detection of atrazine at 0.60 ppb. Resampling of this well confirmed the detection. Fifty-two wells were sampled in corn and sugarcane use areas of Florida including the sandy regions in north central Florida, the Florida panhandle, sugarcane use areas south of Lake Okeechobee and corn growing regions in southern Dade county. Atrazine was detected in four wells (8%) with a maximum concentration of 0.25 ppb. Fifty wells were chosen in Louisiana in representative crop growing regions for rice, sugarcane, cotton and soybeans. Also, twenty wells were monitored in Texas in the Brazos River alluvium area, southern High Plains and Ogollala aquifer area (south of Lubbock), and Gulf Coast. No detections of atrazine or its degradation products were found in either Louisiana and Texas.

Ratio of Atrazine Degradation Products to Atrazine: Ratios of the levels of desethylatrazine, deisopropylatrazine and diaminochlorotriazine as a function of atrazine levels were computed to determine a relationship between these variables. Detections below the LOD were excluded in this analysis. It has been proposed that a ratio of desethylatrazine to atrazine (DAR) greater than unity is an indicator of nonpoint source contamination (*14*). The DAR hypothesis is predicted on the assumption that atrazine degrades slowly in the vadose zone to desethylatrazine and the later has higher mobility than the parent compound. Adams and Thurman (*12*) hypothesized that a small DAR ratio may be an indicator of point-source contamination of an aquifer. Distribution of the desethylatrazine to atrazine ratios for all the detections of atrazine shows that the high values of atrazine (> 3.0 ppb) have a DAR value close to zero. Similarly, when atrazine is near the detection limit of 0.10 ppb, the DAR ratio for some wells were significantly greater than unity. However, these data were extreme values in the distribution and the correlation between the DAR and atrazine concentrations was poor (R square value of 0.051). The DAR hypothesis may be valid as a generalized statement; however, a number of factors may cause variation of the DAR. Some of these factors include preferential flow (such as macropores), surface water interactions, lack of degradation in highly permeable soils, history of atrazine use at the site, etc. The ratios of deisopropylatrazine to atrazine and diaminochlorotriazine to atrazine were calculated to attempt a relational analysis. The conclusions from these distributions were similar to results of DAR ratios.

An evaluation was made of the distribution of desethylatrazine with deisopropylatrazine in all the wells in the nineteen states participating in this program. The ratio of deisopropylatrazine to desethylatrazine (referred to as D2R) has been suggested by Thurman (*12*) as typically less than unity because of preferential degradation of atrazine by desethylation instead of deisopropylation. A D2R ratio greater than unity suggests the formation of deisopropylatrazine from other sources such as cyanazine or propazine instead of atrazine. The distribution of deisopropylatrazine vs. desethylatrazine shows that a large number of wells have D2R ratios less than unity (desethylatrazine levels > deisopropylatrazine). However, a small number of wells in the extreme show D2R ratios > 1 which may be caused by sources other than atrazine. Pesticide use history in the area where these wells were found could assist in determining the source for the greater detection of deisopropylatrazine.

Conclusions and Research Needs

Ciba has completed a large-scale groundwater monitoring program for atrazine and its degradation products in the United States. This monitoring has been conducted in nineteen major atrazine use states with over 1,500 wells selected based on high atrazine use, groundwater vulnerability, and previous atrazine detections. The well selection in this study was highly biased for positive detections because of the targeted well selection process used by these states and hence, these monitoring data cannot be used for accurate extrapolation of exposure estimates for the general population using rural drinking water wells. This study has provided extensive groundwater monitoring data for atrazine and its chloro and hydroxytriazine degradation products in a very large geographic area where these detections are expected. This study was conducted in cooperation with the various state agencies and involving the states very effective for addressing the data gaps on atrazine degradation products in groundwater. The state agencies and their affiliates have also benefited from this program through training in the sample collection and GLPs related to the field phase of groundwater monitoring. Results from this study can be used to facilitate the design and implementation of site-specific water management plans and product stewardship activities to protect groundwater.

Acknowledgment

The authors wish to thank the nineteen state agencies and affiliates who participated in this study, Environmental and Public Affairs of Ciba and Quality Associates, Inc., for the technical support of the field phase. The authors also wish to thank the Biochemistry Resources Department of Ciba for performing the GC/MSD analyses of chlorotriazines; and Alta Analytical Laboratory Inc. and ABC Laboratories Inc. for performing the LC/MS/MS analyses of hydroxyatrazines.

Literature cited

1. Balu, K., Paulsen, R. T. In *Interpretation of Atrazine in Ground Water Data Using a Geographic Information System*; Weigmann, D. L., Ed.; Pesticides in the Next Decade: The Challenges Ahead, Proceeding of the Third National Research Conference on Pesticides, Virginia Water Resources Research Center, Virginia Polytechnic Institute and State University, Blacksburg, VA, November 8-9, 1990.
2. U.S. Environmental Protection Agency, National Survey of Pesticides in Drinking Water Wells, Phase 1 Report, Report No. EPA/570/09-90/015.
3. Holden, L. R., et al. *Environ. Sci. And Technol.* **1992**, Vol. *26*, pp. 935-943.

4. Kross, B. C. et. al. , The Iowa State-wide Rural Well Water Survey - Water Quality Data, Initial Analysis, Iowa Geological Survey Technical Information Series Report No. 19, 1990, pp. 1424.
5. LeMasters, Gary, and D. J. Doyle, Grade A Dairy Farm Well Water Quality Survey, Wisconsin Department of Agriculture, Trade and Consumer Protection, Madison, Wisconsin, 1990, pp. 36.
6. Detroy, M. G., P. K. B. Hunt, and M. A. Holub, Ground-Water Quality Monitoring Program in Iowa: Nitrate and Pesticides in Shallow Aquifers, Proceedings of the Agricultural Impacts in Ground Water, National Water Well Association, Dublin, Ohio, 1988, p. 255-278.
7. Klaseus, T. G., G. C. Buzick and E. C. Scheider, Pesticides and Groundwater: Survey of Selected Minnesota Wells, Minnesota Department of Health and Minnesota Department of Agriculture, Minneapolis, MN, 1988, pp. 95.
8. California Environmental Protection Agency, Sampling for Pesticides Residues in California Well Water - 1993 Update, Department of Pesticide Regulations, Sacramento, CA, 1993, pp. 167.
9. Spalding, R. F.; Burbach, M. E.; Exner, M. E. *Ground Water Monitoring Review.* **1989,** Vol. *9*, pp. 126-133.
10. Maas, R. P., et. al. *J. Environ. Qual.* **1995**, Vol. *24*, pp. 426-431.
11. Domagalski, J. L.; Dubrovsky, N. M. *Journal of Hydrology*, **1993**, Vol. *130*, pp. 299-338.
12. Adams, C. D.; Thurman, E. M. *J. Environ. Qual.* **1991**, Vol. *20*, pp. 540-547.
13. Mills, S. Margaret; Thurman, E. M. *Environ. Sci. Technol.* **1994**, Vol. *28*, pp. 600-605.
14. Winkelmann, D. A.; Kline, S. J. *Environmental Toxicology and Chemistry*, **1991**, Vol. *10*, pp. 347-354.
15. Postle, J., 1994. "Report on the DATCP/Ciba-Geigy 200 Well Sampling Program," Wisconsin Department of Agriculture, Trade and Consumer Protection, Madison, WI.

Chapter 20

Pesticide Movement to Groundwater: Application of Areal Vulnerability Assessments and Well Monitoring to Mitigation Measures

J. Troiano[1], C. Nordmark, T. Barry, B. Johnson, and F. Spurlock

Department of Pesticide Regulation, California Environmental Protection Agency, 1020 N Street, Room 161, Sacramento, CA 95814–5624

An empirical statistical approach has been used to identify areas vulnerable to ground water contamination in California (*1*). The method does not assume any particular pathway for ground water contamination. One-square mile areas of land where pesticide residues have been found in water due to non-point source agricultural applications were categorized statistically first with respect to climate and, then, soil characteristics. The ability of the statistical model to identify potentially vulnerable areas was evaluated by conducting a well sampling study targeted to herbicides used in grape and citrus. Areas for sampling were chosen based on results of the statistical model and on data for pesticide use. Localized mitigation measures for off-site movement of herbicides are being developed using results from the vulnerability modeling and well sampling studies.

A regional approach is being developed in California to prevent contamination of ground water from non-point source application of pesticides. The objective is to construct management options for pesticide use based on local geographic conditions and agronomic practices. Successful implementation of a regional approach requires an understanding of how local climatic and geographic conditions govern residue movement to ground water. The purpose of this paper is: to explain the modeling approach that has been developed to determine spatial vulnerability in California (*1*); to describe results from a well sampling study that was conducted to evaluate the model (*2*); and to outline how the vulnerability assessment will be implemented in mitigating pesticide movement to ground water.

[1] **E-mail: jtroiano@edpr.ca.gov.**

Previous Modeling Approach for Determining Spatial Vulnerability

One predominant approach to identifying vulnerable land areas has been to develop models, *a priori,* and then conduct well monitoring studies to test the veracity of derived land indices (*3*). A common assumption made during model development was that residues reach ground water primarily by leaching through soil from simple percolation (*4*). A few well sampling studies have been conducted to test the relevance of land vulnerability indices (*5-9*). The indices were not reliable because pesticide residues were detected in areas identified as relatively invulnerable. In our experience, identifying leaching from simple percolation as the sole cause of detections in large retrospective well surveys has been problematic because residues move to ground water by other routes. These routes may include movement of surface water into agricultural drainage wells (*9,10*), Karst formations (*11*), or cracks in clay soils (*12*). Measurement and modelling the movement of residues in macropore flow has only recently gained attention (*13,14*).

Development of Vulnerability Assessment for California Conditions

Analysis of previous investigations on the cause of detections in California well water indicated that the finds were associated with a wide range in soil, climatic, and pesticide use conditions. Detections of simazine and diuron in coarse soils of the dry inland valleys were suspected to be due to leaching caused by excess percolation from irrigation. But residues had also been detected in wells that were sampled in areas of hardpan or clay soils where leaching was an unlikely cause of contamination (*10, 15*).

Owing to the broad range in California's climatic and soil conditions, we decided to take an empirical statistical approach to profiling areas of ground water contamination by pesticides (*1*). The method did not assume any particular pathway for ground water contamination, and it did not rely upon deriving relative levels of vulnerability between land areas. One square-mile sectional areas of land (*16*) where pesticide residues had been found in ground water and the detections attributed to legal agricultural use were designated as known contaminated (KC) sections.

Statistical clustering methods were used to identify groups of KC sections that were charcterized first according to climate, then followed by soil variables. Based on rainfall amount, two climatic clusters were identified, one wet and one dry. Five sections that were members of the wet climate cluster were located in Del Norte and Humboldt counties which receive approximately 153 cm of rainfall, annually. The remaining two-hundred-fifty-four KC sections were members of the dry climate cluster, mostly located in inland basins. Irrigation is mandatory in these areas because the climate is Mediterranean with hot, dry summers and little annual rainfall, typically less than 50 cm. Further clustering based on soil variables was conducted in the dry climate cluster. Five distinct clusters of KC sections were identified using two soil variables. One variable reflected soil texture as measured by the percentage of particles passing a number 200 sieve. Average sectional values for soil texture of the five clusters ranged from coarse to fine soil conditions (Table I). The second soil variable indicated the presence or absence of a hardpan which, for the raw soil data, was assigned a weight of 0 if the soil did not have a hardpan or 1 if a hardpan was present. Average cluster values for hardpan ranged from 0.01, indicating practically no soils in a section with a hardpan, to 0.94 where nearly all soils in a section contained a hardpan (Table I).

Table I. Description and Average±Standard Deviation (SD) Sectional Values for Variables that Reflect the Presence of Hardpan and % Soil Particles Passing a No. 200 Soil Sieve in Each of 5 Clusters of Sections with Ground Water Contaminated by Pesticides.

Cluster Description	*# of KC Sections*	*Cluster Variables*	
		Hardpan[1]	*No. 200 Sieve*[2]
KC1. No Hardpan and Coarse Textured	72	0.08±0.11	36± 5.9
KC2. Hardpan and Coarse-Medium Textured	82	0.50±0.14	49± 7.7
KC3. No Hardpan and Medium Textured	26	0.01±0.03	60± 6.4
KC4. Hardpan and Medium Textured	26	0.94±0.13	62±10.1
KC5. No Hardpan and Fine Textured	48	0.03±0.10	82± 4.3

[1] Scale from 0 to 1 with a 0 value representing no soils in section with hardpan and a 1 indicating all soils in that section with hardpan.
[2] Measured by the percentage by weight of soil particles that pass a No. 200 soil sieve. The smaller the percentage, the more coarse textured the soil.
SOURCE: Adapted from ref. 1.

The results of the clustering analysis were incorporated into a method to classify sections that lacked either well sampling data or positive detections into the KC soil clusters. The classification algorithm employed Principal Components Analysis (PCA classification method) and it allowed for a not-classified category. A plot of the classification of sections with soil data in Fresno and Tulare counties indicated that the statistical clusters were associated with discrete geographical areas (Figure 1).

Well Sampling Survey Testing CALVUL Modeling Approach

The results of the PCA classification method were employed in the design of a well sampling study. The objectives of the survey were: 1) to gain experience and confidence in the use of the approach as a tool to define potentially vulnerable areas; and 2) to assess the classification algorithm. The well sampling study was conducted in Fresno and Tulare counties, California (Figure 1). Sections for well sampling were selected from three groups with characteristics similar to: 1) the coarse soil cluster with no hardpan (KC1); 2) the coarse to medium soil texture cluster with approximately 50% of the soils in a section containing a hardpan (KC2); and 3) not-classified into one of the KC soil clusters (Figure 1). Sections were not sampled from the KC3, KC4, or KC5 categories because of the

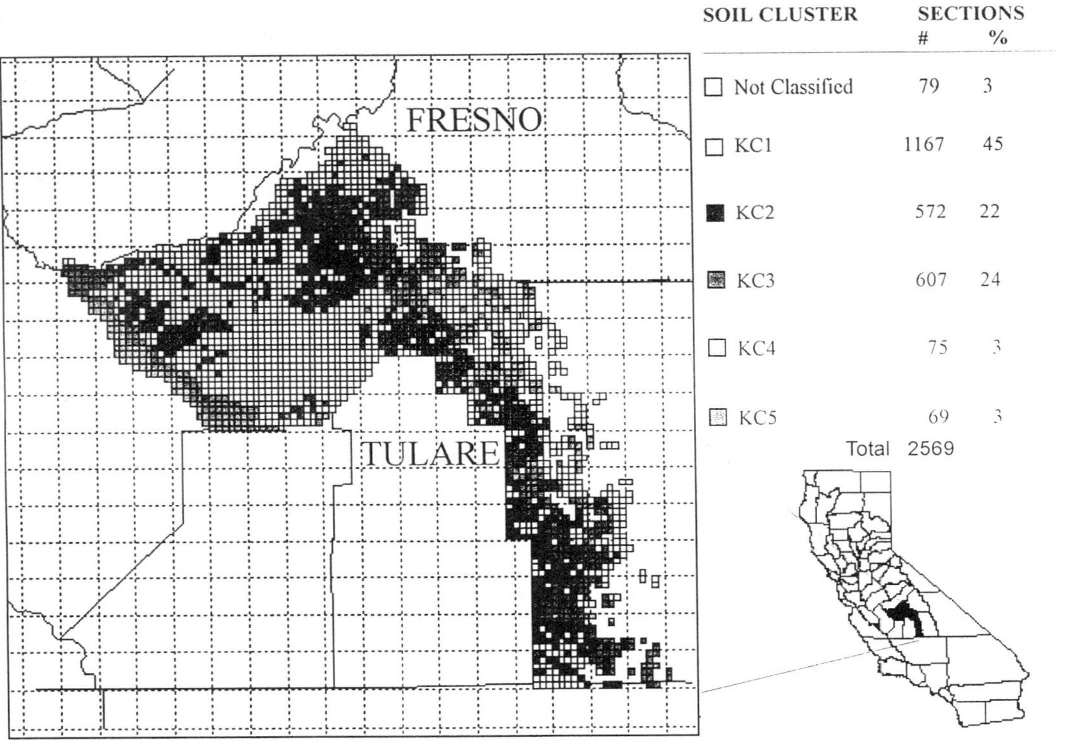

Figure 1. Use of PCA classification to identify sections of land in Fresno and Tulare counties into vulnerable soil clusters.
(Adapted with permission from reference 1. Copyright 1994 Kluwer Academic.) (*See* color photograph on page 461.)

lower number of candidate sections categorized into each. The experimental unit was a section of land with one well sampled per section.

To ensure that pesticide use patterns in candidate sections were similar to those in KC sections, candidate sections were chosen with agricultural cropping and pesticide use patterns that were similar to KC sections in Fresno and Tulare counties. Based on data from the 1990 and 1991 California Department of Pesticide Regulation's (DPR) pesticide use reports, total use of simazine, bromacil, and diuron for the 72 sections in the KC1 cluster was greatest on grapes and orange. Total use for the 82 sections in the KC2 cluster was greatest on orange, grapes, and olive. Thus, candidate sections were chosen as potential experimental units if use of simazine, bromacil, or diruon had been reported on oranges, grapes or olives in either 1990 or 1991.

Sixty candidate sections with one well sampled per section were to be randomly chosen from each of the KC1 and KC2 soil clusters, and from the not-classified sections. Potential wells for sampling were identified by surveying each targeted section for the presence of wellheads, residences, or occupied buildings. When possible, wells situated near vineyards, citrus, or olive orchards were preferentially sampled. Detailed sampling procedures were used to assure the quality and integrity of the well and the quality and interity of the subsequent water samples (*2*). Well water samples were analyzed for the known ground water contaminants, atrazine, bromacil, diuron, prometon, simazine and the triazine breakdown products desethylatrazine and desisopropylatrazine. The primary analytical method employed Solid Phase Extraction and analysis by thermospray triple-stage quadrupole mass spectrometer (TSP)-LC/MS/MS. The confirmation analytical procedure utilized a manifold system or the Zymark Autotrace SPE, C18 Solid Phase Extraction (SPE). Detection was conducted with Liquid Chromatography coupled to a TSQ-700 Mass Spectrometer (TSP-LC/MS/MS). Minimum Reporting Limits and quality control information are reported in Table II.

Table II. Method Reporting Limits (MRL) and QC Data for Sample Analytes.

	Primary Analysis				*Confirmatory Analysis*			
	MRL	*Recovery*[1]			*MRL*	*Recovery*[1]		
Chemical Analyte		*Mean*	*LCL*[2]	*UCL*[2]		*Mean*	*LCL*[2]	*UCL*[2]
	(ppb)	*(ppb)*	*(%)*	*(%)*	*(ppb)*	*(ppb)*	*(%)*	*(%)*
Atrazine	0.05	94	85	102	0.05	89	83	95
Bromacil	0.05	103	87	119	0.05	89	75	103
Desethylatrazine	0.10	89	71	107	0.05	104	77	131
Desisopropylatrazine	0.10	86	71	100	0.05	62[3]	18	106
Diuron	0.05	98	88	109	0.05	101[3]	91	111
Prometon	0.05	91	88	103	0.05	93	73	113
Simazine	0.05	99	81	116	0.05	98	87	109

[1] Percentage of spike levels

[2] LCL=Lower Control Limits and UCL=Upper Control Limits determined as the mean of the percent recovery±2 standard deviations.

[3] Five spiked replicates each at 0.1, 0.5, 2.0, and 10.0 ppb, spike levels for the rest were 0.05, 0.1 and 0.5 ppb.

SOURCE: Adapted from ref. 2.

Objective 1) Experience and Confidence in Use of CALVUL Approach. Seventy-six of the 176 wells sampled were positive, a detection rate of 43% (Table III). More than one residue was detected in 55 wells. The detections paralleled cropping and pesticide use criteria derived for the candidate sections from the 1990 and 1991 pesticide use reports: Simazine and diuron are used on grape and their residues were prevalent in the coarse soil cluster (KC1) where grape was the predominant crop; and simazine, diruon, and bromacil are used on citrus and their residues were prevalent in the hardpan soil cluster (KC2) where citrus was the predominant crop. Detections of atrazine and desethylatrazine were infrequent and probably related to use in non-crop areas.

Table III. Summary Statistics for Detection of Residues in Well Water Samples.

Analyte	*Number of Detections*[1]	*Mean Concentration* (ppb)	*Minimum Value* (ppb)	*Maximum Value* (ppb)
Simazine	50	0.2	0.1	0.8
Disopropylatrazine	59	0.7	0.1	4.0
Diuron	36	0.3	0.1	1.2
Bromacil	20	0.6	0.1	3.0
Atrazine	3	0.1	0.05	0.2
Desethylatrazine	2	0.35	0.3	0.4

[1] Also equal to the number of positive sections out of 176 sampled sections.

We considered the detection rate of 43% high for four reasons: 1) At the time of sampling, the statewide rate of positive detections for these pesticides, as determined from DPR's Well Inventory database, was lower at approximately 10%; 2) In this study only one well was sampled per section, when more then one well is sampled per section the probability of detection and estimate for percentage of positive sections are greater; 3) Use of domestic and irrigation wells to monitor ground water quality introduces variation in results due to differences in construction, depth, and spatial location relative to sources of contamination; and 4) The rate of positive detections in this study was similar to or greater than rates obtained in other targeted well studies that were based on single well water samples (*2*).

Objective 2) Assessment and Modification of Classification Algorithm. The rate of detections in the two vulnerable soil clusters was 41% for KC1 and 57% for KC2, but the rate of detection in not-classified sections was relatively high at 31%. Two explanations for detections in the not-classified sections were: 1) that the not-classified sections should also be considered vulnerable, perhaps forming a new soil cluster; or 2) that the PCA classification method was too restrictive in assigning cluster membership, resulting in not-classified sections that should have been considered members of one of the existing KC soil groups. The possibility of a new soil cluster was investigated by conducting another cluster analysis of the soil variables on the combined data set. Since no new cluster was indicated, use of the complete PCA analysis for the classification algorithm may have been too restrictive because many of the not-classified sections failed the algorithm at Principal Components that probably represented sampling error rather than meaningful experimental information (*17*).

Based on this evaluation, we chose to explore an alternative classification method based on Canonical Variates Analysis (CVA classification method) (*18-20*). The first 2 canonical variates produced by the Canonical Discriminant Analysis accounted for 98% of the variation in the original 254 KC sections. The classification algorithm was based on circular population tolerance intervals constructed around a plot of the first two canonical variates for the mean of each of the 5 KC soil clusters (*2*). A candidate section was considered a member of a KC soil cluster if the Euclidian distance between the canonical variate coordinates of each candidate section and the mean of each KC soil cluster was within the circular tolerance interval. This procedure retained the possibility of producing not-classified sections because the coordinates for a candidate section could fall outside the tolerance interval for all 5 KC soil clusters.

Application of the CVA classification method to all candidate sections in Fresno and Tulare counties produced a geographical distribution of KC soil clusters that was similar to the PCA classification method. However, more candidate sections were classified into the five KC soil clusters with the CVA method, resulting in a dramatic reduction of not-classified sections: only 79 candidate sections were not-classified compared to 868 using the PCA classification method (Figure 2).

Inspection of the CVA classification of the sections used for the well sampling study indicated that 148 of the original 176 sampled sections were classified into KC1 and KC2 soil clusters. Only two of the sampled sections remained members of the not-classified category. The CVA classification procedure appeared more practical in terms of implementation than the PCA classification because a greater number of sections were subject to descriptive terms and, hence, to management practices developed for those conditions.

Application of CALVUL Modeling Approach to Mitigation Measures.

The geographic presentation of the soil clustering results was consistent with our previous observations that contamination could be caused by different hydrologic processes. Leaching was suspected as the pathway for downward movement in coarse, sandy soils in Fresno county where broad bands of simazine had been detected deep in soil cores (*21*). Deep percolation of irrigation water is a major source for ground water recharge in this arid area. In contrast, very little residue was detected deep in cores sampled from hardpan soil in Tulare county, a result contrary to the mechanism of leaching (*22*). A subsequent study in Tulare county indicated that a probable route of non-point source contamination in these soils was movement of pesticide residues in runoff water into agricultural drainage wells (*10*).

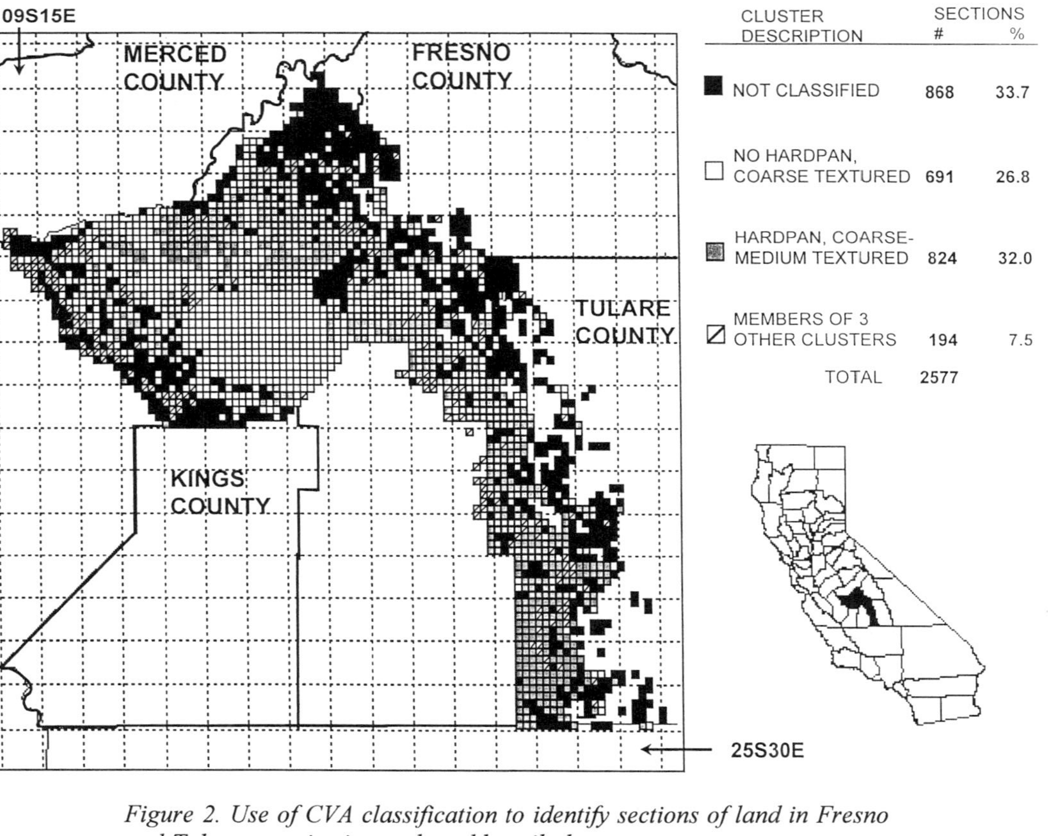

Figure 2. Use of CVA classification to identify sections of land in Fresno and Tulare counties into vulnerable soil clusters.
(Adapted with permission from reference 2. Copyright 1997 Kluwer Academic.) (*See* color photograph on page 462.)

Management practices are now being proposed to address the specific transport mechanisms and local use conditions in Fresno and Tulare Counties. For coarse sandy soils, control of water percolation by increased irrigation management should minimize leaching of pesticides (*23*). For the hardpan soils, runoff of residues could be minimized by more complete incorporation of pesticides during or after soil application. Some preliminary studies have been conducted on the California State University Fresno campus to demonstrate the potential for movement of pesticides residues from row middles to furrows in runoff water in citrus orchard situations. Residues were measured in runoff water collected after application of a simulated heavy rain event (Table 4). Mechanical disruption of the first few inches of soil with a plant cultivator, made immediately after pesticide application, decreased the mass of simazine in runoff through a combined decrease in the amount of runoff water produced and a decrease in the concentration of simazine in the runoff water. Although rainfall is indicated on labels as a method to incorporate many pre-emergence herbicides, compaction of soil in bare row middles, particularly in orchard situations, increases runoff potential and may require special incorporation procedures (*24*).

Table IV. Effect of Mechanical Incorporation on Simazine Movement in Simulated Rain Runoff Water from a Citrus Grove.

Treatment	*Amount of Runoff Water*[1] (L)	*Concentration of Simazine* (mg/L)	*Mass of Simazine* (mg)
No Incorporation	204	0.87	179
Mechanical Incorporation	101	0.14	14

[1] Simulated rain runoff collected from 3.05 x 5.5 m plots with simazine at 2.2 kg/ha and 2.52 cm of rainfall applied through Rainbird macro-sprinklers.

Staff from the DPR, county farm advisors, and staff from the University of California have met with local growers to discuss well monitoring data, the relationship between cropping patterns and detections in wells, transport mechanisms of pesticide movement to ground water as they relate to soil characteristics, and potential mitigation measures that maintain residues on-site after application. In conjunction with DPR, growers and farm advisors have begun to identify management practices for evaluation at on-farm demonstration sites. Participation by other interested parties, including registrants, commodity groups, and/or Pest Control Advisors is being encouraged. Ultimately, the goal is to evaluate management options at the demonstration sites, and use the sites as an educational tool in an effort to encourage voluntary implementation of management practices that prevent movement of pesticides to ground water.

Incorporation of Additional Explanatory Variables

Further analysis of the well sampling study provides an example of how other explanatory variables could be incorporated into the approach. Depth to ground water was initially a logical choice as a factor for defining vulnerable areas, but lack of a statewide database precluded its use in the original clustering procedure. Data were available for the smaller area of Fresno and Tulare counties so the relevance of depth to ground water could be investigated. Depth to ground water data were derived from the Spring 1990 map of the depth from surface to first aquifer (U.S. Bureau of Reclamation, 1989). The rate pf detection appeared related to depth to ground water with a greater probability of detection in sections with shallow ground water. In the KC1 soil cluster, 58% (26 of 45) of sections were positive when ground water was shallower than 15 meters compared to 16% (8 of 51) positive sections when ground water depth was greater than 15 meters. A similar effect was observed in the KC2 soil cluster where 66% (23 of 35) of sections were positive when ground water depth was shallower than 15 meters compared to 12% (2 of 17) positive sections when ground water depth was greater than 15 meters.

The experience obtained from developing and testing the CALVUL modeling approach will aid in the design of future well sampling studies. The probability of detecting residues in a study should be increased through combined use of the soil classification algorithm, pesticide use data, and depth to ground water data.

Conclusions

1. A empirical approach employing multivariate statistical methods has been used to identify areas vulnerable to ground water contamination under California conditions. One important feature of the approach is that vulnerable conditions have been described that have a wide range in climatic and soil conditions.

2. Greater understanding of the transport mechanisms of pesticide movement to ground water has been derived through analyzing combined results from the CALVUL modeling approach and other well sampling and research studies. Two different pathways have been identified in areas with a Mediterranean climate: Leaching in coarse, sandy soils, and movement in winter rain runoff in areas with hardpan soils where runoff water is disposed into subsurface drainage wells.

3. Broad geographic patterns were associated with the statistical clusters, indicating that management practices spatially matched to predominant soil cluster characteristics. For example, greater irrigation management has been suggested as a method to control leaching in geographic areas represented by a coarse soil cluster. In contrast, improvements in incorporation is suggested in geographic areas where hardpan soils are predominant: improved incorporation would mitigate movement of residues in runoff water.

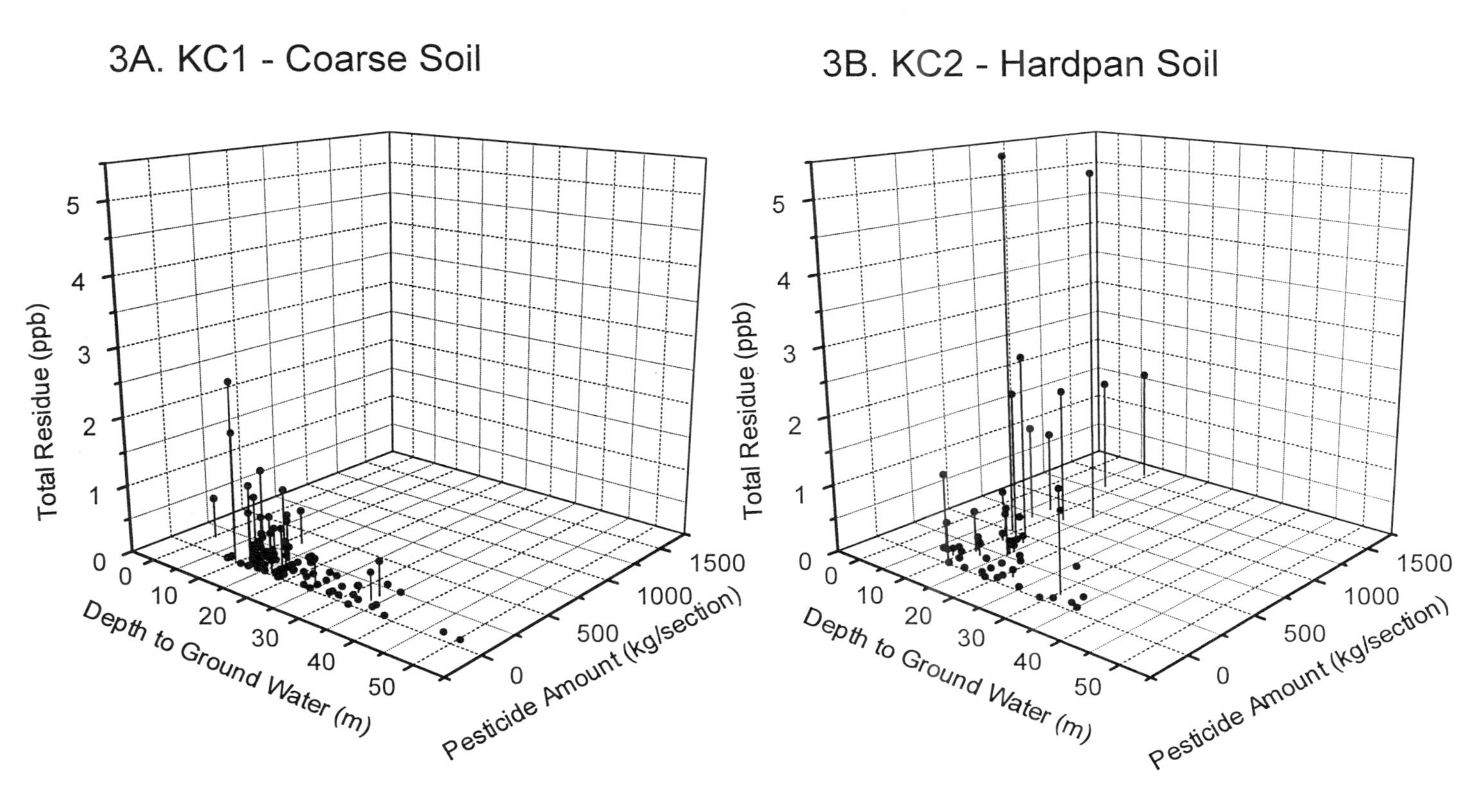

Figure 3. Magnitude of total pesticide residues in well water samples in relation to depth to ground water and amount of pesticide applied in sampled sections. Adapted from ref. 2.

4. The CALVUL modeling approach is flexible and allows the addition of data from other variables to describe additional agronomic and hydrogeologic features of vulnerable sections.

Acknowledgments

The help of all in the Environmental Hazards Assessment Program during the well sampling study is deeply appreciated. We would also like to thank Dallas Johnson of the Department of Statistics, Kansas State University, Kansas for his suggestions on the use of Canonical Variates Analysis as an alternative classification method.

Literature Cited

1. Troiano, J.; Johnson, B. R.; Powell, S.; and Schoenig S. *Environ. Monit. and Assess.* **1994**, *32*, 269-288.
2. Troiano, J.; Nordmark, C.; Barry, T.; and Johnson, B. R. *Environ. Monit. and Assess.,* **1997,** in press.
3. National Research Council, *Ground Water Vulnerability Assessment: Predicting Relative Contamination Potential Under Conditions of Uncertainty;* National Academy Press, 2101 Constitution Avenue, Washington D.C. 20418, 1993.
4. Corwin, D.L., and Wagenet, R.J. (Organizers), *Applications of Gis to the Modeling of Non-point Source Pollutants in the Vadose Zone, 1995 Bouyoucos Conference, Proceedings. May 1-3, 1995, Riverside, California,* USDA-ARS, U.S. Salinity Laboratory 450 Big Springs Road, Riverside, CA, 92507, 1995.
5. *ANOTHER LOOK: National Pesticide Survey Phase II Report; EPA 570/9-91-020,* Office of Water (WH-550), USEPA Washington D.C. 20406, 1992.
6. Balu, K., and Paulsen, R. T.; *In Pesticides in the Next Decade: The Challenges Ahead, Proceedings of the Third National Research Conference on Pesticides, Nov 8-9, 1990,;* Weigmann D.L., Ed.; Virginia Water Resources Research Center, Virginia Polytechnic Institute and State University; Blacksburg, VA, 1991; pp 431-446.
7. Holden, L. R.; Graham, J. A.; Whitmore, R. W.; Alexander, W. J.; Pratt, R. W.; Liddle, S. K.; and Piper, L. L. *Environ. Sci. and Technol.* **1992,** *26*, 935-943.
8. Kalinski, R. J.; Kelly, W. W.; Bogardi, I.; Ehrman, R. L.; and Yamamoto, P.D. *Ground Water* **1994,** *32*, 31-34.
9. Roux, P. H.; Hall, R. L.; and Ross Jr, R. H. *Ground Water Monit. Rev.* **1991**, *XI*, 173-181.
10. Braun, A.L.; and Hawkins, L.S.: *Presence of Bromacil, Diuron, and Simazine in Surface Water Runoff from Agricultural Fields and Non-crop Sites in Tulare County, California;* PM 91-1, Environmental Monitoring and Pest Management Branch, Department of Pesticide Regulation, 1020 N Street, Room 161, Sacramento, CA 95814-5624, 1991.
11. Hallberg, G. R. Agri., Ecosystems and Environ. **1989**, *26*, 299-367.

12. Graham, R. C.; Ulery, A. L.; Neal, R. H.; and Teso, R. R. Soil Sci., **1992**, *153(2)*, 115-121.
13. Bergstrom, L.; McGibbon A.; Day S.; and Snel M. *Environ. Toxicol. and Chem.* **1991**, *10*, 563-571.
14. Chen, C.; Thomas, D. M.; Green, R. E.; and Wagenet, R. J. *Soil Sci. Soc. Am. J.*, **1993**, *57*, 680-686.
15. Miller Maes. C.; Pepple M.; Troiano J.; Weaver D.; Kinmaru W.; and SWRCB Staff. *Sampling for Pesticide Residues in California Well Water 1992 Well Inventory Data Base, Cumulative Report 1986-1992;* EH 93-2, Environmental Monitoring and Pest Management Branch, Department of Pesticide Regulation, 1020 N Street, Room 161, Sacramento, CA 95814-5624, 1992.
16. Davis, R. E.; and Foote, F. F. *Surveying Theory and Practice,* fifth edition, Ch. 23, New York, NY, 1966.
17. Jackson, D. A. *Ecology*. **1993**, *74(8)*, 2204-2214.
18. Johnson, D. Dept. of Statistics, Kansas State University, Manhattan, KS, USA, personnel communication, 1995.
19. Adams, S. M.; Ham, K. D.; and Beauchamp, J.J. *Environ. Toxicol. and Chem.* **1994.** *13(10)*, 1673-1683.
20 *SAS/STAT User's Guide, Release 6.03 Edition*. SAS Institute Inc., SAS Circle, Box 8000, Cary, NC 27512-8000, 1988.
21. Zalkin, F.; Wilkerson, M.; and Oshima, R. H. *Pesticide Movement to Groundwater EPA Grant #E009155-79 Volume II. Pesticide Contamination in the Soil Profile at DBCP, EDB, Simazine and Carbofuran Application Sites*, Environmental Hazards Assessment Program, California Department of Pesticide Regulation, 1020 N Street, Room 161, Sacramento, CA 95814-5624, 1984.
22. Wellling, R.; Troiano, J.; Maykoski, R.; and Loughner, G. In *Proceedings of the Agricultural Impact on Ground Water - A Conference August 11-13, 1986, Omaha, Nebraska;* National Water Well Association, 6375 Riverside Dr. Dublin, OH 43017, 1986; pp 666-685.
23 Troiano, J.; Garretson, C.; Krauter, C.; and Brownell, J. *J. of Environ. Qual.* **1993**, *22,* 290-298.
24. Leonard, R. A. In *Pesticides in the Soil Environment: Processes, Impacts, and Modeling;* Cheng, H. H., Ed.; SSSA Book Series, no. 2: Soil Sci. Soc. of Amer., 677 S. Segoe Rd., Madison, WI 53711, 1990; Chapter 9, pp 303-349

Chapter 21

Exposure to the Herbicides Atrazine and Simazine in Drinking Water

D. P. Tierney[1], J. R. Clarkson[2], B.R. Christensen[3], K.A. Golden[3], and N. A. Hines[3]

[1]Novartis Crop Protection, Inc., P.O. Box 18300, Greensboro, NC 27419
[2]Montgomery Watson, 365 Lennon Lane, Walnut Creek, CA 94598
[3]Montgomery Watson, 545 Indian Mound, Wayzata, MN 55391

A population-linked database was used to assess exposure to the herbicides atrazine and simazine in drinking water provided by community water systems (CWS) in 21 major use states. Herbicide concentration and population data from 1993 through 1995 were paired for each CWS and aggregated for all CWS to construct state and multi-state exposure profiles. The assessed populations were 110 million for atrazine and 107 million for simazine. The majority of the CWS population had no detectable exposure to atrazine and simazine. All simazine and 99.9% of atrazine populations had exposure below their respective drinking water MCLs. Thirteen of 13,688 CWS had atrazine multi-year mean concentrations above the MCL ranging from 3.06 to 6.19 ppb. Exposures to atrazine and simazine both corresponded to a margin of safety of at least 10,000 for 94% and 96% of the assessed population. Ciba Crop Protection (Ciba) is continuing this monitoring program at least through 1997 to provide a 5-year database, and will update this assessment with 1996 and 1997 data.

Atrazine (2-chloro-4-ethylamino-6-isopropyl-amino-s-triazine) and simazine (2-chloro-4,6,bis(ethylamino)-*s*-triazine) are triazine herbicides *(1)*. Both exhibit herbicidal activity on certain annual broadleaf and grass weeds through inhibition of photosynthesis. In the United States, annual atrazine use is the greatest on corn (83%) followed by sorghum (11%) and sugarcane (4%). Atrazine and simazine annual use in the 21 major use states (Figure 1) account for 92% and 91% of the U.S. use, respectively *(2)*. Simazine is used less extensively than atrazine on corn.

The major uses of simazine are on fruit (especially citrus), nuts, and corn crops. In contrast to atrazine, the greatest use of simazine occurs in Florida and California rather than in the midwestern corn states.

Atrazine has been detected in surface water and groundwater in several of the major use states *(3-7)*. Typically, groundwater detections of atrazine occur much less frequently than surface water detections. Also, groundwater detections are usually lower in concentration than that in surface water. In surface water, atrazine concentrations in streams and rivers are episodic, with major peaks in the spring and early summer after field application in April and May. In impounded water bodies (reservoirs), the peak concentrations are usually lower than in rivers and occur at the same time; however, the duration may be longer due to longer hydraulic residence time. Simazine is detected less frequently than atrazine in ground and surface water in the U.S. and at lower concentrations *(3-7)*.

Historically, there have been few studies designed to assess exposure to pesticides through drinking water. The Environmental Protection Agency (EPA) conducted a national survey of rural individual and community water systems (CWS) wells for over 100 pesticides, including atrazine and simazine, in the late 1980s (*8*). This survey included only groundwater sources. Iowa (*9,10*), Minnesota (*11*), and Wisconsin (*12*) conducted private well and CWS pesticide surveys of drinking water in the late 1980s and early 1990s. Again, atrazine and simazine were included. The primary focus was groundwater and the studies were usually limited to only one year of data. A linkage of population to exposure was not made in these studies.

This study expands on these past assessments of atrazine and simazine in drinking water using monitoring data collected for a more recent time period (January 1993-December 1995) from CWS in 21 states. Herbicide exposure and population data from 1993 through 1995 are reported here, although Ciba is continuing this monitoring program at least through 1997 (5-year database). This study provides a more complete assessment of the herbicides' frequency of occurrence and concentrations for CWS populations on both ground and surface water sources through the development of a population-linked exposure (PLEX) database. These results add substantially to the body of knowledge on drinking water exposure to these herbicides and evaluate exposure relative to the established federal drinking water standards.

EPA Drinking Water Standards

The presence of atrazine and simazine in ground and surface raw water sources raised questions regarding possible exposure through drinking water. To provide guidance, EPA developed drinking water health advisory levels (HALs) for both chemicals in 1988 (*13*). By 1993, EPA, through the Safe Drinking Water Act (SDWA), also established maximum contaminant level (MCL) and monitoring requirements for several pesticides, including atrazine and simazine (*14*). The recommended HALs and enforceable MCLs are permissible concentrations in drinking water at which adverse health effects would not be expected to occur for the specified exposure duration. Both HALs and MCLs are based on the no observable effect level (NOEL) in animal toxicity studies. HALs are defined as the concentration of a chemical in drinking water that is not expected to cause any adverse noncarcinogenic effects for up to a certain number of consecutive days of exposure or a certain number of years of exposure, calculated with a margin of safety (Table I).

Beginning in 1993, CWS, initiated compliance monitoring of finished water for atrazine and simazine on a quarterly schedule for surface water supplies and once or twice annually for groundwater supplies. The purpose was to assess atrazine and simazine annual running mean concentration for each CWS for compliance with their respective MCL (Table I).

METHODS

Herbicide Major Use States

A hierarchical protocol was developed to determine the segments of the U.S. population served by CWS with potential exposure to atrazine and/or simazine. Based on agricultural land use data (*2*), company product use data (*15,16*), and summary national herbicide survey information (*17*), the 21 major use states were selected for quantitative exposure assessment (Figure 1). These 21 states represent 68% (175 million) of the total U.S. population (*18*) and 92% and 91% of the annual atrazine and simazine use in pounds, respectively, in the U.S. in 1988-89 (*15,16*). The highest atrazine use states are Illinois, Iowa, Nebraska, Indiana, Kansas, Ohio, and Wisconsin. The highest simazine-use states are California and Florida.

Population-Linked Exposure (PLEX) Database

Drinking water is provided to nearly 243 million people, or 94% of the total U.S. population, by 58,000 CWS (*19*). The other 15 million people (6%) receive drinking water from private wells or other nonregulated systems (*19*). A CWS, as regulated under the SDWA, is defined as a facility which provides piped water for human consumption to at least 15 service connections and provides water to the same population year round. A CWS can use different raw water sources: groundwater, surface water (rivers, lakes, and reservoirs) or blends of both.

There are 34,591 CWS in the 21 major use states. These facilities provide drinking water to 92% (161 million) of the 175 million people in these states (Table II). SDWA quarterly compliance monitoring data for atrazine and simazine from CWS in the 21 major use states were obtained from the state regulatory agencies. These primary data represent a 3-year period (January 1993-December 1995). In addition, data from monitoring studies (secondary data) for the Great Lakes were used to supplement SDWA data (*20*). There are 14,440 CWS (42%) with 58,177 quarterly samples analyzed for atrazine entered into the PLEX database (Table III) and 13,853 CWS (40%) with 53,791 simazine data points (Table IV).

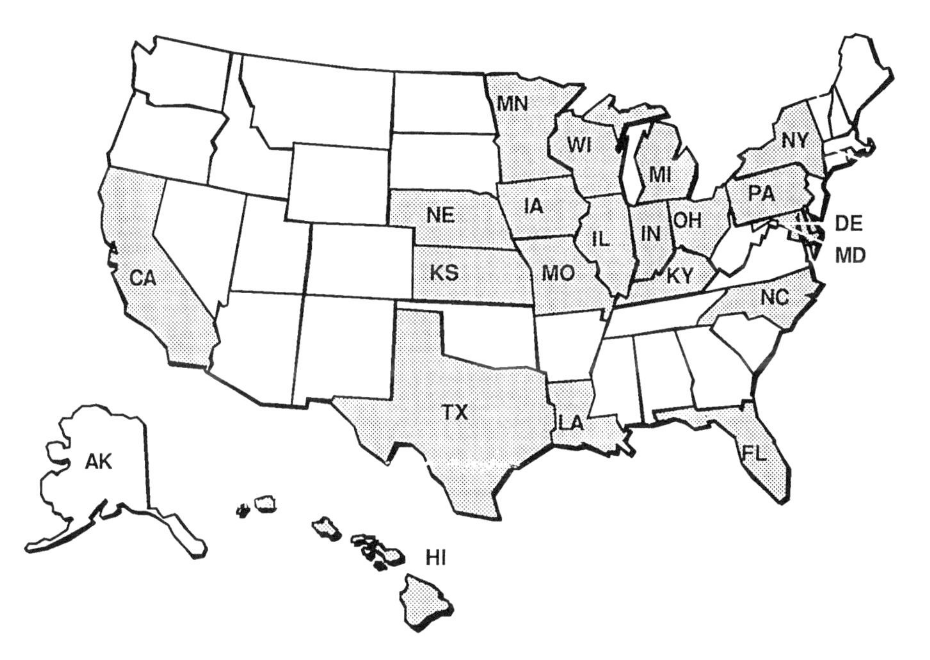

Figure 1. Major Use States for Atrazine and Simazine

Table I. Health Advisory Levels (HALs) and Maximum Contaminant Levels (MCLs) for Atrazine and Simazine

		Atrazine			*Simazine*		
Exposure	*Duration*	*MCL (ppb)*	*HAL (ppb)*	*Safety Factor*	*MCL (ppb)*	*HAL (ppb)*	*Safety Factor*
1-day, child:	5 consecutive days	-	100	100	-	70	100
10-day, child:	14 consecutive days	-	100	100	-	70	100
7-year, child:	Approx. 7 years	-	50	100	-	70	100
7-year, adult:	Approx. 7 years	-	200	100	-	70	100
70-year, adult:	Lifetime	3	3	1000	4	4	1000

Table II. 21 Atrazine/Simazine Major Use States: Numbers of Community Water Systems (CWS), State and CWS Populations

State	State Population	Surface Water CWS		Groundwater CWS		Other CWS		Total CWS		Population Not Served by CWS
		Number	Total Population	Number	Total Population	Number	Total Population	Number	Total Population	
California	31,211,000	499	9,741,226	3,159	22,555,155	60	1,865,294	3,718	34,161,675	-
Delaware	700,000	3	265,800	237	142,074	1	171,800	241	579,674	120,326
Florida	13,679,000	23	1,300,000	2,034	12,300,000	124	690,000	2,181	14,290,000	-
Hawaii	1,172,000	11	26,006	122	1,285,760	5	28,644	138	1,340,410	-
Illinois	11,697,000	462	6,780,297	1,336	2,476,105	367	1,651,351	2,165	10,907,753	789,247
Indiana	5,713,000	87	1,143,586	804	1,762,187	33	1,022,093	924	3,927,866	1,785,134
Iowa	2,814,000	83	567,032	1,036	1,350,946	37	392,528	1,156	2,310,506	503,494
Kansas	2,531,000	274	756,983	614	731,866	32	743,962	920	2,232,811	298,189
Kentucky	3,789,000	328	3,248,001	172	372,614	18	133,780	518	3,754,395	34,605
Louisiana	4,295,000	58	1,899,245	1,346	2,657,450	0	0	1,404	4,556,695	-
Maryland	4,965,000	54	3,670,671	453	526,331	6	70,000	513	4,267,002	697,998
Michigan	9,478,000	279	5,051,393	1,329	1,741,468	12	214,753	1,620	7,007,614	2,470,386
Minnesota	4,517,000	38	1,303,428	919	2,116,636	0	0	957	3,420,064	1,096,936
Missouri	5,234,000	96	3,737,841	1,082	1,432,177	10	270,354	1,188	5,440,372	-
Nebraska	1,607,000	8	10,979	665	859,175	3	389,261	676	1,259,415	347,585
New York	18,031,000	656	11,301,181	2,271	4,502,646	275	743,725	3,202	16,547,552	1,483,448
North Carolina	6,945,000	292	3,415,464	2,386	1,324,220	5	103,291	2,683	4,842,975	2,102,025
Ohio	11,091,000	294	6,434,680	1,246	3,310,067	0	0	1,540	9,744,747	1,346,253
Pennsylvania	12,048,000	309	5,896,236	1,780	1,523,170	214	2,917,688	2,303	10,337,094	1,710,906
Texas	18,031,000	923	10,019,022	4,338	5,397,837	70	1,626,417	5,331	17,043,276	987,724
Wisconsin	5,082,000	38	1,458,572	1,172	1,902,617	3	138,446	1,213	3,499,635	1,582,365
Total	174,630,000	4,815	78,027,643	28,501	70,270,501	1,275	13,173,387	34,591	161,471,531	17,356,621

- Indicates no data

Table III. 21 Major Use States PLEX Data for Atrazine January 1993 - December 1995

	Totals	Groundwater	Surface Water	Other
Data				
Number of Samples	58,177	46,187	9,688	2,302
Number of Detections	5,423	1,359	3,660	404
Percent of Detections	9	3	38	18
CWS				
Number of CWS with Data	14,440	12,170	1,973	297
Percent CWS with Data	42	43	41	23
Number of CWS with No Detections	13,169	11,823	1,116	230
Number of CWS with Detections	1,271	347	857	67
Percent of CWS with No Detections [1]	91	97	57	77
Percent of CWS with Detections	9	3	43	23
Populations				
Major Use State Population	174,630,000			
Population on CWS	161,471,531	70,270,501	78,027,643	13,173,387
Population Served by CWS with Data	113,285,471	44,128,148	60,482,352	8,674,971
Percent Major Use State Population Assessed	65	25	35	5
Percent CWS Population Assessed	70	63	78	66
Population with No Detections	88,951,223	42,305,650	40,903,324	5,742,249
Population with Detections	24,334,248	1,822,498	19,579,028	2,932,722
Percent of population with No Detections[1]	79	96	68	66
Percent of population with Detections	21	4	32	34

[1] Percent CWS and populations with or without detects are based on the number of assessed CWS and populations respectively.

Table IV. 21 Major Use States PLEX Data for Simazine January 1993 - December 1995

	Totals	Groundwater	Surface Water	Other
Data				
Number of Samples	53,791	43,785	7,949	2,057
Number of Detections	730	101	595	34
Percent of Detections	1	0.23	7	2
CWS				
Number of CWS with Data	13,853	11,903	1,641	309
Percent CWS with Data	40	42	34	24
Number of CWS with No Detections	13,589	11,835	1,458	296
Number of CWS with Detections	264	68	183	13
Percent of CWS with No Detections[1]	98	99	89	96
Percent of CWS with Detections	2	1	11	4
Populations				
Major Use State Population	174,630,000			
Population on CWS	161,471,531	70,270,501	78,027,643	13,173,387
Population Served by CWS with Data	110,815,240	44,011,929	57,732,186	9,071,125
Percent State Population Assessed	63	25	33	5
Percent CWS Population Assessed	69	63	74	69
Population with No Detections	101,578,523	43,529,287	49,965,500	8,083,736
Population with Detections	9,236,717	482,642	7,766,686	987,389
Percent of population with No Detections[1]	92	99	87	89
Percent of population with Detections	8	1	13	11

[1] Percent CWS and populations with or without detects are based on the number of assessed CWS and populations respectively.

The majority (82%) of the CWS in the 21 states use groundwater as the raw water source (Table II). Thus, the two herbicides' databases contain more groundwater (4-5 times) than surface water samples (Tables III, IV). The databases for both herbicides were also dominated by samples with analytical results reported as nondetections. The limit of detection (LOD) required for the analysis of atrazine and simazine samples varied among the 21 major use states. Sixteen states used an LOD of ≤0.3 ppb for atrazine and ≤0.4 ppb for simazine. The maximum LOD was 2.5 ppb for atrazine and 8.0 ppb for simazine. The maximum LOD was reported by the state of Louisiana. CWS data from Louisiana were not included in the aggregate atrazine and simazine exposure profiles (Tables V, VI). The Louisiana exposure profile would be an artifact of the high LOD values and could underestimate the expected surface and groundwater concentrations based on either product's use pattern in the state.

Since the PLEX databases for atrazine and simazine were dominated by samples with nondetections, the CWS exposure profiles (Tables V, VI) are primarily driven by the LOD concentration. To develop the PLEX databases, a numerical value had to be assigned to the samples with nondetectable residues. Following EPA guidance, a concentration of one-half the detection limit was assigned to all with nondetectable residue samples (*21*). The substitution value is arbitrary (*22,23*) and provides no actual knowledge of the concentration values below the reporting limit. However, it does provide a conservative estimate of drinking water exposure by assuming all samples have atrazine or simazine present at one-half the LOD.

The CWS with atrazine and simazine monitoring data served populations of 113 and 111 million, respectively (Tables III, IV). The populations associated with atrazine monitoring data represent 70% of CWS and 65% of total populations, respectively, in the 21 states (Table III). The populations for simazine exposure were slightly less: 69% and 63% (Table IV). These CWS populations were also placed in the database.

Herbicide concentration and population data were then paired for each CWS and aggregated for all CWS to construct state and multi-state exposure profiles for each herbicide through the development of the PLEX database. All data were entered into individual state PLEX databases along with CWS population data and source water type (groundwater; surface water including river, reservoir, or lake; or "other" for blended waters). Average concentrations for the two herbicides in finished drinking water were determined for each CWS. When several annual means were available, the exposure concentration for the CWS is the average of all available annual records since 1993. From these state-specific databases, an aggregate or multi-state exposure profile was developed for each herbicide for the three source-water classifications. For each category, the numbers of CWS and populations served by these facilities were totaled. In a similar fashion, the simazine multi-state exposure profile was developed (Table VI). EPA guidance was used to develop protocols for data collection, database preparation, and data analysis (*21*).

Table V. 21 Major Use States: Population Exposure to Atrazine Above and Below the Maximum Contaminant Level (MCL), Three Year Period (1993-1995)[1]

Group MCL = 3.0 ppb	*Number in Group*	*Population Served*	*Percent of Population*
Surface Water >3.0 ppb	13	16,161	0.1
Surface Water ≤3.0 ppb	1,904	58,570,166	99.9
Groundwater >3.0 ppb	0	0	0.0
Groundwater ≤3.0 ppb	11,474	42,564,740	100.0
Other (Blends) >3.0 ppb	0	0	0.0
Other (Blends) ≤4.0 ppb	297	8,674,971	100.0
Total >3.0 ppb	13	16,161	0.01
Total ≤3.0 ppb	13,675	109,809,877	99.99

[1]Database dominated by samples with nondetections (91%); Louisiana data not included.

Table VI. 21 Major Use States: Population Exposure to Simazine Above and Below the Maximum Contamination Level (MCL), Three Year Period (1993-1995)[1]

Group MCL = 4.0 ppb	*Number in Group*	*Population Served*	*Percent of Population*
Surface Water >4.0 ppb	0	0	0.0
Surface Water ≤4.0 ppb	1,585	55,836,161	100.00
Groundwater >4.0 ppb	0	0	0.0
Groundwater ≤4.0 ppb	11,207	42,448,521	
Other (Blends) >4.0 ppb	0	0	0.0
Other (Blends) ≤4.0 ppb	309	9,071,125	100.0
Total >4.0 ppb	0	0	0.0
Total ≤4.0 ppb	13,101	107,355,807	100.0

[1]Database dominated by samples with nondetections (99%); Louisiana data not included.

RESULTS

The actual exposure of the CWS population to atrazine and simazine in each of the 21 states was evaluated using SDWA compliance monitoring data collected between January 1993 and December 1995 (Tables V, VI). These data represent the best available information from state SDWA agencies in the 21 states. Drinking water data entered into the PLEX database provide a direct link between the population and the concentration of atrazine and simazine in the drinking water.

Atrazine

Atrazine was detected infrequently in the CWS samples. It was not detected in 91% of the samples (Table III). As expected, there were more nondetections observed in groundwater than surface water sources. Atrazine was not detected in 97% of the groundwater samples, compared to 57% of the surface water samples (Table III).

Of the assessed population, 89 million had no detectable exposure to atrazine in drinking water (Table III). Overall, 13,675 CWS serving 109,664,420 individuals (99.9%) had average atrazine concentrations less than the 3.0 ppb MCL over the three year period (Table V).

Thirteen of the 13,688 assessed CWS had average concentrations for the 3-year period above 3.0 ppb (Table V). The multi-year average atrazine concentrations ranged from 3.06 to 6.91 ppb. Nine of the 13 CWS had multi-year average concentrations between 3.0 and 4.0 ppb. All 13 CWS obtain raw water from an impounded (reservoir) surface water source.

Simazine

Simazine was detected even less frequently than atrazine in the CWS quarterly samples. It was not detected in 98% of the samples (Table IV). Again, groundwater samples had more nondetections (99%) than surface water (89%).

Of the population served, 102 million (89%) had no detectable concentrations of simazine in drinking water (Table IV). Overall, 13,101 CWS serving 107,355,807 individuals (100%) had average simazine concentrations less than the HAL of 4.0 ppb over the three-year period (Table VI). No CWS had a multi-year mean simazine concentrations above the MCL of 4.0 ppb (Table VI).

Uncertainty

The PLEX database is considered to be representative of potential exposure for populations served by CWS. The population-linked estimates of atrazine and simazine concentrations represent actual exposure to persons consuming potable water from regulated water supplies. However, a few limitations exist and introduce some degree of uncertainty. Not all CWS have finished-water monitoring data. The absence of monitoring data for a CWS is usually due to two factors: 1) a monitoring waiver has been granted to the CWS by the state SDWA agency, or 2) the CWS purchases finished water from another CWS. In addition, populations on private wells within a state were not evaluated since they do not receive water from a CWS. Some CWS, especially groundwater systems, are represented by exposure concentrations based on less than 4 quarterly samples.

CONCLUSIONS

This drinking water exposure assessment is the most comprehensive study conducted to date in the U.S. to evaluate an agricultural product's presence in CWS drinking water. It targets a CWS population of 161 million of 175 million who receive drinking water from 34,591 CWS in the 21 major atrazine and simazine use states (Table II). The objective was to better assess the two products' frequency of occurrence and exposure for CWS populations actually drinking the water.

CWS monitoring data for the 21 major use states represent a reasonably conservative estimate of exposure of U.S. populations to atrazine and simazine through drinking water. This is illustrated for atrazine (Figure 2) and simazine (Figure 3) by comparing the toxicological end points (used to establish a drinking water reference dose) with the individual herbicide's MCL, and the actual concentration profiles (1993-95) for the three types of CWS water source categories. The margin of safety (MOS) is calculated from the no observed effect level (NOEL) used to establish the drinking water reference dose for each chemical (*13*). The lifetime drinking water MCL for atrazine (3.0 ppb) and simazine (4.0 ppb) have a 1,000-fold safety factor incorporated into the calculation. The actual exposure to atrazine for 94% of the assessed population is equal to or less than 0.3 ppb, which corresponds to a MOS of at least 10,000 from the NOEL for the most sensitive species tested in animal toxicity studies. Additionally, the most exposed population for atrazine (mean concentration 3.1-6.9 ppb) had a MOS of approximately 1,000. Similarly, the vast majority of the population (96%) exposed to simazine was at or less than 0.24 ppb, which corresponds to a MOS greater than 10,000. In both Figure 2 and Figure 3, it should be noted that although the bar width is proportional to the percent of the population exposed, both sets of data are dominated by samples with nondetects, thereby representing conservative exposure estimates.

These CWS exposure concentrations represent conservative exposure scenarios for persons in the 21 major use states and could be reasonably extrapolated to the other 29 states with populations using drinking water supplied by CWS. The individual well and CWS populations in the 29 minor use (8-9% of product use annually) states were not assessed. However, it is expected that drinking water exposure to atrazine and simazine would not be greater than, and most likely would be less than, exposure observed at the CWS in the 21 major use states (Tables III, IV, V, and VI). Presumably, for the vast majority of the 84 million people in the 29 minor use states, atrazine and simazine would not be present (nondetectable) in drinking water. Therefore, there is essentially no exposure.

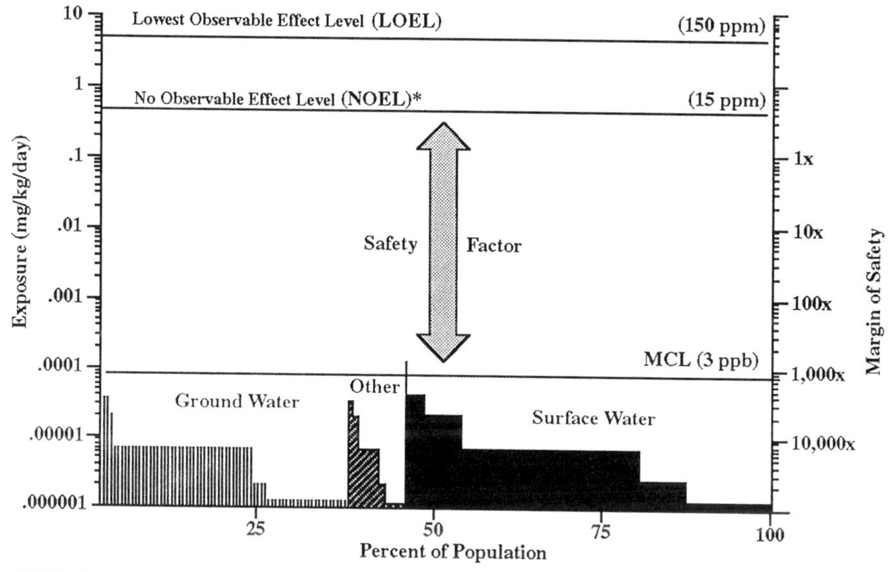

Figure 2. Atrazine exposure profile and associated populations. Bar width is proportional to the percent of the population exposed and data are dominated by samples with nondetects (91%, where 1/2 the LOD is used).

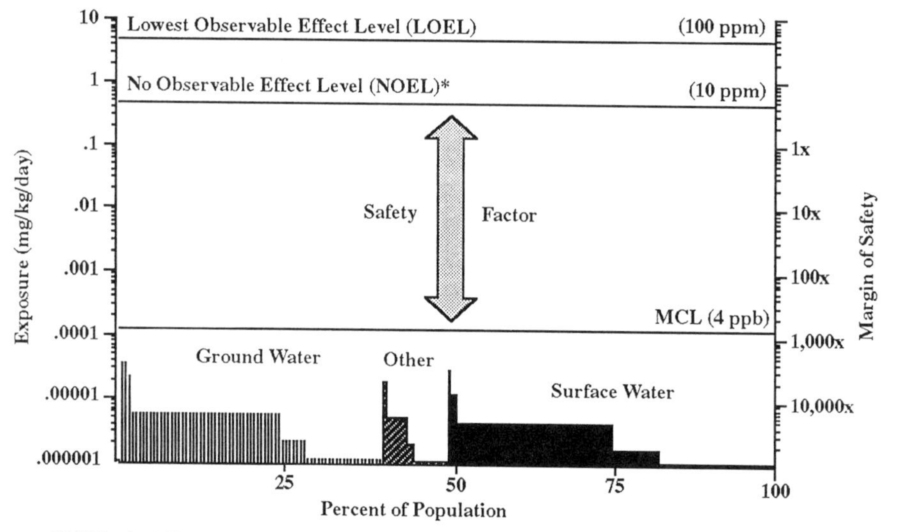

Figure 3. Simazine exposure profile and associated populations. Bar width is proportional to the percent of the population exposed and data are dominated by samples with nondetects (99%, where 1/2 the LOD is used).

Private wells were not included in the PLEX database. Persons receiving potable water from private wells were not quantitatively assessed, but represent 6% of the U.S. population (*19*). National groundwater studies of private wells have shown that over 98% of private wells have atrazine concentrations ≤0.02 ppb (*24,25*), and over 99.8% have simazine concentrations below 0.38 ppb (*8*). These reports indicate that exposure from private wells is minimal.

The PLEX analysis indicates that exposure to these two herbicides in CWS drinking water at concentrations above MCLs is localized. It shows, except for a few surface water supplies in the 21 major use states, the overall population exposure to atrazine and simazine in drinking water is geographically limited, and, in most of the major and minor use states, is below detection limits. Surface water CWS with atrazine multi-year average concentrations above the MCL serve small communities in agricultural areas. These 13 CWS draw raw water from reservoirs. Focus is now on these CWS which require targeted, product-crop watershed use evaluations to determine what mitigative remedies will reduce the finished water concentrations and ensure compliance with the SDWA. Ciba is committed to partnerships with communities on site specific management plans to improve water quality.

LITERATURE CITED

(1) Mills, M.S.; Thurman, E.M. *Environ. Sci. and Technol.* **1994**, *28*, pp. 600-605.

(2) U.S. Department of Agriculture; *Agricultural Chemical Usage*, U.S. Department of Agriculture, National Agricultural Statistics Service, 1992.

(3) Baker, D.B. *Sediment, Nutrient, and Pesticide Transport in Selected Great Lakes Tributaries.* Great Lakes National Program Office, U.S. Environmental Protection Agency, Region 5, 1988.

(4) Thurman, E.M.; Goolsby, D.A.; Meyer, M.T.; Kolpin, D.W. *Environ. Sci. and Technol.* **1991**, *25*, pp. 1794-1796.

(5) Goolsby, D.A., Thurman, E.M.; Kolpin, D.W. *Open-File Rep. U.S. Geol. Surv.*, No. 91-4034, **1991,** pp. 183-188.

(6) Goolsby, D.A., Coupe, R.H.; Markovchick, D.J. *Open-File Rep. U.S. Geol. Surv.*, No. 91-4163, **1991**.

(7) Keck, P. *Missouri River Monitoring Study.* Missouri River Public Water Supplies Association. 1991

(8) U.S. Environmental Protection Agency. November 1990. *National Survey of Pesticides in Drinking Water Wells, Phase I Report.* Office of Water and Office of Pesticides and Toxic Substance. EPA 570-9-90-015.

(9) Iowa Department of Natural Resources. March 1988. *Pesticide and Synthetic Organic Compound Survey.* Report to Iowa General Assembly on the Results of the Water System Monitoring Required by House File 2303.

(10) Kross, B.C., et al. 1990. *The Iowa Statewide Rural Well Water Survey: Water Quality Data: Initial Analysis.* Iowa Department of Natural Resources, Technical Information Services, 19, 142 p.

(11) Klaseus, T.G., et al. February 1988. *Pesticides and Groundwater: Surveys of Selected Minnesota Wells.* Minnesota Department of Health and Minnesota Department of Agriculture. Prepared for Legislative Commission on Minnesota Resources, 93 p.
(12) Wisconsin Department of Agriculture. 1989. *Trade and Consumer Protection.* Grade A Dairy Farm Well Water Quality Survey.
(13) U.S. Environmental Protection Agency. 1989. *Drinking Water Health Advisory 7: Pesticides.* Lewis Publishers, Chelsey, Michigan, 819 pp.
(14) U.S. Environmental Protection Agency. Wednesday, January 30, 1991. *National Primary Drinking Water Regulations: Final Rule.* 40 CFR Part 141, 142, 143.
(15) Ciba-Geigy Corporation. 1988. *Atrazine Use Data by County.* Ciba-Geigy Corporation, Greensboro, NC.
(16) Ciba-Geigy Corporation. 1989. *Simazine Use Data by County.* Ciba-Geigy Corporation, Greensboro, NC.
(17) U.S. Environmental Protection Agency. 1992. *Pesticides in Ground Water Database. A compilation of Monitoring Studies: 1971-1991, National Summary.* U.S. Environmental Protection Agency, Office of Pesticide Programs, U.S. Government Printing Office. Washington, DC.
(18) Bureau of Population Census. 1994. *Estimates of the Resident Populations of States: July 1, 1991 to 1993 and July 1, 1992 to 1993. Components of Change.* Suitland, MD.
(19) U.S. Environmental Protection Agency. 1993. *National Compliance Report.* U.S. Environmental Protection Agency, National Public Water System Supervision Program, U.S. Government Printing Office, Washington, DC.
(20) Schottler, S.P.; Eisenreich, S.J. *Environ. Sci. and Technol.* **1994**, *28*, pp. 2228-2232.
(21) U.S. Environmental Protection Agency. 1989. *Risk Assessment Guidance for Superfund, Volume 1, Human Health Evaluation Manual (Part A).* Office of Emergency and Remedial Response, U.S. Government Printing Office, Washington DC.
(22) Helsel, D.R., R.M. Hirsch. *Statistical Methods in Water Resources.* Elsevier, New York. 1992.
(23) Helsel, D.R. *Environ. Sci. and Technol.* **1990**, *24*, pp. 1768-1774.
(24) Holden, L.R.; Graham, J.A.; Whitmore, R.W.; Alexander, J.W.; Pratt, R.W.; Liddle, S.K.; Piper, L.L. *Environ. Sci. and Technol.* **1992**, *26*, pp. 935-943.
(25) U.S. Environmental Protection Agency. 1992. *Another Look-National Survey of Pesticides in Drinking Water Wells. Phase 2 Report.* U.S. Environmental Protection Agency. U.S. Government Printing Office, Washington, DC. EPA/579/09-91-020.

Chapter 22

Impact of Midwest Farming Practices on Surface and Groundwater Equally

J. L. Hatfield and D. B. Jaynes

[1] **National Soil Tilth Laboratory, Agricultural Research Service, USDA 2150 Pammel Drive, Ames, IA 50011–4420**

Nonpoint source pollution has been linked with agricultural practices across the United States. Farming practices can be modified to improve ground and surface water quality and have a positive impact on sustainability and adaptability by the farmer. Farming practices can alter the hydrologic balance of a landscape and ground and surface water quality problems can be related to the flow of water through the landscape. Surface runoff, which is a major nonpoint source problem, can be reduced by adoption of conservation tillage practices. To fully understand the impact of farming practices on ground and surface water quality requires that each farming practice be examined from a mechanistic view of water management.

Farming is intense in the Midwest with over 80% of the agricultural land area in some aspect of grain crop production. Coupled with the intense crop production is the use of herbicides and nitrogen fertilizer as part of the production system. Herbicide use in the Midwest represents nearly 60% of all the herbicides applied in the United States (*1*). These herbicide use patterns have remained fairly stable over the past 10 years with atrazine being used on nearly 65% of the corn area. Nitrogen fertilizer is used on nearly 97% of the grain crop area of the Midwest. With these intensive levels of inputs there have been detections of herbicides and nitrate-nitrogen in surface and ground water of the Midwest.

In the last 10 years several reconnaissance studies of the ground and surface water resources of the Midwest have been undertaken by a number of different groups (*2-6*). These studies have shown that pesticides, including atrazine (2-chloro-4-ethylamino-6-isopropylamino-*s*-triazine) were detected in almost half of the groundwater samples. The differences among studies were related to analytical reporting limits, well selection criteria, and time of sample collection (*4*). Kolpin *et al.* (*5*) found that in 837 samples from 303 wells across the Midwest, five of the six most frequently detected compounds were herbicide metabolites. They suggested that levels of metabolites may be more frequently detected in groundwater and that degradation pathways of herbicides in the soil need to be better understood. Kolpin *et al.* (*7*) found detectable levels of herbicides or atrazine metabolites in 28.4% of the 303 midwestern wells sampled in 1991. None of the wells sampled had herbicide concentrations that exceeded standards for safe drinking water.

Nitrate also has been found in shallow ground water samples in the Midwest. Burkart and Kolpin (*2*) reported nitrate-nitrogen concentrations above the Federal EPA standard for safe drinking water (10 $\mu g\ L^{-1}$) in 6% of their samples. Madison and Burnett (*8*) found 6.4% of the 123,656 wells sampled throughout the Midwest had nitrate-nitrogen concentrations above 10 $\mu g\ L^{-1}$ and 13% had concentrations between 3 and 10 $\mu g\ L^{-1}$. Hallberg (*3*) suggested that movement of herbicides and nitrate into ground water wells would be dependent upon the intensity of the farming practices and the hydrologic and geologic conditions.

Goolsby and Battaglin (*9*) analyzed data from surface water samples from the Midwest and found that concentrations and mass transport of herbicides follow an annual cycle. In their reconnaissance study, several drainage basins in the Mississippi River watershed were sampled beginning in 1989. They found that less than 3% of the herbicide mass applied to crop land was transported into streams; however, this mass was sufficient to cause atrazine concentrations to exceed 3 $\mu g/L^{-1}$ portions of the Mississippi River for short periods of time. Peak herbicide concentrations were found in storm runoff during May, June, and July with other detections occurring throughout the year. Concentrations were related to the amount applied. Nitrate-nitrogen concentrations throughout the year exhibited a different pattern than herbicides with the highest concentrations in the winter and spring and the lowest during the summer. There is little information about the magnitude of agrichemical loads to surface and shallow ground water from agricultural watersheds.

There has also been little mechanistic investigation of the attributes of farming practices that could be managed to influence ground and surface water quality. Hatfield (*10,11*) described how different attributes of farming practices could be modified to improve environmental quality. This report will show where the potential lies in modifying farming practices for further improvements in ground and surface water quality.

Impact of Farming Practices on Hydrology

Movement of herbicides and nitrate-nitrogen are determined by movement of water. Therefore, to develop a mechanistic model of farming practices and water quality, it is important to understand the patterns of water movement associated with different farming practices. The distribution and amount of precipitation occurring throughout the year are determined by movement of fronts and air masses across a region. The amount of precipitation that a specific area receives is dependent upon its position on the earth and size of the land area, and nearness to mountains. These are basic concepts that underlie the input side of the hydrologic balance. Another important factor is the application of irrigation water to a field from either a ground or surface water source. Precipitation that falls on the earth may infiltrate the soil, or be intercepted by the leaves and stems of the vegetation canopy and subsequently evaporated, or run off from the field into nearby streams or rivers. Of the precipitation that infiltrates the soil, a portion may be returned to the atmosphere by evaporation from both the soil and crop canopy. Any excess may move through the soil profile beyond the reach of the root system and ultimately become part of the ground water. One can imagine the soil as a leaky container that can lose water from the surface if water is applied more quickly than the soil can absorb the water; and from the bottom the amount applied is beyond the ability of the soil to hold water.

Water may be removed from the soil volume through evaporation from the soil surface. Evaporation occurs when the soil is wet and rapidly decreases as the soil dries. Evaporation is a small portion of the water lost from agricultural systems and may represent less than 15% of the annual precipitation (Hatfield and Prueger, unpublished data). Growing crops explore the soil volume with a root system that extracts water from various soil layers depending upon the stage of crop growth and available soil water. Water taken up by the root system is moved to the leaves where evaporation occurs as part of the process of cooling the leaves. This process is referred to as transpiration and is a critical part of plant growth. Without an adequate soil water supply, agricultural plants undergo stress resulting in reduced yield potential. In agriculture, the soil water evaporation process and transpiration are combined into a term, evapotranspiration. Evapotranspiration removes water from the soil profile; the amount being determined by atmospheric conditions. Hatfield (*12*) described this process and the various techniques available to either estimate or directly measure evapotranspiration. These methods have been used to determine the rate of water use by different cropping systems and to compare cropping systems for their efficiency in using water for crop growth.

The critical aspect of evapotranspiration for assessment of nonpoint source pollution and farming practices is the seasonal relationship between precipitation and evapotranspiration. This can be understood by examining the records from three locations shown in Figure 1. Precipitation throughout the year is dependent upon the general climate of the area as shown in this example from Ames, Iowa; Peoria, Illinois; and West Lafayette, Indiana. The precipitation and evapotranspiration distribution patterns throughout the year are similar among locations with the highest rainfall during the summer months. While the distribution patterns are similar, the monthly amounts are different, which leads to a greater excess in the soil profile at Peoria and West Lafayette than at Ames in the early spring. This additional water could lead to either increased leaching or greater potential surface runoff in the eastern part of the Corn Belt. There would be similar soil water use patterns with latitude across the Midwest due to temperature and length of the growing season.

There are differences in evapotranspiration caused by different tillage practices. Hatfield *et al.* (*13*) showed that no-tillage practices that increased the crop residue on the soil surface decreased the soil water evaporation rates in the

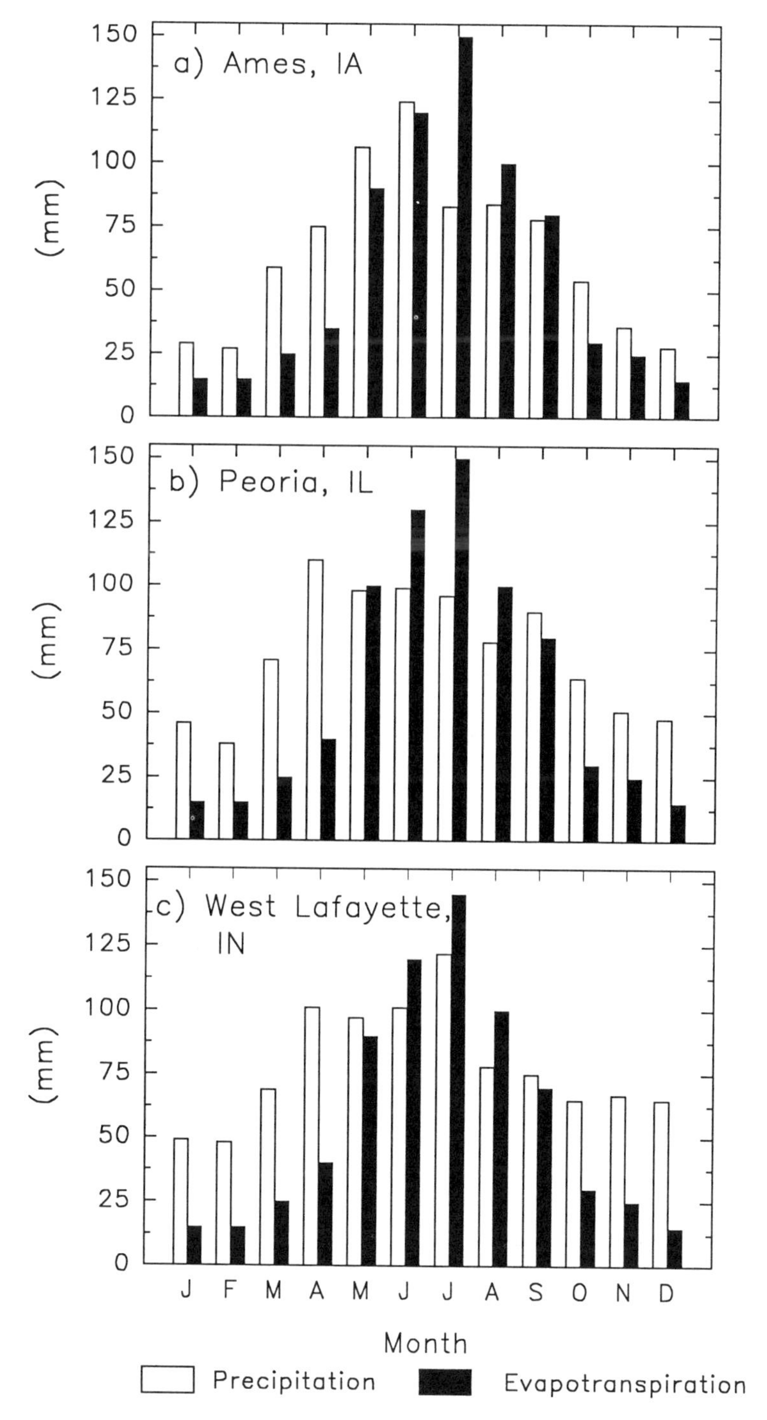

Figure 1. Yearly distribution of precipitation and evapotranspiration for Ames, Iowa, Peoria, Illinois and West Lafayette, Indiana based on a 30 year normal, 1961-1990.

early season and increased evapotranspiration later in the growing season because of the increased soil water in the soil profile. Typically, in central Iowa, the yearly total of evapotranspiration is approximately 650 mm. The largest portion of this amount occurs in the summer, and less than 150 mm occurs from October through April. These seasonal patterns show that the greatest chance of surface runoff from fields would occur in the early spring in the Midwest.

Surface Runoff Resulting from Farming Practices. Baker and Laflen (*14*), after studying the effects of tillage practices on herbicide washoff, concluded that washoff water from crop residue would either run off or infiltrate into the soil based on a combination of factors. These factors were the timing and intensity of rainfall and the infiltration capacity of the soil. Baker and Shiers (*15*) found that the largest concentration of cyanazine [2-[[4-chloro-6-(ethylamino)-1,3,5-triazin-2-yl]amino]-2-methylpropionitrile], alachlor [2-chloro-2'-6'diethyl-N-(methoxy-methyl)-acetanilide], and propachlor [2-chloro-N-isopropylacetanilide] occurred after the first rainfall event following their application. Rainfall events within 12 hours of herbicide application are responsible for considerable loss from washoff from the crop residue. Whether there is movement from the edge of the field will depend upon the amount of rainfall and the infiltration rate into the soil.

Surface runoff of herbicides from the edge of fields is a short-lived event as depicted in Figure 2. In this event, within the Walnut Creek watershed in Iowa, atrazine was transported in the surface runoff which occurred shortly after application. Concentrations were measured between 25 and 30 $\mu g\ L^{-1}$ for atrazine while metolachlor (2-chloro-N-(2-ethyl-6-methylphenyl)-N-(2-methoxy-1-methyl-ethyl)acetamide) was measured at 75 to 80 $\mu g\ L^{-1}$ (*16*). Fawcett *et al.* (*17*) summarized the results of different natural rainfall studies across the United States and concluded that conservation tillage practices reduced herbicide runoff by 70% compared to moldboard plowing. The no-till practice has been shown to increase infiltration and reduce surface runoff and sediment transport because of the effect of the residue on protecting the soil surface. Conversely, Myers *et al.* (*18*) found that no-till practices increased surface runoff. There is a difference in the studies reported in the literature on herbicide runoff when natural rainfall is compared to simulated rainfall. Fawcett *et al.* (*17*) found that rainfall simulation studies that applied very heavy rainfall amounts after herbicide application produced the highest concentrations in the runoff. They also stated that this type of event would rarely occur. Under normal conditions though there would still be some infiltration of water into the soil and subsequent movement of herbicide into the upper soil layer.

Surface runoff will be more likely to occur when the soil is saturated and there is an excess of precipitation compared to water use. The central and eastern part of the Corn Belt, as illustrated by Peoria and West Lafayette, have a greater chance of soil profile saturation in the spring as compared to Ames. Moving from east to west across the Corn Belt, the chances of surface runoff of herbicides to decrease would be expected. However, surface runoff is a result of sloping land and as the slope increases, there is increased chance of a runoff event after rainfall. Hills in western Iowa will experience a greater chance of surface runoff in the

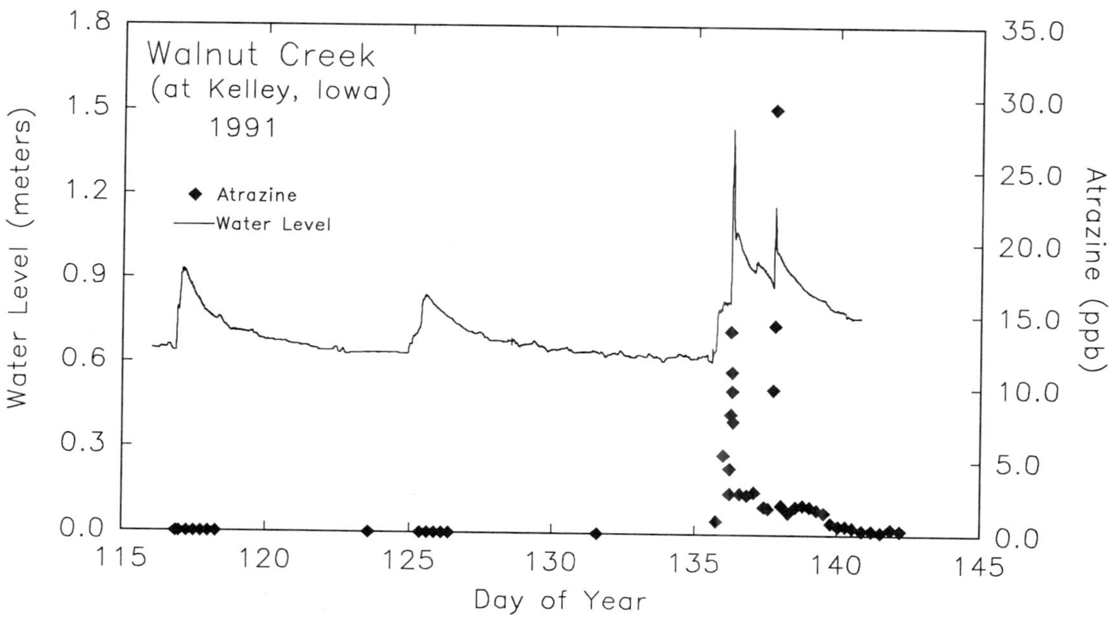

Figure 2. Distribution of atrazine concentration during a surface runoff event on Day-of-Year (DOY) 140 in 1991 from a conventional tilled field in Walnut Creek watershed.

spring than will relatively level central Iowa even with less rainfall. Even though there is an excess of precipitation above evaporation in the late fall and early winter, there is little chance of surface runoff because the soil profile has been depleted of soil water by the crop water use over the summer and there is little herbicide available at the surface to be moved by surface runoff events. All of these factors must be integrated to assess the effect of changing farming practices on water quality.

Leaching Resulting from Farming Practices. The quantity of water that infiltrates the soil may increase the amount of water moved through the root zone and toward the ground water. This movement is likely to occur in the portion of the year immediately after application when there is little water use by the crop and the soil profile is saturated. Soluble herbicides and nitrate-nitrogen are easily transported with moving water, and when water moves below the bottom of the root zone, chemicals are moved out of the root zone. This would be considered offsite movement and subject to the overall water flow patterns within the landscape. Leaching through the soil profile would be dependent upon the amount of crop residue, the presence or absence of cracks and macropores in the soil profile, and the distribution of water within the soil profile. Green *et al.* (*19*) found that increasing the amount of corn residue on the soil surface decreased the time for the peak atrazine concentrations to occur in the leachate from more permeable soil columns. They suggested that crop residue intercepts the herbicide and permits a more steady pattern of infiltration into the soil. Under crop residue, the herbicide was moving more slowly through the soil profile. Gish *et al.* (*20*) evaluated no-till compared to tilled fields and found in the no-till field that 28% of the water samples collected below the root zone had detectable amounts of atrazine ranging from 0 to 10 $\mu g\ L^{-1}$. In the tilled fields, 53% of the samples had detectable levels of atrazine ranging from 3 to 35 $\mu g\ L^{-1}$. In contrast, an earlier study by Isensee *et al.* (*21*) and Gish *et al.* (*22*) reported that increased infiltration under no-tillage conditions enhanced the chemical transport because of increased preferential flow through the root zone. The effects of management practices on preferential flow and chemical movement through the soil profile need to be more completely evaluated across a wider range of soils and soil water management practices.

Herbicide movement through no-till soils is complicated because of the addition of organic matter to the soil with the adoption of this farming practice. Blevins *et al.* (*23*) found that organic matter increased in the surface layer of long-term no-till plots. Dick (*24*) also reported that reducing the intensity of tillage increased the organic matter content. Organic matter has been correlated with changes in herbicide activity, and Upchurch *et al.* (*25*) reported that the soil organic fraction accounted for 66% of the variation in herbicide activity. Increases in soil organic matter have been related to changes in herbicide activity and degradation. Hudson (*26*) reported that increases in soil organic matter content led to an increase in the soil water holding capacity. Changes in the soil organic matter content will lead to changes in water movement patterns through increased storage within the soil profile but will also change herbicide interaction with the soil. The recent work by Gish *et al.* (*20*) suggests an interaction between the movement of herbicides and biological activity in the soil profile caused by changes in soil organic matter content.

Movement of herbicides through the soil profile results in groundwater and subsurface drainage concentrations that are often below the 3.0 $\mu g\ L^{-1}$ Health Advisory Level (HAL). The HAL is a standard for the annual average concentration in drinking water and is set using wide safety margins assuming daily intake over lifetime. Studies conducted within Walnut Creek in Iowa on the quality of subsurface drainage water show a responsiveness to changes in the soil water content as evidenced by the increase in atrazine concentration during periods with increased flow. These data, collected during July and August of 1992, show that peak atrazine concentrations are less than 1 $\mu g\ L^{-1}$ and are typically less than 0.5 $\mu g\ L^{-1}$ (Figure 3). Nitrate-nitrogen concentrations in contrast, decrease with increased subsurface flow and range from 10 to 15 $\mu g\ L^{-1}$ under normal flow. These changes in concentrations with increased flow would suggest that there is considerable preferential movement through the soil profile. Owens *et al.* (*27*) measured nitrate-nitrogen concentrations in suction lysimeters below a continuous corn and corn-soybean rotation at Coshocton, Ohio and found concentrations from 10 to 14 $\mu g\ L^{-1}$. They also found a seasonal variation from 1 to 20 $\mu g\ L^{-1}$ caused by the crop rotation. This led them to conclude that shallow ground water could be protected by adopting a crop rotation pattern over continuous corn production. Kanwar *et al.* (*28*) found in north central Iowa that continuous corn with no-tillage had the largest water loss through the soil profile but the lowest nitrate-nitrogen concentrations in subsurface drainage compared to moldboard plow and chisel plow. In contrast, the results of Hatfield *et al.* (*29*) from Walnut Creek watershed in central Iowa showed no seasonal trend in the nitrate-nitrogen concentrations in the subsurface drainage waterflow. This difference may be due to the scale of the project and the observations of field-sized areas rather than small plots or due to the differences among the soils at the study sites.

Leaching of herbicides and nitrate-nitrogen is a process of offsite movement that can affect either surface water or ground water quality. In areas with subsurface drainage, the discharge is moved across the landscape and is transported to open channels where the water becomes part of the surface water. In areas without subsurface drainage, the water that leaches through the soil profile can become part of the ground water recharge. There is a balance at all locations regarding the partitioning among surface runoff, subsurface drainage, and ground water recharge. The balance will be impacted by the type of farming practice and the patterns of crop water use throughout the year.

Studies in the Midwest on Water Quality

There are many unknowns about the effect of farming practices on water quality. To address these questions and begin to develop a research program that would address the issues associated with farming practices, a research and education effort began in 1990 in the Midwest. The Management Systems Evaluation Areas (MSEA) Program was established in 1990 by the United States Department of

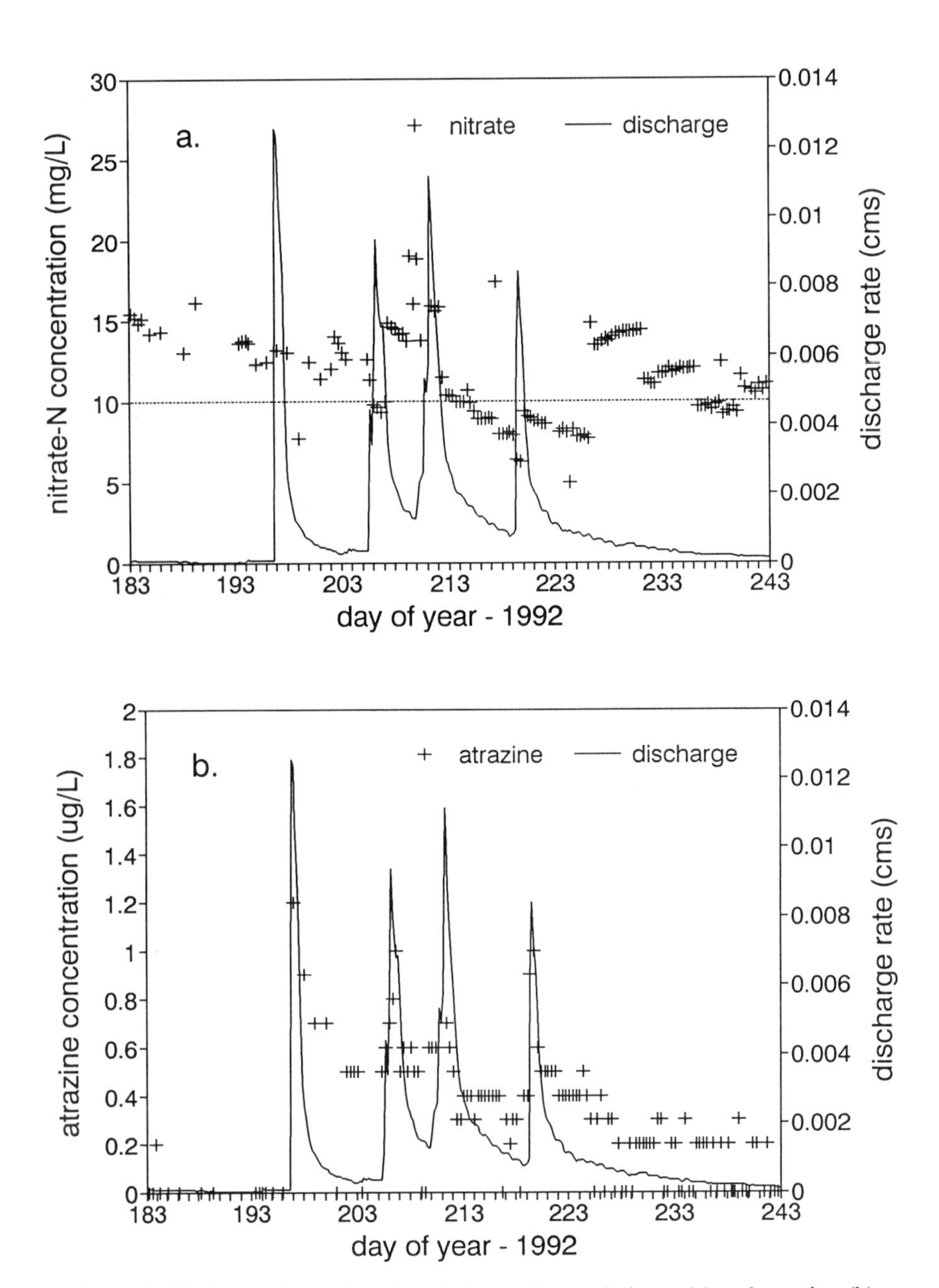

Figure 3. Hydrograph of subsurface drainage flow and nitrate (a) and atrazine (b) concentrations in a single subsurface drainage line from a 10 ha field in Walnut Creek watershed during July and August, 1992.

Agriculture (USDA)-Agricultural Research Service (ARS). As described by Onstad *et al.* (*30*), this program was established to evaluate the effects of agricultural chemicals on ground water quality in areas representing a variety of soil, geologic, and climatic conditions. An additional goal was to develop best management practices (BMPs) that protect ground water from agricultural chemical contamination while addressing the economic, environmental, and social needs of midwest agricultural production. Administration of the program is a cooperative effort between the USDA-ARS, USDA-Cooperative State Research Education and Extension Service (CSREES), State Agricultural Experiment Stations, Natural Resources Conservation Service (NRCS), United States Geologic Survey (USGS), and United States Environmental Protection Agency (USEPA). Research is conducted at five primary sites in Iowa, Minnesota, Missouri, Nebraska, and Ohio and at several additional sites within these and adjacent states.

Current research activities at these sites focus on the protection of ground water in specific aquifers or geographical areas. Research objectives at these MSEA research sites are to: a) measure impacts of prevailing and modified farming systems on chemical constituents of ground water and surface waters; b) identify and increase the understanding of factors and processes that control fate and transport of agricultural chemicals; c) assess impacts of agricultural chemicals and practices on ecosystems; d) assess benefits of using modified farming systems in the Midwest; e) evaluate social and economic impacts of adopting modified farm management systems; and f) transfer appropriate technology to other agricultural areas (*31*).

The Iowa MSEA project goal is to evaluate the effects of agricultural management systems on ground water and surface water quality in three regions of the state (*32*). These sites are located in the northeast part of the state on the Iowan surface landform; in southwest Iowa, near Treynor, on the deep loess deposits; and on the Des Moines Lobe landform region near Ames. Walnut Creek watershed near Ames, Iowa was the site selected for an intensive watershed scale study. Specific research objectives for the Iowa MSEA project are: a) quantify physical, chemical, and biological factors that affect transport and fate of agricultural chemicals; b) determine effects of crop, tillage, and chemical management practices on quality of surface runoff, subsurface drainage, and groundwater recharge; c) integrate information from the first two objectives with data on soil, atmospheric, geologic, and hydrologic processes to assess impacts of these factors on water quality; d) evaluate socioeconomic effects of current and newly developed management practices; and e) understand the ecological effects of agricultural chemicals, distinguishing them from impacts of other agricultural practices (*32*).

Observations from Walnut Creek Watershed

The Walnut Creek watershed has been described in a detailed report on research protocols and descriptions by Sauer and Hatfield (*33*). In this report, observations on surface discharge and shallow ground water will be reported. Farming practices within Walnut Creek watershed are typical of the Midwest with the percent of land area treated with atrazine very similar to the average for the Midwest. Nitrogen fertilizer rates for the watershed are slightly below the statewide average of 150 kg ha^{-1}, however, they are typical of the Des Moines Lobe landform region. Observations have been made with shallow ground water piezometers installed around a series of fields typical of the Midwest with chisel plow, ridge tillage, and no-till and in a corn-soybean rotation.

Shallow Ground Water Observations. Observations through 1993 in the shallow wells showed only one atrazine detection above the 3 µg L^{-1} HAL and this was in the 1.5-3 m depth below the soil surface. These wells were placed at the edge of the fields having a history of atrazine use within the watershed since the mid-1960's. Therefore, the observations should reflect the history of field application. Similar results were found for alachlor, metribuzin (4-amino-6-(1,1-dimethylethyl)-3-(methylthio)-1,2,4-triazin-5(4H)-one), and metolachlor in these piezometers (*27*). The mean atrazine concentration for the 1.5-3 m depth was 0.065 µg L^{-1} and would support the general conclusion of Burkart and Kolpin (2) that there are a few detections above the HAL. Observations made in the deep ground water wells (>150 m) within Walnut Creek watershed revealed no detections of any herbicide, and nitrate-nitrogen concentrations less than 1 µg L^{-1}. Edge of field samples show that there is little historical movement of atrazine under a range of climatic conditions and farming practices.

Surface Water Observations. Surface water discharge from the 5600 ha Walnut Creek watershed has been monitored since 1991 and shows only a small amount of herbicide loss (*29*). Stream discharge from the watershed would represent a composite of subsurface drainage and surface runoff from fields. There is a continual movement of water from the watershed, and nitrate-nitrogen concentrations typically average between 15 and 20 µg L^{-1} in the spring and 5-10 µg L^{-1} in the late summer. Herbicide concentrations are more dependent the precipitation events within the watershed. Loads in 1991 and 1993 were the largest for all herbicides and can be attributed to the spring rains in 1991 and the large summer rains in 1993. These large rain events caused an increase in the surface runoff and subsurface drainage. In 1992 and 1994, the loads were quite low due to the reduced number of large rain events early in the growing season. Atrazine loss ranged from 0.2 to 7.0% of the amount applied during the growing season. Concentrations of atrazine at the discharge point of Walnut Creek have averaged less than 0.5 µg L^{-1} since 1990 (Figure 4). These samples are collected as weekly grab samples from automated samplers located at the discharge point. There is considerable scatter throughout the year in response to the variation in the distribution of rainfall events.

Surface water quality within Walnut Creek is influenced by the subsurface drainage system and the hydrology of the watershed. There is little evidence of an effect of crop rotation on nitrate-nitrogen or herbicide movement into the subsurface drainage systems. The concentrations within selected subbasins of the watershed reflect the overall water movement within the watershed.

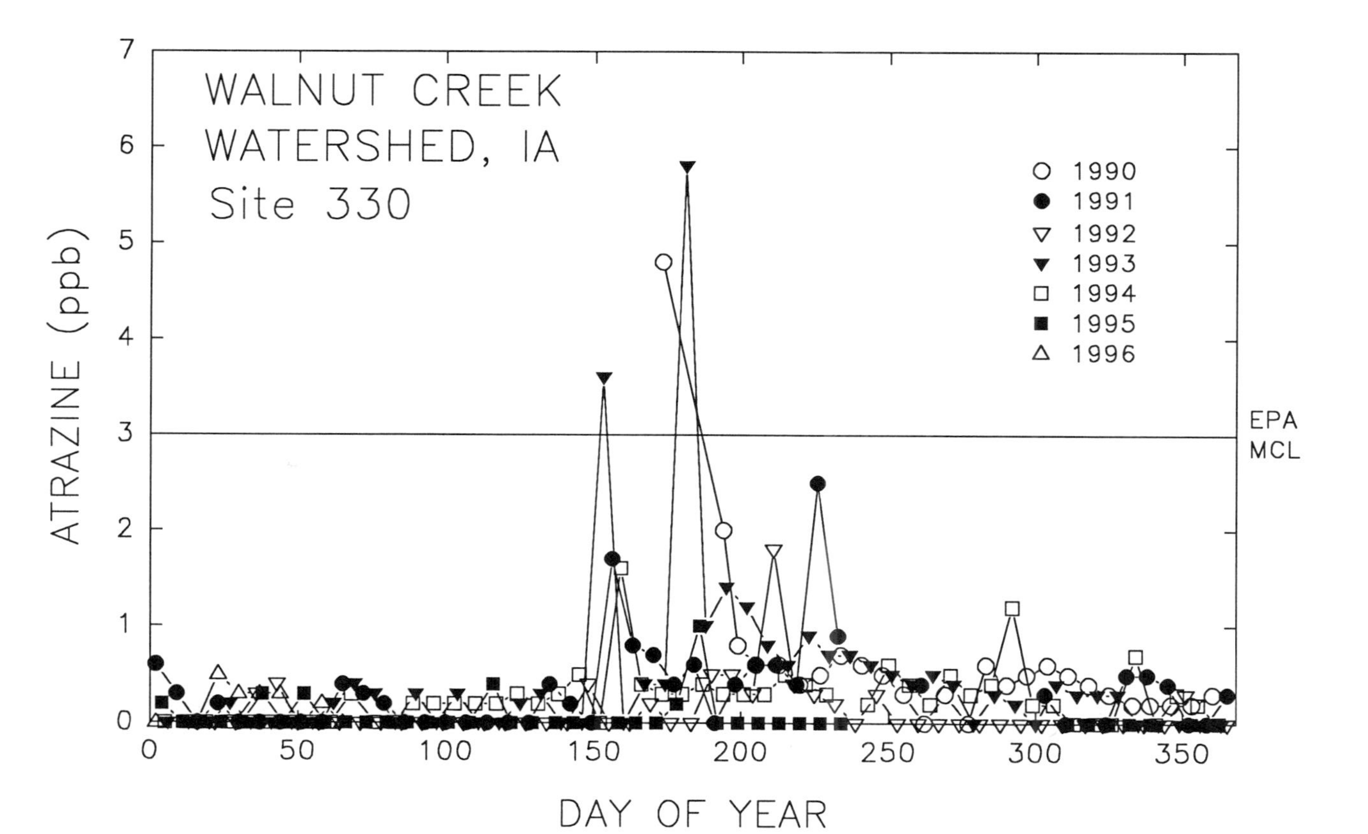

Figure 4. Weekly concentrations of atrazine (µg/L) at the discharge point of Walnut Creek watershed measured since 1990.

Concentrations of atrazine detected in the subsurface drain lines may be residual atrazine from previous growing seasons since atrazine was introduced in the mid-1960's. The amounts that are lost, however, are quite small. Runoff concentrations from the edge of single fields in excess of 25 µg L^{-1} in 1991 are not reflected in the concentrations leaving the watershed because only a small portion of the stream flow consists of surface runoff in Walnut Creek watershed (Figure 4). Atrazine application across the watershed is evenly distributed and field records show that about 50% of the land area receives some atrazine. Metolachlor is more widely used as a herbicide in crop production within Walnut Creek. Both of these herbicides are soluble in water and easily moved with water flow. Atrazine losses with current farming practices in Walnut Creek are less than the amounts reported by Goolsby and Battaglin (*9*). Nitrate-nitrogen losses are higher than reported by Goolsby and Battaglin (*9*) since there is a greater proportion of subsurface drainage and grain crop production in Walnut Creek compared to the Midwest. Nitrate-nitrogen concentrations are higher in subsurface drainage discharge than in surface runoff.

Modifications of Farming Practices to Improve Water Quality

Water movement is the primary transport mechanism for herbicides and nitrate-nitrogen, and efforts to modify farming practices to reduce the loss of herbicides below current levels must start with this mechanism. Surface runoff can be reduced through the adoption of conservation tillage practices as demonstrated by Fawcett *et al.* (*17*). Surface runoff of herbicides occurs when the soil is saturated and there is heavy rainfall shortly on fields after broadcast applications. Total surface water discharge from a watershed does not exhibit the concentrations or loads from an individual field since herbicides are not uniformly applied across a watershed. This can be greatly reduced by incorporation of herbicides at the time of planting as described by Baker and Mickelson (*34*). Incorporation of a herbicide into the upper soil layer allows for more rapid infiltration and soil-bind, thus reducing the potential for movement with surface runoff. In conservation tillage systems, incorporation is not always possible due to the need to leave as much residue on the surface as possible. In these cases, delaying application when the soil is saturated and there is heavy rain predicted could potentially reduce surface transport of herbicides. This concept needs to be evaluated for adoption into herbicide management practices.

Leaching through the soil profile is considered to be the opposite effect of decreasing surface runoff. There are few studies that have examined the interrelationships between these two processes at the field and watershed scale for herbicide and nitrate-nitrogen transport. Leaching may increase with conservation tillage as demonstrated by Kanwar *et al.* (*28*); however, they found a decrease in the herbicide and nitrate-nitrogen concentrations in the leachate so that there is a net decrease in total load. Nitrogen management could include a change in the time of application to more closely link the application with peak crop needs. This would reduce the potential for leaching since the nitrogen would be placed in the soil at a time when there is both rapid uptake of water and nitrate and little potential for leaching. There are many offsetting processes that occur within the soil profile as changes in farming practices are made, and these are related to the organic matter content and preferential flow pathways through the soil. There is little information to compare these mechanisms across soils and management practices, although we can assume that increasing the organic matter content in the surface not only increases infiltration, but also increases the degradation rate of herbicides due to increased biological activity.

Much of the Corn Belt has subsurface drainage; a situation which offers the potential for management of the tile system to reduce offsite movement. Management of subsurface drainage to reduce movement in the spring could impact both nitrate-nitrogen and herbicide losses from fields and improve water management for the crop later in the growing season.

Modification of farming practices to achieve environmental goals is not a new research area. Stewart *et al.* (*35,36*) developed the first report on the interrelationships of nonpoint source pollution and farming practices. The principles detailed in those reports are valid today, as they were 20 years ago. The concepts of modifying farming practices have the potential to reduce both the onsite and offsite impacts. However, as the studies from Walnut Creek show, herbicide losses are small and modifications in farming practices will provide only slight reductions in herbicide loads.

Field management practices coupled with landscape changes around fields, e.g., vegetative filter or buffer strips, can provide environmental benefits. Surface runoff appears to be the largest contributor to herbicide loss across the Midwest and some reductions will be possible with the adoption of conservation tillage. Baker *et al.* (*37*) evaluated the effect of filter strips on herbicide removal and showed that removal efficiency is based on antecedent soil water in the vegetative filter strip, runoff volume, and herbicide concentration.

Summary

Farming practices can be modified to achieve water quality goals. These modifications need to be developed for each particular region of the United States and evaluated locally in cooperation between local research communities and growers. The research community will gain insights into the newest farming practices and the modifications can be quickly transferred to and adopted by local growers. These changes can be integrated into current practices by farmers and landowners without major changes in their farming operations. This partnership can yield positive results for soil management, profit of farming operations, and environmental quality.

Literature Cited

1. Gianessi, L. P.; Puffer, C. M. Herbicide use in the United States. Resources for the Future, Qual. of the Environ. Div., 1991, Washington, D.C.
2. Burkart, M. R.; Kolpin, D. W. J. Environ. Qual. 1993, 22, pp. 646-656.
3. Hallberg, G. R. Agric. Ecosys. Environ. 1989, 26, pp. 399-367.

4. Kolpin, D. W.; Goolsby, D. W.; Thurman, E. M. J. Environ. Qual. 1995, 24, pp. 1125-1132.
5. Kolpin, D. W.; Thurman, E. M.; Goolsby, D. A. Environ. Sci. Technol. 1996, 30, pp. 335-340.
6. Thurman, E. M.; Goolsby, D. W.; Meyer, M. T.; Kolpin, D. W. Environ. Sci. Technol. 1992, 26, pp. 2440-2447.
7. Kolpin, D. W.; Goolsby, D. A.; Aga, D. S.; Iverson, J. L.; Thurman, E. M. In Selected Papers on Agricultural Chemicals in Water Resources of the Midcontinental United States. U.S. Geological Survey Open File Report 93-418, Denver, CO, 1993, pp. 64-74.
8. Madison, R. J.; Burnett, J. O. Water Supply Paper 2275. U.S. Geological Survey, Dept. of the Interior, Washington, DC, 1985, pp. 93-105.
9. Goolsby, D. A.; Battaglin, W. A. In Selected papers on agricultural chemicals in water resources of the Midcontinental United States; Goolsby, D. A.; Boyer, I. L.; Mallard, G. E., Eds.; U.S. Geological Survey, Open File Report 93-418, Department of the Interior, Washington, DC, 1993, pp. 1-25.
10. Hatfield, J. L. Wat. Sci. Tech. 1993, 28, pp. 415-424.
11. Hatfield, J. L. Second Intl. Weed Control Cong. 1996, pp. 1339-1348.
12. Hatfield, J. L. In Irrigation of Agricultural Crops; Stewart, B. A.; Nielsen, D. R., Eds.; Agronomy, 1990, 30, pp. 435-474.
13. Hatfield, J. L.; Prueger, J. H.; Sauer, T. J. In Evapotranspiration and Irrigation Scheduling; Camp, C. R., Sadler, E. J., Yoder, R. E., Eds.; ASAE. 1996, pp. 1065-1070.
14. Baker, J. L.; Laflen, J. M. J. Soil Water Conserv. 1983, 38, pp. 186-193.
15. Baker, J. L.; Shiers, L. E. Trans. ASAE. 1989, 32, pp. 830-833.
16. Hatfield, J. L.; Jaynes, D. B. In Water Resources Engineering; Espey, Jr., W., Combs, P. E., Eds.; ASCE. 1995, pp. 1370-1374.
17. Fawcett, R. S.; Christensen, B. R.; Tierney, D. P. J. Soil Water Conserv. 1994, 49, pp. 126-135.
18. Myers, J. L.; Wagger, M. G.; Leidy, R. B. J. Environ. Qual. 1995, 24, pp. 1183-1192.
19. Green, J. D.; Horton, R.; Baker, J. L. J. Environ. Qual. 1995, 24, pp. 343-351.
20. Gish, T. J.; Shirmohammadi, A.; Vyravipillai, R.; Weinhold, B. J. Soil Sci. Sci. Am. J. 1995, 59, pp. 895-901.
21. Isensee, A. R.; Nash, R. G.; Helling, C. S. J. Environ. Qual. 1990, 19, pp. 434-440.
22. Gish, T. J.; Isensee, A. R.; Nash, R. G.; Helling, C. S. Trans. ASAE. 1991, 34, pp. 1745-1753.
23. Blevins, R. L.; Thomas, G. W; Smith, M. S.; Frye, W. W; Cornelius, P. L. Soil Tillage Res. 1983, 3, pp. 135-145.
24. Dick, W. A. Soil Sci. Soc. Am. J. 1983, 47, pp. 102-107.
25. Upchurch, R. P.; Selman, F. L.; Mason, D. D.; Kamprath, E. J. Weed Sci. 1966, 14, pp. 42-49.
26. Hudson, B. D. J. Soil Water Conserv. 1994, 49, pp. 189-194.

27. Owens, L. B.; Edwards, W. M.; Shipitalo, M. J. Soil Sci. Soc. Am. J. 1995, 59, pp. 902-907.
28. Kanwar, R. S.; Stoltenberg, D. E.; Pfeiffer, R. L.; Karlen, D. L.; Colvin, T. S.; Honeyman, M. In Proc. Conference by ASCE on Irrigation and Drainage; Ritter, W. F., Ed.; ASCE, New York, NY, 1991, pp. 655-661.
29. Hatfield, J. L.; Jaynes, D. B.; Burkart, M. R. In Proc. of National Agricultural Ecosystem Management Conference, New Orleans, LA, Conservation Technology Information Center, West Lafayette, IN, 1995, pp. 127-153.
30. Onstad, C. A.; Burkart, M. R.; Bubenzer, G. D. J. Soil and Water Conserv. 1991, 46, pp. 184-188.
31. USDA-ARS. Water Quality Research Plan for Management Systems Evaluation Areas (MSEA's), an Ecosystems Management Program. ARS-123, Springfield, VA, 1994, 45 pp.
32. Hatfield, J. L.; Baker, J. L.; Soenksen, P. J.; Swank, R. R. In Proc. of Conf. on Agricultural Research to Protect Water Quality, Minneapolis, MN, Soil and Water Conservation Soc., Ankeny, IA, 1993, pp. 48-59.
33. Sauer, P. J.; Hatfield, J. L. Walnut Creek Watershed: Research Protocol Report. National Soil Tilth Laboratory Bulletin 94-1, USDA-ARS, Ames, IA, 1994, 418 pp.
34. Baker, J. L.; Mickelson, S. K. Weed Tech. 1994, 8, pp. 862-869.
35. Stewart, B. A.; Woolhiser, D. A.; Wischmeier, W. H.; Caro, J. H.; Frere, M. H. USDA-ARS, Washington, DC, Report ARS-H-5-1, 1975, 111 pp.
36. Stewart, B. A.; Woolhiser, D. A.; Wischmeier, W. H.; Caro, J. H.; Frere, M. H. USDA-ARS, Washington, DC, Report ARS-H-5-2, 1976, 187 pp.
37. Baker, J. L.; Mickelson, S. K.; Hatfield, J. L.; Fawcett, R. S.; Hoffman, D. W.; Franti, T. G.; Peter, C. J.; Tierney, D. P. Brighton Crop Protection Conference, Brighton, England, 1995, pp. 479-487.

Triazine Water Issues: Regulatory, Risk, and Ecotoxicology

Chapter 23

The Role of Groundwater Surveys in Regulating Atrazine in Wisconsin

Gary S. LeMasters

Trade and Consumer Protection, Division of Agricultural Resource Management, Wisconsin Department of Agriculture, P.O. Box 8911, Madision, WI 53708–8911

The Wisconsin Department of Agriculture, Trade and Consumer Protection has used surveys of atrazine in groundwater to guide regulations to protect the resource. The first surveys led to restrictions and prohibitions on atrazine use in 1991. Subsequent surveys yielded new atrazine prohibition areas and tightened use restrictions. Surveys have positive and negative effects. Surveys give participants ownership of their groundwater quality and can push the regulatory process. Surveys can generate controversy when participants are not fully aware of the regulatory consequences of participation. Negative consequences for farmers can include the loss of atrazine and increased weed control costs, lowered agricultural land value, and alienation from neighbors affected by the prohibition area. For all participants negative effects include costs of improvements in their water supply and possible reductions in property values. Survey objectives must be tempered by the limits of survey methodology.

In 1958, Geigy Chemical Corporation made available research samples of atrazine [2-chloro-4-(ethylamino)-6-(isopropylamino)-*s*-triazine]; at that time it was registered for industrial and non-selective uses only (*1*). In one of the major understatements in the history of weed science, the author noted that "Atrazine appears very promising for weed control in corn..." (ibid., pp. 32). It was first used in Wisconsin to control weeds in corn in 1960. By 1970, the majority of corn acres were treated with atrazine (*2*). The first groundwater samples were analyzed for atrazine in the early 1980s. A series of groundwater surveys played a fundamental role in driving (and being driven by) the process that led to the

Wisconsin Department of Agriculture, Trade and Consumer Protection (WDATCP, "the department") regulating the use of atrazine for the 1991 growing season. Additional survey results have contributed data for annual revisions to the rule. For the 1996 growing season, the Atrazine Rule (Ch. ATCP 30, Wis. Adm. Code, "the atrazine rule") sets atrazine application rates below federal label rates and prohibits atrazine use on approximately one million land acres within 91 atrazine prohibition areas, encompassing some of the most productive agricultural soils in the state.

The interactions between the groundwater surveys and the regulatory process are the subject of this paper. Background information on atrazine use in Wisconsin and the Wisconsin Groundwater Law (Ch. 160, Wis. Stats., "the groundwater law") will be presented. The interactions between regulating the use of atrazine and surveying its distribution in private water supply wells will be discussed, beginning with a voluntary moratorium on its use in the lower Wisconsin River valley for the 1990 growing season, and ending with the atrazine rule for 1996. The lessons learned in Wisconsin may be of value to other states as they address groundwater contamination from pesticides.

Atrazine Use in Wisconsin

Atrazine was first approved for use on corn in 1960 and it was rapidly adopted by Wisconsin farmers. The first random survey of pesticide use in Wisconsin was conducted in 1969 (*2*). Atrazine alone or in combination with other herbicides was applied to 949,000 acres of corn, 91 percent of the corn acreage treated preemergence. About 790,000 acres of corn were treated postemergence with atrazine, either alone (412,000) or with crop oil (378,000), 83 percent of the total corn acreage treated postemergence. Atrazine was especially popular to control quackgrass in old alfalfa fields being rotated into corn, a common practice on Wisconsin dairy farms. Annual and cumulative atrazine use in Wisconsin is shown in Figure 1, based on pesticide use surveys conducted in 1970 (*2*), 1978 (*3*), 1985 (*4*) and 1990 (*5*).

The Wisconsin Groundwater Law

The Groundwater Law (Chapter 160, Wis. Stats.), adopted in 1984, guides regulatory responses to detections of pesticides in groundwater. All groundwater receives equal protection; there is no aquifer classification system to permit differential regulation based on present or future use of the aquifer. Therefore, a detection of atrazine in a shallow well in the sandy, lower Wisconsin River valley carries the same weight as a detection from a much deeper well finished in bedrock.

The Groundwater Law also established a process for setting numerical, health-based standards for substances in groundwater, based on the belief that Wisconsin residents should not be exposed to a significant health risk by drinking

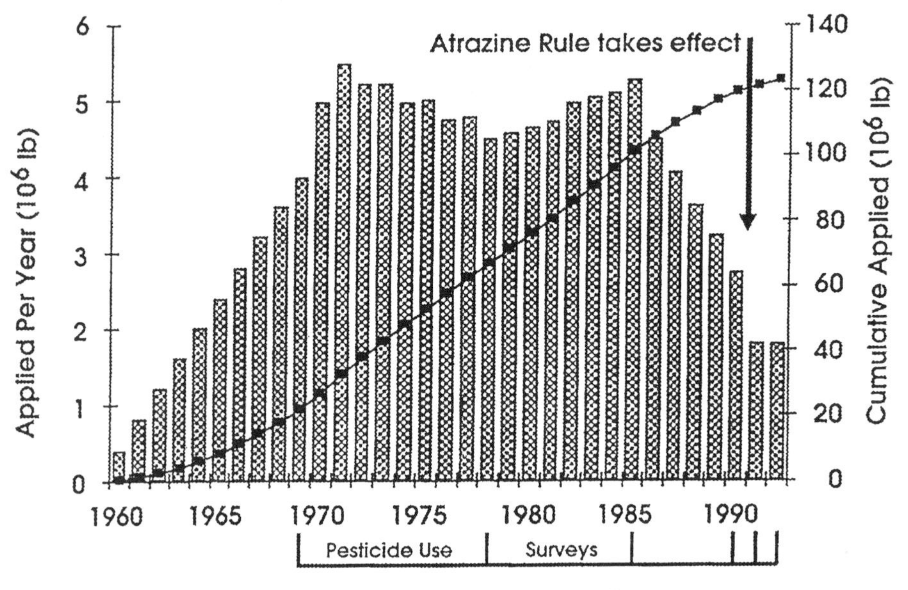

Figure 1. Atrazine use in Wisconsin.

their water. The numerical standards replace earlier concepts of "non-degradation" or "no detrimental effect" with enforceable, measurable concentrations (*6*). Two standards are set for each substance, the enforcement standard (ES) and the preventive action limit (PAL).

The ES serves as a regulatory red light. If it is exceeded at a Point Of Standards Application (POSA), the department that regulates the activity or practice causing the contamination must prohibit the activity or practice.

The Wisconsin Groundwater Law states "Preventive action limits shall serve as a means to inform regulatory agencies of potential groundwater contamination problems. A preventive action limit is not intended to be an absolute standard at which remedial action is always required." (ss 160.001 (*8*), Wis. Stats.) The PAL is set at a percentage of the ES (10% for atrazine, due to its potential carcinogenicity classification). In responding to a substance exceeding the PAL at a POSA, the regulatory agency "...shall implement responses for a specific site designed to:

a) Minimize the concentration of the substance in the groundwater at the point of standards application where technically and economically feasible;
b) Regain and maintain compliance with the preventive action limit, unless, in the determination of the regulatory agency, the preventive action limit is either not technically or economically feasible, in which case, it shall achieve compliance with the lowest possible concentration which is technically and economically feasible; and
c) Ensure that the enforcement standard is not attained or exceeded at the point of standards application." (ss. 160.23 (*l*), Wis. Stats.).

Under the Groundwater Law, both the Wisconsin Department of Natural Resources and the Wisconsin Department of Agriculture, Trade and Consumer Protection were required to "...establish regulations to assure that regulated facilities and activities will not cause the concentration of a substance in groundwater affected by the facilities or activities to exceed the enforcement standards and preventive action limits at a point of standards application." (Ch. 160.001(*4*), Wis. Stats.). These are given in Chapter ATCP 3 1, Wis. Adm. Code. The WDNR administrative rule is Chapter NR 140, Wis. Adm. Code.

Points of Standards Application (POSA) means the specific location, depth or distance from a facility, activity or practice at which the concentration of a substance in groundwater is measured for purposes of determining whether a preventive action limit or an enforcement standard has been attained or exceeded (ss. 160.01(*5*), Wis. Stats.). WDNR considers a monitoring well to be a POSA, while WDATCP excludes them unless the landowner agrees otherwise. WDATCP chose this route to make it easier to install monitoring wells on private property as part of its Monitoring Project for Pesticides. This exemption places great importance on samples from private water supply wells, which both WDNR and WDATCP consider to be a POSA.

The regulatory process is driven by the groundwater standards. There were no numerical standards for atrazine when it was first detected in Wisconsin groundwater in the early 1980s. There was an unofficial health advisory level of 215 micrograms per liter (μg/L), which had only been exceeded near commercial pesticide mixing and loading sites. The standard-setting process for atrazine began in 1985 with the WNDR transmitting a list of substances, including atrazine, which had been found in groundwater to the Wisconsin Department of Health and Social Services (DHSS) and asking DHSS to recommend groundwater standards. Atrazine had been detected in monitoring wells maintained by WDATCP and in a few private water supply wells in the lower Wisconsin River valley. On June 25, 1986, DHSS submitted to WDNR a recommended enforcement standard (ES) and preventive action limit (PAL) of 0.35 and 0.035 micrograms per liter (μg/L), respectively, calculated using procedures in the Groundwater Law.

$$\frac{3.5 \text{ mg atrazine/kg body weight/day} \times 10 \text{ kg body weight}}{\begin{array}{c}1000 \times 10 \times 10 \times 1 \text{ L drinking water/day} \\ \times\ 100\% \text{ of exposure from drinking water}\end{array}} = 0.35\ \mu\text{g atrazine/L}$$

The 1000-fold uncertainty factor (UF) accounts for inter- and intra-species differences. One 10-fold UF accounts for the fact that the most sensitive toxicological endpoint has not been determined, and the second 10-fold UF is used for United States Environmental Protection Agency (USEPA) Category C carcinogens when performing Acceptable Daily Intake - based quantitative risk assessment approach rather than linearized multi-stage model for low dose extrapolation (*7*).

The proposed standards were considered at public hearings conducted by the WDNR in 1987. A number of people commented that the 100,000-fold uncertainty factor seemed unwarranted. Another person commented that the proposed standards probably were too high, since they did not consider pesticide breakdown products. The only data the department had collected on atrazine in groundwater due to field use was from the WDATCP Monitoring Project for Pesticides. Every field had exceeded the proposed PAL for atrazine. In a memorandum to the WDATCP Secretary, the Administrator of the Agricultural Resource Management Division noted that "If these proposed standards become part of the WDNR regulations our department will be faced with determining if these products can be safely used in Wisconsin. The scope of the problem will be far greater than anything we have faced so far with aldicarb." (Robert, WDATCP memorandum, 1986).

DHSS reviewed new information on atrazine toxicology from the registrant and USEPA and decided that the 10-fold UF applied due to the lack of a toxicological endpoint was no longer warranted. The revised proposal for an ES of 3.5 μg/L and a PAL of 0.35 μg/L was submitted to WDNR on March 10, 1988 (*8*). The standards were approved by the WDNR Board in June of 1988 and were taken to public hearings to state senate in August, 1988. The department provided testimony on the possible regulatory implications of the new standards (Ehart, unpublished WDATCP policy paper). Atrazine had been detected above the proposed ES at several sites in the Lower Wisconsin River Valley, but not in the Central Sands area, where the aldicarb problem was centered. Therefore, with the possible exception of "karst" regions in southwest Wisconsin, "less susceptible" landscapes should not have been greatly affected. The department also told the senate committee that it was conducting a survey of pesticides in well water from Grade A dairy farms. The standards for atrazine took effect in October, 1988, over 28 years since atrazine was first applied to corn fields in Wisconsin.

Surveys of Atrazine in Groundwater

The department uses monitoring wells to determine the impact of pesticide use practices on groundwater quality in susceptible areas of the state. However, regulatory programs must be built on samples from points of standards application, which includes monitoring wells. Therefore, the department conducted a series of groundwater surveys of atrazine contamination in private water supply wells to define the scope of the problem and guide regulations. The most important surveys will be summarized, in chronological order, including the lessons learned from each survey.

Grade A Survey

The Grade A Dairy Farm Well Water Quality Survey (*9*) was the first statistical survey of pesticides in Wisconsin groundwater. Between August, 1988 and February, 1989, 534 randomly selected Grade A dairy farms were visited and a water sample was collected. A total of 66 wells contained detectable (LOD = 0.15 µg/L) residues of atrazine, either alone (64), with alachlor (1) or metolachlor (1). The proportion of wells on Grade A farms that contain detectable levels of atrazine was estimated to be between 9 and 15 percent. Between 5 and 9 percent contained atrazine at or above the PAL of 0.35 µg/L.

There were about 23,500 Grade A dairy farms in Wisconsin at the time of the survey. The department was careful to state from the beginning that the results could only be applied to this population, not the population of all private water supply wells in the state. The report was well received in the scientific and regulatory community, but posed as many questions as it answered.

The Grade A survey was successful from a statistical perspective because the objectives were clearly defined, and methods were used that could achieve the objectives. The department maintains a list of Grade A dairy farms, so the population was known and finite and the sampling frame could be selected easily. The department has statutory authority to collect water samples at Grade A dairy farms, because the water is an ingredient in milk. The department has a staff of trained inspectors who visit each farm periodically and who were able to collect the water samples. The Grade A dairy farmers were cooperative and concerned about their groundwater quality.

These results demonstrated that atrazine was the pesticide posing the greatest concern regarding groundwater. Laboratory analysis can recover residues of the most commonly used herbicides and insecticides, yet, besides atrazine, the only analytes detected were alachlor (5 samples), metribuzin (1 sample), and metolachlor (1 sample).

The Grade A survey demonstrated that the problem was not confined to "susceptible" landscapes and required the department to rethink its notion on susceptibility. Many of the detections were on farms in areas of medium-textured soils, with no irrigation, and groundwater at some depth in limestone or sandstone bedrock.

The results also generated public interest in water testing. The department did not have the resources to collect samples from the general public, yet there was a public outcry for more information about the atrazine problem. This led in 1990 to the Rural Well Survey.

The Grade A survey suffered in the eyes of some because the department did not consider in advance the difficulty it would face in trying to determine whether the atrazine detections were caused primarily by the sins of past farming practices, by current use practices, or both. If historical practices were to blame, the department was not complying with its duties under the Groundwater Law by proposing limits on current use practices. This criticism is still being raised today with the Grade A survey and its successors.

The department was told many times that the atrazine problem on Grade A dairy farms was due to historically high use rates, especially for quackgrass control. Atrazine was readily adopted by dairy farmers. It gave them effective control of quackgrass in first-year corn following several years of alfalfa. The field would be fall-plowed and treated with atrazine at rates of 2 to 4 pounds active ingredient per acre (lb. a.i.). A second treatment would be made in the spring before planting, again using 2-4 lb. a.i. Atrazine also was widely used with crop oil to rescue the corn crop when weeds escaped the planned treatments. Since these practices were no longer legal, even on the federal label, the department did not need to regulate atrazine use.

The atrazine problem at dairy farms was also attributed to historical mishandling in the form of spills and backsiphoning incidents. Farmers routinely did their own mixing, loading and spraying. The wettable powder and granular formulations could be difficult to mix and early spray equipment could be quite crude. The liquid formulation was not available until the early 1970s. A factor that may have contributed to this legacy of careless handling was the common perception during much of atrazine's use history that it posed no threat to human health. The department was told by farmers and agribusiness people that representatives of the chemical companies would drink a glass of what they claimed to be atrazine at dealer meetings to demonstrate its safety.

During the 1960s, when atrazine was being rapidly adopted, the scientific community, agribusiness and farmers had little understanding of groundwater contamination from pesticide handling. For example, in 1964 the United States Department of Agriculture Agricultural Research Service published recommendations for the safe disposal of empty pesticide containers and surplus pesticides, The publication was targeted to farmers, commercial pesticide applicators, city, state and federal pest control officials, and others who use large quantities of pesticides.

The recommendations for disposing of pesticide containers begin with this step:

"Drain any pesticide remaining in the container into a pit dug in sandy soil." (*10*).

The recommendations for disposing of surplus pesticides state the following: "If surplus pesticides cannot be given to a responsible person in need of such material, they should be poured into a hole dug in the ground and covered with dirt to a depth of at least 18 inches. Leftover spray mixture should be poured into a pit dug in sandy soil." *(ibid., pp. 6)*.

The department attempted to resolve the role of point- versus non-point sources of contamination by funding research at a Grade A dairy farm (*11*). After a detailed study, the researchers concluded that "Although some evidence of point-source pollution exists, most groundwater residues probably result from normal field applications..." *(ibid., p.119)*.

Ciba-Geigy Corporation also attempted to determine the source(s) of the atrazine in groundwater at the Grade A dairy farms. A consultant investigated detections at 19 farms, and conducted similar investigations at 13 farms with no detects of atrazine. Detailed site maps were made and the farm operator was interviewed about atrazine mixing and loading practices and use history. From this information a determination on the most likely pathway for the transport of atrazine to groundwater was made. The consultant determined several potential pathways for the transport of atrazine to groundwater. These included normal field use, spills in the field, backsiphonings, and improper disposal of rinsate and containers (*12*).

Some members of the agricultural community felt that the Grade A survey results were the telltale fingerprint of the legacy of almost 30 years of high application rates and mishandling of atrazine. Since the department had addressed problems in pesticide handling with the certified applicator program and rules on pesticide mixing and loading, and farmers were already using much lower rates of atrazine due to changes in the federal label, state regulations on the use of atrazine were unneccessary. The department has devoted considerable resources to this issue but was never able to convince some members of the agricultural community that restrictions on the use of atrazine were needed.

Lower Wisconsin River Valley Survey

Between 1985 and 1989, WDNR and the WDATCP staff sampled 65 private water supply wells in the Lower Wisconsin River Valley, the majority in 1989. This landscape is considered highly susceptible to groundwater contamination and atrazine had been detected above the ES in several WDATCP monitoring wells in the valley. However, the department does not consider a monitoring well to be a POSA, so no regulatory action could be taken. In the private wells, atrazine exceeded the ES in 7 wells, 30 exceeded the PAL and 40 contained detectable residues. (In 1986 the state laboratories lowered their limit of detection (LOD) for atrazine from 1 µg/L to between 0.1-0.4 µg/L.) The Groundwater Law requires the department to prohibit the activity or practice which uses or produces the substance that exceeds the ES. Therefore, the department prohibited atrazine use on about 2,000 acres, affecting 15 landowners. Ciba-Geigy supported the prohibition (*13*). This was the first restriction in Wisconsin on the use of atrazine and was a direct outgrowth of sampling the 65 wells in the valley. At the same time, the department created the Atrazine Ad Hoc Committee to help prepare rules on the use of atrazine for the 1991 growing season.

The Rural Well Survey and the 1991 Atrazine Rule.

The publicity surrounding the Grade A and Lower Wisconsin Valley surveys generated a great deal of public interest in water testing. To meet this demand, and to gather more data on atrazine in groundwater outside of Grade A dairy farms, the department, in cooperation with WDNR and Ciba-Geigy Corporation, designed and conducted the Rural Well Survey. Phase 1 used an immunoassay test kit to screen 2187 water samples submitted by well owners. In Phase 2, follow-up samples were collected from over 400 wells with elevated levels of nitrate-nitrogen or a triazine test above the PAL (0.35 µg/L) and analyzed by conventional methods. A subset of these samples was split with Ciba-Geigy for analysis by their laboratory. Survey details are in Williams (*14*) and Brady et al. (*15*).

The survey began in January, 1990. By April, the capacity of 2400 participants had been reached. The results were announced at the first meeting of the Atrazine Ad Hoc Committee in April, 1990. The samples had come largely from well owners in Dane County and surrounding agricultural counties in southern Wisconsin. For the 209 samples analyzed up to that point from Dane County, 52 percent, showed a positive response to the immunoassay. The publicity surrounding this announcement put great pressure on the ad hoc committee to recommend that the use of atrazine be restricted beyond the Lower Wisconsin River Valley. "The issue has become more complex because the results show the problem is not confined to any particular area," (*16*).

Phase 2 sampling began in March of 1990 and continued through October. A total of 431 samples were analyzed for atrazine and other pesticides at the department laboratory. A total of 236 samples from wells with triazine test results over the PAL were split with Ciba-Geigy for analysis. The Phase 2 results confirmed the presence of atrazine and provided more data to drive the demand for further restrictions on atrazine use, the so-called "county restrictions" (see below).

First Draft of the 1991 Atrazine Rule. The first draft of proposed regulations on atrazine use was presented to the ad hoc committee at their first meeting in April, 1990. The major rule provisions were: maximum annual application rates from 1.0 to 2.0 lb. a.i., based on texture of the surface soil; atrazine applied only between April 15 and July 15; no irrigation for two years after atrazine use without an irrigation management plan; use only by certified applicators; no more than 0.75 lb. ai. on coarse soils in the Lower Wisconsin River Valley, no irrigation for two years following atrazine applications to coarse soils in the Lower Wisconsin River Valley; a process for creating atrazine prohibition areas; and establishment of seven prohibition areas in the Lower Wisconsin River Valley, the same areas that were created by voluntary cooperation in 1990.

The major problem facing the committee was how to deal with the contamination between the PAL and the ES. The Groundwater Law requires the department to minimize groundwater contamination, but applies the test of technical and economic feasibility to restrictions on use in response to contamination below the ES. In an economic analysis presented to the ad hoc committee in June, 1990, the department suggested that while it was technically feasible to grow corn without atrazine, it probably was not economically feasible. This was especially the case for weed control in a dairy rotation, where, due to increased weed pressure rotating from alfalfa to corn, losing atrazine would mean switching to more expensive herbicides and more trips across the field.

One committee member felt that farmers had voluntarily cut back on atrazine use. "The dairy well survey has had a tremendous impact in the farming community and everything we've done since has added to the concern" (*17*). However, others felt that the publicity surrounding the early results of the Rural Well Survey made delaying the atrazine rule until 1992 unacceptable (*18*).

At the next committee meeting in July the department presented a plan to further restrict atrazine use in counties with "significant" contamination between the PAL and ES. The proposal also modified statewide use rates somewhat. The controversy centered on the county restrictions. As proposed, if the department detected atrazine in groundwater at levels above the PAL in five or more civil townships (36-square-mile parcels of land identified in the Public Land Survey), atrazine could only be used in no-till or as a rescue treatment and the maximum application rate would be 2 lb. a.i. per acre. In the committee draft, 9 counties were initially affected. The department would amend the rule annually to add new counties based on additional groundwater testing.

Committee reaction was diverse. A representative of Ciba-Geigy Corporation said the proposal "bordered on irresponsibility" because the proposed method of targeting counties failed to distinguish contamination due to labeled use at current rates from point-sources due to mishandling and historically higher use rates. A representative of the Sierra Club said the proposal "is way too generous when you consider we have overwhelming evidence that the chemical is polluting ground water in this state." (*19*).

The Board of ATCP approved the rule for public hearings at its August 8 meeting, but several members voiced their concerns "I have a lot of questions about this proposal", said James Harsdorf, and Board Chairman Louis Wysocki said "There will be changes" (*20*). By that time, the number of restricted counties had grown to 12.

In addition, the Board adopted an amendment under which, in the 12 restricted counties, atrazine may be used at one-half the maximum amounts statewide. This provision replaced the provision allowing use only for rescue or no-till, at statewide rates. This change meant that maximum rates on coarse soils would be 0.5 lb. ai/A, which is well below the federal label rates. Although the intent of the amendment was to lessen restrictions, this provision was viewed by many as a de facto ban on atrazine use, based on atrazine levels above the PAL rather than the ES. This action was perceived to be beyond the department's authority in the Groundwater Law.

Public hearings were held in October, 1990. Most aspects of the rule were criticized by one side or the other, but the harshest criticism was directed at the county restrictions. Based on the public outcry and lack of political support, this component was dropped from the final rule draft. In its place was the concept of Atrazine Management Areas. The Lower Wisconsin River valley was the only AMA established.

When the dust settled, the final rule adopted for the 1991 growing season was quite similar to the first draft. The application window was extended to July 31, an additional 0.5 lbs. ai/A of atrazine could be used on coarse and medium soils where it had not been used the previous year, and the maximum rate on fine soils where atrazine had been used the previous year was lowered by 0.5 lb. ai/A. The department was mandated to evaluate the effectiveness of the rule and report to the legislature after five years. Contingent on funding, this evaluation was to include the results of two statistically designed surveys of atrazine in groundwater.

The results of the Rural Well Survey were the catalyst for the extended debate on further use restrictions in counties with more extensive contamination, but those provisions were not included in the final rule for 1991. A department official stated when the rule was adopted that, based on department surveys of current atrazine use rates, the final rule would do little to alter the use of atrazine in Wisconsin (*21*). While perceived as an admission that the rule was inadequate, the fact is that farmers had reduced their reliance on atrazine considerably since the mid-1980s, due to the registration of glyphosate, the need for more rotational flexibility, and federal agricultural policy that took land out of corn production.

The immunoassay method to determine atrazine residues in water is a cost-effective screening technique but it must be used properly. A regulatory agency must decide if the immunoassay method requires confirmation by conventional analysis. In Wisconsin, the department has determined that a detection above the ES must be confirmed with conventional laboratory analysis before considering a prohibition on atrazine use. In one case from the Rural Well Survey, the atrazine test showed 27.3 μg/L atrazine equivalents. However, the conventional analysis showed 0.21 μg/L atrazine, 1.5 μg/L deisopropylatrazine, 0.84 μg/L simazine, 93.0 μg/L prometon, and 15.0 μg/L bromacil. Upon investigation, the department determined that staff at the facility, who were untrained in pesticide use, had routinely used a granular herbicide containing prometon and simazine to treat the gravel parking lot adjacent to the test well.

The 1992 Atrazine Rule

The rule created three additional Atrazine Management Areas and eight Prohibition Areas. The Atrazine Management Areas were delineated using maps showing groundwater sample results for the Grade A survey, Grade A follow-up, and Rural Well Survey. The Prohibition Areas were based on exceedences of the ES in samples collected in Phase 2 of the Rural Well survey.

Beginning in January 1991, the State of Wisconsin Laboratory of Hygiene (SLOH) began offering the public an inexpensive test for triazines. This program was a result of the tremendous response to the Rural Well Survey. During 1991,

SLOH would use the immunoassay to analyze over 3000 samples. These data provided additional evidence to support the need for further restrictions on atrazine use in certain parts of the state. However, for the 1992 rule, only sample results received by April 1, 1991 were considered; the effects of this testing would not be felt until the 1993 growing season.

At the same time, the department would add these data to the Grade A and Rural Well survey results to counteract a legislative effort to preempt the Atrazine Rule and prohibit the use of atrazine statewide. The department interpreted these data to show that atrazine contamination is regional in nature, versus statewide, and is largely below the ES, therefore a statewide ban is not warranted.

A toll-free number encouraged participation. Both our agency and WDNR could refer concerned citizens who want their water tested to the toll-free number. The large number of samples helped determine the extent of contamination.

The 1993 Atrazine Rule

The Atrazine Rule for 1993 was much more restrictive. Three reasons for this: new groundwater standards for atrazine and its chloro metabolites; the results from Ciba-Geigy for Phase 2 of the Rural Well Survey; and extensive use of the triazine immunoassay test.

In February, 1992, the Atrazine Total Chlorinated Residue (TCR) groundwater standard was adopted by WDNR. The ES (3 μg/L) and PAL (0.3 μg/L) are applied to the sum of atrazine and three chloro metabolites, deethylatrazine [2-chloro-4-amino-6-(isopropylamino)-*s*-triazine], deisopropyl-atrazine [2-chloro-4-(ethylamino)-6-amino-s-triazine] and diaminoatrazine [2-chloro-4,6-diamino-s-triazine]. The department had only recently begun to analyze groundwater samples for the three metabolites, so the department believed that the new standard would not affect the existing database very much.

However, on March 6, 1992, the department received from Ciba-Geigy Corporation the results of its analysis of 236 water samples from the follow-up phase of the Rural Well Survey (*22*). There were 54 samples that had been below the ES based on the department's analysis for parent atrazine only that now exceeded the new ES for TCR. As a result, the department notified each well owner that their water should be considered unsafe to drink. The department established 54 new prohibition areas statewide.

The Ciba-Geigy report also showed that the PAL was exceeded in the majority of samples with any detectable residue of atrazine and/or its breakdown products.

Between April 1, 1991 and April 1, 1992 the department received over 3500 results from triazine tests. These results showed that atrazine is present in aquifers in most areas where it has been used. Furthermore, based on a statistical analysis of the Rural Well Survey report from Ciba-Geigy, a sample with a positive response to the triazine test would very likely exceed the PAL for TCR.

Based on the new TCR standard, the report from Ciba-Geigy, and the extensive watershed testing, the department proposed that the entire state be declared an Atrazine Management Area, with the same application rates allowed in

Atrazine Management Areas in 1992. Based on public hearing testimony, up to 1.5 lb. ai/A of atrazine could be applied on medium- and fine-textured soils where atrazine had not been used the year before. This allowed the continued use of popular herbicide pre-mixes containing slightly over 1 lb. of atrazine

The triazine immunoassay was used by WDNR in groundwater quality programs in its priority watersheds during 1992. At some point in the planning process, the staff working at the local level were given the mistaken impression that the test results would be confidential and that there would be no regulatory response to detections. The samples were analyzed at the State Laboratory of Hygiene and are public. Therefore, WDATCP must consider them just as it does any other groundwater results. A great deal of animosity was created in some counties when the department began investigating the exceedances and proposing atrazine prohibition areas and atrazine management areas. It was due in large part to this experience that triazine testing was discontinued the following year in priority watersheds.

Atrazine Rule Revisions for 1994-1996

Annual rule revisions have consisted of defining more prohibition areas based on additional exceedances of the ES. The department completed several surveys during this time.

Between August, 1992 and June, 1993, the department collected samples from 200 wells as part of a national study sponsored by Ciba-Geigy Corporation (*23*). The results confirmed the presence of atrazine metabolites in samples from the Central Sands region, showed that, in 44 wells from the Rural Well Survey, atrazine levels had declined in 35 and increased in 9, and showed that low triazine detections were generally confirmed by gas chromatography analysis for atrazine and its chloro metabolites. The survey was helpful in promoting the need for statewide restrictions on atrazine use based on the large number of triazine test results received during this rule revision cycle. Many of the detects were of low concentrations and the department needed to know if these were true detections.

In 1996, the department completed the Atrazine Rule Evaluation Survey (*24*), one component of the overall rule evaluation strategy. The objective of the survey was to estimate the proportion of private water supply wells containing detectable residues of atrazine and to estimate the concentration of atrazine residues in the population of contaminated wells. Groups of wells were sampled in 1994 and 1996. The proportion of contaminated wells did not change between 1994 and 1996, but there was a significant decrease in the concentration of atrazine residues in the contaminated wells during this time period. The department relied heavily on these survey results to argue that the current atrazine regulation was sufficiently protective of Wisconsin groundwater (*25*).

In 1995 the department conducted the Exceedance Survey (*26*). Owners of all wells that had tested above an ES for atrazine or any other pesticide were contacted and offered a free, follow-up sample analysis. Of the 195 candidate wells, samples were collected from 122, 111 of which had exceeded the ES for atrazine. Forty-three (43) percent of these wells still exceeded the ES, even though the majority were in atrazine prohibition areas.

Role of Surveys in Revising the Atrazine Groundwater Standards

The groundwater standard for TCR was the end result of protracted discussions between DHSS, WDNR, WDATCP, USEPA and Ciba-Geigy. In the Agricultural Resource Management (ARM) Division at WDATCP, WDNR, and DHSS Division of Health proposed that the atrazine metabolites, deethylatrazine and desiopropyl atrazine, be included in routine pesticide analytical methods, and that the sum of the analytical results for these three compounds be used to determine compliance with existing NR140 groundwater standards for atrazine. At that time, the only data on the two metabolites came from research (*27-28*). Neither the department laboratory nor the State Laboratory of Hygiene (SLOH) included these analytes.

The first data on metabolites in groundwater came from monitoring wells at a Grade A dairy farm being studied as an outgrowth of the Grade A survey (*11*).

In the fall of 1990, WDNR and WDATCP resampled 50 wells that were in Phase 2 of the Rural Well Survey. Although these results showed that metabolites were commonly found, their presence would not affect atrazine regulations until the NR140 standard was revised to include them.

General Problems With Statistical Surveys and Groundwater Protection

The Atrazine Rule Evaluation Survey provided estimates of the proportion of wells that contain detectable levels of atrazine which exceed the PAL and ES. The results from this survey are quite similar to those from the Grade A survey, and both compare well with the Iowa Rural Well Survey (*29*), as shown Table I. If these estimates are valid, our sampling to date has indicated that only a small percentage of the wells that exceed the ES for atrazine and these unknown wells are not being protected from further contamination. The current approach is reactionary in dealing with violations of the ES. Participants whose wells contain levels of atrazine residues above ES, both farmers and non-farmers, may spend more money than non-participants, either to secure a safe water supply or control weeds with more costly substitutes to atrazine. Yet, many contaminated wells could potentially remain unidentified and unprotected. From the standpoint of protecting the groundwater resource, site-specific responses to exceedances of the ES from small random samples would seem to leave the majority of contaminated groundwater vulnerable to continued contamination.

Using private water supply wells to protect the groundwater resource may not provide a realistic picture of the contamination. While the Groundwater Law applies protection equally to all groundwater, regardless of current use, the Atrazine Rule has relied almost exclusively on data from private water supply wells, which are invariably finished some depth into the saturated zone. Atrazine levels at the water table may be higher, so the data from private wells may underestimate the problem. Chesters *et al.* measured higher levels of atrazine at

Table I. Estimates of Atrazine in Groundwater from the Grade A Dairy Farm Well Water Quality Survey (9), the Iowa State-Wide Rural Well-Water Survey (*29*) and the USEPA National Pesticide Survey (*30*).

	Grade A Survey	*Iowa Rural Well Survey*	*USEPA National Pesticide Survey*
Date Issued	*April 1989*	*Nov. 1990*	*Nov. 1990*
Number of Wells	534	686	783
Detection Proportion Estimate	10 - 16 %	6 - 10%	0.1 - 2.0%
Median Detection	0.45 ug/L	0.41 ug/L	0.28 ug/L
Maximum Detection	19.4 ug/L	7.71 ug/L	7.0 ug/L
Method Detection Limit	0.15 ug/L	0.13 ug/L	0.12 ug/L

the water table than at depth in monitoring wells on a Grade A dairy farm (*11*), while some monitoring wells in the lower Wisconsin River Valley have shown the opposite trend (J. Postle, personal communication).

Data Management and Groundwater Regulations

A program such as the department groundwater program is driven by sample results and data management is critical to its success. Requests for groundwater information from private citizens, realtors and consultants are made frequently. A typical request would be for any information about atrazine in drinking water in a certain part of a rural county. Comprehensive data management is also important to convince the public that the regulations are warranted.

Our data management has matured over the past eight years. In the first years of the program, sample results were maintained in DBase on a personal computer and results were presented on maps using mylar overlays and dots of different colors to indicate no-detects below the PAL, and so on. Preparing such maps required considerable lead time and the results were not very professional. As the program matured and computer technology improved, a relational database was developed, using the Wisconsin Unique Well Number (WUWN) as the key field to link tables of well locations, well ownership, atrazine results, results for other analytes, and compliance actions. There currently are about 17,000 wells, 15,000 well owners, and 27,000 of the results were in the database.

In a perfect world, every sample result we receive from every program would be identified with a WUWN, which either would already be in our database or its location would accompany the test result. In reality, much time is devoted to keeping track of well locations. Some private citizens are reluctant to provide their well locations when requesting a triazine test, and are not required to do so. Of more concern are improper or missing WUWNs on samples from other state agencies. The WUWN is a good concept, but any state considering using such a program needs to ensure cooperation among all those who collect groundwater samples. In Wisconsin, one solution would be to have a policy that no laboratory receiving state support can analyze a groundwater sample unless the well is properly located.

The most significant step forward in our data management has been developing a geographic information system. The system consists of Arc Info software (Environmental Systems Research Institute, Redlands, CA.) running on a workstation, with output produced on a large format color inkjet printer. This has been of great value in preparing materials for public hearings and meetings with the Board of ATCP and other interested parties. It is important in planning such a program to provide enough funding to hire professional, trained staff to operate the equipment and produce professional maps.

Conclusions

The data on atrazine in groundwater from surveys conducted by the department have been integral in the regulatory program that has evolved. In fact, surveys are being used to evaluate the effectiveness of the regulations themselves in reducing the levels of atrazine in groundwater.

The kinds of surveys used in Wisconsin reflect the regulatory framework provided by the Wisconsin Groundwater Law. Each state should determine what role, if any, similar surveys could play in programs to protect groundwater.

Atrazine had been used extensively in Wisconsin for about 28 years before the first groundwater standards were adopted. The groundwater had already been contaminated, so the regulatory program by necessity was largely reactive. Monitoring wells, possibly supported by computer models, would be a better choice than private water supply wells to prevent contamination from new compounds.

Survey objectives must be clearly defined and methods must be available to meet the objectives. Despite the most careful preparations, surveys may have unanticipated and possibly undesirable consequences. Wisconsin has relied on samples from private water supply wells, many on farms, to define the extent of contamination. Unfortunately, it is not possible to determine with confidence that these wells are contaminated from current use practices, rather than from historical mishandling and higher application rates.

Literature Cited

1. Bartley, C. E. Late research report: triazine compounds. *Farm Chemicals* **1959**, *122(5)*, 28-34.
2. *General Farm Use of Pesticides 1969, Wisconsin and Illinois, Indiana, Michigan and Minnesota:* Wisconsin Statistical Reporting Service: Madison, WI, 1970
3. *Pesticide Use on Wisconsin Field Crops, 1978*; Wisconsin Agriculture Reporting Service: Madison, WI, 1979.
4. *Wisconsin 1985 Pesticide Use*; Wisconsin Agricultural Statistics Service: Madison, WI, 1986.
5. *Wisconsin 1990 Pesticide Use*; Wisconsin Agricultural Statistics Service: Madison, WI, 1991.
6. Bochert, L. H. The framework for groundwater management in Wisconsin: an historical perspective. *In Working together to manage Wisconsin's groundwater.*
7. Anderson, H. A.; Belluck, D. A.; S. K. Sinka. **1986**. *Public health related groundwater standards - 1986: summary of scientific support documentation for NR 140.10.* Wisconsin Department of Health and Social Services, Division of Health.
8. Anderson, H. A.; Belluck, D. A.; S. K. Sinka. **1988**. *Public health related groundwater standards - 1988: summary of scientific support documentation for NR 140.10.* Wisconsin Department of Health and Social Services, Division of Health.

9. LeMasters, G.; Doyle, D. J. *Grade A Dairy Farm Well Water Quality Survey*; Wisconsin Department of Agriculture, Trade and Consumer Protection and Wisconsin Agricultural Statistics Service: Madison, WI, 1989.
10. *Safe disposal of empty pesticide containers and surplus pesticides.* U.S. Department of Agriculture, Agricultural Research Service; U.S. Government Printing Office: Washington, DC, 1964: 0-750-455(10).
11. Chesters, G.; Levy, J.; Gustafson, D.; Read, H.; Simsiman, G.; Liposcak, D.; Xiang, Y. *Sources and extent of atrazine contamination of groundwater at Grade A dairy farms in Dane County, WI.* Final report to Wisconsin Department of Agriculture, Trade and Consumer Protection and Wisconsin Department of Natural Resources. Water Resources Center, Univ. of Wis. - Madison, 1975 Willow Drive, Madison, WI 53706. 1991.
12. Paulsen, R. T.: *A summary of atrazine data in groundwater in Wisconsin:* Study number Roux #02856Y, submitted to Ciba-Geigy Corporation, Greensboro, NC, 1990.
13. *Atrazine Maker Agrees to Limited Moratorium:* The Country Today, February 21, 1990.
14. *Williams, R .K.: Immunoassay Analysis and GC/MS Confirmation for Residues of Atrazine in Water Samples from a Field Study Conducted by the State of Wisconsin.* Ciba-Geigy Corporation Agricultural Division, Residue Chemistry Department, Report No. ABR-91069. Greensboro, NC, **1982**.
15. Brady, J. F.; LeMasters, G. S.; Williams, R. K.; Pittman, J. H.; Daubert, J. P.; Cheung, M. W.; Skinner, D. H.; Turner, J.; Rowland, M. A.; Lange, J.; Sobek, S. Immunoassay analysis and gas chromatography confirmation of atrazine residues in water samples from a field study conducted in the State of Wisconsin. *J. Agric. Food Chem.* **1995**, *43, 268-274.*
16. *Atrazine Pops Up in More Private Wells*: The Country Today, June 13, 1990.
17. *Some on Ad Hoc Committee Favor State Ban on Atrazine*: The Country Today, June 20, 1990.
18. *State Officials Groping for Atrazine Rule:* Agri-View, June 21, 1990.
19. *Ban on Atrazine Sought:* Wisconsin State Journal, July 19, 1990.
20. *Ag Board OKs Atrazine Rules:* Wisconsin State Journal, August 9, 1990.
21. *State Admits Atrazine Proposal Ineffective:* Milwaukee Sentinel, November 29, 1990.
22. Williams, R. K. *Residues of atrazine and chloro-triazine metabolites in follow-up well water samples from a rural well sampling program conducted by the State of Wisconsin.* Ciba-Geigy Corporation Agricultural Division, Residue Chemistry Department, Report No. ABR-91085. Greensboro, NC, **1985**.
23. Postle, J. *Report on the DA TCP/Ciba 200 well sampling program.* Wisconsin Department of Agriculture, Trade and Consumer Protection: Madison, WI 1994.

24. LeMasters, G. S.; Baldock, J. *A Survey of Atrazine in Wisconsin Groundwater: Final report.* ARM Pub 26A. Wisconsin Department of Agriculture, Trade and Consumer Protection: Madison, WI, 1997.
25. LeMasters, G. S.; Morrison, L.; Postle, J.; Vanden Brook, *J. Groundwater protection: an evaluation of Wisconsin's atrazine rule.* ARM Pub 26B. Wisconsin Department of Agriculture, Trade and Consumer Protection: Madison, WI, 1997.
26. Postle, J. *Resampling wells that previously exceeded a pesticide enforcement standard.* ARM Pub 27. Wisconsin Department of Agriculture, Trade and Consumer Protection: Madison, WI, 1996.
27. DeLuca, D. B. *Analytical Determination of Atrazine, Alachlor and Their Selected Degradation Products in Contaminated Groundwater: Implications for Wisconsin Groundwater Standards.* M.S. Thesis, Univ. of Wis.-Madison, 1990.
28. Belluck, D.; Benjamin, S.; Dawson, T. Groundwater monitoring for pesticide degradates and formulation materials: lessons from Wisconsin. *J. Pesticide Reform.* **1990**, 9, 28-31.
29. Kross, B. C.; Hallberg, G. R.; Bruner, D. R.; Libra, R. D. *The Iowa state-wide rural well-water survey, water-quality data: initial analysis.* Technical information series 19; Iowa Department of Natural Resources: Iowa City, IA 1990.
30. U.S. EPA. *National survey of pesticides in drinking water wells.* Phase I report; U.S. Environmental Protection Agency, Office of Water and Office of Pesticides and Toxic Substances; U.S. Government Printing Office: Washington, DC, 1990; EPA 570/9-90-015.

Chapter 24

Herbicides in Drinking Water: A Challenge for Risk Communication

David B. Baker

Water Quality Laboratory, Heidelberg College, Tiffin, OH 44883

To narrow gaps between public perceptions of human health risks posed by herbicides in drinking water and scientific perspectives of those same risks, it is necessary to build public confidence in the risk assessment process. To build such confidence, it is essential that the public understand the basic principles of risk assessment and trust those agencies charged with conducting risk assessments. Because of the wealth of information available on the toxicities of triazine herbicides, and their exposure patterns through drinking water, these compounds provide useful examples for public education regarding (1) risk assessment, (2) the setting of drinking water standards and their interpretation, and (3) the features of the safe drinking water act that permit ongoing consideration of new toxicological and exposure information. This paper illustrates several approaches that have been useful in improving public understanding of risk assessment and of the human health risks associated with the occurrence of herbicides in drinking water.

Is our water safe to drink? Concerned citizens frequently direct that question to water supply officials, agricultural and environmental agencies, industries, and environmental research and monitoring organizations. One of their concerns is the occurrence of herbicides in drinking water.

It has been known for many years that herbicide residues occur in midwestern drinking water supplies (*1-3*). Until recently this topic has received only limited attention because herbicide concentrations in drinking water derived from both groundwater and surface water sources are generally well below existing federal drinking water standards or lifetime health advisories. The types of water supplies where standards are sometimes exceeded are well known, and programs are being developed and implemented to address those problems. However, in recent years, some environmental advocacy organizations have contended that herbicides pose significant human health threats, even at concentrations well below drinking water standards. These claims have been advanced most strongly by the Environmental Working Group, an environmental advocacy organization based in Washington, D.C. They have recently published two reports on this topic -- *Tap Water Blues: Herbicides in Drinking Water* (*4*) and *Weed Killers by the Glass* (*5*).

Both of these reports contain useful information regarding the concentrations of herbicides in midwestern water supplies. However, in my view, both reports are highly biased in their presentation of information and, in addition, contain much misinformation (*6,7*). Using the services of a public relations firm, they release their reports through media events that are often effective in generating newspaper, television, and radio headlines and in alarming some of those members of the public concerned with environmental and public health issues. Unfortunately, media representatives often lack the background to recognize the biases and misinformation in the reports and, although they often seek and present alternative viewpoints, end up giving the advocacy group undeserved credibility (*8*). Thus, it is not surprising that the issue of herbicides in drinking water has become a focus of considerable public discussion, as well as an important public policy and political issue.

Scientists often lament the wide gap between public perceptions of risk and scientific assessments of risk. Frequently, the public perceives certain risks to be far greater than supported by scientific risk assessment. Since public policies and expenditures generally track public perceptions of risks, rather than scientific assessments of risk, considerable potential exists for inefficient use of resources available for advancing environmental protection and public health (*9*). To narrow the gap between public perceptions of risk and scientific risk assessments, it will be necessary to build public confidence in the risk assessment process. Such confidence requires that the public understand the procedures, benefits, and limitations of risk assessment, and trust those entities charged with conducting risk assessments. Building such understanding should be a major objective of risk communication.

The triazine herbicides, and especially atrazine, provide an excellent opportunity for public education about risk assessment because much is known regarding both the toxicity of these compounds and their concentrations in drinking water. The triazine herbicides are currently undergoing Special Review by the EPA in association with possible excess cancer risks associated with their occurrence(*10*). The knowledge base for triazine risk assessment includes many "state of the science" studies that have recently been completed in connection with the Special Review (*11*). The challenge in risk communication is to accurately reflect the processes of risk assessment and the data which support current assessments of health risks. It also becomes necessary to counter misinformation regarding herbicide health effects that is communicated by some advocacy groups.

In this paper, I will illustrate the approach to risk communication and risk assessment education that I use in our laboratory's environmental extension program.

The Concentration Makes the Poison

To answer the question "Is our water safe to drink?" scientists use the procedures of risk assessment. This procedure involves comparing two major sets of factors that affect risks -- the concentrations of particular contaminants and the toxicities of those contaminants (Figure 1). Even for herbicides with relatively high toxicity, based on laboratory animal studies, if the concentrations are low enough, no significant adverse health effects will occur. Conversely, even for herbicides with very low toxicities, if the concentrations are high enough, adverse health impacts will occur. Thus, it is always necessary to compare concentrations and toxicities in making risk assessments.

The concentrations of herbicides in drinking water are usually measured and reported in micrograms per liter (µg/L). One µg/L is the same as one part per billion (ppb). If a person was to drink two liters of water per day containing a pesticide at a concentration of 1 µg/L for 365 days per year for 70 years, that person would consume a total of 51.1 mg of pesticide over the 70-year period. This amount of pesticide is equivalent in weight and size to about 14% of one aspirin tablet. Thus, for atrazine, the question is whether or not atrazine is sufficiently toxic that consumption of an amount equivalent to less than an aspirin tablet over a lifetime poses significant human health risks.

Normally, to assess the risks of herbicides in drinking water, it is not necessary to directly evaluate the toxicological literature. Instead, we can compare drinking water concentrations with federal drinking water standards for the herbicides (Figure 1). The U. S. Environmental Protection Agency (EPA) has been charged with evaluating the toxicological literature and setting drinking water standards such that consumption of drinking water containing herbicides at concentrations equal to or less than their drinking water standards should not adversely impact human health. Since the adequacy of current drinking water standards has been questioned, the public needs to be familiar with the methods used by the EPA in setting drinking water standards.

How Drinking Water Standards Are Set

As part of the registration process, either to bring a new pesticide onto the market or to maintain registration of an existing pesticide, the EPA requires that a battery of toxicological tests be completed. The EPA specifies the testing protocols that are to be used. The tests look at both acute effects, which are associated with short term exposures at relatively high concentrations, and chronic effects, which are associated with long-term exposures to relatively low concentrations.

Just as in the development of new medicines, toxicological testing for pesticides starts with testing on laboratory animals, such as mice, rats, rabbits, and dogs, and on various bacterial or cell cultures. In contrast with medicines, where clinical trials on human subjects generally follow the animal testing, toxicological testing for pesticides stops with the animal testing. The toxicity of a pesticide to humans is then predicted based on the toxicity of that pesticide to animals. Because of the uncertainty in extrapolating animal test results to humans, safety factors are incorporated into estimates of human pesticide doses that should pose no significant human health risks.

Virtually all toxicity testing incorporates dose-response testing (Figure 2). In such testing, the relationships between the sizes of pesticide doses contained within food and adverse health impacts are investigated. A control group of animals receives food lacking any of the pesticide, while other groups of animals receive food with increasing concentrations of the pesticide (Figure 2). Doses are reported in mg pesticide per kg body weight of the test animal per day. This method of reporting doses facilitates comparisons of doses among animals of differing sizes, such as mice, rats, and dogs, and even extrapolation to humans.

The high doses are chosen such that obvious adverse health effects are apparent. Two particular doses are important for the setting of drinking water standards. These are the lowest observed adverse effect level (LOAEL), and the next lower dose, the no observed adverse effect level (NOAEL). Somewhere between these two doses lies a *threshold* dose, a dose at which the onset of adverse effects occurs. Often there is a five- to ten-fold difference between the LOAEL and NOAEL doses. Because of the high costs associated with such testing, no attempts are made to zero in on the threshold value. Instead, the NOAEL is generally used as the starting point for setting drinking water standards.

A wide variety of "adverse health effects" are examined in the various toxicological studies (*12*). These include determination of lethal doses, dermal/ocular effects, growth rates, organ weights, blood chemistry, multi-generational reproductive studies, developmental effects, mutagenicity, and carcinogenicity. As a first step in setting drinking water standards, the EPA's Office of Drinking Water decides which adverse effect appears to pose the greatest threat to human health. Subsequent standards are set in two different ways, depending on whether the greatest threat is associated with carcinogenic effects (cancer causing effects) or non-carcinogenic effects.

Standards Based on Non-Carcinogenic Effects. If non-carcinogenic effects are thought to pose the greatest threat, then the EPA identifies the NOAEL for the most sensitive animal species and adverse effect, and that NOAEL becomes the starting point for the incorporation of a variety of safety factors that lead to the drinking water standard. This procedure is illustrated in Figure 3 for chronic effects from atrazine. A 100-fold safety factor is incorporated into the drinking water standard for virtually all pesticides. This factor includes a 10-fold safety factor based on the uncertainty in extrapolating from one animal species to another, and a second 10-fold safety factor to allow for variable sensitivities among individuals of the human population. Because the triazine herbicides are classified as Class C carcinogens (possible human carcinogens), the EPA's Office of Drinking Water incorporates an additional 10-fold safety factor into the drinking water standard. To allow for alternate pathways of pesticide entrance into humans, such as via food, an additional 5-fold safety factor is added. These separate safety factors yield a combined 5,000-fold safety factor for drinking water.

The drinking water dose deemed safe for chronic exposure to humans is therefore 5,000 times smaller than the dose which has no observed adverse effect

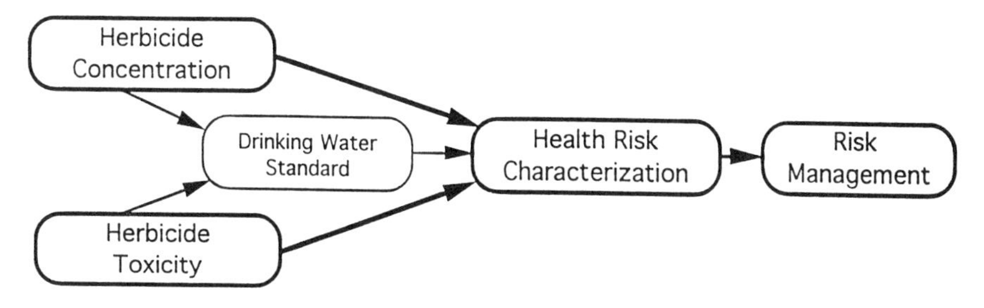

Figure 1. Basic components of risk assessment and management for herbicides in drinking water.

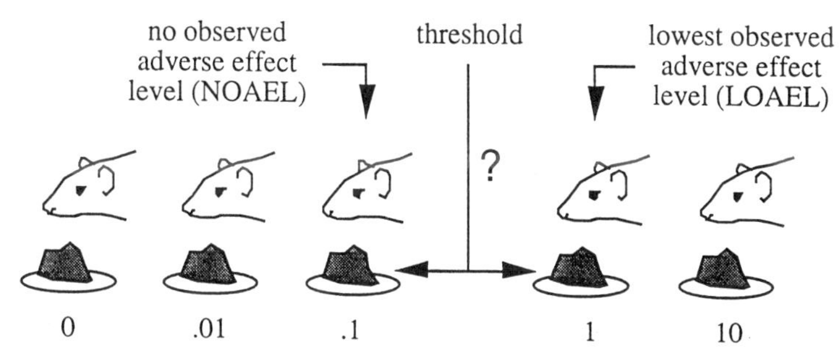

Figure 2. Animal feeding studies for evaluation of relationships between pesticide doses and animal responses.

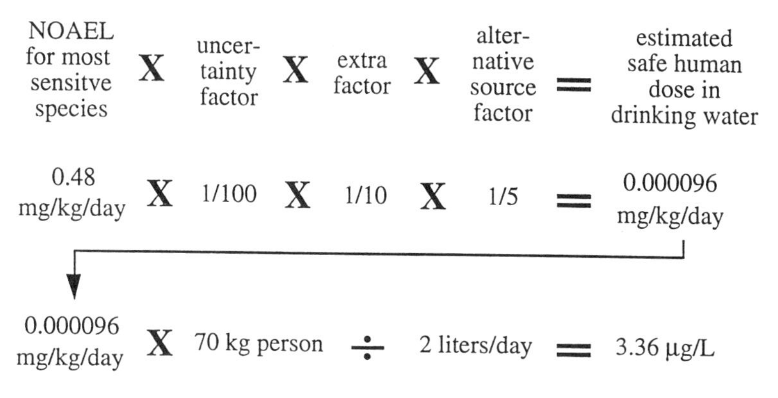

Figure 3. EPA's calculation of drinking water standard (lifetime health advisory) for atrazine.

on the most sensitive animal for which toxicological data are available. This dose, in mg per kg body weight per day, is converted into a drinking water concentration for a person of average size (154 lbs) who consumes 2 liters of water per day. The rounded-off value of 3 µg/L is the chronic or lifetime drinking water standard for atrazine. The Office of Drinking Water deems the combined safety factor sufficiently large that they consider consumption of drinking water containing atrazine at 3 µg/L or less over a lifetime to pose no significant adverse health effects. This standard is called the lifetime health advisory (LHA) for atrazine. After the LHA for atrazine was set at 3 µg/L, the EPA subsequently reevaluated the toxicity data for atrazine and concluded that the NOAEL for atrazine should be 3.5 mg/kg/day rather than the 0.48 mg/kg/day used in setting the standard. The manufacturer of atrazine requested that the drinking water standard be proportionally adjusted upward but the EPA declined (*13*). Consequently, the current LHA for atrazine is actually about 41,000 times lower than the current NOAEL.

In addition to the LHA for atrazine, the EPA also sets health advisories (HAs) related to shorter term exposures at higher concentrations for both children and adults (*12*,*14*). These are shown for five commonly used herbicides in Table I. In all cases, no significant adverse human health effects are thought to be associated with exposure at or below the health advisories for the durations of time indicated for each health advisory. For calculation of HAs for children, body weights and water consumption of children are used, rather than adult weights and water consumption (*12*).

Table I. Health Advisories for Selected Herbicides

Health Advisories Concentrations in µg/L	*10 kg Child*			*70 kg Adult*		
	One Day	*Ten Day*	*Long Term*	*Long Term*	*Life Time*	*MCL*
Atrazine	100	100	50	200	3	3
Cyanazine	100	100	20	70	1	
Simazine	70	70	70	70	4	4
Alachlor	100	100				2
Metolachlor	2000	2000	2000	5000	70	

While the Office of Drinking Water has calculated LHAs for most pesticides that are used in substantial quantities, not all of these have been extended to regulatory status, at which time they are referred to as maximum contaminant levels (MCLs). MCLs have been set for three of the five herbicides listed in Table I (alachlor, atrazine and cyanazine). MCLs are generally set at the same value as LHAs.

Standards Based on Carcinogenic Effects. In cases where carcinogenicity is deemed to pose the greater health threat, drinking water standards are calculated in a different manner. For cancers, no threshold concentrations are assumed to exist.

Instead, any dose, no matter how small, is assumed to carry with it some additional risk of cancer. To estimate the magnitude of that risk, the Office of Drinking Water uses what is called a conservative multi-stage, linear model for extrapolating from the high doses used in animal testing to the much lower doses associated with human exposures to pesticides (*12*). The model produces a cancer potency factor called a Q* (Q star). This cancer potency factor can be used to calculate the added lifetime cancer risk associated with a given drinking water concentration of a pesticide or other chemical. In calculating the cancer potency factor, the EPA uses a conservative approach in that they use the 95th percentile upper bound limit. In effect, this means that they expect the cancer potency factors to overestimate cancer occurrences 95% of the time and underestimate cancer occurrences only 5% of the time.

Drinking water standards for carcinogens are set on a case by case basis, such that the added lifetime cancer risk with consumption at the standard should not increase cancer risks by more that one-in-10,000 to one-in-1,000,000. This procedure is used for chemicals that are classified as Class A or Class B carcinogens. Class A carcinogens are called known human carcinogens and include such compounds as benzene, cigarette smoke and asbestos. Class B carcinogens are called probable human carcinogens. Chemicals are classified as Class B carcinogens when they induce cancers in multiple animal species and in multiple tissues, even though there may be no direct evidence that they cause cancer in humans. The herbicide acetochlor is an example of a Class B carcinogen. Chemicals are classified as Class C carcinogens (possible human carcinogens) when evidence of carcinogenicity is limited to a single animal species and there is no evidence of cancer induction in humans. The triazine herbicides are Class C carcinogens. As noted above, the Office of Drinking Water adds an additional 10-fold safety factor into the drinking water standards for Class C carcinogens.

For triazine herbicides, evidence of cancer induction is limited to female rats of the Sprague-Dawley strain, where triazines induce the formation of mammary tumors at an earlier age than in controls. The atrazine doses observed to trigger increased mammary tumors in this strain of rats are 41,000 times higher than the doses received by humans at the drinking water standard. Triazines do not induce cancers in mice or rats of the Fisher strain. Much research is underway to determine the biochemical and physiological mechanisms by which triazines induce mammary tumors in female Sprague-Dawley rats (*15,16*). It is hoped that a better understanding of these mechanisms will indicate whether or not the same mechanisms might operate in humans and whether or not the induction mechanism in these rats involves threshold effects.

For some Class C carcinogens, including the triazine herbicides, the Office of Drinking Water has calculated cancer potency factors. Thus, cancer risks can be calculated for these compounds using their drinking water concentrations and their cancer potency factors. In Table II, cancer potency factors are shown for five commonly used herbicides. Table II also includes the added lifetime cancer risk that accompanies consumption of water at the MCL or LHA for each compound, and the concentration of each herbicide that carries with it a one-in-1,000,000 added lifetime risk. At such low cancer risk levels, the risks are considered to be additive (*17*). Thus consumption of all five herbicides at their drinking water standards over a lifetime would result in an aggregate lifetime risk of 71 in 1,000,000.

Table II. Cancer Potencies and Associated Lifetime Cancer Risks for Selected Herbicides at Their MCLs or Lifetime Health Advisory Levels

Herbicide	*Cancer Potency Factor (Q*)*	*Drinking Water Standard MCL/LHA, μg/L*	*Theoretical Lifetime Cancer Risk at Standard, x 10^{-6}*	*Concentration at 10^{-6} lifetime cancer risk, μg/L*
Atrazine	0.22	3	19	0.16
Cyanazine	1.00	1	29	0.03
Simazine	0.12	4	13	0.31
Alachlor	0.08	2		0.44
Metolachlor	0.01	70	4.6	18
Aggregate Risk			3.9	

Herbicide Concentrations in Drinking Water

Much is known regarding the concentrations of triazine herbicides in drinking water (*18-22*). Except for the few cases where water supplies are using some type of activated carbon filtration, the concentrations of herbicides in treated tap water are the same as the concentrations in the surface or ground water used as the raw water supplies (*2,23*). Consequently, concentrations in streams, rivers and reservoirs generally provide good estimates of the concentrations in drinking water.

The concentration patterns of herbicides in streams and rivers draining midwestern agricultural watersheds are well documented (*18*). Concentrations are highly seasonal, with peaks during spring storm runoff events following herbicide applications (*18*). By late summer and fall, stream concentrations are much lower, even during runoff events. Likewise, runoff events during the winter, and early spring prior to application are very low. These patterns are illustrated with data from the Maumee River of northwestern Ohio (Figure 4). Monthly average concentrations during May, June and July often exceed 3 μg/L. However, annual average concentrations, also shown in Figure 4, have yet to exceed the drinking water standard of 3 μg/L. Such is apparently the case for the vast majority of midwestern rivers (*19*).

Since the MCL for atrazine is a chronic standard, the EPA judges compliance with that standard by comparing it with annual average concentrations. Even though daily, and even monthly average concentrations, often exceed the MCL in midwestern rivers, such excursions do not represent violations of drinking water standards unless annual averages also exceed the standard. The National Research Council, in *Pesticides in the Diets of Infants and Children*, also recommends that compliance with chronic standards be based on average concentrations (*17*). The monthly and annual averages shown in Figure 4 are based on detailed storm event sampling during the mid-April through mid-August period, coupled with two samples per month during other seasons. Federal drinking water regulations state that compliance with the standard is to be based on a running average of four quarterly samples. Unfortunately, quarterly samples are inadequate to accurately characterize average annual concentrations in most midwestern rivers.

Statewide herbicide exposure patterns have recently been developed for several states (*19*). Table III represents a summary of atrazine concentrations in Ohio's public and private drinking water supplies (*24*). The supplies are listed by rank order from those with the highest concentrations to those with the lowest concentrations. For each supply or group of supplies, the percentage of the state's population served is also listed. Highest concentrations occurred in a small proportion of private rural wells, which upon investigation generally reflected point sources of contamination. The most vulnerable public water supplies are those derived from surface waters draining from watersheds where land use is dominated by row crop agriculture.

Table III. Ohio Drinking Water Sources, Populations Served and Average Atrazine Concentrations[1]

Water Source	*% of Population*	*Concentration (μg/L)*
Private wells >3.0 ppb	0.05	5.00
Upground Reservoirs	2.05	2.19
Scioto River	2.32	2.02
Private wells 1.0 - 3.0 ppb	0.03	2.00
Sandusky River	0.40	1.73
Maumee River	0.57	1.50
Big Walnut Creek	4.89	0.91
Huron and Vermilion Rivers	0.02	0.89
Other reservoirs	3.95	0.75
Alum Creek	0.31	0.62
Private wells 0.2 - 1.0 ppb	0.32	0.60
Lake Rockwell	3.68	0.56
Other Surface Water	5.12	0.48
Ohio River	8.59	0.24
Private wells 0.05 - 0.2 ppb	0.46	0.10
Lake Erie	23.09	0.07
Private wells <0.05 ppb	17.67	0.025
Public wells	28.50	0.025

[1]Adapted from ref. *24*.

The information shown in Table III can be plotted as a ranked, variable width histogram, as shown in Figure 5. Such graphs provide a convenient way to depict herbicide concentrations in drinking water supplies. The graphs and supporting

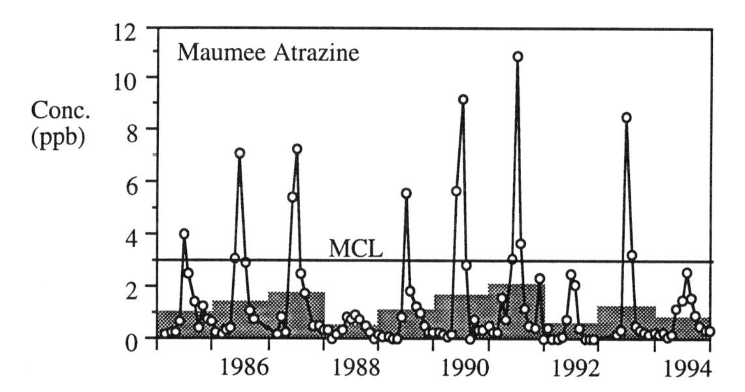

Figure 4. Monthly (dots) and annual average (bars) concentrations of atrazine in the Maumee River, 1985-1994. Adapted from ref. *24*.

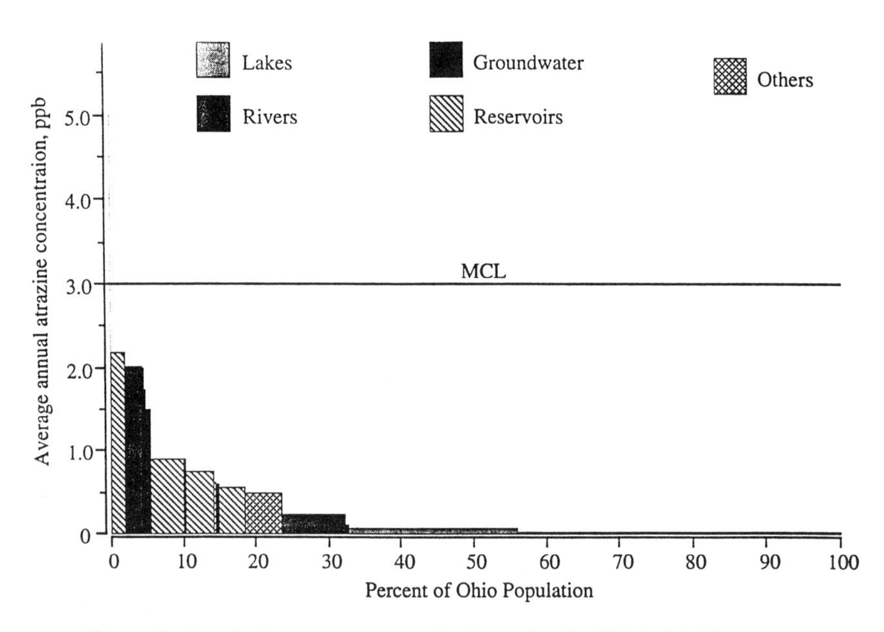

Figure 5. Population exposure graph of atrazine in Ohio's drinking water. Adapted from ref. *24*.

tables show those supplies which have the highest concentrations, and thus can support targeting of risk management programs, should such programs be deemed appropriate. The graphs also indicate that most drinking water supplies are in compliance with the atrazine MCL. The amounts by which the actual atrazine concentrations fall below the MCL represent additional safety factors beyond those incorporated in the drinking water standard. Statewide atrazine population exposure graphs for Illinois and Iowa are very similar to that of Ohio (*19*). In the midwest, highest average annual atrazine concentrations often occur in reservoirs having relatively small, agricultural watersheds. Atrazine in groundwater supplies is typically very much lower than in surface water supplies.

In Figure 6, data from a recent testing program conducted by the Environmental Working Group (*5*) are plotted in a similar fashion as in the statewide population exposure graphs. In addition to the average concentrations, the peak concentration for each supply is also shown. The Environmental Working Group collected samples at three-day intervals during the May 15 though July 31 period (1995) for 29 midwestern cities (*5*). The cities were selected as representing supplies likely to be contaminated, based on previous monitoring data, with all but one utilizing surface water. These results are noteworthy in that in only three of the 29 supplies did average atrazine exceed 3 µg/L, even though the sampling period was restricted to the season of the year having the highest herbicide concentrations. In presenting the results of their study to the public, the Environmental Working Group claimed that herbicide concentrations throughout the midwest frequently exceeded federal drinking water standards (*5,7*). They based their claims on improper comparisons of peak daily concentrations and short-term average concentrations with the chronic standard, in direct conflict with the recommendations of the National Research Council (*17*) and the EPA (*25*). In fact, their data provide strong evidence that drinking water standards would not be exceeded in these cities. It is important to note that the peak atrazine concentrations they observed did not exceed any of the EPA's short term health advisories for children.

Risk Characterization

A crucial part of risk assessment is risk characterization. It is the component of risk assessment that involves comparison of toxicity and concentrations. This is the step that answers the question, "Is our water safe to drink?"

Non-Cancer Effects. One way of comparing toxicity and drinking water concentrations is to plot both toxicity and concentrations on the same graph and see if they overlap. This requires expressing toxicity and drinking water concentrations in the same units. It is most convenient to convert drinking water concentrations into units of dose expressed in mg/kg body wt/day, assuming a 154-lb person consumes 2 liters of water per day. Figure 7 illustrates such a graph for Ohio. It is necessary to use a log scale for doses since the graph must cover 9 orders of magnitude to encompass the midrange of doses used in animal toxicity testing and the doses that most people receive in their drinking water. The gap

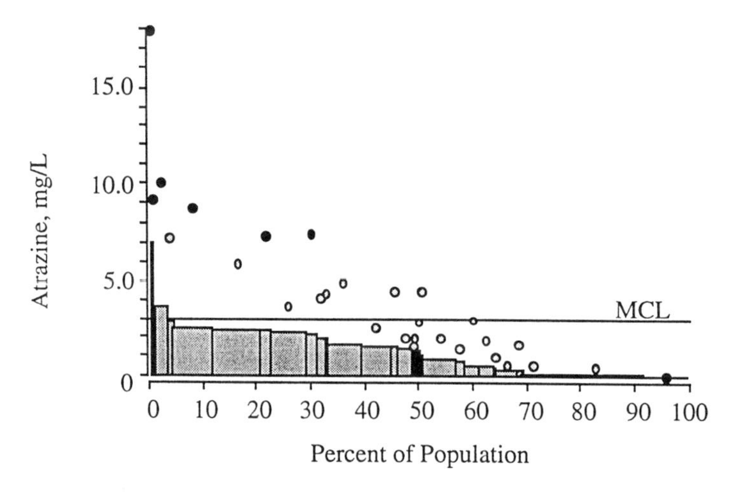

Figure 6. Summary of data from Weed Killers by the Glass. Bars shown mean concentrations and dots show peak concentrations. Adapted from ref. *5*.

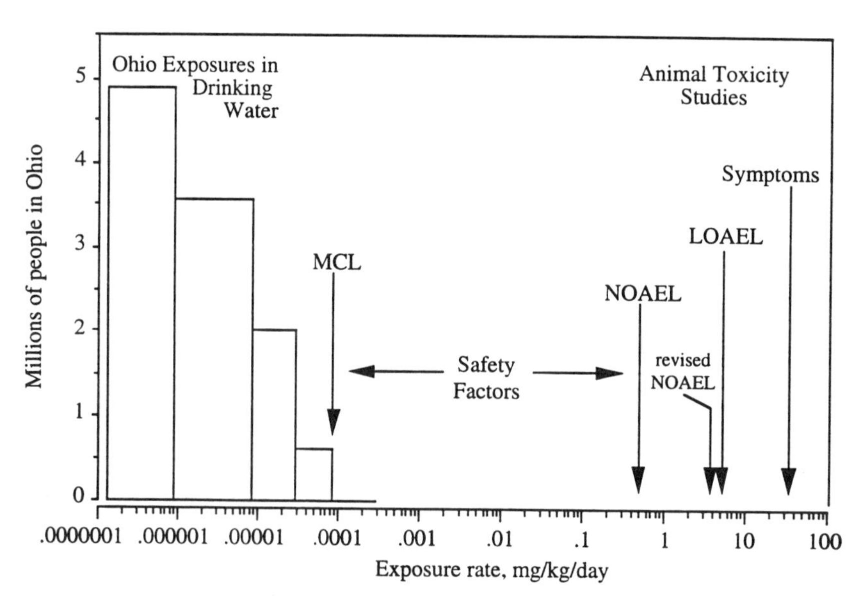

Figure 7. Comparison of atrazine toxicity data and atrazine exposure data for Ohio. Adapted from ref. *24*.

between the NOAEL for the most sensitive animal and the MCL represents the safety factors incorporated into the drinking water standard. It is important that the public not confuse the MCL with the threshold for adverse human health effects. While it is uncertain what the threshold level for adverse human health effects would be for atrazine, the EPA considers the safety factor sufficiently large that no adverse human health effects would be expected at concentrations below the MCL and furthermore that the MCL does include a margin of safety. Consequently, even the small portion of Ohio's population that consumes water slightly above the MCL would be very unlikely to experience adverse health effects from atrazine in drinking water. Since atrazine concentrations for the vast majority of Ohio's residents are well below the MCL, most residents enjoy safety factors much greater than those incorporated into the drinking water standard.

Because infants and children consume more water per unit body weight than adults, the doses for infants and children would be shifted slightly to the right from the adult doses shown in Figure 7. In Table IV, the revised NOAEL for atrazine is shown, along with the doses and associated safety factors for adults, children and infants who consume drinking water containing atrazine at 3 μg/L and, for infants, also at 15 μg/L. It is noteworthy that at 15 μg/L, the safety factor for an infant is 1,000-fold relative to the revised NOAEL. In *Pesticides in the Diets of Infants and Children*, the National Research Council recommended that an additional 10-fold safety factor be added to the 100-fold safety factor normally present in order to account for the possibility that children and infants might be more sensitive to pesticides than adults (*17*). The recommended 1,000-fold safety factor is present for infants, even when the atrazine concentration is at 15 μg/L. Thus, atrazine concentrations in drinking water should not pose adverse health effects for infants, even if infants would be more sensitive to atrazine than adults.

Table IV. Safety Factors in Drinking Water Standards for Adults, Children and Infants

NOAEL or drinking water concentration	*Dose mg/kg body wt./day*	*Safety Factor relative to NOAEL*
NOAEL (EPA revised value)	3.5	---
3 ppb, adult	0.000086	41,000
3 ppb, child	0.0003	11,700
3 ppb, infant	0.0007	5,000
15 ppb, infant	0.0035	1,000

Are safety factors in the range of 1,000-fold to 41,000-fold adequate to protect health? It is instructive to compare the size of safety factors present for pesticides with the size of safety factors for other chemicals that we commonly consume. In Table V, the lethal doses which result in 50% mortality in mice (i.e., the LD50s) are shown for atrazine, aspirin, and caffeine. Atrazine is the least toxic of all of these substances, since it takes a larger dose of atrazine to result in 50% mortality than for the other substances. The doses for adults consuming 2 liters/day of drinking water containing 3 μg/L atrazine, 8 aspirin tablets per day, and 5 cups of coffee per day are shown in Table V. The safety factors between the above doses and their LD50s are 20,000,000 for atrazine, 29 for aspirin and 17 for caffeine. Clearly, safety factors for herbicides are very large relative to safety factors for many chemicals that we voluntarily consume.

Table V. Safety Factors for Atrazine, Aspirin and Caffeine Relative to Lethal Doses

	Atrazine	*Aspirin*	*Caffeine*
LD50 for mice mg/kg body wt./day	1,800	1,200	150
human adult doses mg/kg body wt./day	0.00009 (3 ppb, 2 L/day)	41 (8 tablets/day)	9 (5 cups coffee/day)
safety factor relative to LD50 for mice	20,000,000	29	17

Cancer Effects. Since cancer potency factors are available for the triazine herbicides, it is possible to calculate both the individual and aggregate cancer risks associated with herbicide concentrations present in various drinking water supplies. One such effort was undertaken by the Environmental Working Group and published in *Tap Water Blues: Herbicides in Drinking Water* (*4*). They examined data on the concentrations of five herbicides (alachlor, metolachlor, atrazine, cyanazine and simazine) in the drinking water for 121 cities, mostly from vulnerable midwestern supplies. Using the cancer potency factors for each herbicide, they calculated the cancer risk associated with each herbicide and added the risks for the five herbicides to obtain an aggregate risk for each city. They also calculated average risks for the assessed communities in each state. Rather than refer to the combined risks as aggregate risks, the Environmental Working Group referred to the combined risks as multiples of a federal standard of 1×10^{-6} (one-in-a-million) added lifetime risk. Thus, for the residents of Bowling Green, Ohio, they stated that cancer risks were 30.3 times higher than the federal standard. They clearly imply that there is a federal standard of 1×10^{-6} that is applicable to the combined risks for 5 herbicides. There are no federal regulations indicating that the combined risk for 5 herbicides should be 1×10^{-6}. The Office of Drinking Water considers risks of 1×10^{-4} to 1×10^{-6} to be acceptable for single compounds. To the public, risks labeled 30 times higher than a federal standard likely sound more frightening than concentrations that yield a 30×10^{-6} added lifetime chance of cancer, particularly when such statements are accompanied with calls for an immediate banning of the herbicides because of high cancer risks.

A more understandable way to inform the public about the magnitude of cancer risks associated with herbicides in drinking water is to indicate the predicted numbers of additional cancers per year associated with given aggregate risks in the populations of interest. Table VI indicates the assessed populations for each state and the associated average aggregate cancer risks, taken from *Tap Water Blues*. In Table VI, I have added an estimate of the additional cancers per 70 years (lifetime) and the additional cancers per year that would be expected in the assessed populations of each state, based on these same aggregate risks (*6*). In no state would these five herbicides be expected to result in even a single cancer per year in the assessed populations. For the approximately 11,000,000 residents of the assessed portions of these states, herbicides in drinking water would be estimated to cause an additional 2.34 cancers per year. The bulk of this estimated risk is associated with atrazine and cyanazine, which are classified as possible human carcinogens (i.e., they may or may not cause cancers in humans). The estimate also represents the 95th percentile upper bound limit of cancer risk. Consequently, the 2.34 cancers per year may significantly overestimate the actual risks posed by these compounds.

Table VI. Theoretical Cancer Occurrences Due to Herbicides in Worst Case Midwestern Water Supplies

State	*Population of Affected Communities*[1]	*"Statewide Average" Risk*[1]	*Herbicide Cancers per 70 yrs*[2]	*Herbicide Cancers per yr*[2]	*Expected New Cancers per yr, All Causes*[2]
Illinois	1,400,000	12.6	17.6	0.25	6,258
Indiana	1,300,000	27.9	36.3	0.52	5,811
Iowa	760,000	17.4	13.2	0.19	3,397
Kansas	830,000	8.1	6.7	0.10	3,710
Kentucky	650,000	3.8	2.5	0.04	2,906
Louisiana	1,500,000	8.5	12.8	0.18	6,705
Minnesota	930,000	30.6	28.5	0.41	4,157
Missouri	2,125,000	14.3	30.4	0.43	9,499
Nebraska	450,000	9.0	4.1	0.06	2,012
Ohio	1,044,200	17.3	18.1	0.26	4,668
Total	10,989,200	15.4		2.43	49,122

[1]Adapted from ref. *4*.
[2]Adapted from ref. *6*.

The assessed population in the ten states covered about 11,000,000 residents, or about 21% of the total population of these states. If this assessed population is typical of the United States population as a whole, in terms of cancer rates from all sources, then more than 49,000 new cancers would be expected per year among these 11,000,000 people. Obviously, 2.34 cancers is very small relative to 49,000 new cancers per year. Most of the unassessed population of these states would have much lower herbicide concentrations in their drinking water because their water supplies are withdrawn from groundwater, the Great Lakes or rivers having largely nonagricultural watersheds.

How protective is a standard set at a 1×10^{-6} lifetime risk of cancer? Often, 1×10^{-6} added lifetime risk is set forth as a negligible cancer risk. It is also frequently viewed as the maximum acceptable risk, although the basis for this interpretation is not clear (*26*). To illustrate the degree of protection offered by a 1×10^{-6} lifetime risk standard, consider the number of cancers per year that would occur in the United States if everyone were exposed continuously to a chemical that conveyed that level of risk. With a United States population of 258,000,000, there would be 258 cancers over a 70 year period (a "lifetime"). This is equivalent to 3.7 new cancers per year. In 1994, there were 1,250,000 new cancers identified in the United States. Thus the total cancer risk faced by citizens of the United States is about $338{,}000 \times 10^{-6}$. Since a standard represents the maximum allowable concentration, the average concentration in drinking water supplies would generally be well below the standard. Any supplies above the standard would have to institute programs to lower the concentration below the standard. Supplies in compliance would already have concentrations below the standard. Consequently, a 1×10^{-6} risk standard actually assures a much lower average risk than 1×10^{-6}.

As noted in Table II, given atrazine's cancer potency factor, the atrazine MCL of 3 ppb is accompanied by a cancer risk of 19×10^{-6} and concentration of 0.16 µg/L results in a 1×10^{-6} risk. As noted above, atrazine concentrations are actually well below the standard, even in vulnerable midwestern supplies. The population weighted average atrazine concentration in drinking water of the United States is certainly less than 0.16 µg/L. If one assumes average atrazine concentrations of 0.3 µg/L among the 52,000,000 midwestern residents and 0.05 µg/L for the remaining 206,000,000 residents in the United States, then the population weighted average atrazine concentration in drinking water would be about 0.1 µg/L, and the theoretical estimates of the number of cancers possibly attributable to atrazine would be 2.3 cancers per year. The above concentration estimates likely overestimate average drinking water concentrations of atrazine. Clearly, banning atrazine does not appear to be an effective way to reduce overall cancer occurrences in the United States.

Additive and Synergistic Effects. An issue that frequently arises in connection with the occurrence of herbicides in drinking water is the fact that multiple herbicides, together with several herbicide breakdown products, often occur simultaneously in drinking water supplies. Does this simultaneous occurrence change the picture relative to possible adverse health effects from herbicides? To answer this question, it is again necessary to start out considering the concentrations and toxicities of the individual chemicals.

A strong case can be made that atrazine, cyanazine and alachlor represent "worst case" drinking water risks, relative to the effects of pesticides. Within the Great Lakes regions, these three herbicides rank in the top five in terms of quantities applied (*27*). Since pesticide concentrations are generally proportional to pesticide use, it is not surprising that monitoring programs show these compounds to be present at concentrations much higher than most other pesticides (*18*). Not only are other pesticides generally present at much lower concentrations, but they also have higher drinking water standards. Consequently, the theoretical health risks accompanying those compounds would be much less than for atrazine, cyanazine and alachlor. As noted above, the risks for atrazine, cyanazine and alachlor are, themselves, very small.

What about additive or synergistic non-cancer effects? It is well known that care must be exercised when taking multiple medicinal drugs. Both prescription and non- prescription drugs are taken at doses that do have significant effects on human physiology and biochemistry. It is these effects that can interact in ways that have adverse impacts on our health. If drugs were taken at doses below the therapeutic ranges, both the beneficial effects and the complicating additive or synergistic effects would usually disappear. Herbicide doses in drinking water are thousands to hundreds of thousands of times lower than doses which have no adverse effect on the most sensitive animal species tested. Consequently, effects on human physiology and biochemistry of sufficient size to interact either additively or synergistically seem highly unlikely. This conclusion is supported by animal testing of mixtures of five herbicides (alachlor, atrazine, cyanazine, metolachlor and metribuzin) and nitrogen fertilizers, all at concentrations up to 100 times their median concentrations in Iowa ground water (*28*). Neither reproductive nor developmental toxicity were observed at any of the dose ranges.

The doses of herbicides that we receive through drinking water are very small relative to the doses of medicines that we frequently take, even though the medicines are often more toxic than the herbicides. Adverse reactions to medicinal drugs have been shown to be responsible for from 3 - 7% of all hospital admissions (*29*).

Incorporation of New Data. The regulations set forth in the Safe Drinking Water Act specifically allow for consideration of any new toxicological or exposure data that may become available. Should new data suggest that large health risks may be present, use of a particular pesticide could immediately be suspended. Likewise, as new areas of concern arise, such as that of endocrine disruption (*30*), pesticides can be subjected to new types of testing to evaluate possible effects via those mechanisms. Thus, the setting of drinking water standards is a dynamic process that allows for incorporation of new, relevant data that may become available. Because of the large safety factors built into drinking water standards, examples of known adverse human health impacts from herbicide exposure through drinking water are, to my knowledge, absent from the literature.

Risk Management

From the above analyses, adverse health risks associated with the occurrence of herbicides in drinking water appear to be very small. Consequently, any management efforts aimed at reducing herbicide concentrations in drinking water that are accompanied by high costs to farmers or consumers need to be evaluated very carefully. Since the health risks appear small, the possible health benefits of reducing those risks are also small. Care must be taken when small benefits are accompanied by high costs. A fundamental outcome of risk assessment should be improved efficiency in the use of resources to improve public health and the environment.

Fortunately, there are many agricultural best management practices that are either profitable or cost neutral that do result in reduced concentrations of herbicides in drinking water supplies. Some of these result in reduced use of herbicides, some result in changes in the herbicides that are used, and some minimize off-field transport of herbicides. Aspects of agricultural nonpoint source pollution other than herbicide runoff probably represent greater environmental threats and costs than does herbicide runoff. These include phosphorus, nitrogen and sediment runoff. Significant reductions in herbicide runoff can be economically accomplished within the context of comprehensive farm plans aimed at minimizing the varied adverse impacts of food production on water resources.

Literature Cited

1. Environmental Protection Agency. Federal Register. 1983, vol 48, No 194, pp 45505.
2. Baker, D. B. J. *Soil and Water Cons.* **1985**, *vol* 40, pp 125-132.
3. Frank, R.; Logan, L.; Clegg, B. S. *Arch. Environ. Contam. Toxicol.* 1990, *vol* 19, pp 319-324.
4. Wiles, R.; Cohen, B.; Campbell, C.; Elderkin, S. *Tap Water Blues: Herbicides in Drinking Water;* Environmental Working Group: Washington, DC, 1994, pp 276.
5. Cohen, B.; Wiles, R.; Bondoc, E. *Weed Killers by the Glass;* Environmental Working Group: Washington, D.C., 1995, pp 83.
6. Baker, D. B.; Richards, R. P.; Baker, K. N. *A Review of the Science, Methods of Risk Communication and Policy Recommendations in Tap Water Blues: Herbicides in Drinking Water;* Water Quality Laboratory, Heidelberg College, Tiffin, OH, 1994, pp. 47.
7. Baker, D. B. *A Review of Weed Killers by the Glass and Related Reports by the Environmental Working Group*; Water Quality Laboratory, Heidelberg College, Tiffin, OH, 1994, pp. 19.
8. Schultz, M. *Forbes Media Critic.* **1996**, Winter, pp 86-89.
9. U. S. Environmental Protection Agency. *Reducing Risk: Setting Priorities and Strategies for Environmental Protection*; U. S. EPA, Science Advisory Board. 1990, SAB-EC-90-021.

10. U. S. Environmental Protection Agency. *Atrazine, Simazine and Cyanazine; Notice of Initiation of Special Review*; 1994. OPP-3000-60; FRL-491905, pp 63.
11. (This Publication - ACS Symposium on Triazines)
12. U. S. Environmental Protection Agency. *Drinking Water Health Advisory: Pesticides*. 1989; Lewis Publishers: Chelsea, MI, pp 819.
13. Browner, C. U.S. EPA - Letter to Emilio Bontempo, Ciba-Geigy Corporation. Re: Petition for reconsideration and administrative stay of drinking water regulations. October 17, 1994.
14. U. S. EPA. *Drinking Water Regulations and Health Advisories*; EPA, Office of Water: Washington, DC, November **1994**.
15. Eldridge, J. C.; McConnell, R. F.; Tisdel, M. O.; Wetzel, L. T. (This symposium volume)
16. Anderson, M. E.; Barton, H. A.; Clemen, H. G. III; Gearhart, J. M.; Allen, B. C. (This symposium volume)
17. National Research Council. *Pesticides in the Diets of Infants and Children*; National Academy Press: Washington, D.C. **1993**, pp 386.
18. Richards, R. P.; Baker, D. B. *Environmental Toxicology and Chemistry*. **1993**, *vol* 12, pp 13-26.
19. Richards, R. P.; Baker, D. B.; Christensen, B. R.; Tierney, D. P. *Env. Science and Technology*. **1995**, *vol* 29, pp 406-412.
20. Goolsby, D. A.; Thurman, E. M.; Kolpin, D. W.; Battaglin, W. A. (This symposium volume.)
21. Richards, R. P.; Baker, D. B. (This symposium volume.)
22. Barrett, M. R. (This symposium volume.)
23. Miltner, R. J.; Baker, D. B.; Speth, T. F.; Fronk, C. A. *J. Amer. Water Works Assoc*. **1989**, *vol* 81, pp 43-52.
24. Baker, D. B.; Richards, R. P. In *Proceedings of the Fourth National Conference on Pesticides: New Directions in Pesticide Research, Development, Management, and Policy*; Weigmann, D. L., Ed., Virginia Water Resources Center: Blacksburg, VA, 1994, pp 200-221.
25. Pontius, F. W. *J. Amer. Water Works Assoc*. **1990**, *vol* 82, pp 32-52.
26. Kelly, K. A.; Cardon, N. C. *EPA Watch*. The DeWeese Company, Chantilly, VA, **1994**, *vol* 3, pp 4-8.
27. United States General Accounting Office. Pesticides - Issues Concerning Pesticides Used in the Great Lakes Watershed. **1993,** GAO/RCED-93-128.
28. Heindel, J. J.; Chapin, R. E.; Gulati, D. K.; George, J. D.; Price, C. J.; Marr, M. C.; Myers, C. B.; Barnes, L. H.; Fail, P. A.; Grizzle, T. B.; Schwetz, B. A.; Yang, R. S. H. *Fundamental and Applied Toxicology*. **1994**, vol 22, 605-621.
29. *The Merck Manual of Diagnosis and Therapy;* Berkow, R. Ed.; 16th Edition; Publisher: Merck Sharp & Dohme Research Laboratories, Rahway, NJ; pp 2642.
30. Safe, S. H. *Environmental Health Perspectives*. **1995**, *vol* 103, pp 346-351.

Chapter 25

Estimated Ecological Effects of Atrazine Use on Surface Waters

Steven D. Mercurio

Department of Biological Sciences, Mankato State University, Mankato, MN 56002–8400

A model for ecological impacts of atrazine has been developed for surface waters in the Upper Midwest based on the seasonal intensive use of triazine herbicides utilizing the following parameters: 1) ecologically sensitive periods of stream ecosystems; 2) soil type and slope effects on atrazine runoff; 3) toxicity to various species in a flowing system; and 4) mitigation measures that are economically beneficial. Based on Southern Ontario data, stream concentrations of atrazine are inversely log-linearly related to soil water infiltration rates for soils with slopes <1-3% and are significantly increased by higher slopes. Using the Canadian data, mean Spring and Summer atrazine stream concentrations at the peak of primary algal production exceed the annual concentration for clay soils or highly sloping soils and for peak concentrations for all soils. Laboratory and mesocosm toxicity data suggest increased atrazine toxicities in flowing systems and decreased estimates of minimum effect levels.

The Federal Insecticide, Fungicide and Rodenticide Act (FIFRA) has required the U.S. Environmental Protection Agency (EPA) to conduct ecological risk assessments on pesticides since 1975 (*1*). The tiered risk assessment approach varies from single aquatic species analysis (Tier 1) to aquatic field testing (Tier 4). Currently, the Office of Pesticide Programs (OPP) does not routinely require either mesocosm or field studies (Tier 4) for pesticide registration. Instead, risk assessments are generally based upon a comparison of laboratory toxicity values to computer modeling estimates of pesticide concentrations in surface water. For a complete review of current policy directions, the work of the Aquatic Risk Assessment and Mitigation Dialogue Group (2) should serve as a reference point. Exposure levels are determined by factors such as the active ingredient of the pesticide including metabolites and degradation products, labeled use pattern including area of use, maximum and usual application rates, application methods, frequency of application, and fate and transport characteristics. Exposure levels are also determined by watershed factors, including the timing, intensity and duration of post-application rainfall, the percentage of the drainage area treated, the extent and type of vegetative cover, the distribution of slopes, and the distribution of soil types with different infiltration rates. Pesticides are generally modeled from maximum application rates as a worst case scenario using programs such as Pesticide Root Zone Model (PRZM) to estimate pesticide loadings to surface water and Exposure Analysis Modeling System (EXAMS) to estimate pesticide concentrations in surface water. Actual measured concentrations are used where available. Risk is then estimated from these simulated runoff profiles by the determination of a single value, the Risk Quotient based on standard LC_{50} assays.

The use of atrazine (2-chloro-4-ethylamino-6-isopropylamino-1,3,5-triazine) is the most widespread and intensive of any agricultural pesticide in North America, especially in the cornbelt of the Midwest. Residues in surface waters have been seasonally measured at phytotoxic concentrations (*3*). Current regulations represent a uniform attempt to assess and control environmental inputs by limiting application rates, providing for buffer zones and specifying protected areas. Due to differences in climate, soil type and slopes by region and the prevalence of use of a given pesticide, such as atrazine, simple runoff or receiving water models and impact predictions may be compromised, especially in a localized small stream or watershed. A reinvestigation of the ecotoxicity of atrazine is herein proposed using the following criteria: 1) determination of the most sensitive period of Northern Midwest stream ecosystems to herbicide application; 2) using field data to model stream atrazine concentrations ranges in the most sensitive period including correlations to quantitative information on soil compositions and slopes; 3) selection of the toxicity data which most closely approximate a flowing, multispecies aquatic system; and 4) suggesting remediation options that yield reductions in atrazine runoff into streams while providing economic benefits to farmers.

Determination of Period of Peak Sensitivity of Northern Stream Ecosystems

The difficulty with interpretation of most herbicide runoff data is that the modeling of peak and average concentrations should coincide with the relevant period of greatest impact determined by the individual ecoregion's production cycle. There are a number of factors impacting primary (algal) production (*4*), such as nutrient loading and energy inputs. The minimum number of tests needed to monitor water (lake) systems was determined to be 1) Secchi disk transparency (indirectly measures algal density), 2) chlorophyll, a direct, reliable determinant of algal density, and 3) total phosphorus as a measure of water fertility (*5*). Nutrients may be delivered by soil or fertilizer carried in runoff from rain events, sewage discharges or from the prevalence of a disruptive fish species, such as carp. However, the first event in a normal yearly cycle is the requirement for a certain water temperature or light supply in large lakes and ocean areas in temperate climates (*6,7*). The number of days having the longest photoperiods in a given season usually just precede or coincide with highest atmospheric or water temperatures, but water depth, rainfall, turbidity, etc., must be considered to pinpoint the true season of greatest production for a given river or stream. The connection between photoperiod (degree-days/hours or meter-candle-hours), temperature and primary production of streams has been described in the scientific literature (*8*). The river continuum concept adds a caveat that energy and nutrient input, species diversity, numbers of developing versus adult organisms, etc., are interrelated and differ from headwaters to the mouth of the river due to flow, water source, and forest cover (*9,10*). Given the normal cycle of most streams in the northern part of the U.S., the productive season can be taken as the agricultural growing season as a first approximation. For small, first-order streams with forest canopy cover, the highest primary productivity coincides with the Spring planting time, a period with high light intensity and low water temperatures (*11*). The most productive period for streams or lakes in the upper Midwest ranges from June to September in Minnesota (July and August peaks) or late April to late September further south (*12*). These time estimates represent an integration of a variety of aquatic ecosystem data, including ecoregion, production as monitored by chlorophyll and dissolved oxygen concentrations, and nutrient loadings (especially phosphorus) to transparency measurements as a measure of non-point source pollution impacts.

Effect of Soil Slope and Composition on Stream Atrazine Concentrations

A summary table in the synoptic review by R. Eisler (*3*) appeared to indicate: 1) that high concentrations of atrazine can occur in runoff water shortly after application (4,900 μg/L in Iowa), 2) that higher concentrations occur in runoff from an area with low permeable clay soils (maximum 25 μg/L) when compared to more permeable loam (maximum 14 μg/L) or highly permeable sand-dominated soils [maximum 4 μg/L from a study in Ontario, Canada; (*13*)], and 3) there can be substantial differences in atrazine concentrations based on time of sample collection, intensity of application in a given region and site of sampling in similar regions. The effect of soil texture and slope were found to be significant in these studies, but no quantitative relationship was established. More recent studies have focused on the total atrazine loss in a given area or watershed as runoff into surface waters. From all available data, it appears that large rivers that drain substantial areas of the Midwest have similar concentrations of atrazine, reflecting atrazine usage or application rates rather than a specific characteristic of that stream. A study reported in 1994 that as a percentage of use, atrazine transport in the Mississippi River averages 1.53% of total agricultural use drainage basin (*14*). Although much of the data in this study is from sites along the Mississippi watershed that have similar percentages of atrazine river transport levels compared to soil application amounts (e.g., Baton Rouge, Louisiana and Thebes, Illinois), other sites with different climatic conditions or flow rates may not be directly comparable (e.g., Platte River at Louisville, Nebraska) as suggested by the data of White *et al.* (*15*). A 1989 USGS analysis of pesticide concentrations in the

Midwest (*16*) indicated no significant differences in postplanting stream concentrations for atrazine based on basin size, while the highest peak concentrations of simazine and alachlor occurred in the smallest basins. In this study, it was clear that pulses of atrazine occurred during storm events, but lasted at measurable concentrations at the majority of sample sites through the harvest period (91%, 98% and 76% of the preplanting, postplanting and harvest samples, respectively). This long period of elevated atrazine concentration may be a concern for aquatic life, and groundwater and reservoir contamination.

Considerations for Developing a Quantitative Soil Runoff Model. Difficulties with prediction of runoff potential from an intensively applied agricultural chemical onto various soil textures in watersheds of different sizes has been a continuing problem in the area of nonpoint source pollution. Problems with predicting atrazine runoff by current models was indicated by an atrazine risk assessment panel assembled by the Ciba-Geigy Corporation (*17*). Their analysis showed that current "uncalibrated" runoff models were not considered predictive of atrazine fate and transport using soil absorption and degradation rates from laboratory data. PRZM-2 modeling of atrazine surface concentrations overestimated runoff nearly by an order of magnitude, while GLEAMS underpredicted by 1/5-1/2 an order of magnitude from a single test site of Tennessee silt loam using an application rate of 0.92 kg/ha. Similar behavior was observed in a Georgia sandy loam site.

Data supporting the effect of soil textures on atrazine runoff and stream contamination appear in the previously cited Canadian study of Frank and Sirons (*13*). An examination of 11 agricultural watersheds (average 4279 ha) in Southern Ontario indicated mean annual atrazine concentrations (time-weighted but not flow-weighted; flow monitored for total loss calculations) in stream water of 1.8, 0.5 and 0.3 μg/L for mostly clay, loam and sandy soil-containing basins having a mean application rate of 1.7 kg/ha. The Frank and Sirons study indicated that the flow-weighted mean loss of atrazine (mg/ha) came from storm runoff pulses that were highest for clay soils (66.5% of total atrazine lost), less for loam soils (55.8%) and a minority for sandy soils (29.4%). This figure is amplified if one considers that 27.2% of total atrazine yearly loss for clay soils (versus 4.8% and 20.6% for loam and sandy soils respectively) in the Southern Ontario region comes from storm runoff in the application period from May-August when the stream flow volume was 252 m^3/ha/yr (versus 659 and 1002 m^3/ha/yr for loam and sandy soils). The rest of the storm runoff, that accounts for the overwhelming majority of the loss for loam soils, was during the January-April period of snowfall and melt. The actual weighted mean loss for clay soils was over an order of magnitude higher (>12.6-fold) than that of sandy or loam soils during the application period in this region. If the concentrations of atrazine in streams are calculated from the total atrazine and desethyl atrazine loss data and stream flow volumes, they would have been 8.42, 0.74 and 0.39 mg/m^3/yr or (μg/L/yr) for clay, loam and sandy soils, respectively, during the four month period during and immediately following the time of application (May-August). Note then that the clay soil application would result in surface water contamination at a rate 21.5 times that of sandy soils. The totals lost of atrazine (input) into streams as a percentage of the application rate on a yearly basis for the periods of 1975-1976 and 1976-1977, respectively, were 1.93% and 1.55% for clay soils, 0.61% and 0.57% for loam soils, and 0.33% and 0.26% for sandy soils. Unfortunately, the lack of runoff in permeable sandy soils has been shown in a USGS study in Topeka, Kansas causes residues to leach into the soil and ground water instead (*18*). The leaching rate appears to be linear with application rate (doubling water concentrations with twice the application rate). A recent Canadian study further confirms that immediate rainfall following atrazine application is the largest concern (*19*), with a 6% immediate loss of atrazine for simulated rainfall immediately after application and 3% for rainfall one week following application.

Soil textures are also apparently taken into account in the label directions for effective application of atrazine (AAtrex 4L [Registered Trademark of Ciba-Geigy Corp.]). Broadcast rates of 2 pts./A (in a 1:1 mixture with Princep 80W [Registered Trademark of Ciba-Geigy] for corn) for sandy soils, 2.4 pts/A for low organic matter loam soils and 3 pts./A for high organic matter or clay soils are recommended in the instructions. Thus, the soils with the highest runoff potential to surface waters would receive the highest application rate.

Atrazine Runoff Model Description. Combining the above list of parameters, the following approach is suggested as an example of a first step in assessing the potential impact of atrazine runoff on peak production of aquatic ecosystems. The source of the data was taken from the Frank and Sirons (*13*) study of atrazine runoff in Southern Ontario. Estimates of site locations (latitude and longitude) were made by Steven Clegg of the Ontario Ministry of Food and Agriculture from a 1:250,000 scale map and 1979 site descriptions. Soil parameters were matched to those sites on a 1:1,000,000 map (obtained courtesy of K. Bruce MacDonald of Agriculture and Agri-Food Canada) and soil data (*20*) determined by William Effland of the Environmental Fates and Effects Branch of the Office of Pesticide Programs in the U.S. EPA. A summary of the soil characteristics is presented in Table I. Note that map data are relied upon for this analysis. Discrepancies between map and study presentations of soil characteristics are indicated in the column labeled surface texture (largest difference estimated for site AG-13). Ted Oke of the Food and Rural Affairs Office of the Ontario Ministry of Agriculture confirmed that the AG-13 site had a mixture of Brookston clay, Brookston clay-sand spot phase, Plainfield sand and Berrien sand to sandy loam. For the sake of this analysis, the soil characteristics for AG-13 were considered a mixture closer in properties to the loam soils.

Estimated soil properties from Table I are found in Table II. Since individual site-specific infiltration rates were determined from the map's general description of soil type (rather than site-specific soil analyses), only a range of possible values were indicated. To determine whether these data have any quantitative value, a selection of an infiltration rate was made per range. If the infiltration rate was calculated to be <0.5 cm/hr, 0.5 was conservatively taken as the parameter. Similarly, infiltration rates >7.6 cm/hr were noted as 7.6 for this analysis. An average of 4.0 cm/hr was taken for the remaining soils. The soil infiltration rate was then correlated with mean atrazine concentrations by linear regression to the

Table I: Soil Landscapes and Associated Soil Properties for 11 Southern Ontario Agricultural Watersheds (based on Frank and Sirons, 1979 (*13*))

Site	*Map Symbol*[a]	*Soil Development*[b]	*Surface Texture Class*[c] *(Parent Material)*	*Total Map Area, kha (study area)*	*Slope Class*	*Surface Form*	*Drainage Class*	*Depth To Water Table*
AG-1	Ucy Ml1 9	Gleysolic	clay (clay)	210.2 (5.1)	1-3%	level	poor	0-1
AG-2	Ucl Ml1 44	Gleysolic	fine sandy loam (silty clay loam)	9.0 (7.9)	1-3%	level	poor	0-1
AG-3	Ecl Mr4 27	Gray Brown Luvisolic	silt loam (silty clay loam)	126.1 (6.2)	4-9%	ridged	well-mod. well	2-3
AG-4	Elm Mu4 69	Gray Brown Luvisolic	loam (loam)	81.2 (1.9)	4-9%	undulating	imperfect	2-3
AG-5	Elm Mm4 69	Gray Brown Luvisolic	loam (loam)	37.3 (3.0)	4-9%	rolling	well-mod. well	2-3
AG-6	Esd Fu1 63	Gray Brown Luvisolic	silt loam (gravelly sand)	25.9 (5.5)	1-3%	undulating	rapidly	>3
AG-7	Esd Lu1 183	Gray Brown Luvisolic	sandy loam (sand)	9.8 (5.6)	1-3%	undulating	imperfect	1-2
AG-10	Ucy Ll1 47	Gleysolic	silty clay (heavy clay)	236.5 (3.0)	1-3%	level	poor	0-1
AG-11	Ecy Mu4 164	Gray Brown Luvisolic	clay (clay)	54.3 (2.4)	4-9%	undulating	imperfect	1-2
AG-13	Ucy Ml1 9	Gleysolic	clay (clay)	210.2 (2.0)	1-3%	level	poor	0-1
AG-14	Esd Lu1 100	Gray Brown Luvisolic	sandy loam (sand)	15.0 (4.5)	1-3%	undulating	well-mod. well	2-3

[a]Dominant soil landscape (*20*); map scale = 1:1,000,000; map reliability: high to medium; subdominant soil landscapes and inclusions not considered for this table.
[b]Tentative correlation to U.S. soil classification system: Gleysolic - Aquic (i.e., wet) suborders; Gray Brown Luvisolic - Hapludalfs.
[c]Comparison to sites listed by Frank and Sirons (*13*) - Clay soils: AG-1, 3, 10, 11; Loam soils: AG-4, 5, 6, 14; Sandy soils: AG-2, 7, 13. Site Ag-13 was described as sandy soil in the original reference.
Variations in map scale and level of detail could result in different interpretations.

mean atrazine concentrations reported by site (11 total) using SYSTAT Version 5.0 (SYSTAT, Inc., Evanston, IL). The data showed no significant correlation due to an important confounding variable, namely slope. If moderate slopes are present, as in sample AG-3, higher runoff concentrations would be expected. When moderate slopes were eliminated (4-9% slope range), a significant negative correlation between the mean of the annual time-weighted water concentrations (µg/L) and the logarithm of the infiltration rate expressed in cm/hr was observed (R^2=0.660, *P*=0.026, [mean atrazine concentration] = -1.22*log [infiltration rate] + 1.346]. These data indicate that as water infiltration decreased, stream atrazine concentrations increased. It also confirms the importance of moderate slopes on runoff, regardless of soil type (i.e. highest atrazine stream concentrations in the study were observed in an area of loam and clay soil textures and a moderate slope).

Table II: Estimated Soil Properties from 11 Southern Ontario Sites[a]

Site	*Bulk density, mg/m^3*	K_{sat}[b]	*Index Surface Runoff Class*	*Infiltration Rate, Cm/hr*
AG-1	1.3	low	high	<0.5
AG-2	1.4	mod. high	low	0.5-7.6
AG-3	1.2	mod. low	high	0.5-7.6
AG-4	1.3	mod. high	medium	0.5-7.6
AG-5	1.3	mod. high	medium	0.5-7.6
AG-6	1.2	mod. high	low	0.5-7.6
AG-7	1.4	high	very low	>7.6
AG-10	1.3	low	high	<0.5
AG-11	1.3	low	very high	<0.5
AG-13	1.3	low	high	<0.5 (actually 0.5-7.6)
AG-14	1.4	high	very low	>7.6

[a]Estimated through data of Table I and a USDA soil manual (*21*).
[b]K_{sat} Classes - High: 86.4-864.0 m/day; Mod. High: 8.64-86.4 m/day; Mod. Low: 0.864-8.64 m/day; Low: 0.0864-0.864 m/day.

To show how changes in infiltration rate increased atrazine concentrations, the data are expressed in mean annual atrazine concentration versus the logarithm of 1/infiltration rate in Figure 1 (open circles and lower line in plot; [mean atrazine

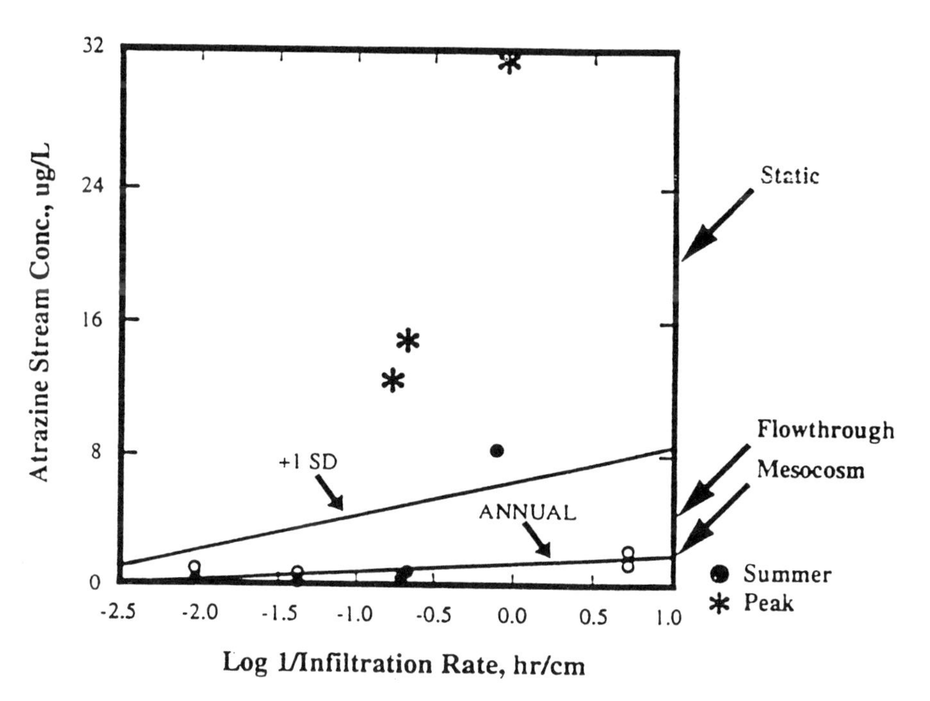

Figure 1. Atrazine stream concentration estimates are indicated by regressions of mean annual stream concentrations (**o**) and +1 SD above the annual values. Mean summer and peak concentrations are indicated for average values of sand, loam and clay soils. Minimum toxicity values are indicated on the right margin for Static algal impacts, Flowthrough toxicity to algae and Mesocosm data.

conc.] = 0.53*log [1/infiltration rate] + 1.346). The standard deviation line in the figure (+1 SD) was calculated by a linear regression of the mean atrazine concentration for each site added to the standard deviation of that site as the dependent variable and mean atrazine concentration as the independent variable [R^2 = 0.870, P<0.001, (mean atrazine conc. + 1 SD) = 4.085* (mean atrazine concentration) + 0.845]. Alternatively, the SD versus mean atrazine concentration had a significant correlation (R^2 = 0.788, P<0.001). This indicates that the variation of data was reasonably consistent when expressed as a percent of the annual mean for any given atrazine concentration, or that the true data range for +1 SD based on the regression is approximately four times the mean yearly atrazine concentration for this given data set.

The average summer atrazine stream concentrations (filled circles) and the highest reported (storm peak as asterisks) values are also plotted on this figure based on the average infiltration rate for that given soil texture across watersheds from the Frank and Siron's data set (average infiltration rates for all sites studied = 1.13, 4.9 and 5.2 cm/hr for clay, loam and sand, respectively). The summer values for sand and loam fell well within the mean annual concentration +1 SD area. However, the stream atrazine concentrations for clay soils exceeded the +1 SD area. Therefore, this analysis suggests that even when one standard deviation variation is included into the annual time-weighted mean atrazine stream concentrations, this range of values fails to account for the true impact of summer concentrations for clay soils. This is an important factor in surface water ecological risk determination, especially considering that the peak productivity of the streams in the study area occurs in this time period and many of those streams have runoff from clay soils.

Acute risk should be assessed by the storm pulses or peak values expressed in the figure. Chronic risk to primary production should be assessed in the figure as a range above the average summer atrazine stream concentrations, but below the peak values. The area between the summer averages and the yearly averages really represents the background concentrations for determining risk, since these levels are achieved outside of the time of peak productivity of the streams in this region. A linear adjustment for application rate may be made to predict the effects this variable has on the above parameters in this region (*18*).

In summary, it appears that 1) the true acute effects of atrazine likely occur during rain events (the closer the rain(s) to application time, the worse the event); 2) the chronic data really represent the late Spring to late Summer/early Fall period of high extended levels of atrazine runoff and peak productivity of these ecosystems (especially in central to northern U.S. regions); 3) yearly averages represent background levels that may have little true relevance to toxicity to stream ecosystems; and 4) clay soils or high slopes pose significant problems in controlling atrazine runoff and, therefore, toxicity to aquatic species in streams.

Toxicity to Aquatic Organisms in Flowing Water

The arrows on Figure 1 labeled static, flowthrough and mesocosm express the expected toxicity to the aquatic ecosystems by atrazine. The arrow labeled "static" at 20 μg/L atrazine estimates that only concentrations above this level affect individual species as suggested from analyses of mainly laboratory studies in confined vessels, and is a level recommended by publications such as Solomon *et al.* (*17*) and Huber (*22*). The "flowthrough" arrow at 4 μg/L atrazine suggests that in flowing systems, individual phytoplankton experience toxic effects starting at this concentration [based on analyses in a flowthrough apparatus by Schafer *et al.* (*23*)]. Note that even the mean summer atrazine stream concentrations for clay soils exceeded at least two-fold that of the flow-through and mesocosm calculations of effects on primary and therefore secondary, etc., production of these streams (*24*). Peak (pulse) levels for clay or highly sloping soils exceed the static calculations. This model suggests that acute and chronic damage may be occurring to stream ecosystems during their peak productive periods based on toxicity determinations of laboratory and mesocosm flow-through systems. The effect of rain pulses on the streams' turbidity and impacts of atrazine on aquatic organisms for unique watersheds should be confirmed with field analyses for validation purposes.

Primary Toxicity to Aquatic Organisms. Effects of atrazine on aquatic phytoplankton communities are not readily detected at levels <0.5 μg/L; reduced photosynthesis occurs in sensitive species at 1.0-5.0 μg/L (*25*); and at concentrations >100 μg/L causes permanent changes in the algal community structure in a two-week period (*26*). Algae have the widest range of response, having EC_{50} values of 8-1500 μg/L as compared to 30-163 μg/L for macrophytes (*27,28*). Most LOEC values for invertebrates appear at concentrations >100 μg/L (*29*). However, egg hatchability of the caddisfly *Triaenodes tardus* Milne had an EC_{50} of 22 μg/L (*30*), while chronic studies of leeches and snails led to an estimation of adverse effects to invertebrates at approximately 2% of the LC_{50}s (*31*). Atrazine concentrations as low as 20 μg/L caused initial inhibitions of photosynthesis in phytoplankton, aquatic insect perturbations and other direct and/or secondary effects on the ecosystem (*32-34*). Atrazine concentrations in experimental pond mesocosm ecosystems ranging from 20-500 μg/L caused direct adverse effects on aquatic vegetation and gizzard shad and indirect influences on almost all species excepting macroinvertebrate total abundance (*24*). The lowest concentration causing direct effects of atrazine on higher aquatic species was reported for the carp, *Cyprinus carpio*. Following 6-12 hours of exposure to 10 μg/L atrazine, significant alterations in serum glucose and enzyme activity were noted (*35*). Additional information concerning the impact of atrazine on aquatic life and esturaine productivity is described by Stevenson (*36*) and Ward (*37*). Most of the toxic effects mentioned above for atrazine at concentrations <20 μg/L represent the effect of subchronic exposures (more than 24 hours but less than two weeks of continual exposure at a stable concentration). Although concerns about the validity of the lower estimates of atrazine concentrations that yield direct toxicity exist (*17*). Effects on algae composition have been observed, even a year following water dosed at concentrations as low as 20 μg/L (*38*).

Secondary Effects on Dissolved Oxygen and Nutrients. Although secondary effects on other organisms have been referred to above, some of the most recent data indicates effects of atrazine that appear to be related to the ability of algae to oxygenate streams and utilize nutrients. A study by Detenbeck, Hermanutz and Swift (*39*) indicates that the lowest level of atrazine contamination that they applied (15 µg/L) dropped maximum dissolved oxygen concentrations 23% in a flow-through wetland mesocosm in Monticello, MN, while enhancing phosphate and ammonia concentrations. Stay and coworkers (*40*) reported that the effect of atrazine on dissolved oxygen levels, primary productivity, etc., of mesocosms is not readily apparent until the zooplankton bloom stresses the system. Static (nonflowing) or one species systems may not be the best estimate of the true impact of atrazine in the field.

Possible Remediation Measures

A promising method currently used by some successful farmers for decreasing atrazine runoff is just changing the bandwidth of application (i.e., not treating the entire field). Changing to a 50-cm band reduced atrazine runoff by 69% (*41*). Clearly this would enhance the effects of decreased application rate, no-till conservation plowing or buffer strips to reduce runoff, while maintaining a herbicidal concentration on the critical areas.

Another possibility for reducing atrazine runoff or leaching into groundwater at the present reduced application rates without loss of efficiency is altered formulation or incorporation. A sucrose-encapsulated form of atrazine (20-40 mesh size) applied 1 hour before a simulated rainfall (2 hours at rate of 40 mm/h followed by 1 hour rest and then 2 hours at 25 mm/h) resulted in 1) a constant release versus a large surge in atrazine concentration for the liquid formulation and 2) a 43% and 38% reduction in effluent loss for no-till and chisel plow, respectively (*42*). Similar successes for various mesh sizes, clay content and extruder processing speeds were reported for starch-encapsulated atrazine, with an average of 68% reduction in leaching from the first 5 cm of soil compared with dry flowable formulations (*43*).

Acknowledgements

Appreciation for help on this project go to those of the U.S. Environmental Protection Agency, Pesticide Programs/Environmental Fates and Effects Division/Ecological Effects Branch - Anthony F. Maciorowski (overview), William Effland (soil analysis), Candace Brassard (incident data), David Farrar (statistics and graphics), Harry Winnik (graphics), Henry Nelson (atrazine surface runoff), Rick Petrie (provided atrazine literature), Dave Jones (modeling) and Bob Pilsucki (downloaded Canadian GIS system). Mary Moffett, Frank S. Stay and Naomi Dettenbeck of the U.S. EPA Research Laboratory-Duluth provided current toxicity data. Thanks also go to Don Goolsby and Steve Larson at the U.S. Geological Survey (recent studies of atrazine in streams), Art Buikema of Virginia Polytech (seasonal stream productivity), Steve Hieskary of the Minnesota

Pollution Control Agency (primary production in Minnesota's lakes), and the Canadian scientists Matthew Craig (GIS), Bryan Monette (GIS), Steve Clegg (map points), K. Bruce MacDonald (map) and Ted Oke (site soil confirmation).

Literature Cited

1. Urban, D. In *Aquatic Mesocosm Studies in Ecological Risk Assessment*; Graney, R. L.; Kennedy, J. H.; Rodgers, J. H. Jr., Eds.; Lewis Publishers: Boca Raton, FL, 1994; pp 7-16.
2. Aquatic Risk Assessment and Mitigation Dialogue Group. *Aquatic Dialogue Group: Pesticide Risk Assessment and Mitigation*; SETAC Press: Pensacola, FL, 1994; pp 1-220.
3. Eiser, R. *U.S. Fish Wild. Serv. Biol. Rept.* **1989,** *85*, 1-53.
4. Boynton, W. R.; Kemp, W. M.; Keefe, C. W. In *Estuarine Comparisons;.* Kennedy, V. S., Ed.; Academic Press: New York, NY, 1982; pp 69-90.
5. *Volunteer Lake Monitoring: a Methods Manual*; EPA/440/4-91/002; U.S. Environmental Protection Agency: Washington, DC., 1991.
6. Gieskes, W. W. C.; Kraay, G. W. *Neth. J. Sea Res.* **1975,** *9*, 166-196.
7. Cadee, G. C.; Hegeman, J. *Neth. J. Sea Res.* **1986,** *20*, 29-36.
8. Cummins, K. W. *BioSci.* **1974,** *24*, 631-641.
9. Vannote, R. L.; Minshall, G. W.; Cummins, K. W.; Sedell, J. R.; Cushing, C. E. *Can. J. Fish. Aquat. Sci.* **1980,** *37*, 130-137.
10. Minshall, G. W.; Petersen, R. C.; Cummins, K. W.; Bott, T. L.; Sedell, J. R.; Cushing, C. E.; Vannote, R. L. *Ecol. Microgr.* **1983,** *53*, 1-25.
11. Marzolf, E. R.; Mulholland, P. J.; Steinman, A. *Can. J. Fish. Aquat. Sci.* **1994,** *51*, 1591-1599.
12. Heiskary, S. A. *Lake Reservoir Management* **1989,** *5*, 85-94.
13. Frank, R.; Sirons, G. J. *Sci. Total Environ.* **1979,** *12*, 233-239.
14. Battaglin, W. A.; Goolsby, D. A.; Mueller, D. K. In *Effects of Human-Induced Changes on Hydrologic Systems*; Marston, R. A.; Hasfurther, V. R., Eds., Proceed. Ann. Summer Symp. Amer. Water Res. Assoc., Jackson Hole, WY, 1994; pp. 1073-1085.
15. White, A. W.; Barnett, A. P.; Wright, B. G.; Holladay, J. H. *Environ. Sci. Technol.* **1967,** *1*, 740-744.
16. Thurman, E. M.; Goolsby, D. A.; Meyer, M. T.; Mills, M. S.; Pomes, M. L.; Kolpin, D. W. *Environ. Sci. Technol.* **1992,** *26*, 2440-2447.
17. Solomon, K. R.; Baker, D. B.; Dixon, K. R.; Giddings, J. M.; Geisy, J. P.; Hall, L. W. Jr.; Klaine, S. J.; La Point, T. W.; Richards, R. P.; Weisskopf, C. P.; Williams, W. M.; Kendall, R. J. *Ecological Risk Assessment of Atrazine in North American Surface Waters*; Ciba Crop Protection: Greensboro, NC, 1995, pp 1-234.
18. Perry, C. A. *U.S.G.S. Water-Resources Invest. Report* **1991,** *91-4017*, 1-61.
19. Bowman, B. T.; Wall, G. J.; King, D. J. *Can. J. Soil Sci.* **1994,,** *74*, 59-66.

20. Agriculture Canada. *Soil Landscapes of Canada: Ontario - South*; Canadian Soil Inventory, Land Resource Research Centre, Research Branch, Publ: 1988 (cartography), 1989 (numerical data), pp. 52:101B.
21. Soil Survey Division Staff. *Soil Survey Manual*; U.S. Department of Agriculture Handbook No. 18, 1993, pp. 89-113.
22. Huber, W. *Environ. Toxicol. Chem.* **1993,** *12*, 1865-1811.
23. Schafer, H.; Hettler, H.; Fritsche, U.; Pitzen, G.; Roderer, G.; Wenzel, A. *Ecotoxicol. Environ. Safety* **1994,** *27*, 64-81.
24. Huggins, D. G.; Johnson, M. L.; deNoyelles, F. Jr. In *Aquatic Mesocosm Studies in Ecological Risk Assessment*; Graney, R. L.; Kennedy, J. H.; Rodgers, J. H. Jr., Eds.; Lewis Publishers: Boca Raton, FL, 1994; pp. 653-697.
25. deNoyelles, F., Jr.; Kettle, W. D. and Sinn, D. E. Ecol. 63: 1285-1293.
26. Hamala, J. A.; Kollig, H. P. *Chemosphere* **1985,** *14*, 1391-1408.
27. Huggins, D. G. *Ecotoxic Effects of Atrazine on Aquatic Macro invertebrates and Its Impact on Ecosystem Structure*; Ph.D. dissertation, Univ. Kansas: Lawrence, KS, 1990.
28. Larsen, D. P.; deNoyelles, F. Jr.; Stay, F.; Shirogama, T. *Environ. Toxicol. Chem.* **1986,** *5*, 179-190.
29. Macek, K. J.; Buxton, S.; Sauter, S.; Gnilka, S.; Dean, J. 1976. *Chronic Toxicity of Atrazine to Selected Aquatic Invertebrates and Fishes*; EPA/600/3-76-047; U.S. Environ. Protect. Agency, Environ. Res. Lab.: Duluth, MN, 1976.
30. Belluck, D.A. *Pesticides in the Aquatic Environment*; M.S. Thesis, Univ. Illinois: Champaign-Urbana, IL, 1980.
31. Streit, B.; Peter, H. M. *Arch. Hydrobiol. Suppl.* **1978,** *65*, 235-267.
32. deNoyelles, F. Jr.; Kettle, W. D.; Fromm, C. H.; Moffett, M. F.; Dewey, S. L. *Entomol. Soc. Amer. Misc. Publ.* **1989,** *75*, 41-56,
33. Dewey, S.L. *The Effects of the Herbicide, Atrazine, on Aquatic Insect Community Structure and Emergence.* M.S. Thesis, Univ. Kansas: Lawrence, KS, 1983.
34. Kettle, W. D.; deNoyelles, F. Jr.; Heacock, B. D.; Kadoun, A. M. *Bull. Environ. Contam. Toxicol.* **1987,** *38*, 47-52.
35. Hanke, W.; Gluth, G.; Bubel, H.; Muller, R. *Ecotoxicol. Environ. Safety,* **1983,** *7*, 229-241.
36. Stevenson, J. C.; Jones, T. W.; Kemp, W. M.; Boyton, W. R.; Means, J. C. In *Proceedings of the Workshop on Agrichemicals and Estuarine Productivity*; Beauford, North Carolina, September 18-19, 1980, Natl. Ocean. Atmos. Admin., Off. Mar. Pollut. Assess.: Boulder, Colorado, 1982..
37. Ward, G.S.; Ballantine, L. *Estuaries* **1985,** *8*, 22-27.
38. Neugebaur, K.; Zieris, F. J.; Huber, W. *Z. Wasser-Abwasser-Forsch.* **1990,** *23*, 11-17.
39. Detenbeck, N.E.; Hermanutz, R; Allen, K.; Swift, M.C. *Environ. Toxicol. Chem.* **1996,** *15*, 937-946.

40. Stay, F. S.; Larsen, D. P.; Katko, A.; Rohm, C. M. In *Special Technical Testing Publication 865*; Amer. Soc. Test. Mater.: Philadelphia, PA, 1985; pp 75-90.
41. Gaynor, J. D.; Van Wesenbeeck, I. J. *Weed Technol.* **1995,** *9*, 107-112.
42. Hickman, M.; Schreiber, M. M. *Controlled Release Herbicide Formulations, Their Efficacy and Role in Reducing Ground Water Contamination*: Fedrip. Database, Nat. Tech, Inform. Serv., 1992.
43. Fleming, G. F.; Wax, L. M.; Simmons, F. W. *Weed Technol.* **1992,** *6*, 297-302.

Chapter 26

Triazines in Waters of the Midwest: Exposure Patterns

R. Peter Richards and David B. Baker

Water Quality Laboratory, Heidelberg College, 310 E. Market Street, Tiffin, OH 44883

Concentrations of triazine herbicides in waters receiving runoff from agricultural lands are seasonal in nature, with highest concentrations in the six weeks to two months following application, and lower to non-detectable concentrations during the rest of the year. In rivers and streams, concentrations are highest during runoff from storm events in this post-application period, and lower during base flow periods. This pulsed exposure pattern has important implications for possible impacts on aquatic ecosystems, since algae may have time to recover from the highest concentrations during intervening low flow periods. Typically, atrazine dominates the triazines: atrazine concentrations usually exceed the sum of cyanazine, simazine, and the atrazine breakdown products DIA and DEA, particularly during times of the year characterized by elevated concentrations.

With improvements in analytical technology, agricultural pesticides are being found more and more frequently in surface water throughout the midwestern United States. Concentrations are highly variable in time and space. While many different herbicides and insecticides are occasionally detected in natural waters, only four or five of the pesticides quantified by standard GC/MS techniques are found above detection limits consistently enough to provide useful information on patterns of occurrence in time and space. All five are herbicides; they are atrazine, alachlor, metolachlor, cyanazine, and metribuzin. Two of these are S-triazines: atrazine and cyanazine. Atrazine is by far the most widely studied of these herbicides. Consequently it is the compound most frequently used as an example of pesticide behavior in the environment. A third S-triazine, simazine, is sometimes detected, usually in low concentrations. In addition, certain breakdown products are often detected, particularly the atrazine breakdown products des-ethyl atrazine (DEA) and de-isopropyl atrazine (DIA). In specific situations these breakdown products can have higher concentrations than the parent compounds. Triazine herbicide use is generally most intense in the midwestern states of Ohio, Indiana, Illinois, and Iowa. This paper illustrates patterns of triazine concentrations in midwestern surface waters, drawing on detailed datasets (*1-3*), and creates a context in which the possible impacts of triazines on aquatic ecosystems may be better understood and evaluated.

Spatial Patterns of Triazine Use

County-level herbicide use estimates (*4*) for the period 1978-1989 have been widely used for depicting spatial patterns of herbicide use; relatively complete sets of maps for herbicides and crops are available in several recently published documents (*5,6*). Similar datasets of county-level herbicide use have been prepared by the Agricultural Research and Statistical Service for the years 1991, 1993, and 1994.

Atrazine is used almost exclusively on corn, with minor applications to sorghum. Its area of most intense use is the corn belt of the upper Mississippi River watershed, particularly the states of Iowa, Illinois, Indiana, and Ohio. The use distribution, properties, and environmental occurrence of atrazine are thoroughly discussed by Solomon *et al.* (*7*). Cyanazine is used on corn and cotton. The use patterns for cyanazine in the late 1980s were very similar to those of atrazine, but the quantities used were much smaller (*4-6*) and have declined since, because cyanazine is being phased out of use. Simazine is used primarily on corn and alfalfa and in orchards. It is used in much smaller quantities than atrazine in the midwest, but is also used extensively along the southern east coast and along the west coast(*4-6*)

Temporal Patterns of Herbicide Concentration in Surface Water

Patterns in Streams and Rivers. Various aspects of temporal patterns of triazine concentrations in rivers and streams have been described in a number of papers in the literature. Early works (e.g. *8-11*) focused on atrazine runoff from fields into rivers and streams. Recent works (e.g. *1, 2, 7, 12, 14*) have broadened the focus to include other triazine and non-triazine herbicides, and possible ecological and human health effects from exposure to these compounds.

Triazine herbicides move to streams and rivers primarily in runoff from rainfall events. As a consequence, concentrations typically increase and decrease more or less in parallel with stream discharge (*1*). In some systems, triazines may also move to streams and rivers through shallow groundwater pathways (*15*); this pathway is more important during low flow conditions than during storm runoff. The relative importance of direct runoff and groundwater baseflow is a function of surficial and bedrock geology and a number of other factors. Triazines are transported in rivers and streams primarily in the dissolved state.

On an annual basis, triazine herbicides tend to occur in highest concentrations during the two or three months following application, declining to low to non-detectable concentrations by mid to late fall, and remaining at low concentrations until the following application season (*1, 16*). In most of the midwest, the period during which elevated concentrations occur typically begins in May and extends into July.

Cyanazine concentrations are usually below detection limit by late summer or early fall, whereas it is not uncommon to find small but detectable quantites of atrazine in storm runoff throughout the year, in basins where agricultural land use is dominant. Simazine concentrations in midwestern rivers are generally too small to reveal a very clear annual pattern, except that detections are almost entirely limited to the two months following application.

The typical patterns described above and in the following sections are the result of non-point processes. Occasionally, high concentrations which are not associated with storm runoff have been observed in streams and rivers monitored by the Water Quality Lab. Sometimes these occur simultaneously for two compounds known to be applied together. These concentrations rarely persist long enough to appear in two successive samples (which are usually three days or less apart). These are most likely the result of point-source inputs, either accidental spills or emptying excess herbicide from a spray tank. While apparently infrequent, of short duration, and of unpredictable timing, these point-source events often have higher concentrations than most non-point storm-runoff events.

Storm Runoff and Pulsed Exposures. Because of the importance of storm runoff in the transport of triazine herbicides, rivers and streams are characterized by pulses of elevated concentration which result from the passage downstream of the runoff from different storms (Figure 1). Smaller streams tend to have wider extremes of concentration - higher maximum concentrations during storm runoff and longer periods of low concentration between storms - than larger rivers, which carry water resulting from contributions from different tributaries whose maximum concentrations enter at different times. All rivers and streams, however, show the pulsed exposure pattern.

The pulsed nature of triazine runoff has important implications for possible ecosystem effects of herbicide exposures, because these herbicides act by inhibiting photosynthesis rather than by inflicting direct damage on cells, and plants such as algae may recover to a large extent between pulses of high concentration. Pulsed concentrations also hinder assessments of long-term human exposures through drinking water, because it is difficult and expensive to carry out a sampling program sufficiently detailed to obtain a reliable estimate of the average over time of the rapidly fluctuating concentrations. The quarterly sampling program required by the Safe Drinking Water Act (*13*) is clearly inadequate to deal with such extreme fluctuations.

Major differences in average and extreme concentrations (Table I) occur among different compounds and in watersheds of different sizes and geographic locations, due to differences in crop distributions and pesticide use rates, differences in hydrologic residence times, and compound-specific differences in rates of breakdown.

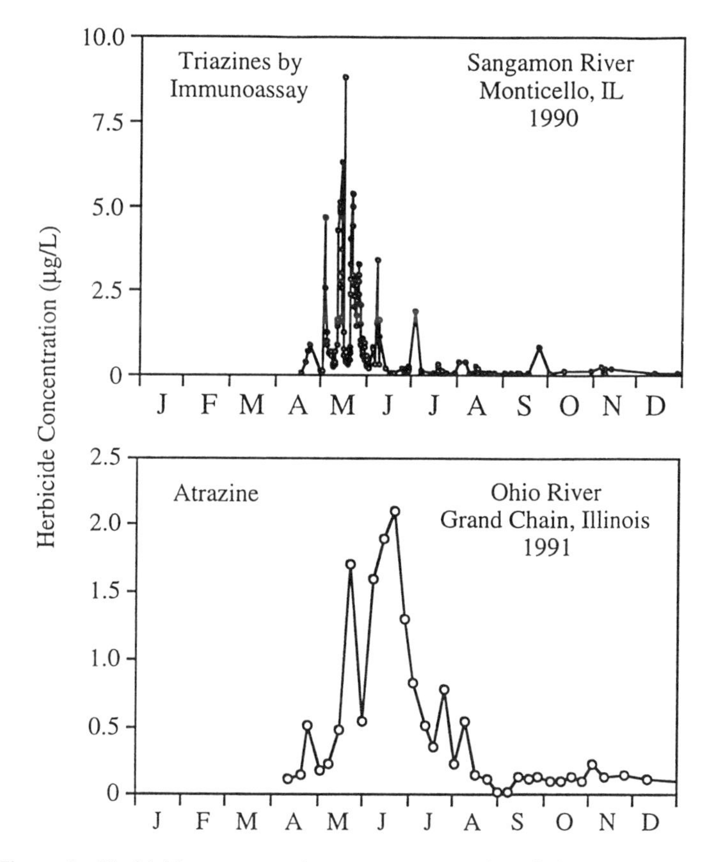

Figure 1. Herbicide concentrations over an annual cycle in representative midwestern rivers and streams.

Table I. Annualized time weighted mean, median, and 90th percentile concentrations (µg/L) for triazines in representative midwestern rivers.

River	Atrazine			Cyanazine			Simazine		
	Mean	Med	90th	Mean	Med	90th	Mean	Med	90th
Data of Baker and Richards (*1*)									
Maumee River, OH	1.14	0.22	3.46	0.30	0.05	0.77	0.11	0.03	0.24
Sandusky River, OH	1.17	0.19	3.26	0.20	0.05	0.35	0.08	0.03	0.13
Lost Creek, OH	0.91	0.10	1.71	0.26	0.05	0.21	0.03	0.03	0.04
Honey Creek, OH	1.35	0.10	3.48	0.22	0.05	0.42	0.07	0.03	0.10
Rock Creek, OH	0.97	0.10	2.28	0.13	0.05	0.14	0.16	0.03	0.05
Data of Goolsby and colleagues (*2*)									
Huron River, OH 1990	1.39	0.10	4.16	0.63	0.05	1.58	0.11	0.03	0.22
Roberts Creek, IA 1990	2.17	0.10	0.75	1.24	0.05	0.10	0.04	0.03	0.03
Old Man Creek, IA 1990	1.13	0.10	1.63	0.72	0.05	0.09	0.03	0.03	0.03
Iroquois River, IL 1990	0.91	0.10	2.89	0.43	0.05	1.23	0.04	0.03	0.03
Sangamon River, IL 1990	0.83	0.10	2.66	0.27	0.05	0.53	0.05	0.03	0.03
Sangamon River, IL 1991	0.15	0.10	0.28	0.11	0.05	0.21	0.01	0.03	---
Silver Creek, IL	1.89	0.10	4.83	0.78	0.05	2.88	0.23	0.03	0.54
W. Fk. Big Blue, NE 1990	1.65	0.10	6.88	0.29	0.05	0.47	0.06	0.03	0.08
W. Fk. Big Blue, NE 1991	0.47	0.10	1.41	0.24	0.05	0.71	0.01	0.03	---
Delaware River, KS	1.36	0.10	5.76	0.08	0.05	0.10	0.03	0.03	0.03
Cedar River, IA	0.55	0.10	1.45	0.27	0.05	0.43	0.03	0.03	0.03

In calculating the statistics in Table I, an observed concentration was assumed to represent a period of time equal to half the time between it and the preceeding sample, or 3.5 days, whichever was less, plus half the time between it and the following sample, or 3.5 days, whichever was less. Time not represented by any sample was assigned a concentration of 0.1 µg/L in the case of atrazine, 0.05 µg/L for cyanazine, and 0.025 µg/L for simazine. The great majority of this unrepresented time occurred in the fall, winter, and early spring months when concentrations are known to be low. Observations below detection limit were assigned a concentration of half the detection limit (generally 0.025 µg/L).

Annual and Monthly TWMCs. Substantial differences in average and peak concentrations also occur from year to year within a given watershed, as a function of intensity and timing of rainfall. Figure 2 shows a 10-year pattern of monthly time-weighted mean concentrations (TWMCs) of atrazine in the Maumee River in northwest Ohio. The maximum monthly average concentration occurs about the same time every year, but the year-to-year differences are considerable. For example, 1988 was a drought year with very little runoff and correspondingly low average concentrations. By contrast, 1991 was a year with several large storms shortly after herbicide application, and this history is reflected in the magnitude of its peak concentration.

The annual TWMCs, shown as bars, are much less variable than the monthly average concentrations. Of interest from a regulatory standpoint is the observation that, while monthly average concentrations often exceed the atrazine Maximum Contaminant Level (MCL) for drinking water of 3 µg/L, none of the annual average concentrations do so, and it is the annual average concentrations which are used to evaluate compliance with drinking water standards. This pattern is typical of other midwestern rivers used as water supplies: they often exceed the MCL for short periods of time but rarely do so on an annual average basis (*14*).

Patterns in Lakes and Reservoirs. Lakes and reservoirs also typically show an annual pattern, but one in which the short-term pulses of concentration which characterize rivers are less conspicuous or absent. Lakes and large reservoirs with hydraulic residence times on the order of several years or more can be expected to show only gradual changes in concentration over several years in response to trends in herbicide runoff. Smaller lakes, ponds, and many reservoirs, with residence times on the order of a half-year to several years, show annual pulses of concentration with varying degrees of damping. Because inflowing water with high concentrations mixes with resident water with lower concentrations, maximum concentrations observed in lakes and reservoirs are generally substantially smaller than those in their tributaries, and minimum concentrations may be greater, if the resident herbicide is not eliminated in the winter months through breakdown or wash-through.

Relatively high average concentrations have been found in some (usually small) reservoirs in the midwest. These typically have mostly agricultural watersheds, and a sufficiently high water demand that little water is released from the reservoir during much of the year. These reservoirs may have concentrations high enough to pose a potential threat to their ecosystems and/or to exceed drinking water standards, and are excellent examples of what can happen when multiple uses of a watershed are not properly coordinated.

Concentration Distributions and Ecological Risk. In ecological risk assessments of herbicides, the upper percentiles of concentration are generally of more interest than the average concentrations. The distribution of 90th percentile (upper 10th percentile) instantaneous concentrations in atrazine data for 58 midwestern rivers, streams, lakes, and reservoirs is shown in Figure 3. This dataset is probably biased high by factors involved in selecting a site for monitoring for herbicides, including selecting sites at which herbicide detections are expected, selecting small watersheds more frequently than large ones, and selecting ag-dominated watersheds more frequently than those dominated by other land uses. Nonetheless, 43 of the 58 water bodies had concentrations which were below 5 µg/L 90% of the time. Mesocosm studies summarized in Solomon *et al.* (7) suggest an ecological no effect level of 20 µg/L.

It may make more sense to compare toxicological data with environmental concentrations which are averaged over time frames similar to those used in toxicological experiments. Solomon *et al.* (7) used 4-day and 21-day running-average concentrations to characterize the exposure distribution in preference to instantaneous concentrations. However, they also found that the values of these averaged concentrations were very similar to and slightly lower than those of instantaneous concentrations at percentiles (90th and 95th) typically used for ecological risk assessment. This result stems from the relationship between the autocorrelation structure of the data and the length of the running-average window (R.P. Richards, unpublished). This is a useful finding, because it indicates that distributions of instantaneous concentrations can be used in these assessments without serious loss of representativeness, when running-average concentrations are difficult or impossible to calculate because of the nature of the raw data (e.g. widely and unevenly spaced observations).

Multi-parameter Patterns

Chemographs for the same time period for atrazine and the triazine herbicide breakdown products DEA and DIA are shown in Figure 4. River flow is shown in gray in the background. Before the first runoff event following application, atrazine concentations are less than 1 µg/L, and the breakdown products comprise about half of the sum atrazine+DIA+DEA. Spikes in the curve for atrazine as a percent of the sum correspond to storm runoff events, and occur shortly after peak flow. Detailed examination of the data shows that the spikes in percent atrazine are related to reduced concentrations of DIA and DEA rather than increased concentrations of atrazine.

Following the application period, storm events bring new atrazine into the system at higher concentrations, and atrazine becomes dominant over the

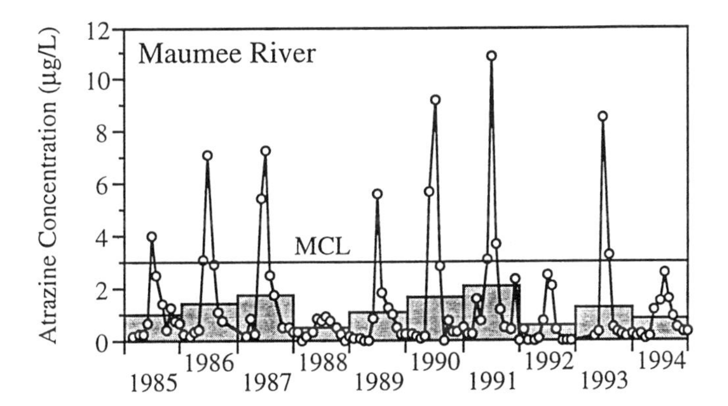

Figure 2. Monthly average concentrations (dots) and annual average concentrations (shaded bars) of atrazine in the Maumee River at Bowling Green, Ohio.

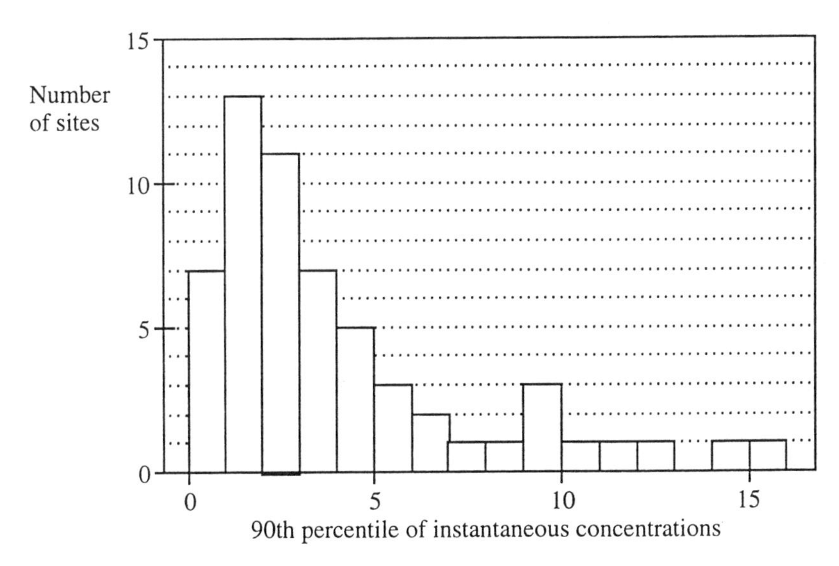

Figure 3. Histogram of 90th percentiles of instantaneous concentrations of atrazine in 58 midwestern streams, rivers, lakes, and reservoirs.

breakdown products. Thereafter, atrazine gradually breaks down into DEA and DIA (and other breakdown products which were not analyzed). By the end of the July, atrazine is about equal to its breakdown products in concentration. During the rest of the year, the ratio of atrazine to breakdown products will not change greatly, but their concentrations will gradually decline to the near-zero levels seen at the beginning of the graph.

Both the the increasing transformation of atrazine to breakdown products and the general decrease in concentrations which characterize the post-application period represent conditions of decreasing ecosystem risk, since toxicity to aquatic organisms is not only concentration-dependent but generally decreases in the order atrazine > DEA > DIA > other products (*17*).

The generality of these patterns is indicated by Figure 5, in which TWMCs of atrazine, cyanazine, simazine, DEA, and DIA for the period 1982-1993 are plotted as stacked bar graphs for five midwestern rivers studied by Baker and Richards in the Lake Erie basin (*1*), and nine rivers studied by Goolsby and colleagues in the Mississippi drainage system in 1990 and 1991 (*2*). In these graphs, the data are normalized by dividing all TWMCs at a station by the atrazine TWMC at that station. In all rivers except one, the sum of the three parent triazines plus DEA and DIA is less than twice the atrazine concentration alone. The sole exception is for the 1991 data of Goolsby and colleagues. 1991 was a drought year with very little runoff at these stations, and apparently little applicatiion of new pesticide. As a consequence, the atrazine breakdown products dominate the runoff. The actual concentrations for this year, however, are among the lowest of any site. Thus the pattern is more a consequence of an unusual scarcity of parent compounds rather than an unusually large quantity of breakdown products.

Conclusions

• Triazine concentration patterns have a strong seasonal component, with elevated concentrations during a 4-12 week period following application. The duration of elevated concentrations depends on compound-specific properties such as soil half-life and on the timing and intensity of rainfall following application.

• Triazine concentration patterns in streams and rivers are storm-runoff driven, and pulsed exposures alternating with periods of recovery are typical. This may be particularly important for understanding ecosystem effects, since triazines work by inhibiting photosynthesis, and complete recovery may occur between pulses of exposure.

• High concentrations of triazines in surface waters are associated with small streams and rivers, with watersheds dominated by agricultural land use, and with reservoirs which impound such streams. Reservoirs are particularly at risk if the outflow is intermittant or seasonal only.

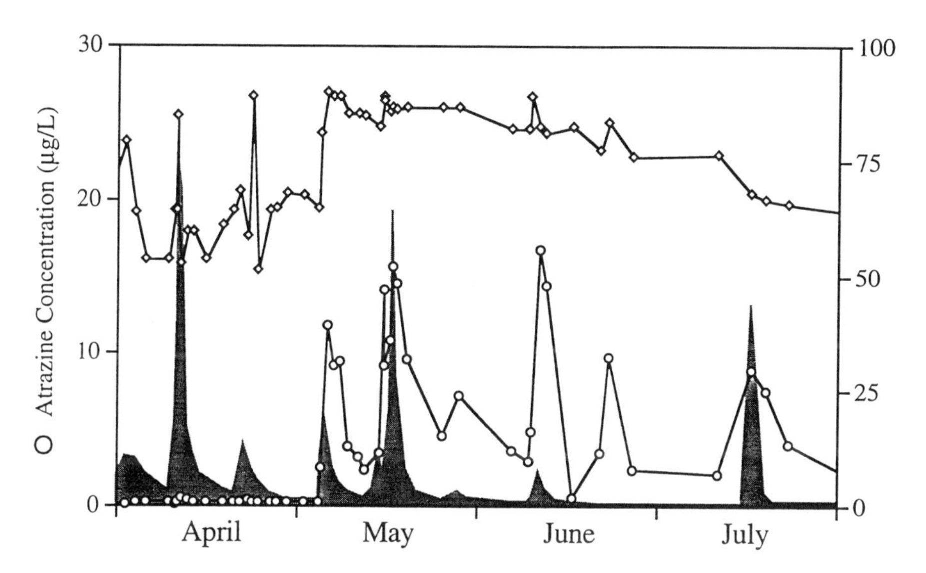

Figure 4. Chemograph for atrazine, Huron River, April to July 1990. The gray shaded plot displays the mean daily flows. Atrazine concentrations are shown as circles. Atrazine concentrations expressed as a percent of atrazine+DIA+DEA are shown using diamonds.

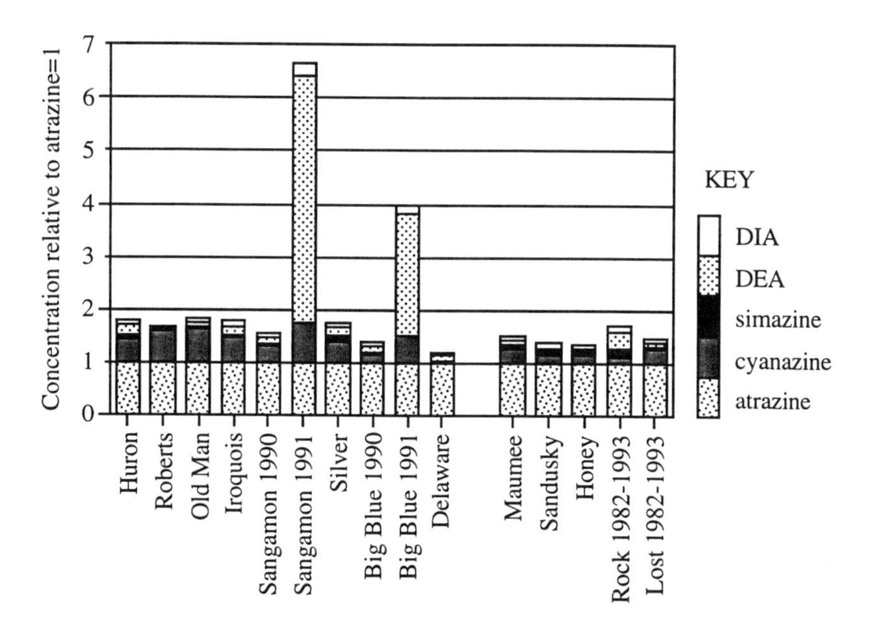

Figure 5. Relative amounts of atrazine, cyanazine, simazine, DIA, and DEA in midwestern rivers: annual or multi-year averages.

• Atrazine dominates total triazine concentrations in the midwest. Atrazine concentrations generally are larger than the sum of cyanazine, simazine, DEA, and DIA, both in individual samples and as averages over the course of a year.

Literature Cited

1. Richards, R. P.; Baker, D.B. *Environ. Toxicol. Chem.* **1993**, *12*, 13.
2. Scribner, E.A.; Goolsby, D.A.; Thurman, E.M.; Meyer, M.T.; Pomes, M.L. *Concentrations of Selected Herbicides, Two Triazine Metabolites, and Nutrients in Storm Runoff from Nine Stream Basins in the Midwestern United States, 1990-1992*; Open-File Report 94-396; U.S. Geological Survey: Denver, Colorado, 1994 .
3. Illinois EPA. *Pesticide monitoring: Illinois EPA's Summary of Results, 1985-1988;* IEPA/WPC/90-297; Illinois EPA, Division of Water Pollution Control: Springfield, Illinois, 1990.
4. Giannessi, L.P.; Puffer, C. 1991. *Herbicide Use in the United States*; Resources for the Future: Washington, DC, 1991.
5. Battaglin, W.A.; Goolsby, D.A. *Spatial Data in Geographic Information System Format on Agricultural Chemical Use, Land Use, and Cropping Practices in the United States;* Water-Resources Investigations Report 94-4176; U.S. Geological Survey: Denver, Colorado, 1995.
6. Larson, S.J.; Capel, P.D.; Majewski, M.S. *Pesticides in Surface waters: Distribution, Trends, and Governing Factors*; Ann Arbor Press: Chelsea, Michigan, 1996.
7. Solomon; K.R.; Baker, D.B.; Richards, R.P.; Dixon, K.R.; Klaine, S.J.; LaPoint, T.W.; Kendall, R.J.; Weisskopf, C.P.; Giddings, J.M.; Giesy, J.P.; Hall, J.W., Jr.; Williams, M.W. *Environ. Toxicol. Chem.* 1996, *15*, 31.
8. Wauchope, R.D. *J. Environ. Qual.* **1978**, *7*, 459.
9. Muir, D.C.G.; Yoo, J.Y.; Baker, B.E. *Arch. Environ. Contam. Toxicol.* **1978**, *7*, 221.
10. Frank, R.; Sirons, J. Sci. Total Environ. 1979, 12, 223.
11. Frank, R.; Sirons, G.J.; Thomas, R.L.; McMillan, K. *J. Great Lakes Res.* **1982**, *5*, 131-138.
12. Bodo, B.A. *Environ. Toxicol. Chem.* **1991**, *10*, 1105-1121.
13. Pontius, F.W. *J. Am. Water Works Assoc.* **1990**,*82*, 32.
14. Richards, R.P.; Baker, D.B.; Christensen, B.R.; Tierney, D.P. *Environ. Sci. Technol.* **1995**, *29*, 406-412.
15. Squillace, P.J.; Thurman, E.M.; Furlong, E.T. *Water Resources Research* **1993**, *29*, 1719-1729.
16. Thurman, E.M.; Goosby, D.A.; Meyer, M.T.; Kolpin, D.W. *Environ. Sci. Technol.* **1991**, *25*, 1794-1796.
17. Eisler, R. *Atrazine Hazards to Fish, Wildlife, and Invertebrates: A Synoptic Review*; Biol. Report 85; U.S. Fish Wildlife Serv.: Washington, DC, 1989.

Chapter 27

The Aquatic Ecotoxicology of Triazine Herbicides

Jeffrey M. Giddings[1] and Lenwood W. Hall, Jr.[2]

[1]Springborn Laboratories, Inc., 790 Main Street, Wareham, MA 02571
[2]Wye Research and Education Center, University of Maryland, Queenstown, MD 21658

The effects of triazine herbicides on aquatic species and ecosystems are reviewed. Effects on aquatic plants are reversible; photosynthesis resumes when the herbicide disappears from the water, and sometimes even while it is still present. Effects on aquatic plant communities are further ameliorated by species replacements, so the communities as a whole are less sensitive than their most sensitive species. Atrazine, a representative triazine herbicide, is acutely toxic to aquatic plants (algae and macrophytes) at concentrations in the range of 20 to 200 μg/L. Chronic toxicity to plants occurs at concentrations ten times lower than acute toxicity. Aquatic invertebrates and fish are much less sensitive than plants, with acute toxicity occurring at 1000 to 200,000 μg/L. Ecologically significant effects in aquatic ecosystems are likely only if plant communities are severely damaged by prolonged exposure to high atrazine concentrations.

The objective of this paper is to review the data on triazine herbicide toxicity to aquatic organisms and ecosystems. The focus is on atrazine because atrazine is the most widely used and widely studied triazine, and because a detailed ecological risk assessment of atrazine has recently been completed (*1*). Besides reviewing the laboratory data on triazine toxicity, we will discuss the ecological implications of triazine effects in real ecosystems, drawing particularly on evidence from microcosm and mesocosm studies.

Triazine herbicides are photosynthetic inhibitors. Their primary physiological effect is to block electron transport in Photosystem II (*2*). The effect is reversible: when the herbicide is removed from the plant cell, photosynthesis resumes, and the plant's normal biochemistry is quickly restored. The effect is non-lethal to aquatic plants unless exposure continues for a very long time—weeks or months, long enough for the plant literally to starve. In fact, the effect of triazines on aquatic plants is similar to the effect of reduced light, such as occurs when suspended solids shade the plants in a muddy stream after a rainstorm. Because triazines act by blocking a specific photosynthetic mechanism, they are not highly toxic to animals.

Acute Toxicity Distributions

The acute toxicity of a substance to aquatic organisms is generally expressed as the LC50 or EC50 concentration, which is the concentration that kills half of the test population or reduces plant growth by 50%. A lower LC50 concentration implies a more sensitive species. Based on studies submitted to EPA for product registrations (*3*), atrazine is generally intermediate in toxicity compared with other triazines (Table I). As expected, aquatic plants (algae and duckweeds) are considerably more sensitive to atrazine than animals, with values in the range of 20 to 500 μg/L. Invertebrates are less sensitive than plants, with LC50 values from 1000 to 7,000 μg/L (and, in one extreme case, a crab, nearly 200,000 μg/L). Fish are less sensitive still, with most values between 10,000 and 100,000 μg/L, roughly two to four orders of magnitude higher than for aquatic plants. The same trends are evident in the data for other triazines as well. The acute toxicity values for even the most sensitive aquatic animals are always greater than 1,000 μg/L, whereas some plants are sensitive at concentrations less than 100 μg/L. In a risk assessment of triazine herbicides, we are therefore concerned with (a) the potential for reduced productivity of the aquatic plant community due to direct toxic effects, and (b) indirect effects on aquatic invertebrates and fish due to loss of food supply, alteration of habitat, or changes in water quality caused by reduced photosynthesis.

In our risk assessment of atrazine (*1*), we used a probabilistic approach (*4*) to characterize the sensitivity of aquatic species. Acute toxicity data for 52 species were compiled from several sources, sorted in order of sensitivity, and plotted as a cumulative log-normal distribution (Figure 1). The horizontal axis represents the LC50 or EC50 (in μg/L, on a log scale), and the vertical axis represents the ranking of species sensitivity, expressed on a probability scale. The line through the points is a least-squares regression, assuming a log-normal distribution of species sensitivity. In our risk assessment, we used the regression line to estimate the LC50 of the tenth percentile of species sensitivity—that is, the concentration that would be expected to cause acute effects to one-tenth of the species for which we have data. For atrazine, the concentration that would protect 90% of the aquatic species from acute toxic effects was estimated to be 37 μg/L. Of course, the affected species would all be plants; it would take much higher concentrations to cause effects on animals. This approach assumes that protecting ten percent of the species will also protect the ecosystem as a whole, an assumption that turns out to be conservative, as will be discussed below.

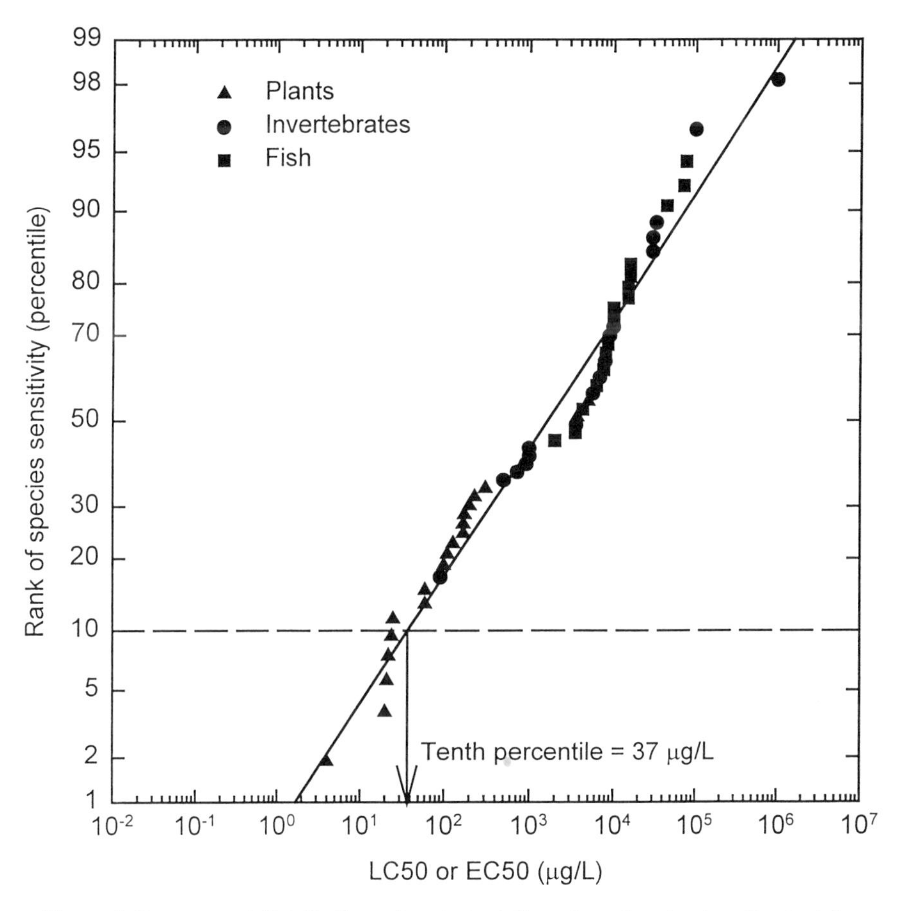

Figure 1. Log-normal distribution of acute toxicity values for atrazine. See text for explanation. Adapted from ref. 1.

Table I. Summary of acute toxicity data (LC50 and EC50 concentrations, in μg/L) for five triazine herbicides to aquatic species. Data from Ref. 3.

Species	Atrazine	Cyanazine	Simazine	Ametryn	Prometryn
		Plants			
Isochrysis galbana	22		500	10	
Skeletonema costatum	24	18	600		8
Selenastrum capricornutum	53	6	100	4	12
Chlamydomonas sp.	60				
Monochrysis lutheri	77			14	
Neochloris sp.	82			36	
Platymonas sp.	100			24	
Chlorococcum sp.	100		2000	10	
Thallassiosira fluviatilis	110			58	
Microcystis aeruginosa	129				
Chlorella sp.	140			320	
Lemna gibba	170	64	140		12
Phaeodactylum tricornutum	200		500	20	
Chlorella pyrenoidosa	282				
Nitzschia closterium	290			62	
Porphyridium cruentum	308			36	
Dunaliella tertiolecta	431		5000	20	
Navicula inserta	460			97	
Navicula pelliculosa		5	90		1
Anabaena flos-aquae		24	36		40
Achnanthese brevipes				19	
Stauroneis amphoroides				26	
Cyclotella nana				55	

Table I (continued). Acute toxicity data for five triazine herbicides to aquatic species.

Species	Atrazine	Cyanazine	Simazine	Ametryn	Prometryn
		Invertebrates			
Chironomus tentans	1000				
Penaeus aztecus	1000				
Mysidopsis bahia	5000			2300	1700
Gammarus fasciatus	6000	2000			
Daphnia magna	7000	45500	1100	28000	18590
Uca pugilator	198000				
Crassostrea virginica		20000			1000
Palaemonetes pugio		56000			
Pteronarcys californica			1900		
Cypridopsis vidua			3700		
Gammarus lacustris			13000		
Penaeus duorarum			113000		1000
Mercenaria mercenaria				11000	21000
		Fish			
Salvelinus fontinalis	5000				
Cyprinodon variegatus	13000	18000		5800	5100
Oncorhynchus mykiss	14667	9000	53900	3200	7200
Pimephales promelas	15000	18500	5700	5700	
Notropis atherinoides	16000				
Lepomis macrochirus	39400	23000	50333	5433	10000
Carassius auratus	60000			14000	4000
Ictalurus punctatus		12667	85000		
Leiostomus xanthurus				1000	1000
Lepomis gibbosus			27000		
Micropterus salmoides			46000		
Lepomis macrolopus			54000		
Pimephales notatus			66000		
Ictalurus natalis			110000		

Chronic Toxicity Distributions

The chronic toxicity of a substance to aquatic organisms is typically expressed as a No Observed Effect Concentration (NOEC). Because NOECs apply to longer exposure times and more sensitive toxicity endpoints, the concentrations are lower than for acute toxicity. The tenth percentile of the chronic toxicity distribution for atrazine is 3.7 μg/L (Figure 2), exactly one-tenth the tenth percentile for acute toxicity.

Model Ecosystem Studies

The results reviewed so far are based on standard laboratory toxicity tests with single species. Numerous studies have also been conducted to measure atrazine effects on model aquatic ecosystems—microcosms and mesocosms. These studies help us to understand and evaluate the significance of the laboratory toxicological data, because the microcosm and mesocosm studies address aggregate responses of multiple species in intact communities. They also allow observation of indirect effects and ecological recovery. Most of these studies involved continuous doses or repeated pulses of atrazine, or took place in static systems in which concentrations remained at fairly steady levels for weeks at a time; thus, they represent the effects of chronic exposure.

Based on results of more than 20 microcosm and mesocosm studies, atrazine exposures below 20 μg/L generally cause no effect on aquatic plants, and where an effect occurs there is always recovery (Figure 3). Between 10 and 100 μg/L there is sometimes an effect but still always a recovery. For example, atrazine at 10 μg/L reduced macrophyte productivity in wetland microcosms, but productivity returned to pretreatment levels within 7 days (while atrazine was still present); macrophyte biomass was unaffected (*5*). In a study with laboratory streams, periphyton productivity—again, not biomass—was affected at 10 μg/L and recovered within 3 weeks (*6,7*). The productivity of pond phytoplankton in microcosms exposed to 15 μg/L recovered within 2 weeks (*8*). Wetland macrophytes exposed to 20 μg/L showed reduced productivity but no effect on biomass, and recovered in 6 weeks (*9*). Stream periphyton exposed to 24 μg/L recovered after 12 days (*10*). Pond periphyton exposed to 32 μg/L recovered after 3 weeks (*11*). Pond phytoplankton exposed to continuous input of 50 μg/L recovered within one day after atrazine input ceased (*12*). Benjamin et al. (The Institute of Wildlife and Environmental Toxicology and Clemson University, unpublished data) showed the same phenomenon in a species of green algae: growth of *Selenastrum capricornutum* was severely inhibited during 32 days of exposure to 10 μg/L atrazine, but when the cells were transferred to clean medium their growth resumed at a normal rate, equal to controls.

Based on these results, we conclude that atrazine exposure of 20 μg/L or less, even for extended periods of time (one of these studies continued for three years), causes no lasting harm to aquatic plant communities. Fifty μg/L is taken, conservatively, as the lowest effect concentration, even though recovery still occurs. Above 100 μg/L there is always an effect and often no recovery.

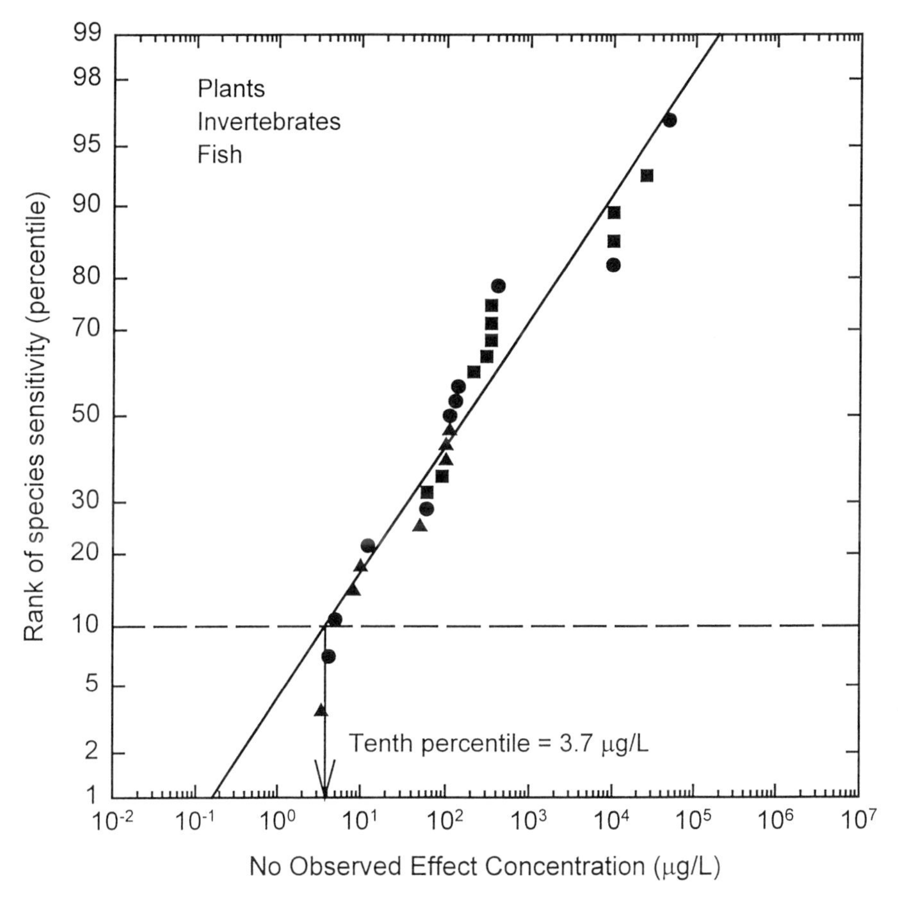

Figure 2. Log-normal distribution of chronic toxicity values for atrazine. Adapted from ref. 1.

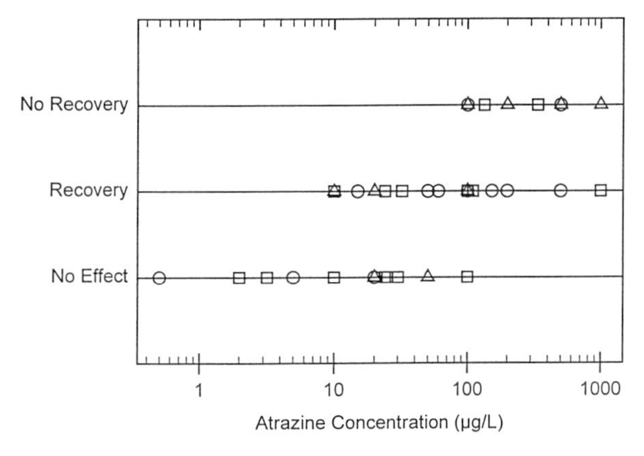

Figure 3. Summary of effects of atrazine on plants in mesocosm and microcosm studies. Each point represents an observed response at one exposure level in one study. Circles: Phytoplankton. Squares: Periphyton. Triangles: Macrophytes.

Thus we have 3 ecotoxicological endpoints for atrazine: 37 μg/L as the tenth percentile of acute toxicity, 3.7 μg/L as the tenth percentile of chronic toxicity, and 20 μg/L as a conservative no effect level for the plant community as a whole.

Recovery, Resistance, and Replacement

Why are plant communities unaffected by prolonged exposure to 20 μg/L atrazine, even though a substantial fraction of plant species are affected by atrazine concentrations as low as 3.7 μg/L? In terms of total biomass and productivity (and these are the ecological endpoints we are usually interested in protecting, rather than the success of particular species of algae) an aquatic plant community is less sensitive to atrazine than its most sensitive species. Three mechanisms contribute to this.

The first is *RECOVERY*. As demonstrated above, even while exposure continues, and immediately after it ceases, aquatic plants are typically able to return to normal levels of productivity. The exact physiological mechanisms that allow this aren't clear, but it is sometimes observed that atrazine-exposed plants produce increased amounts of chlorophyll to compensate for the reduced efficiency of that chlorophyll in photosynthesis. A similar phenomenon is observed when plants respond to shading.

A second mechanism that may be at work is *RESISTANCE*. deNoyelles *et al.* (*13*) observed that algal communities in mesocosms treated with atrazine developed a physiological tolerance, such that a greater concentration of atrazine is needed to cause photosynthetic inhibition. Other investigators have been unable to reproduce this effect in laboratory studies (*6,7,12,14*), but resistance was demonstrated in field studies by Fromm (*15*).

A third mechanism is *REPLACEMENT*. Aquatic plants vary greatly in their sensitivity to atrazine, and the more resistant species are generally able to replace those that are affected. The overall structure and function of the plant community is unchanged, even though the proportions of the species may shift (*13-16*). Unless there is concern about a particular species of plant, these changes would not generally be considered significant to the ecosystem.

Indirect Effects

Even if atrazine causes no direct toxic effects on fish and invertebrates, and only temporary inhibition of plants, the possibility of indirect effects must be considered. One type of indirect effect that could occur is a change in water quality due to reduced photosynthesis: reduced dissolved oxygen and pH, and increased alkalinity, conductivity, and nutrient levels, all due to lower rates of CO_2 uptake, nutrient uptake, and oxygen production. These effects have been observed in microcosm and mesocosm studies. For example, dissolved oxygen in pond mesocosms was 1 mg/L less than controls for 7 days following treatment with 20 μg/L atrazine (*13*). At higher treatment levels (500 μg/L), dissolved oxygen was reduced by 1 to 3 mg/L for up to 22 days, and pH fell by 0.3 units (*13*); total alkalinity increased by 5 to 10 mg/L (*17*). In non-flowing laboratory systems, dissolved oxygen decreased after exposure to 100 μg/L atrazine and higher, but not after exposure to 20 μg/L (*18*). Several investigators have reported increases in inorganic nitrogen, phosphorus, conductivity and alkalinity (*5,8,12,19*) at atrazine concentrations high enough to cause severe effects on the plant community (100 µg/L and greater). Such changes would be most significant in static macrophyte-dominated systems, in which water chemistry is strongly influenced by biological activity. In flowing water, especially small streams, water chemistry is generally controlled by physical processes (advection, diffusion), and changes in rates of photosynthesis and nutrient uptake would be expected to have little or no impact on water quality.

A second type of potential indirect effect would be reduced growth or survival of fish caused by changes in their food supply, especially small invertebrates—assuming that the invertebrates were themselves reduced by a reduction in plant productivity. Indirect effects on fish production have been observed in some studies (*8,13,20*), but only at atrazine exposure levels that cause major impacts on the plant community. They do not occur at levels that cause only subtle effects on plants.

Summary

In interpreting the ecological significance of the toxicity data for atrazine and other triazines, several factors must be taken into account: (1) Effects of triazines on aquatic plants are transient and reversible. (2) Aquatic plant communities are less sensitive than individual species due to the potential for recovery, resistance, and replacement. (3) Indirect effects occur only at high levels of exposure (high enough to cause major damage to the plant community). (4) Other stressors (such as nutrients, and shading caused by suspended solids) often accompany triazine exposure and could be of greater ecological significance than triazines.

This review has been essentially qualitative. Our published risk assessment on atrazine (*1*) put this information into a quantitative framework, and determined the probability that atrazine concentrations measured in US surface waters actually cause significant ecological effects. The conclusion of the probabilistic risk assessment was that atrazine residues in surface waters do not present a significant risk to the aquatic environment, though risk is higher in some small watersheds with extensive pesticide use, and in reservoirs which receive drainage from those watersheds. Site-specific risk assessments were recommended for the ecosystems at highest risk.

Acknowledgments

This review was adapted from the report of the Atrazine Ecological Risk Assessment Panel (Keith Solomon, David Baker, Peter Richards, Kenneth Dixon, Stephen Klaine, Thomas La Point, Ronald Kendall, Carol Weisskopf, Jeffrey Giddings, John Giesy, Lenwood Hall, and Marty Williams). The Panel was supported by Ciba Crop Protection and coordinated by Richard Balcomb.

Literature Cited

1. Solomon, K.R.; Baker, D.B.; Richards, R.P.; Dixon, K.R.; Klaine, S.J.; La Point, T.W.; Kendall, R.J.; Weisskopf, C.P.; Giddings, J.M.; Giesy, J.P.; Hall, L.W., Jr.; Williams, W.M. *Environ. Toxicol. Chem.* **1996**, *15*, 31-76.

2. Ebert, E.; Dumford, S.W. *Res. Rev.* **1976**, *65*, 2-60.
3. U.S. Environmental Protection Agency. *Pesticide Toxicity Database.* **1995**. Office of Pesticide Programs, U.S. EPA, Washington, DC.
4. *Aquatic Risk Assessment and Mitigation Dialogue Group. Final Report.* Society of Environmental Toxicology and Chemistry Foundation for Environmental Education: Pensacola, Florida, 1994.
5. Johnson, T.B. *Environ. Toxicol. Chem.* **1986**, *5*, 473-485.
6. Kosinski, R.J. *Environ. Pollut. (Ser. A)* **1984**, *36*, 165-189.
7. Kosinski, R.J.; Merkle, M.G. *J. Environ. Qual.* **1984**, *13*, 75-82.
8. Hoagland, K.D.; Drenner, R.W; Smith, J.D.; Cross, D.R. *Environ. Toxicol. Chem.* **1993**, *12*, 622-637.
9. Huckins, J.N.; Petty, J.D.; England, D.C. *Chemosphere*. **1986**, *15*, 563-588.
10. Krieger, K.A.; Baker, D.B.; Kramer, J.W. *Arch. Environ. Contam. Toxicol.* **1988**, *17*, 299-306.
11. Pratt, J.R.; Bowers, N.J.; Niederlehner, B.R.; Cairns, J., Jr. *Arch. Environ. Contam. Toxicol.* **1988**, *17*, 449-457.
12. Brockway, D.L.; Smith, P.D.; Stancil, F.E. *Bull. Environ. Contam. Toxicol.* **1984**, *32*, 345-353.
13. deNoyelles, F., Jr.; Kettle, W.D.; Sinn, D.E. *Ecology*. **1982**, *63*, 1285-1293.
14. Hamala, J.A.; Kollig, H.P. *Chemosphere.* **1985**, *14*, 1391-1408.
15. Fromm, C.H. *Effects of the Herbicide Atrazine on Eutrophic Plankton Communities.* MS Thesis, University of Kansas, 1986.
16. Hamilton, P.B.; Jackson, G.S.; Kaushik, N.K.; Solomon, K.R.; Stephenson, G.L. *Aquatic Toxicol.* **1988**, *13*, 123-140.
17. Kettle, W.D. *Description and Analysis of Toxicant-Induced Responses of Aquatic Communities in Replicated Experimental Ponds.* Ph.D. Dissertation, University of Kansas, 1982.
18. Stay, F.S.; Katko, A.; Rohm, C.M.; Fix, M.A.; Larsen, D.P. *Arch. Environ. Contam. Toxicol.* **1989**, *18*, 866-875.
19. Hamilton, P.B.; Lean, D.R.S.; Jackson, G.S.; Kaushik, N.K.; Solomon, K.R. *Environ. Pollut.* **1989**, *60*, 291-304.
20. deNoyelles, F., Jr.; Kettle, W.D.; Fromm, C.H.; Moffett, M.F.; Dewey, S.L. In *Using Mesocosms to Assess the Aquatic Ecological Risk of Pesticides: Theory and Practice*; Voshell, J.R., Ed.; MPPEAL 75; Entomological Society of America: Lanham, MD, 1989, pp 41-56.

Chapter 28

Triazine Herbicides: Ecological Risk Assessment in Surface Waters

Keith R. Solomon and Mark J. Chappel

Centre for Toxicology, University of Guelph, Guelph, Ontario N1G 2W1, Canada

Triazine herbicides are widely used pesticides in North America. Residues of these substances are found in many surface waters and ecological effects in these ecosystems are a possible concern. A probabilistic risk assessment technique was used to assess the risks associated with these substances in surface waters. The exposure characterization concentrated on monitoring data from US and Canadian watersheds with a focus on high-use areas. The effects characterization showed that phytoplankton and aquatic plants were the most sensitive organisms followed by, in decreasing order of sensitivity, arthropods and vertebrates such as fish. Based on an integrative risk assessment using laboratory bioassay data and environmental monitoring data from watersheds in high-use areas, it was concluded that, in general, the triazines do not pose a significant risk to the aquatic environment. Although some inhibitory effects on algae, phytoplankton or macrophyte production may occur in small streams vulnerable to agricultural runoff, these effects are likely to be transient and quick recovery of the ecosystem is expected. A subset of surface waters, principally small streams in areas with intensive use of triazines, may be at greater risk. In these cases, site-specific risk assessments should be conducted to assess possible ecological effects in the context of the uses to which these ecosystems are put and the effectiveness and cost-benefits of any risk mitigation measures that may be applied.

Trace amounts of pesticides have been found in a number of aquatic systems in North America. These pesticide residues are primarily those of compounds that have the propensity for movement and some persistence in the environment. Residues are more often detected in aquatic ecosystems that are located close to areas of high pesticide use and where environmental factors, such as intensity of precipitation, increase the likelihood for runoff into aquatic ecosystems. The risks associated with the use of atrazine in North American surface waters have been assessed using probabilistic techniques (*1*) and this approach has been applied for other substances as well (*2*). This paper is based on procedures recommended by the US EPA (*3*) and focuses on the use of probabilistic procedures for assessing risks from the herbicides; atrazine [6-chloro-N-ethyl-N'-(1-methylethyl-1,3,5-triazine-2,4-diamine], simazine [6-chloro-N,N'-diethyl-1,3,5-triazine-2,4-diamine] and cyanazine [2-[[4-chloro-6-(ethylamino)-1,3,5-triazin-2-yl]amino]-2-methylpropionitrile] in aquatic ecosystems in the US and Canada.

Assessing Risks

Most risk assessments are carried out using a tiered approach. The use of tiered approaches has several advantages. The initial use of conservative criteria allows substances that truly do not present a risk to be eliminated from the risk assessment process, thus allowing the focus of expertise to be shifted to more problematic substances. These tiers begin with a simple "worst-case" estimation of environmental concentration which is compared with the effect level for the most sensitive organism (the hazard quotient approach). If the result of this comparison suggests no risk, no regulatory action would be necessary. If the result suggests a potential risk, further tiers of risk assessment with more realistic and more complete exposure and effects data can be applied to the problem. Emulating (*4*) we have used the following tiers in the assessment: 1) hazard quotient, 2) probabilistic risk characterization and, 3) characterization of ecological relevance.

The Quotient Approach. Traditionally, characterizing hazards at the level of the organism has been conducted by comparison of the concentration of the stressor/s found in the environment to the responses reported for the stressor/s in laboratory tests. The simplest approach to this is the use of hazard quotients. Hazard quotients are simple ratios of exposure and effects. For example:

$$Hazard \approx \frac{Exposure\ concentration}{Effect\ concentration}$$

The application of hazard quotients to ecotoxicological risk assessment has been accomplished by comparing the effect concentration of the most sensitive organism or group of organisms to the highest exposure concentration (a worst-case hazard scenario). Depending on the response used to characterize the effect (No Observed Effect Concentration, LC50, etc.), this hazard quotient may be made more conservative by the use of a safety or uncertainty factor such as, for example, division of the effect concentration by 20 (*5*). This is done to allow for unquantified uncertainty in the estimations or measurements of effect and exposure. The hazard quotient approach is based on similar procedures used in human health risk assessment and therefore fails to acknowledge the very significant differences between human health and ecosystem risk assessment (*6*). In contrast to human health protection, individual organisms in the ecosystem are regarded as transitory and, because they are usually part of a food chain, are, in an ecological sense, expendable (*7*). In addition, ecosystem functions are usually highly conserved. The absence of one or more species may have no effect on ecosystem function and ecosystems are, in general, less sensitive than their most sensitive component (*1*).

The hazard quotient approach with uncertainty factors is conservative and is useful where little data on effect or exposure concentrations are available (*4*). However, where more data are available and the variation in the response of organisms in the environment is better defined, the use of large uncertainty factors may be unnecessarily overprotective. The hazard quotient approach also fails to consider the range of variation that may exist in terms of real-world exposures to the substance in question.

The Probabilistic Approach. Expressing the results of a refined risk characterization analysis as a distribution of toxicity values rather than a single point estimate is an approach presently being used in higher tiers of risk assessment by several regulatory agencies (*8*) and others (*1, 2, 4*). A major advantage of this approach is that it uses all relevant single species toxicity data and, when combined with exposure distributions, allows quantitative estimations of risks to ecosystems.

The principle of the probabilistic approach is illustrated in Figure 1. It is well known that many parameters and measures are distributed in a consistent manner (*9*) and that, from these distributions, it is possible to estimate the likelihood that any particular measure will be observed in subsequent sampling of the same population (for example, height of individual humans, Figure 1-A). This also applies to concentrations of substances in the environment. However, in this case, data are often censored by the limits of analytical detection (Figure 1-B) and they are frequently log-normally distributed. When plotted as a cumulative frequency distribution using a probability scale on the Y axis (Figure 1-C), these distributions approximate a straight line which can be used to estimate the likelihood that a particular concentration of the substance

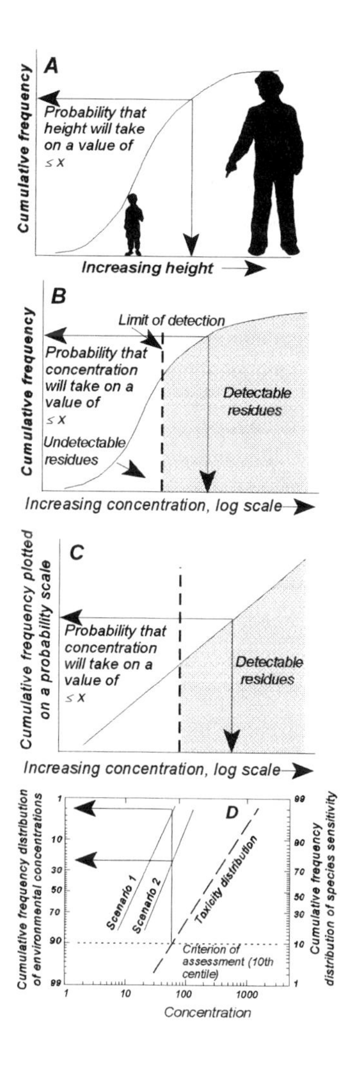

Figure 1. Illustration of the use of the probabilistic approach. In D, horizontal arrows show the intercept of the 10th centile of the toxicity distribution with the environmental concentration distribution and indicates the probability that this concentration will be exceeded.

will be exceeded in an environment where the circumstances are similar. A similar approach can be taken with susceptibility of different organisms in the ecosystem to the substance. The use of these distributions in the probabilistic approach is illustrated in Figure 1-D. In this procedure, it is assumed that the distributions of sensitivity represent the range of responses that are likely to be encountered in the ecosystems where the exposures occur (*8*). If the exposure data were collected over time, the degree of overlap of the exposure distribution with the effects distribution can be used to estimate the joint probability of exposure and toxicity, leading to estimates of exceedence probabilities for responses at a fixed exposure assessment criterion, such as the concentration equivalent to the 10th centile of the species distribution (Figure 1-D). This can be applied to a number of data sets (scenarios, Figure 1-D) and the resulting probabilities used for priority setting or in assessing ecological relevance.

In using distributional approaches, there is an implied assumption that protecting a certain proportion of species for a certain proportion of stressor exposure events will also preserve population and community function. This assumption is consistent with ecotoxicological theory and observations of responses of communities to stress (*10*), and is the reason why assessment endpoints in ecotoxicological risk assessment are directed to the level of population or community function (*7*).

There are some limitations to this approach (*4, 8*). For example, the choice of protection level (e.g., 90% of species) may not be socially acceptable, especially if the 10% of potentially affected species includes organisms of great ecological, commercial, or recreational significance. However, the procedure allows these species to be identified from the distribution. With knowledge of the mode of action of the substance, this question can be readily addressed in higher tiers of risk assessment where the relevance of these species to ecosystem function and inclusion of any keystone species can be assessed.

Where effects data from field studies are available, the protection offered by the chosen assessment criterion may be judged against observations at the functional level of the ecosystem, i.e., the responses of populations and communities (*6*). In the case of atrazine, the 10th centile of the sensitivity distribution of EC5s for sensitive organisms such as algae was judged to be conservative when compared to community response measured in mesocosms (*1*). For this reason the 10th centile of the sensitivity distribution was also used in these studies.

Characterizing the Exposure Data

Description of Data Sets. The data sets (Table I) used to determine the exposure concentrations of pesticides in surface waters from Quebec were obtained from Environment Quebec (Berryman, D, Environment Quebec, Quebec, PQ, personal communication, 1995). The data comprised analyses conducted in a number of rivers and streams in close proximity to intensive agricultural practices in watersheds that drain into the St Lawrence River in South Eastern Quebec. These watersheds feed a range of water bodies from small streams with no tributaries to major rivers. Data from Ontario were obtained from Environment Canada (Struger, J and S. Painter, Environment Canada, Ecosystem Health Effects Branch, Burlington, ON, personal communication, 1996). The simazine data were obtained from analyses conducted on samples from the San Joaquin and Sacramento Rivers in California by the US Geological Survey (*11*).

Selection of Data. Only data from complete data sets were used in this analysis. As the approach assesses probability, all analyses were included, even those where the substance was not detected. Only data sets with more than 10 data points were included in the analysis.

Analysis of Data. Data were analysed as given. Concentrations reported as non-detects (ND, below the limits of detection [LOD] of the analytical method) were assigned a valueof zero for the purposes of the distributional analysis. This procedure differs from the common practise of assigning a value of ½ LOD to the non-detects. It is highly unlikely that all of the NDs are exactly equal to half the LOD. The probabilistic procedure therefore assumes that the NDs are distributed from the LOD to zero, in a continuation of the distribution of the detected concentrations, a much more realistic postulate.

The data were analysed as instantaneous measurements rather than 4-day time weighted mean concentrations (*1*). Atrazine and other triazines have been shown to have a reversible mechanism of action and algae and other aquatic plants will recover from short-term exposures. The data sets were relatively small and, for those collected in Canada, samples were only taken during the growing season. As the crop growing season is also the most productive period in the rivers and streams of these environments, the use of seasonally biassed sampling data is not inappropriate.

The exposure concentration data points in this analysis were assumed to be log-normally distributed. This distribution has been observed in other situations (*1*) and, where deviations from log-normality occurred in our data, regressions based on the log-normal assumption were conservative. For all these data sets, distributions of measured data were plotted directly using a ranked set of data. Percentages were calculated from the formula 100 x n/(N+1) [from (*4*) and (*12*)]. Data were plotted using a log-Pearson Type III distribution (*13*) with the aid of SigmaPlot graphics package (*14*). A linear regression analysis was then performed on the data sets with the aid of SigmaPlot graphics package (*14*).

Characterizing the Toxicity Data

The LC/EC50 and LC/EC5 toxicity data for the assessment of atrazine concentrations were taken from (*1*). Toxicity data for simazine and cyanazine in aquatic organisms were obtained from the EPA one-liner database (*15*). Only studies judged to be core (C) or supplemental (S) were selected from the database. Toxicological endpoints were further selected to include mortality (LC50) and morbidity (EC50). EC50 data were included in the risk assessment to account for responses in organisms in which mortality was difficult to evaluate. Although instantaneous concentration data were used to characterize exposure, 48-96 hour acute toxicity studies (LC50 and EC50) were used in the characterization of sensitivity. The exposure duration in this commonly used acute toxicity test provided an additional degree of conservatism.

Analysis of Toxicity Data. Sensitivity concentrations were analysed using the same methodology as the exposure data. When assessing data of this type, it is frequently necessary to consider multiple tests conducted on the same species. In these cases, only those reporting the lowest effect concentration were included in the study. This approach is also conservative. All data sets had relatively few data points (less than 100), and distributions were plotted directly using a ranked set of data on the assumption that the data represented the universe of measurements. Rank percentages were calculated from the formula [100 x n/(N+1)].

Risk Characterization of Pesticide Residues Found in Surface Waters

The method utilized in this risk assessment was outlined by (*4*) and has been used by other authors (*1, 2*). They have suggested that the 10th centile of the sensitivity distribution can be used as an assessment criterion for risk characterization of pesticides. The overlap of the exposure distribution with the 10th centile of the sensitivity distribution can be used to estimate the likelihood that the exposure concentrations will exceed this assessment criterion (*1*). Because the probabilistic approach is purely numerical, it cannot consider the ecological importance of the potentially affected organisms. This must be judged on the basis of many other factors, including the results of higher tiers of risk assessment where ecological relevance is characterized. As discussed in (*1*), and by other authors in this volume, the aquatic organisms most sensitive to the triazine herbicides are algae, plants and macrophytes. Impact assessment studies in microcosms have demonstrated both functional redundancy and recovery from longer exposures to higher concentrations of these substances than were encountered in our study (*1*), further supporting the use of this assessment criterion. However, estimates of exceedences can be used to rank sites, locations and scenarios on the basis of the probability of adverse effects (*1*). This is useful for risk prioritization, as input for higher tiers of risk assessment, and subsequent risk management decisions (*6*).

Results and Discussion. The results of this assessment are presented in Figures 2 and 3 and, in summary form, in Table I.

Atrazine. The previous risk characterization of atrazine in surface waters of North America (*1*) presented assessments based on data from a number of locations in Canada and the US Midwest. Our results focussed on areas not assessed previously. Analyses from Lake Erie and the Niagara River (Figures 2-A to 2-C, Table I) showed that the probability of exposure concentrations exceeding the 10th centile assessment criteria of both the EPA LC/EC5s and the LC/EC50s of all organisms (*1*) was very low (<0.1%) in all cases, confirming the results of the tier-one hazard quotient assessment. These results would be expected as the dilution factors in large bodies of water, such as the Great Lakes, would substantially reduce higher concentrations resulting from runoff and other sources in rivers flowing into these systems. Similar results were observed in the large rivers of the US Midwest (*1*). However, some sites in Ontario and Quebec (Figure 2-C, Table I) showed concentrations that exceeded the 10th centile of the susceptibility distribution for the EPA LC5s. These sites are confined to smaller tributaries, rivers and streams from areas with high atrazine use and, once again, are similar to observations in the US Midwest (*1*). Thus,

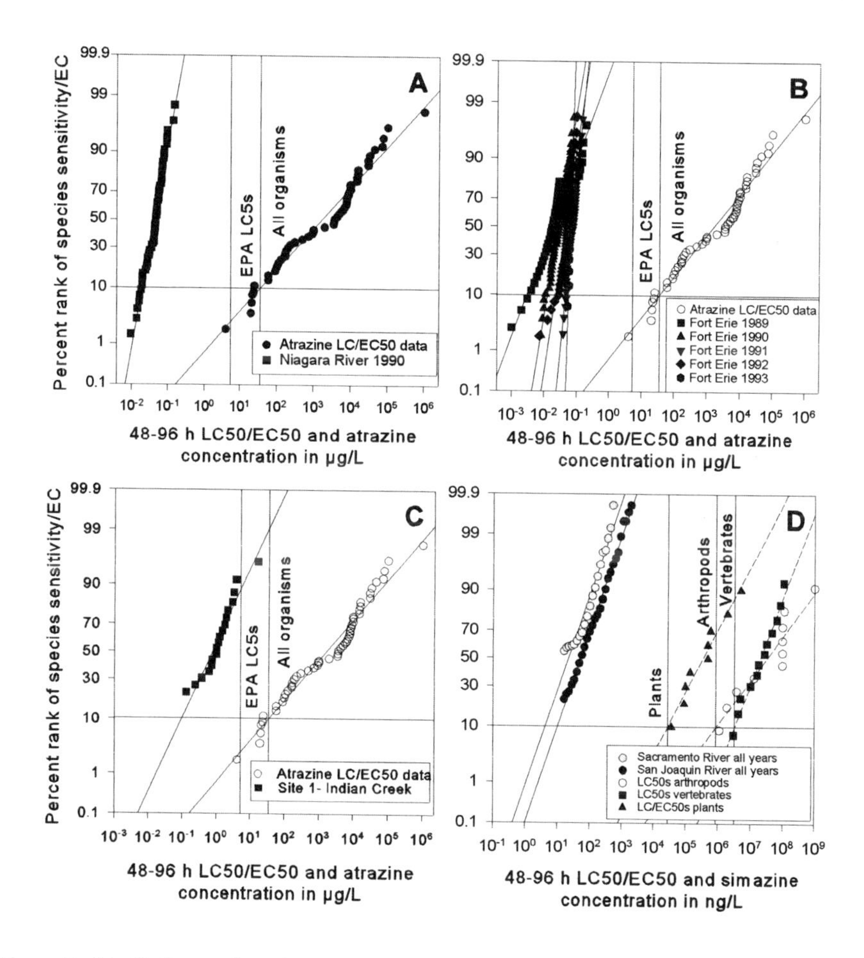

Figure 2. (A-C) Comparison between the distributions of atrazine concentrations and the acute sensitivity to atrazine in the Niagara River, Lake Erie at Fort Erie and in Indian Creek, ON. The vertical lines show the intercept of the 10th centile of the acute LC/EC5 (from *1*) and LC/EC50 distributions. (D) Comparison between the distribution of acute sensitivity to simazine and concentrations in the Sacramento and San Joaquin Rivers, CA. The vertical lines show the intercept of the 10th centile of the acute LC/EC50 distributions for plants, arthropods and vertebrates.

for the large rivers and lakes assessed in this paper, the ecological risks from atrazine concentrations in surface water were minimal. However, for some sites, concentrations may exceed the 10th entile of the sensitivity distribution for the EPA LC5s data — mostly from algae, the most sensitive group. In these cases, site-specific risk assessments should be conducted to assess possible ecological effects in the context of the uses to which these ecosystems are put and the effectiveness and cost-benefits of any risk mitigation measures that may be applied.

Simazine. Data for simazine concentrations from the Sacramento and San Joaquin Rivers were analysed in this paper. The data set for simazine in the EPA onc-liner database *15*) was large (Table I, Figure 2 D) and could be separated into groups of organisms vertebrates, arthropods and plants). As expected, plants (algae) were more sensitive than vertebrates or arthropods. The number of observations of concentrations in the Sacramento and San Joaquin Rivers (Table I) was large and represented samples taken over 3-year period. The highest concentration observed was lower than the lowest sensitivity reported and a tier-one risk assessment would have suggested low hazard. The likelihood of any of the exposure concentrations exceeding the 10th centile of the sensitivity distribution of LC/EC50s for all organisms was less than 0.1% Table I). When the distributions for vertebrates, arthropods and plants (including algae) were separated, the likelihood of any exposure concentrations exceeding the 10th centiles of any of the sensitivity distributions was again less than 0.1% (Figure 2 D, Table I). Thus, the likelihood of either direct or indirect effects of simazine on aquatic organisms in the Sacramento and San Joaquin Rivers was judged to be extremely small.

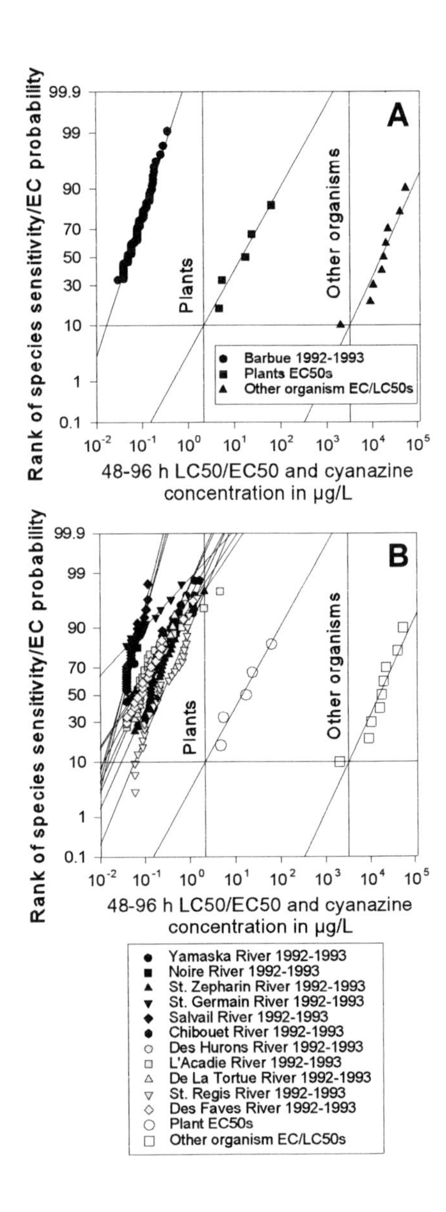

Figure 3. Comparison between the distributions of cyanazine concentrations and the acute sensitivity to cyanazine in a number of sites in Quebec. The vertical lines show the intercept of the 10th centile of the acute LC/EC50 distributions for plants and other organisms.

Cyanazine. Data for cyanazine concentrations from the rivers and streams of South Eastern Quebec were analysed in this paper. The susceptibility data set for cyanazine was not as large as that for simazine (Table I) but, for the purposes of the analysis, the data were separated into plants and other organisms (fish and arthropods). As expected, plants were more sensitive than other organisms. The likelihood of any exposure concentration exceeding the 10th centile of the sensitivity distribution for fish and arthropods at any of the sites in Quebec was less than 0.1%. Exposure data from some sites (Figure 3-A to 3-C, Table I) showed a likelihood that the 10th centile for plant sensitivity distribution would be exceeded, however, in no case, was this greater than 2.5%. These data do not suggest that an ecologically significant risk exists. However, for some sites, this assessment is based on small sample sizes and more data should be gathered and assessed to confirm these observations.

Table I. Regression coefficients and intercepts for the toxicity data distributions for acute exposures of aquatic organisms to various pesticides and concentrations of these in surface waters

Data source	*y = ax + b*				*Regression intercepts (μg/L)*		*Probability of exceeding the 10th percentile of the toxicity distribution*	
	a	*b*	r^2	*N =*	*10%*	*90%*		
Atrazine Fort Erie ON - Large volume samples								
Atrazine LC/EC50 data	0.77	2.51	0.97	52	36			
Atrazine EPA LC5s	1.33	2.74	0.75	8	5.4		All organisms	EPA LC5s
Fort Erie 1989	1.66	7.92	0.94	34		0.1	<0.1	<0.1
Fort Erie 1990	3.84	11.02	0.99	48		0.06	<0.1	<0.1
Fort Erie 1991	6.43	12.3	0.98	45		0.12	<0.1	<0.1
Fort Erie 1992	4.21	10.65	0.78	50		0.09	<0.1	<0.1
Fort Erie 1993	23.02	32.6	0.96	14		0.07	<0.1	<0.1
Atrazine Niagara-on-the-Lake ON - Large volume samples								
N-on-the-L - 1989	4.26	8.96	0.88	30		0.23	<0.1	<0.1
N-on-the-L - 1990	4.19	11.42	0.98	36		0.06	<0.1	<0.1
N-on-the-L - 1991	6.95	12.96	0.94	46		0.11	<0.1	<0.1
N-on-the-L - 1992	9.09	16.73	0.96	51		0.07	<0.1	<0.1
N-on-the-L - 1993	5.62	12.19	0.79	13		0.09	<0.1	<0.1
Atrazine from Rondeau Bay								
Indian Creek June 89-Aug 1990	1.42	5.16	0.93	24		6.16	0.9	11.6
Indian Creek May - Aug, 1990	2.24	5.08	0.96	17		3.44	<0.1	4.3
Atrazine from Quebec PQ								
Yamaska	2.06	5.82	0.97	10		1.7	<0.1	1
Noire	1.72	5.96	0.96	10		1.5	<0.1	1.4
Noire Tem	1.17	6.96	0.93	10		0.3	<0.1	0.3
Blanche	1.29	5.6	0.97	10		3.3	0.5	6.1
St. Zepharin	2.22	4.81	0.98	40		4.6	<0.1	7.6
St. Germaine	1.6	5.65	0.99	57		2.5	<0.1	3.5
Salvail	1.79	5.2	0.98	57		4	0.2	6.6
Chibouet	1.8	4.85	0.96	68		6.3	0.5	12.2
Des Hurons	1.92	5.16	0.99	68		3.8	<0.1	5.9
L'Acadie	2.36	4.31	0.97	40		6.9	0.2	15
De La Tortue	1.85	4.72	0.96	30		7	0.5	14.3
St. Regis	2.21	4.64	0.94	30		5.5	0.2	10.4

Table I. Regression coefficients and intercepts for the toxicity data distributions for acute exposures of aquatic organisms to various pesticides and concentrations of these in surface waters

Data source	*y = ax + b*				*Regression intercepts (µg/L)*		*Probability of exceeding the 10th percentile of the toxicity distribution*			
	a	*b*	*r^2*	*N =*	*10%*	*90%*				
Des Faves	1.67	4.85	0.98	26		7.2	0.8	14.3		
Barbue	2.05	5.25	0.97	108		3.2	<0.1	4		
Cyanazine from Quebec PQ										
Cyanazine EC/LC50s Plants	1.55	3.18	0.94	5	2.2					
Cyanazine EC/LC50s Other organisms	1.84	-2.74	0.9	9	3171		Plants	Other organisms		
Yamaska	3.29	9.77	0.83	10		0.1	<0.1	<0.1		
Noire	3	9.51	0.94	10		0.1	<0.1	<0.1		
St. Zepharin	1.89	6.61	0.97	40		0.7	1.3	<0.1		
St. Germaine	0.92	7.22	0.98	57		0.1	0.6	<0.1		
Salvaile	3.31	9.98	0.93	57		0.1	<0.1	<0.1		
Chibouet	1.33	6.67	0.98	68		0.5	1.7	<0.1		
Des Hurons	1.88	7.07	0.98	68		0.4	0.4	<0.1		
L'Acadie	1.29	6.54	0.89	40		0.6	2.5	<0.1		
De La Tortue	1.98	7.11	0.97	30		0.4	0.3	<0.1		
St. Regis	2.06	6.27	0.95	30		1	2.5	<0.1		
Des Faves	1.62	6.84	0.96	26		0.4	0.9	<0.1		
Barbue	2.61	8.36	0.98	108		0.2	<0.1	<0.1		
Simazine data from the Sacramento and San Joaquin Rivers CA										
Simazine LC50s (EPA oneliner)	0.81	-0.56	0.97	31	188051					
Simazine arthropod LC50s (EPA oneliner)	0.82	-1.17	0.89	10	885157					
Simazine vertebrate LC50s (EPA oneliner)	1.61	-6.77	0.98	12	3229331		All organisms	Vertebrates	Arthropods	Plants
Simazine plant EC50s (EPA oneliner)	1.17	-1.47	0.96	9	27580					
Simazine in the Sacramento River (Oct 91-Apr 94)	1.79	2.59	0.94	431		115	<0.1	<0.1	<0.1	<0.1
Simazine in the San Joaquin (Jan 91-Oct 94)	1.8	1.9	0.99	639		272	<0.1	<0.1	<0.1	<0.1

Ecological Relevance

The risks from the residues of triazine herbicides in aquatic environments must be qualified by our knowledge of the mode of action of the triazines and field observations of their effects. The triazines all act as herbicides by blocking photosynthesis (*16*), however, this process is reversible and the effects will be transient if exposure is short (*1*). The probabilistic risk assessment used in this paper is conservative because it was based on the responses of organisms exposed to a continuous, maximum concentration and did not consider recovery. Because exposures to triazines in the environment are driven by transient rainfall events, exposures will be short and recovery will be possible in most cases.

As pointed out (*1*), resiliency has been demonstrated in the function of phytoplankton exposed to atrazine in ponds and microcosms. Triazine-tolerant or resistant species will likely be able to maintain levels of primary productivity, even in the presence of residues of these herbicides. Field experiments have shown that laboratory toxicity tests tend to overestimate the potential response of aquatic plant communities to triazines (*1*). Through mechanisms involving physiological adaptation and species shifts, plant communities appear to compensate for the effects of atrazine and this reduces impacts on productivity and biomass (*17*). In addition, confounding stressors, such as sediment loading, that accompany greater triazine concentrations associated with runoff events can also inhibit photosynthesis (*1*) and, in fact, may protect chlorophyll from triazine-induced damage.

Acknowledgments

The authors wish to thank David Berryman of Environment Quebec for access to their analytical data. Use of this data does not necessarily imply the agreement of Environment Quebec with the conclusions. The authors also wish thank John Struger and Scott Painter of Environment Canada for access to their analytical data. This work was supported by grants from Ciba Crop Protection (Canada and US) and Environment Canada.

Literature Cited

1. Solomon, K.R., D.B. Baker, P. Richards, K.R. Dixon, S.J. Klaine, T.W. La Point, R.J. Kendall, J.M. Giddings, J.P. Giesy, L.W. Hall, Jr., C.P. Weisskopf, and M Williams. *Env. Tox. Chem.* **1996**, *15*, 31-76.
2. Klaine, S. J., G.P. Cobb, R.L. Dickerson, K.R. Dixon, R.J. Kendall, E.E. Smith and K.R. Solomon. *Env. Tox. Chem.* **1996**, *15*, 21-30.
3. U.S. Environmental Protection Agency. *Framework for Ecological Risk Assessment.* 1992 Risk Assessment Forum, Washington, DC, EPA/630/R92/001. pp 41.
4. SETAC Foundation for Education, Report of the Aquatic Risk Assessment and Mitigation Dialogue Group for Pesticides. **1994,** SETAC, Pensacola, FL, 1994.
5. Canadian Water Quality Guidelines and updates. Task Force on Water Quality Guidelines of the Canadian Council of Resource and Environment Ministers. **1987** (as updated) Environment Canada, Ottawa ON Canada .
6. Solomon, K.R. *Risk Anal.* **1996,** *16*:627-633.
7. Suter II, G., with L.W. Barnthouse, S.M. Bartell, T. Mill, D. Mackay and S. Patterson. *Ecological risk assessment.* Lewis Publishers, Boca Raton, FL, 1993; pp 538 .
8. Health Council of the Netherlands. *Network,* **1993**, *6*(3)/7(1), 8-11
9. Carrington, C.D. *Human and Ecol. Risk Assess.* **1996**, *2*:62-78.
10. Tillman, D. *Ecology* **1996,** *77*:350-363.
11. MacCoy, D., K.L. Crepeau and K.M. Kuivila. *Dissolved pesticide data for the San Joachim River at Vernalis and the Sacramento River at Sacramento, California, 1991-1994.* U.S. Geological Survey, 25286, MS 517, Denver CO. Open-File Report 95-110 1995 pp i-28.
12. Parkhurst, B.R., W. Warren-Hicks, T. Etchison, J.B. Butcher, R.D. Cardwell and J. Volison. Methodology for Aquatic Ecological Risk Assessment. **1995** Final Report prepared for the Water Environment Research Foundation, Alexandria, VA. RP91-AER.

13. McBean, E.A. and F.A. Rovers. *Ground Water Monitoring Rev.* **1992**, *12*, 115-119.
14. SigmaPlot Scientific Graphing System. **1995,** Version 2 for Windows, Jandel Corporation, Inc., San Rafael, CA.
15. EPA. Environmental Effects Branch One-liner Toxdata Electronic File (1991-1995). Personal communication, Brian Montague, US Environmental Protection Agency, 401 M Street, Washington, DC, 20460
16. WSSA. *Herbicide Handbook of the Weed Society of America.* Weed Science Society of America, Seventh Edition - **1994,** Champaign, IL. 323 pp.
17. Herman, D., N.K. Kaushik and K.R. Solomon. *Can. J. Fish. Aquat. Sci.* **1986**, *43*, 1917-1925.

Mammalian Toxicology and Human Risk Assessment

Chapter 29

Toxicity Characteristics of the 2-Chlorotriazines Atrazine and Simazine

J. W. Hauswirth[1] and L. T. Wetzel[2]

[1]Jellinek, Schwartz and Connolly, Inc., 525 Wilson Boulevard, Suite 600, Arlington, VA 22209
[2]Novartis Crop Protection, Inc., P.O. Box 18300, Greensboro, NC 27419

Atrazine and simazine are chlorotriazine herbicides used broadly in agriculture to control annual grasses and broadleaf weeds. An extensive database on the toxicity of these triazines has been developed to support the safety of their use in agriculture. Atrazine and simazine have very low acute toxicity, with oral LD_{50}s of >3000 mg/kg in rats. The results of a total of 38 mutagenicity studies on atrazine and 35 studies on simazine were included in a weight-of-the-evidence evaluation of the mutagenicity data leading to the conclusion that neither triazine possesses genotoxic activity. Oncogenicity studies in three strains of mice for both atrazine and simazine are negative. Neither triazine is oncogenic to male Sprague-Dawley (SD) rats nor is atrazine oncogenic to male and female Fischer 344 rats. However, in female SD rats both chlorotriazines induce the early occurrence and/or increased incidence of mammary gland tumors. Results of additional studies suggest that endocrinologic changes related to triazine administration are likely responsible for the mammary gland effects in female SD rats, and that a threshold dose exists for these findings.

Atrazine (2-chloro-4-ethylamino-6-isopropylamino-s-triazine) and simazine (2-chloro-4,6-ethylamino-s-triazine) are members of a group of s-triazine herbicides used in agriculture to control annual grasses and broadleaf weeds. Atrazine is the most broadly used s-triazine herbicide and its major crop uses are corn, sorghum and sugarcane. The major crop uses of simazine are corn, citrus, grapes, apples and several other fruit and nut crops, as well as non-food uses such as controlling weeds in nurseries. Both s-triazines exert herbicidal activity through inhibition of photosynthesis by preventing electron transfer at the reducing site of the photosynthesis complex II in the chloroplasts. The structures of atrazine and simazine are shown in Figure 1.

The metabolism, general toxicity, mutagenicity, and oncogenicity (carcinogenicity) of atrazine and simazine are discussed herein. In addition, a proposed hypothesis for a mode of action for mammary tumor induction based on results of mechanistic studies is presented.

Metabolism and Metabolite Toxicity

Both atrazine and simazine are metabolized in plants and animals via dealkylation and conjugation to glutathione and cysteine (Figure 2).

They are further metabolized to mercapturates, sulfides, disulfides, and aminotriazines in animals and to N-cysteine conjugates, lanthionine conjugates, aminotriazines, and hydroxytriazines in plants. In plants, the major metabolite of

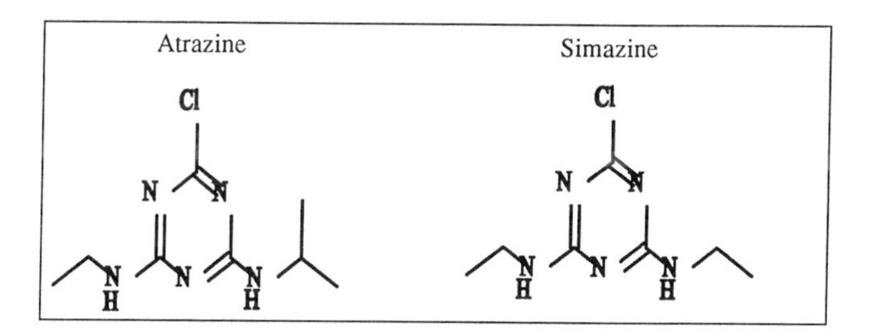

Figure 1. Structure of Atrazine and Simazine

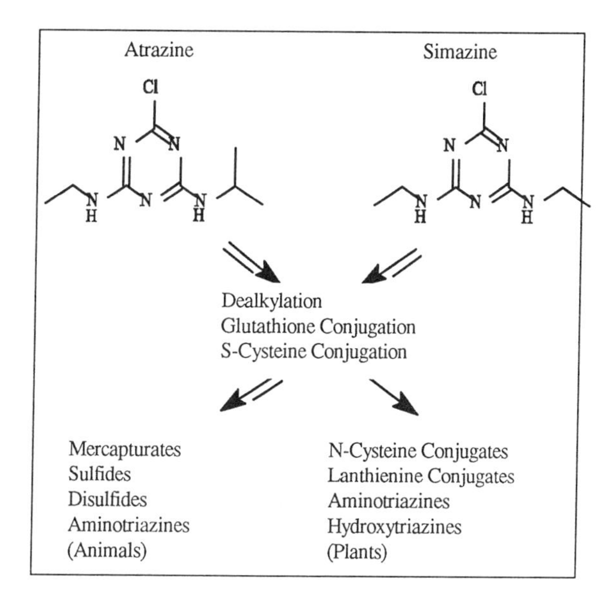

Figure 2. Atrazine and Simazine Metabolism

atrazine is 2-hydroxyatrazine, which does not apear to be present in animals. Both s-triazines are rapidly absorbed, metabolized, and eliminated in the rat (*1-5*).

Toxicity testing has been conducted separately on three dealkylated metabolites of atrazine (diaminochloro-s-triazine, and deethylated and deisopropylated atrazine). Two of these metabolites are common to simazine. Results indicate that the metabolites tested are not more toxic than atrazine itself, and they are not mutagenic or teratogenic. These and the other animal metabolites of atrazine and simazine have been tested through autoexposure (through in vivo metabolism of the test material) in experimental animals—especially the rat, for which extensive metabolism information is available.

The major atrazine plant metabolite, 2-hydroxyatrazine, tested negative for mutagenicity and teratogenicity. A two-year oncogenicity in the Sprague-Dawley (SD) rat at feeding levels up to 400 ppm showed no oncogenic effects in males or females (*6*) Administration of this metabolite was not associated with an increased incidence of any tumor type (benign or malignant).

General Atrazine and Simazine Toxicity

Both atrazine and simazine have been extensively tested for acute, chronic, developmental, and reproductive toxicity, and for potential mutagenic effects. The results of these studies are summarized in Table I and Table II.

Atrazine and simazine have also been tested for developmental toxicity, reproductive toxicity and chronic toxicity in several species and for oncogenic effects in rats and mice; the results of these studies are discussed in the following section. Neither is acutely toxic by the oral route of exposure in rats with LD_{50}s of >3000 mg/kg. The results of these toxicity studies, judged either core minimum or core guideline for acceptability by the USEPA, are discussed separately below.

Developmental and Reproductive Toxicity. Developmental or teratology studies have been conducted for atrazine and simazine in both rats and rabbits. In each case, the pregnant animal was administered the test material during the period of organogenesis (days 6 though 15 of gestation for rats, and days 6 though 19 for

Table I. Hazard Assessment Summary: Atrazine and Simazine

Acute Toxicity:	Low Order (e.g., Oral LD_{50} > 3000 mg/kg)
Reproduction:	Not Reproductive Toxins
	Not Developmental Toxins
Chronic Toxicity and Target Organs for Atrazine	
	Dog—Cardiotoxicity at 1500 ppm
	Mouse—Hematopoietic System Effects at 1500 ppm
	Rat—Mortality at 1000 ppm
	Body Weight Hematopoietic Effects at 500 ppm
Chronic Toxicity and Target Organs for Simazine	
	Dog—Hematology Effects, Body Weight Effects at 100 ppm
	Mouse—Body Weight Effects at 4000 ppm
	Rat—Mortality, Body Weight Effects at 1000 ppm

Table II. Mutagenicity: Atrazine and Simazine

Test	*Reported Response* Atrazine	Simazine
Mammalian Cells	In Vitro Studies	
Sister Chromatid Exchange		
Human Lymphocytes	1 Neg	
CHO Cells	1 Neg	
Hamster Kidney Cells		1 Neg
Chromosomal Aberrations		
Human Lymphocytes	1 Neg	1 Neg
CHO Cells	1 Neg	
CHL Cells		1 Neg
UDS/DNA Damage		
Rat Hepatocytes	1 Neg	3 Neg
Human Fibroblasts	1 Neg	1 Neg
WI-38 Cells		1 Neg
Unknown		3 Neg
Mouse Lymphoma	1 Neg	1 Neg/1 Pos
HGPRT in V79 Cells	1 Neg	
Hamster Nuclear Anomaly	1 Neg	1 Neg
Bacterial Cells & Other		
E. Coli Rec Assay	1 Neg	4 Neg
Ames Salmonella	4 Neg	7 Neg
Host Mediated		
Salmonella	2 Neg	
Yeast	1 Pos	2 Neg
E. Coli	1 Pos	
Mammalian	In Vivo Studies	
Mouse Micronucleus	1 Neg	2 Neg
Dominant Lethal—Mouse	1 Pos/3 Neg	
Chinese Hamster Bone Marrow Chromos Aberrations	1 Neg	
SCE Chinese Hamster	1 Neg	
Mouse Sperm Abnormalities	1 Neg	
Alkaline Elution—Rats	1 Pos	
Mouse Spot Test	1 Neg	
Mouse Bone Marrow Metaphase Analysis	1 Pos/3 Neg	
Chromosomal Aberrations Mouse Germ Cells	2 Neg	
Other		
Drosophila		
Dominant Lethal		2 Pos
Sex-Linked Recessive Lethal	1 Pos/2 Neg	2 Pos
Chromosome Loss/Nondisjunction		2 Neg

rabbits). Fetuses were examined for potential effects on development and for malformations.

In two-generation reproduction studies both the male and female experimental animal (usually rat) were treated with atrazine or simazine for at least 10 weeks before mating to produce the first generation. Females were treated throughout gestation and lactation. Offspring were, therefore, exposed to the test material through the mother's milk and through the diet when they begin feeding. The offspring from the first mating were allowed to mature and mate, producing the second generation of offspring. Reproductive parameters, such as fertility, live births, and post natal mortality were evaluated. Offspring were evaluated for survival, body weight, and external malformations and developmental delays until weaning.

Atrazine. Atrazine was tested for developmental toxicity in New Zealand White rabbits at dose levels of 0, 1, 5, and 75 mg/kg/day (*7*). Effects related to treatment were only seen in the 75 mg/kg/day dose group. In the pregnant animal, administration of atrazine resulted in reduced body weight gain and reduced food consumption. In the fetuses, reduced body weights and reduced ossification of the skeleton were observed. No malformations were seen in the fetuses; therefore, it was concluded that atrazine was not teratogenic in the rabbit. The no-observed effect level (NOEL) was 5 mg/kg/day.

Dose levels of 0, 5, 25, and 100 mg/kg/day were tested for developmental toxicity in SD rats (*8*). Effects seen in this study were similar to those seen in the rabbit. Maternal body weight gain and food consumption were reduced at 100 mg/kg/day, only. Fetal effects consisted solely of reduced ossification at 100 mg/kg/day. Atrazine was not teratogenic in the rat and the NOEL was considered to be 25 mg/kg/day.

In the atrazine two-generation reproduction study, rats were administered the compound in the diet at dose levels of 0, 10, 50, and 500 ppm (approximately 0.5, 5, and 50 mg/kg/day) (*9*). Decreased body weight gain was seen in both parental generations at 500 ppm. The only effect seen in the pups was decreased body weight at weaning and it was seen only in the second generation at 500 ppm. This effect can be attributed to the pups being exposed to dietary atrazine as they begin to be weaned and therefore is not considered a reproductive effect. Therefore, it was concluded that atrazine posed no reproductive hazard. The NOEL was at least 50 ppm or approximately 5 mg/kg/day.

Simazine. In a rabbit (New Zealand White) developmental toxicity study, simazine administered by gavage at dose levels of 0, 5, 75, or 200 mg/kg/day (*10*). Severe maternal toxicity was seen in this study at 75 and 200 mg/kg/day.. At these same dose levels, the embryo and/or fetal toxicity observed were considered to be secondary to the severe maternal toxicity. Simazine was not teratogenic to the rabbit and the NOEL was at least 5 mg/kg/day.

Pregnant Crl:COBS CD(SD)BR rats were administered a simazine suspension by gavage at dose levels of 0, 30, 300, and 600 mg/kg/day (*11*). At 300 and 600 mg/kg/day, reduced body weight, and in some cases body weight loss, was observed in the dams. Developmental toxicity, seen at these same dose levels were considered to be secondary to the maternal toxicity seen at those dose levels. No treatment-related malformations were seen. Simazine was not teratogenic to the rat and the NOEL was at least 30 mg/kg/day.

A rat [Crl:COBS CD(SD)BR] reproduction study was conducted with simazine at dietary dose levels of 0, 10, 100, and 500 ppm (approximately 0.5, 5, and 50 mg/kg/day) (*12*). Administration of simazine did not cause any reproductive toxicity. Toxicity was seen in the parental generations, which consisted of decreased body weight gain and decreased food consumption at the two high doses tested. Simazine was not a reproductive toxicant and the NOEL was at least 10 ppm (~0.5 mg/kg/day).

Chronic Toxicity. The chronic toxicity of atrazine and simazine was tested in three different test species: rat, mouse, and dog.

Atrazine. Atrazine was tested for chronic toxicity in a 2-year rat (SD) study at feeding levels of 0, 10, 70, 500, and 1000 ppm (*13*). These levels correspond to approximately 0, 0.5, 3.5, 25, and 50 mg/kg/day. Survival was significantly reduced at 1000 ppm in females. Body weight gain was adversely affected in males at 1000 ppm and females at 500 and 1000 ppm. Histopathologic lesions were seen in both males (1000 ppm) and females (500 and 1000 ppm). These consisted of bone marrow myeloid hyperplasia, femoris muscle degeneration, retinal degeneration, liver necrosis, epithelial prostate hyperplasia, and kidney pelvic calculi and microcalculi. No treatment-related effects were seen at 0.5 and 3.5 mg/kg/day.

In an 18-month study, atrazine was administered in the diet to CD-1 mice at dose levels of 0, 10, 300, 1500, and 3000 ppm (*14*). These dose levels correspond to approximately 0, 15, 45, 225, and 450 mg/kg/day. Treatment-related effects seen included decreased body weight gain (males and females at 1500 and 3000 ppm), decreased survival (females at 3000 ppm), and histological effects on the bone marrow (1500 and 3000 ppm). No treatment-related effects were seen at 300 ppm which was the NOEL for the study.

Atrazine was tested in Beagle dogs for a period of one year (*15*). The feeding levels tested were 0, 15, 150, or 1500 ppm (approximately 0, 0.5, 5.0, and 34 mg/kg/day. Cardiotoxicity (electrocardiogram alterations and cardiac lesions) was seen at 34 mg/kg/day. No effects related to treatment were seen at 0.5 and 5.0 mg/kg/day.

Simazine. A 2-year chronic feeding study was conducted in CrL CD(SD)BR rats at dose levels of 0, 10, 100, and 1000 ppm simazine (*16*). These dose levels correspond to approximately 0, 0.5, 5.0, and 50 mg/kg/day. Mortality was increased during the course of the study compared to controls at 100 and 1000 ppm in female rats only. Body weight gain was decreased in male and female rats at 1000 ppm and in females at 100 ppm. Food consumption was decreased at 1000 ppm for most of the study. Treatment-related reductions in RBC count, hematocrit, and mean hemoglobin levels were observed in females at feeding levels of 100 ppm and higher with concomitant increases in mean platelet counts. No non-neoplastic histopathologic lesions were observed. No effects related to treatment were seen at 0.5 mg/kg/day.

Groups of Crl:CD-1(CR)BR mice received simazine in their diet at concentrations of 0, 40, 1000, and 4000 ppm corresponding to approximately 0, 5.3, 131.5, and 543.8 mg/kg/day for males and 0, 6.2, 160.0, and 652.1 mg/kg/day for females (*17*). The only significant treatment-related effects seen in this study were reduced body weight gain for mid- and high-dose male and female mice, and in food consumption for mid-dose males and high-dose males and females. The NOEL for the study was 40 ppm (5.3 mg/kg/day).

Groups of Beagle dogs were administered simazine in the diet for one year (*18*). The dose levels tested were 0, 20, 100, and 1250 ppm corresponding to approximately 0, 0.5, 2.5 or 30 mg/kg/day. Reduction in body weight gain was recorded in the high dose males and mid- and high-dose females. High-dose males lost weight during the first two weeks of the study, but cumulative body weight gain after one year was comparable to controls. Slight decreases in hematology parameters were seen in females at these dose levels. No effects of treatment were seen at 20 ppm or approximately 0.5 mg/kg/day.

Mutagenicity. Mutagenicity or genotoxicity is the ability of a chemical to cause DNA damage and thereby cause DNA mutations. These mutations can lead to heritable genetic defects and/or cancer.The mutagenicity data bases on both atrazine and simazine are large and are briefly summarized in Table II. The EPA requirements for the battery of mutagenicity tests required for pesticide registration have been fulfilled for both chemicals. All EPA acceptable studies have been judged to be negative.

Atrazine. Information on the genotoxicity of atrazine has been published in several reviews (*19-21*). This herbicide has been extensively studied in a wide variety of test systems ranging from microbial assays to tests measuring genotoxicity in both somatic (undifferentiated) and germ (immature or mature ovum and sperm) cells of animals (as shown in Table II). An overwhelming majority of the tests conducted (32 negative responses out of 38 tests) have shown no evidence that atrazine damages DNA; however, sporadic positive/equivocal responses were reported in various non-guideline tests (i.e., tests not performed under Good Laboratory Practice (GLP) regulations and specific EPA guidelines) (mouse dominant lethal, mouse bone marrow metaphase analysis, and Drosophila sex-linked recessive lethal). However, the majority of similar or identical tests consistently demonstrated convincing negative results for atrazine at similar exposure levels. The 38 studies and results are listed in Table II. A weight-of-the-evidence (WOE) approach to evaluate the results of these studies indicates that atrazine does not possess genotoxic activity (*22*).

Simazine. Simazine has been subjected to a battery of genotoxicity tests similar to that of atrazine (as shown in Table II). Out of a total of 35 tests, simazine was positive in only 5. In one case, the mouse lymphoma assay, both a positive and a negative response were found. The positive test was conducted before the time that proper controls for this assay were in place. The other four positive responses were seen in tests using the fruit fly, Drosophila melanogater. The results seen in this test species may have resulted from the unique metabolism and DNA repair processes in this organism, leading to a response not likely relevant to human risk considerations.

A WOE approach to evaluate the results of these studies indicates that simazine does not possess genotoxic activity (*22*).

Oncogenicity. Atrazine and simazine have been tested for their potential to induce tumors, either benign and malignant, in rats and mice in several different studies. These studies are briefly summarized in Table III for mice and Table IV for rats.

Atrazine. Atrazine has been tested for oncogenicity in three strains of mice in four separate studies (*14, 23, 24*). Dose levels of up to approximately 450 mg/kg/day (3000 ppm) were tested. Atrazine did not induce the incidence of benign or malignant tumors in any of these mouse studies (Table III).

Atrazine has also been tested for oncogenicity in two strains of rats (SD and Fischer 344; Table IV) (*13, 25-31*). Dose levels of up to 50 mg/kg/day (1000 ppm) were tested. All of the studies, except for the 1995 SD study (1 year), were lifetime studies (24 months). Atrazine did not increase the incidence of benign or malignant

Table III. Mouse Oncogenicity Studies: Atrazine and Simazine

Atrazine:	Negative for Oncogenicity in 4 Studies	**Atrazine:**
	• CD-1 Mouse (1987) up to 3000 ppm	*(14)*
	• CD-1 Mouse (1981) up to 1000 ppm	*(24)*
	• (C57BL/6XC3H/Anf)F_1 Mouse (1969) up to 603 ppm	*(23)*
	• (C57BL/6XAKR)F_1 Mouse (1969) up to 603 pm	*(23)*
Simazine:	Negative for Oncogenicity in 3 Studies	
	• CD-1 Mouse (1988) up to 400 ppm	
	• (C57BL/6XC3H/Anf)F_1 Mouse (1969) up to 82 ppm	
	• (C57BL/6XAKR)F_1 Mouse (1969) up to 82 ppm	

Table IV. Rat Oncogenicity Studies: Effect on Mammary Gland Tumor Incidence

		ATRAZINE	
		Feeding Level Tested (ppm)	
Study	*Results*	*No Increased Incidence*	*Increased Incidence*
Sprague-Dawley Rats			
1986	Positive	10	70, 500, 1000
1987	Negative	10, 50, 500	
1991	Positive*	70	400*
1995	Positive*	15, 30, 50, 70	400*
Fischer-344 Rats			
1991	Negative	10 to 400	
		SIMAZINE	
Sprague-Dawley Rats			
1988	Positive	10	100, 1000

*Early onset only

tumors in male SD rats or in male and female Fischer 344 rats. The results of the F-344 study are discussed in the following paper (*32*). In female SD rats, atrazine administration was associated with an increased incidence and/or early onset of mammary gland tumors. Tumors were seen primarily at dose levels at and above 400 ppm. Only in one study (1986) was an increased incidence of mammary gland tumors seen below 400 ppm and that was at 70 ppm (~3.5 mg/kg/day). In that study, the increase was not associated with a dose response because the incidence at 70 ppm was similar to that seen at 500 ppm. The relevance of this finding is questionable because it could not be reproduced in other studies (1991, 1992 and 1995).

Dose levels of 400 ppm and greater were considered to be excessively toxic or above a maximum tolerated dose (MTD) in this strain of rat. At these dose levels, increased mortality was seen in comparison to the control group in female rats. In addition, these animals did not gain weight as quickly as the control animals. Body weight gain was decreased by >10% in these animals over the course of the studies. The relevance of findings at such high dose levels must be carefully considered as to relevance to possible human exposure scenarios which are orders of magnitude less.

Simazine. Simazine has been tested for oncogenicity in three mouse studies and three strains of mice up to dose levels of 3000 ppm, which is equivalent to approximately 450 mg/kg/day (*17, 23*). Simazine did not induce the incidence of any tumor type—benign or malignant—in these studies (Table III).

An oncogenicity study in SD rats has been conducted on simazine at dose levels of 0, 10, 100, and 1000 ppm (*16, 30*). Those feeding levels correspond to 0 and approximately 0.5, 5 and 50 mg/kg/day. Simazine did not induce the incidence of any tumor type in male rats; however, in female rats the incidence of mammary gland tumors was increased at 100 and 1000 ppm and there was evidence of early onset of tumors. The highest dose level in this study exceeded an MTD in females based on decreased survival (34% in the control group versus 20% at 1000 ppm) and decreased body weight gain in excess of 30% compared to the control group.

Potential Mechanism for Mammary Gland Tumor Formation

Atrazine induces early onset of mammary gland tumors in SD, but not F-344, female rats. In Figures 3 and 4, the incidence with time of palpable mammary gland masses is plotted for both the 1991 and 1992 SD and Fischer F344 rat studies.

In the SD study, only early onset of mammary gland tumors was observed, not increased incidence. As can be seen from the cumulative incidence of palpable masses (confirmed as mammary gland tumors histologically), atrazine induced an early onset of these tumors at 400 but not at 70 ppm (Figure 3). Atrazine did not increase the incidence or decrease the latency time for mammary gland tumors in Fischer 344 rats (Figure 4).

Both SD and Fischer 344 female rats are responsive to genotoxic mammary gland carcinogens. Because the Fischer 344 female rat was not responsive to atrazine and the WOE indicates that atrazine is not genotoxic, atrazine is likely acting through a mechanism other than direct interaction with DNA. Any carcinogenic effect by a chemical that is not attributable to genotoxicity should have a threshold—a dose level

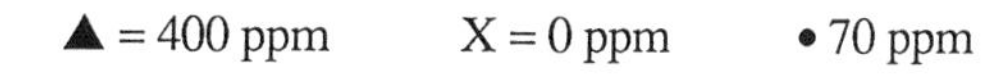

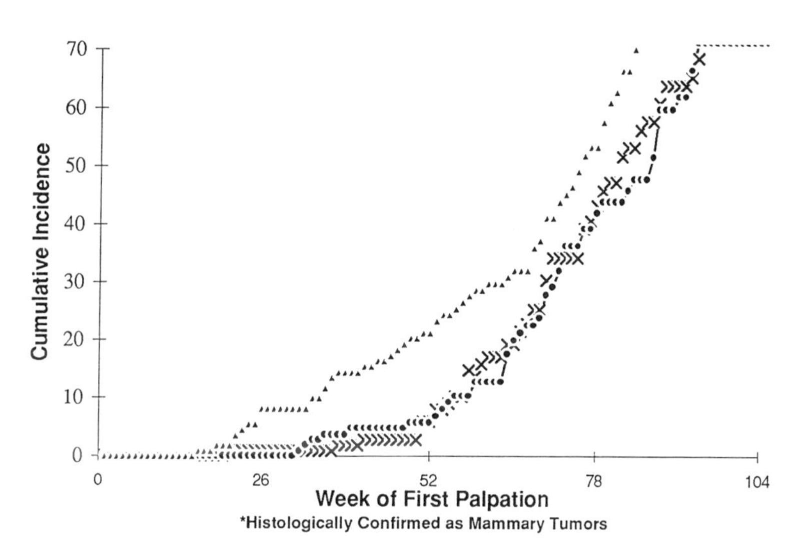

Figure 3. Cumulative Incidence of Palpable Masses* in Lifetime Female SD Rat Feeding Studies With Atrazine

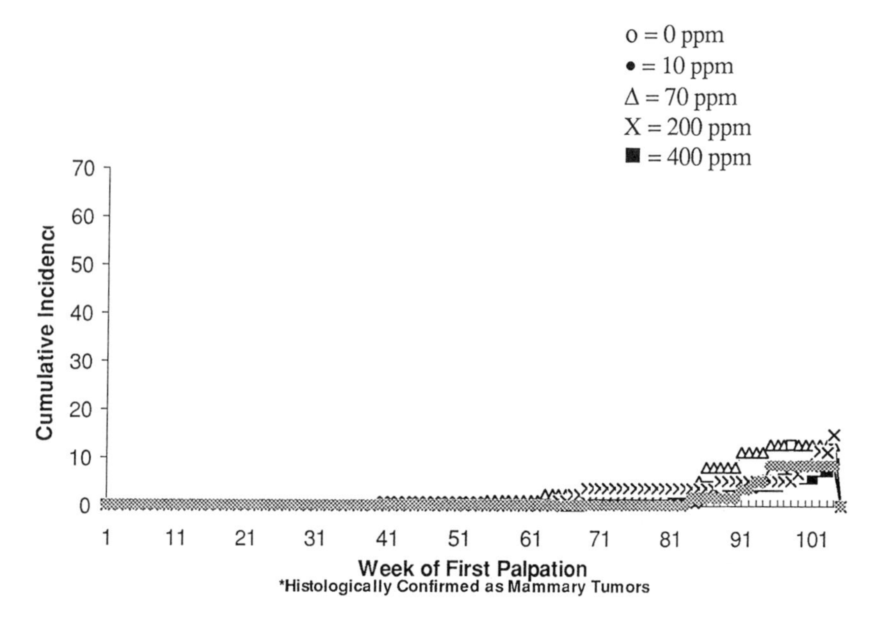

Figure 4. Cumulative Incidence of Palpable Masses* in Lifetime Female Fischer 344 Rat Feeding Studies With Atrazine

at which no tumors will be induced. The threshold for the induction of early onset of mammary gland tumors by atrazine in the 1991 and 1992 studies were 70 ppm and $\leq$ 70 ppm in the 1986 atrazine study.

Because atrazine is not genotoxic, another mode of action must be involved in inducing tumors and/or their earlier onset in SD female rats. Because mammary gland development and tumor formation in the SD rat is hormonally dependent, studying the effects of atrazine on the endocrine system of the SD female rat was appropriate. Atrazine and simazine appear to be acting in a similar manner on the mammary gland and, because more oncogenicity data are available for atrazine, attention was focused primarily on this chlorotriazine.

The first mechanistic study was conducted to determine whether atrazine caused a disruption of reproductive cyclicity in female SD rats and/or Fischer 344 rats (33,*34*). Atrazine was administered in the diet to female SD and Fischer 344 rats at levels of 0, 70 and 400 ppm (SD) and 0, 10, 70 , 200 and 400 ppm (F-344) for two years. Effects of atrazine on various reproductive cycle parameters were studied. Atrazine increased the percent of days the 400-ppm treated SD rats spent in estrus, particularly during the first 12 months of the study (Figure 5). The 70 ppm SD rats were unaffected.

The percent of days spent in estrus was unaffected by atrazine in Fischer rats.

These results indicate that atrazine is affecting the normal estrous cycling pattern of SD, but not F-344, female rats (*31*). Likewise, the graphs of the estrous cycle data (Figure 5) and the mammary tumor onset data (Figure 3) in SD rats exposed to 400 ppm atrazine are comparable in shape. That is, the curve is shifted to the left of control by about 3 to 6 months for the treated animals for each parameter. In addition, the increase in the percent days spent in estrus precedes mammary tumor formation by 3 to 6 months, in either the control or treated animals. Thus, there appears to be a link between a change in normal cycling pattern and the onset of mammary tumors.

Additional experiments were designed to determine whether atrazine possesses estrogenic activity. Estrogen is a known mammary gland carcinogen in both SD and Fischer 344 female rats and in female mice. Logically it can be reasoned that atrazine (or simazine) could not be acting as an estrogen because if it were, it would have the potential to induce mammary gland tumors in both strains of rat or in the mouse. In addition, if either were estrogenic, teratogenic and/or reproductive effects would have been expected in tests for those effects. Additional studies have been conducted and were planned to test for the potential estrogenicity of atrazine and the results are discussed in more detail in following papers (35,*36*).

Summary and Conclusions

In summary the following conclusions can be drawn from the data/information presented on atrazine and simazine:

- both possess low acute toxicity;
- neither are reproductive or developmental toxicants, nor are they teratogenic;
- a WOE approach indicates that they are not mutagenic;

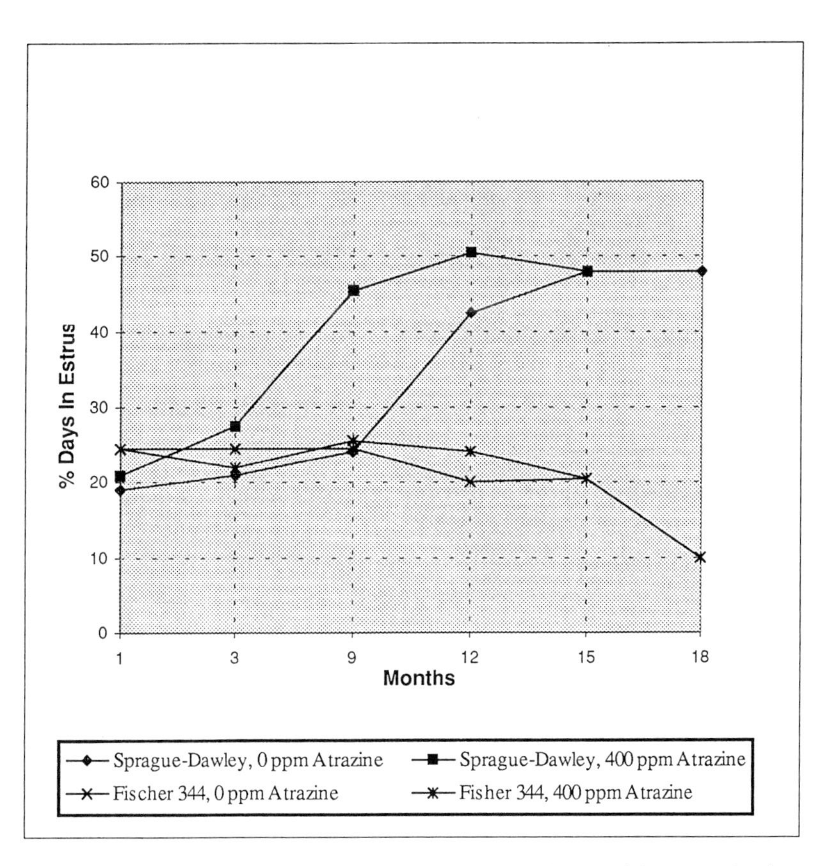

Figure 5. Effect of Feeding Atrazine on the Percent of Days of the Reproductive Cycle Spent in Estrus

- both induced an increased incidence and/or early onset of mammary gland tumors in female SD rats;
- neither are oncogenic in three strains of mice and in male SD rats; and
- results of reproductive tests indicate that atrazine (and simazine) are not estrogenic.

Atrazine specific studies:

- atrazine does not induce any tumor type in male or female Fischer 344 rats;
- exposure to high levels of atrazine is associated with estrous cycle disruption and mammary gland tumorigenesis in SD but not Fischer 344 rats; and
- mammary gland tumors are produced through a non-genotoxic mechanism which has a threshold ($\leq$70 ppm).

Studies have been and are being conducted to further define the mode of action of triazine-induced mammary gland tumorigenesis in female SD rats. Results to date indicate that the chlorotriazines induce mammary gland tumors in SD rats through a non-genotoxic and non-estrogenic mechanism which has a threshold. Therefore, exposure to atrazine and simazine should be regulated through a non-linear reference dose-based approach.

Literature Cited

1. Ikonen, R., Kangas, J., Savolainen, H. (1988). Toxicology Letters 44:109-112.
2. Orr, G. R. (1985). Ciba-Geigy Corporation, unpublished data. EPA acceptable.
3. Orr, G. R.; Simoneaux, B. J. (1986). Ciba-Geigy Corporation, unpublished data. EPA acceptable.
4. Miles, J.; Orr, G. R. (1987). Ciba-Geigy Corporation, unpublished data. EPA acceptable.
5. Thede, B. (1988). Ciba-Geigy Corporation, unpublished data. EPA acceptable.
6. Chow, E. (1995). Ciba-Geigy Corporation, unpublished data. EPA acceptable.
7. Infurna, R. N. (1984). Ciba-Geigy Pharmaceuticals SEF Project No. MIN 832110, unpublished data. EPA acceptable.
8. Giknis, M. L. A. (1989). Ciba-Geigy Pharmaceuticals SEF Project No. MIN 882049, unpublished data. EPA acceptable.
9. Maeniero, J.; Youreneff, M.; Giknis, M. L. A.; Yau, E. T. (1987). Ciba-Geigy Pharmaceuticals SEF Project No. MIN 852063, unpublished data. EPA acceptable.
10. Arthur, A. T. (1984). Ciba-Geigy Corporation, unpublished data. EPA acceptable.
11. Yau, E. T. (1986). Ciba-Geigy Corporation, unpublished data. EPA acceptable.
12. Epstein, D. L. (1991). Ciba-Geigy Corporation, unpublished data. EPA acceptable.

13. Mayhew, D. (1986). American Biogenics Corporation report, unpublished data. EPA acceptable.
14. Hazelette, J. R.; Green, J. D. (1987). Ciba-Geigy Corporation Study No. 842129, unpublished data. EPA acceptable.
15. Hazelette, J. R.; Green, J. D. (1987). Ciba-Geigy Pharmaceuticals SEF Project No. MIN 882049. EPA acceptable.
16. McCormick, G. C. (1988). Ciba-Geigy Corporation. EPA acceptable.
17. Hazelette, J. R. (1988). Ciba-Geigy Corporation. EPA acceptable.
18. McCormick, G. C. (1988). Ciba-Geigy Corporation. EPA acceptable.
19. Plewa, M. J.; Wagner, E. D.; Gentile, G. J.; Gentile, J. M. *Mutation Research.* **1984**, *136*, pp. 233-245.
20. Franekic, J.; Hulina, G.; Kniewald, J.; Alacevic, M. *Environ. Mol. Mutagen.* **1989**, *14*, pp. 62.
21. Brusick, D. J. *Mutation Research.* 1994, *317,* pp. 133-144.
22. Brusick, D. J.; Lohman, P. H. M.; Mendelsohn, M. L.; Moore, D.; Waters, M. D. *Mutation Research.* **1992**, *266,* pp. 1-6.
23. Innes, J. R. M.; Ulland, B. M.; Valerio, M. G.; Petrucelli, L.; Fishbein, L.; Hart, E. R.; Pallotta, A. J.; Bates, R. R.; Falk, H. L.; Gart, J. J.; Klein, M.; Mitchell, I.; Peters, J. *J. Natl. Cancer Inst.* **1969**, *4,* pp, 1101-1114.
24. Sumner, D. D. (1981). Industrial Biotest Laboratories No. 8580-8906. Report prepared by Ciba-Geigy Corporation.
25. Petterson, J.C.; Turnier, J.C. (1995). Ciba Crop Corporation.
26. Rudzki, M.W.; McCormick, G.C.; Arthur, A.T. (1990). Ciba-Geigy Pharmaceuticals Project No. MIN 852214.
27. Thakur, A.K. (1992). Ciba-Geigy Corporation Hazleton Washington Report 483-275. EPA Supplemental.
28. Thakur, A.K. (1992). Ciba-Geigy Corporation. Hazleton Washington Report 483-277, unpublished data. EPA acceptable.
29. Stevens, J. T.; Breckenridge, C. B.; Wetzel, L. T.; Gillis, J. H.; Luempert III, L. G.; Eldridge, J. C. *J. Toxicol. Environ. Health.* **1994**, *43,* pp. 139-153.
30. Wetzel, L. T.; Luempert III, L. G.; Breckenridge, C. B.; Tisdel, M. O.; Stevens, J. T.; Thakur, A. K.; Extrom, P. J.; Eldridge, J. C. *J. Toxicol. Environ. Health.* **1994**, *43*, pp. 169-182.
31. Thakur, A.K. et al., *ACS Symposium Series*, in press.
32. Thakur, A.K. (1991). Ciba-Geigy Corporation. Hazleton Washington Report 483-278.
33. Thakur, A.K. (1991). Ciba-Geigy Corporation. Hazleton Washington Report 483-279.
34. Eldridge, J.C., *ACS Symposium Series*, in press.
35. Simpkins, J.W., *ACS Symposium Series*, in press.

Chapter 30

Results of a Two-Year Oncogenicity Study in Fischer 344 Rats with Atrazine

Ajit K. Thakur[1], Lawrence T. Wetzel[2], Richard W. Voelker[1], and Amy E. Wakefield[1]

[1]Covance Laboratories, Inc., 9200 Leesburg Pike, Vienna, VA 22182
[2]Novartis Crop Protection, Inc., P.O. Box 18300 Greensboro, NC 27419

Atrazine is a widely used chlorotriazine herbicide. Several two-year dietary studies have indicated the potential for atrazine-induced-threshold-mediated-strain-specific-hormonally-invoked mammary tumorigenesis in female Sprague-Dawley (SD) rats. Male SD rats did not show any oncogenic response in these studies. Weight-of-evidence analysis indicates that atrazine is not genotoxic. The present studies with male and female Fischer 344 (F344) fed 0, 10, 70, 200, and 400 ppm of atrazine for 104 weeks were undertaken to compare the strain and sex specificity of the atrazine response. The results from the studies indicate a negative dose-response in mean body weight gain with significant depression at the 200 and 400 ppm levels. In contrast to the female SD rats, there was no indication of either earlier onset or increased number of mammary tumors in F344 rats. The toxicological no-adverse-effect level for atrazine from these results is 70 ppm. Furthermore, the lack of mammary oncogenic response in F344 females provides credence to the hypothesis of atrazine-induced-hormonally-mediated oncogenicity is specific to the SD strain.

Atrazine has been a major herbicide in the United States and other parts of the world for more than 35 years. It exerts its phytotoxic effects through inhibition of the Hill Reaction (*1*) in photosynthesis and correspondingly shows only low-level toxicity to organisms which do not perform photosynthesis.

Atrazine is not a mutagen (*2*), a teratogen, and it does not pose any reproductive hazard (*3*). The chemical is not oncogenic in the male Sprague-Dawley (SD) rat (*4*) and several strains of mice (*5-7*). In a previous study (*8*), there were indications of significant earlier onset (decreased latency) of mammary tumors (benign and malignant combined) in female Sprague-Dawley rats administered atrazine at 400 ppm dietary level for 104 weeks, although the number of tumor-bearing animals was not significantly increased over the control in that group.

In an earlier study (*4*), mean body weights and gains, as well as median survival rates, were significantly decreased at dietary levels of 500 and 1000 ppm in the SD rats indicating that the maximum tolerated dose (MTD) was exceeded at those levels. In that study, there were indications of earlier onset and/or increased frequency of mammary tumors at 70, 500, and 1000 ppm. There was no effect on individual or combined mammary tumor incidence rates or latency at the 10 ppm levels.

In a time-to-death (approximately 136 weeks in some animals) study in Fischer 344/LATI rats, it was reported (*9*) that atrazine caused an increased mammary tumor incidence rate in the high dose (initially 1000 ppm, reduced to 750 ppm after 8 weeks because of indications of high toxicity) males. The results of this particular study indicate that the outcome of atrazine treatment was compromised by effects on survival, i.e. the 750 ppm group males outlived the control and the 375 ppm group males by a significant amount of time.

The present paper reports the results from a 104-week dietary feeding study on male and female Fischer 344 (F344) rats with dose levels of 0, 10, 70, 200, and 400 ppm conducted under the U.S. Environmental Protection Agency (EPA) Good Laboratory Practice Guidelines (*10*). Additionally, there was a concurrent F344 study in females at the aforementioned dose levels with serial sacrifices for hormonal measurements as well as for detecting onset times for pituitary and mammary tumors (*11*).

Methods

Test Materials

Atrazine Technical (2-chloro-4-ethylamino-6-isopropylamino-s-triazine) was used in all studies. The compound was mixed with Purina Certified Rodent Chow #5002 (which also served as the control feed) for treated groups.

Experimental Animals

Charles River Fischer Crl:CDF (F344) male and female rats were randomly assigned to the various treatment groups after a three-week acclimation, and were housed individually throughout the course of the studies.

Experimental Design

Atrazine was fed to F344 rats at 10, 70, 200, and 400 ppm for a maximum of 104 weeks. The high level of 400 ppm and the 70 ppm level were selected based on dose range finding study as well as for comparison of results between the F344 and SD rats. Interval sacrifices were scheduled at 1, 3, 9, 12, 15, 18, and 24 months for 10 females/group (total of 70/group). In a second study run concurrently, 60 male and 60 female rats/group were maintained on the same diets for 104 weeks with no interval sacrifices.

The rats were examined twice daily for mortality, moribundity, and for indications of toxic effects. Palpation for tissue masses was performed prior to the initiation of dosing and weekly thereafter. Body weights were recorded prior to the initiation of dosing, and body weights and food consumption were recorded weekly for weeks 1-16 and every fourth week thereafter. Mean compound consumption was calculated for each body weight/food consumption interval.

Complete necropsies were performed on all animals. For the females sacrificed at the stated intervals, the pituitary, ovary, uterus, and mammary tissues were histologically examined. In the second study, all tissues from all the animals were examined.

In order to avoid bias in evaluation of the microscopic lesions, the slides were coded in regard to dose level prior to histopathological examination. The same pathologist read the slides from all studies to ensure consistency in evaluation.

Statistical Analyses

The body weight gains for the first year of the study (intervals of weeks 0-13, 14-28, and 29-52) for both sexes were evaluated by one-way analysis of variance (*12*) and Dunnett's control versus treatment mean comparison (*13-14*) techniques. A second analysis was performed for the females where the study number was used for blocked analysis of variance (*15*) to obtain overall estimates of treatment effect.

Survival and time-adjusted tumor incidences were analyzed for each study by techniques described in detail in an earlier publication (*16*).

For all of these incidence data, overall significance from both studies for female F344 rats was determined by combination of results over experiments (*17*).

Since the palpation times for the mammary tumors in the study by Pinter et al (*9*) were not available, they were analyzed by exact logistic regression for prevalence of survival adjusted incidences (*18*). Such an analysis is vital for meaningful comparison of tumor rates in the males because of extreme differential survival in the atrazine-treated groups compared to control.

Results

Mean Compound Consumption

Mean compound consumption for the studies are shown in Tables IA (males) and IB (females) for intervals 0-13 weeks, 14-28 weeks, 32-52 weeks, and 56-80 weeks. Since the two studies in the female F344 did not have any differences, they were combined. Figure 1 shows the means at different intervals for the 70 and 400-ppm-female groups for both the SD and F344 strains for comparison. As the figure indicates, these means were practically overlapping throughout the course of the study.

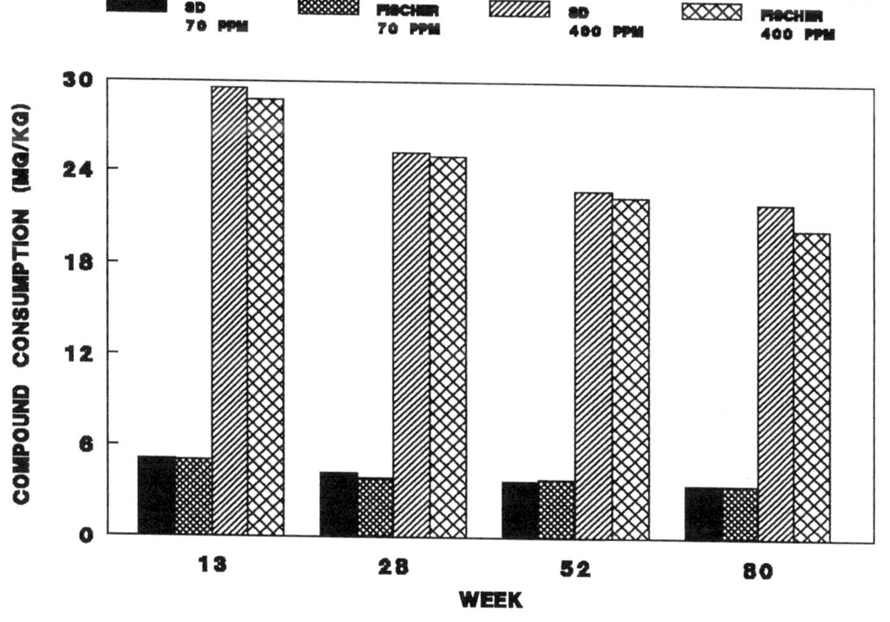

(*See* color on page 463.)

Table IA. Mean Compound Consumption (mg/kg/day) (Males) Fisher-344

Interval (Weeks)	*10 ppm*	*70 ppm*	*200 ppm*	*400 ppm*
0 - 13	0.64 ± 0.116	4.50 ± 0.859	13.13 ± 2.227	26.28 ± 4.483
14 - 28	0.52 ± 0.024	3.68 ± 0.221	10.68 ± 0.318	21.76 ± 1.353
32 - 52	0.46 ± 0.032	3.26 ± 0.233	9.18 ± 0.655	18.98 ± 1.068
56 - 80	0.44 ± 0.012	3.09 ± 0.115	8.88 ± 0.406	18.07 ± 0.787

Table IB. Mean Compound Consumption (mg/kg/day) (Females) Fisher-344 -Both Studies Combined

Interval (Weeks)	*10 ppm*	*70 ppm*	*200 ppm*	*400 ppm*
0 - 13	0.77 ± 0.098	5.46 ± 0.675	15.66 ± 1.923	31.40 ± 3.673
14 - 28	0.69 ± 0.021	4.94 ± 0.118	14.36 ± 0.221	28.69 ± 0.794
32 - 52	0.61 ± 0.041	4.39 ± 0.262	12.84 ± 0.770	26.18 ± 1.519
56 - 80	0.55 ± 0.013	3.92 ± 0.139	11.45 ± 0.534	23.19 ± 1.336

Mean Body Weight Gain

The mean body weight gains for both sexes by collection intervals are shown in Figures 2A and 2B. To obtain a true reflection of effect on the body weight gain for each group, animals with mammary tumors were excluded from the calculation of the means after the initial palpation of a mass in an individual animal.

For the female F344 rats, there was significant negative trend in mean body weght gains in both studies during the Week 0-13 interval because of effects in the 200 and 400-ppm groups. For the Week 14-28 interval, the trend continued in the hormonal study but not in the oncogenicity study. The combined body weight gain means showed a negative trend for both intervals. The 200-ppm and 400-ppm groups showed significant depression of mean body weight gains ranging 10 to 15% during interval 0-13 and in some instances during the interval 14-28 in individual studies as well as when both were combined. The maximum tolerated dose (MTD) was thus achieved at the 200 and/or 400-ppm females based on this finding. The depressions in the 10 and 70-ppm groups ranged approximately between 2-4%. As a result, the 70-ppm group is considered to be the no-effect level (NOEL) for this parameter.

The male 200 and 400-ppm groups of F344 rats showed similar body weight gain depressions during the intervals 0-13 weeks, 14-28 weeks, 29-52 weeks, and 53-76 weeks. The high dose males showed consistent depression in body weight gain during the entire course of the study. The decreases in the 10-ppm group ranged between 0-3%, in the 70-ppm group between 1-6%, in the 200-ppm group

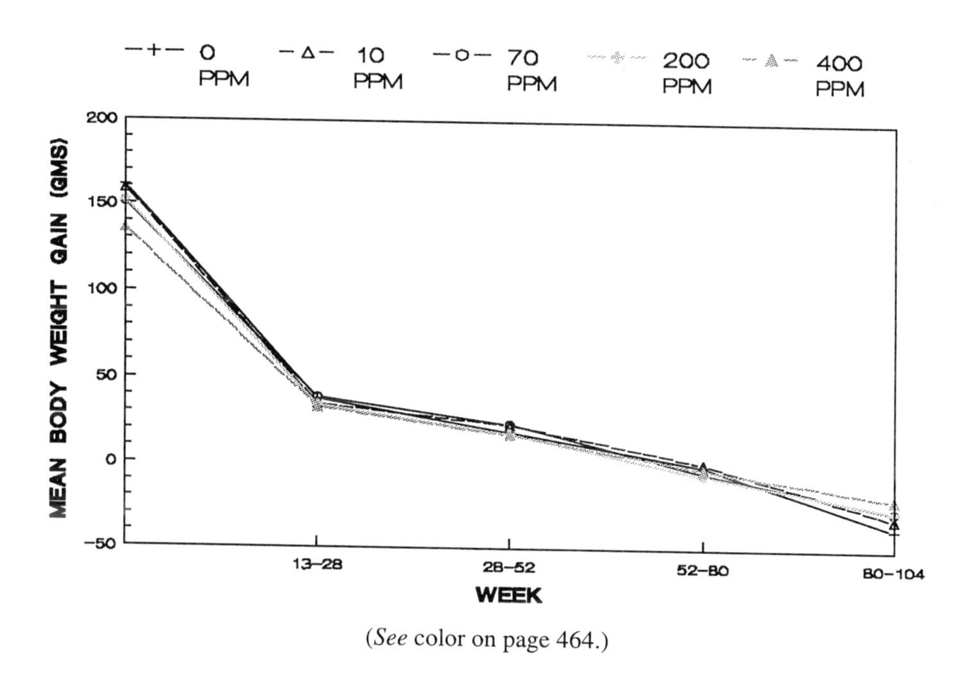

(*See* color on page 464.)

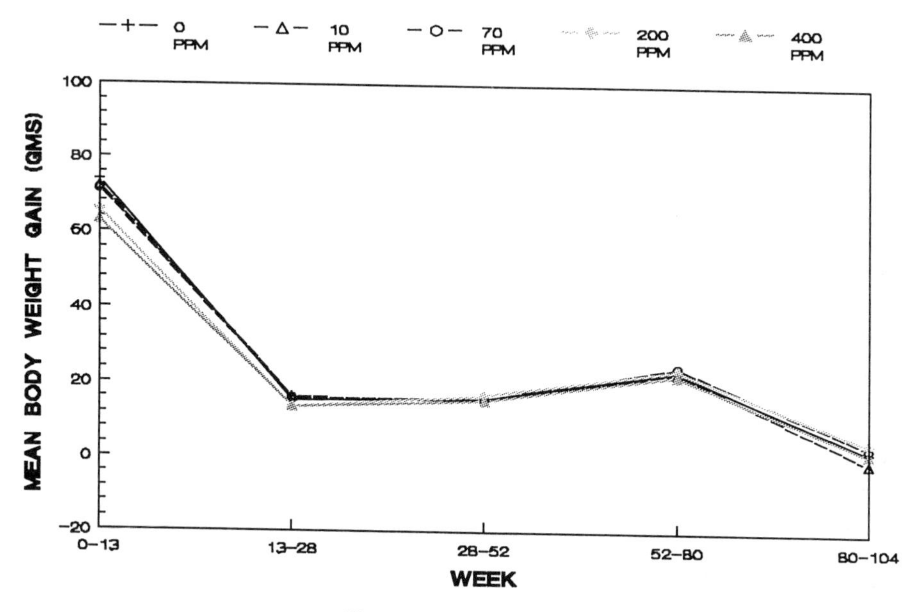

(*See* color on page 464.)

between 6-10%, and in the 400-ppm group between 11-16% of the control. These results indicate that the MTD was achieved in the males at 400-ppm. The 70-ppm level is designated as the NOEL for this parameter in the males.

Survival

Results of the survival analyses for the F344 studies are depicted in Figures 3A and 3B in the form of Kaplan-Meier product limit estimate adjusted mortality curves for the males and the females combining both studies. From Figure 3B and statistical analysis, there was no significant treatment effect (trend p=0.3602, two-sided, combined studies) on female survival. As Figure 3A shows, there was no significant survival effect on the male F344 (trend p=0.2911, two-sided) in the current study. This is in contrast to the results from the work by Pinter et al (*9*) in which the animals were treated until death, resulting in extreme differential survival patterns (Table II). In that study, a number of the 750-ppm males, and a lesser extent, of the 375-ppm males survived longer than the last control male.

Table II. Summary of Mammary Tumors - Fishcher 344 Males (Pinter et al (*9*))

Dose	*0 ppm*	*375 ppm*	*750 ppm*
Weeks of death 0-92	1(F)	0	1(AC)
Weeks of death 93-111	1(FA)	0	1(F)
Weeks of death 112-130	All dead	1(FA)	2(F) + 2(A)
Weeks of death 131-136	All dead	All dead	2(A)
Total	2/49	1/55	8/53

Note: F = Fibroma; FA = Fibroadenoma; A = Adenoma; AC = Adenocarcinoma

Tumor Incidences

The only remarkable tumors in the current studies were pituitary and mammary tumors. Tables IIIA and IIIB show the pituitary and other tumor incidences for the male and female F344 rats. The incidences are broken down by the individual and combined studies. As Table IIIA indicates, there was actually treatment-related significant negative trend in survival-adjusted pituitary tumor (adenoma and/or carcinoma) incidences in the two sexes (males, p=0.0486; females, p=0.0256) in the oncogenicity studies. The trend in the female pituitary tumors in the hormonal study was in the positive direction, but not significant. Because of the opposite directions of the trend statistics in this case, no combined evidence from both studies is appropriate. The pituitary tumors were primarily adenomas in all groups. In any case, the results from these two studies reveal the nature of biological variability in such tumors in this strain. In contrast, the Pinter et al (*9*) data show slight but not significant increases in pituitary adenomas in both sexes. The slight increases are probably a manifestation of increased survival in the treated groups, particularly in the males.

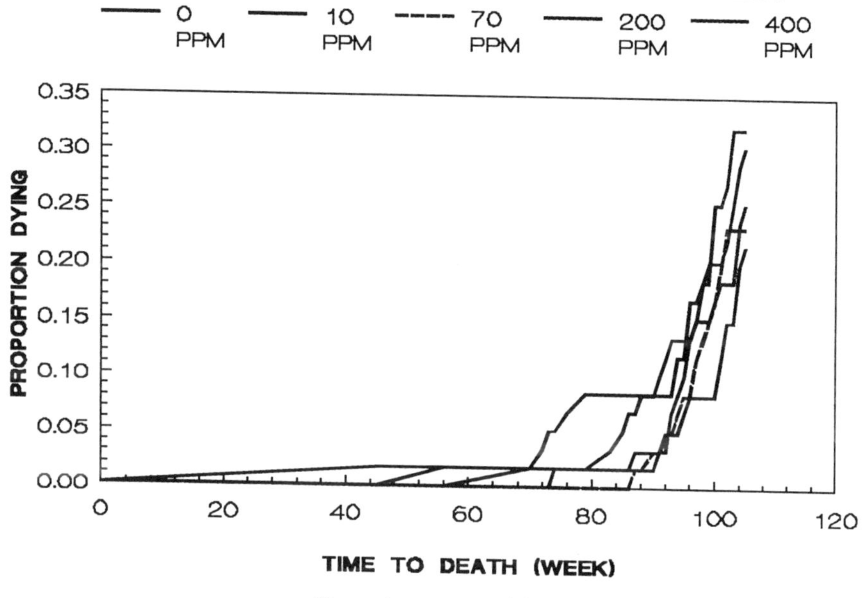

(*See* color on page 465.)

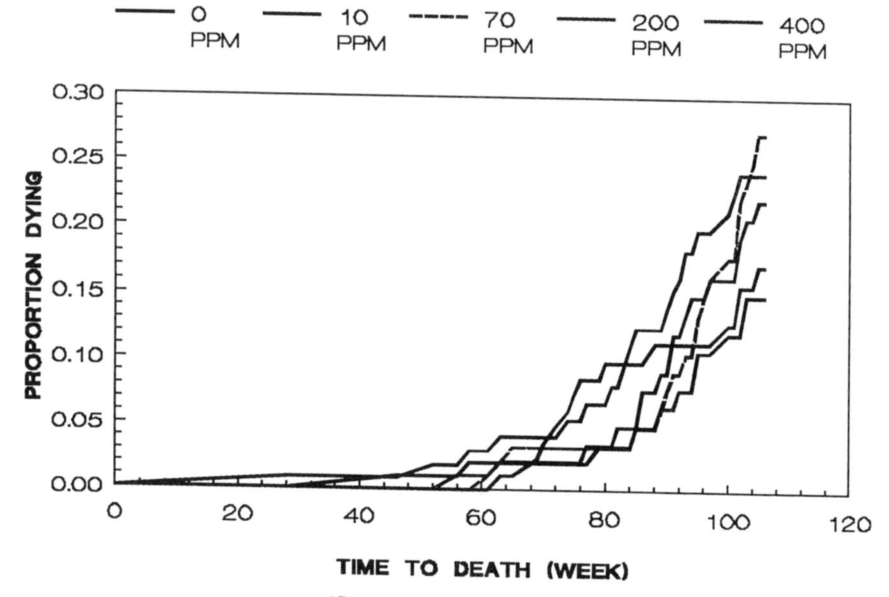

(*See* color on page 465.)

Table IIIA. Summary Of Pituitary Tumors, Mammary Tumors, Leukemias, and Lymphomas - Fischer 344 Males

Dose	*0 ppm*	*10 ppm*	*70 ppm*	*200 ppm*	*400 ppm*
	Pituitary Tumors				
Adenoma	15/59	14/60	15/60	11/60	8/59
Carcinoma	0/59	0/60	0/60	0/60	0/59
	Mammary Tumors				
Fibroadenoma	2/55	0/54	0/52	0/52	1/58
Adenocarcinoma	0/55	1/54	0/52	0/52	0/58
Total	2/55	1/54	0/52	0/52	1/58
	Leukemia and Lymphoma				
Mononuclear Cell Leukemia	21/60	22/60	16/60	26/60	22/60

Table IIIB. Summary Of Pituitary Tumors, Uterine Adenocarcinomas, Leukemias and Lymphomas - Fischer 344 Females

Dose	*0 ppm*	*10 ppm*	*70 ppm*	*200 ppm*	*400 ppm*
Pituitary Tumors - Hormone Study					
Adenoma	10/70	4/70	6/68	6/70	5/69
Carcinoma	0/70	0/70	0/68	0/70	0/69
Adenoma and/or Carcinoma	10/70	4/70	6/68	0/70	5/69
Pituitary Tumors - Onco Study					
Adenoma	22/60	26/60	20/58	19/59	13/59
Carcinoma	1/60	2/60	0/58	1/59	2/59
Adenoma and/or Carcinoma	23/60	28/60	20/58	20/59	15/59
Uterine Adenocarcinoma					
Adenoma	0/130	0/99	0/100	0/97	1/130
Leukemia and Lymphoma					
Mononuclear Cell Leukemia	10/62	18/60	17/64	11/65	16/65
Malignant Lymphocytic Lymphoma	0/62	0/60	1/64	0/65	0/65
Histiocytic Sarcoma	0/62	1/60	3/64	2/65	0/65

The same tables show the mammary tumor incidences in both sexes. Mammary gland adenomas were the primary tumor type observed in the F344 females (Table IIIB). Figures 4A and 4B show the Kaplan-Meier product limit adjusted percentages of known mammary tumors (adenoma and/or carcinoma bearing animals) for both sexes. For comparison with the Pinter et al (*9*) data, they are broken down into intervals (Table II). No onset data were available for the

FIGURE 4A: KAPLAN-MEIER PRODUCT LIMIT ESTIMATES OF MAMMARY TUMORS : F344 MALES

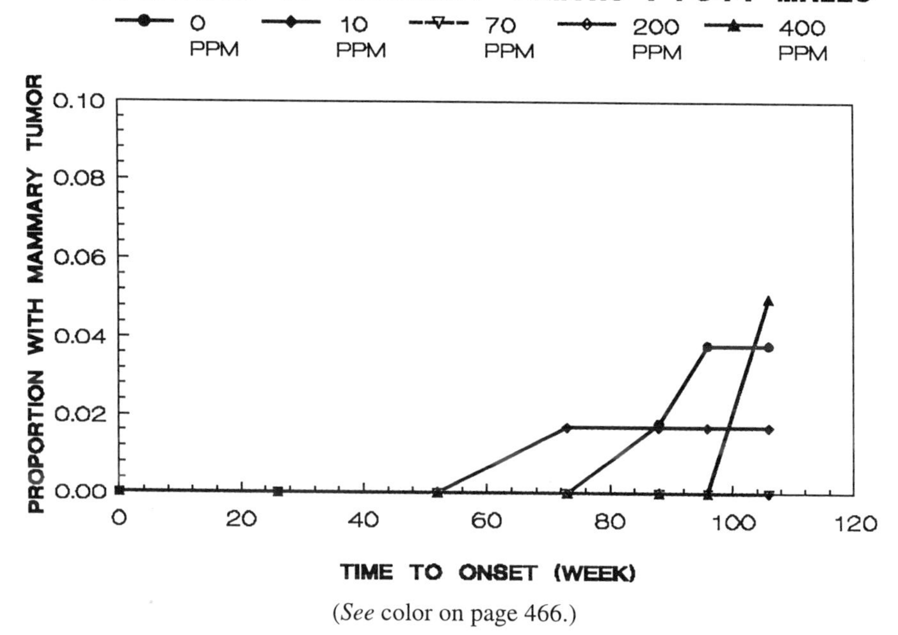

(*See* color on page 466.)

FIGURE 4B: KAPLAN-MEIER PRODUCT LIMIT ESTIMATES OF MAMMARY TUMOR: F344 FEMALES

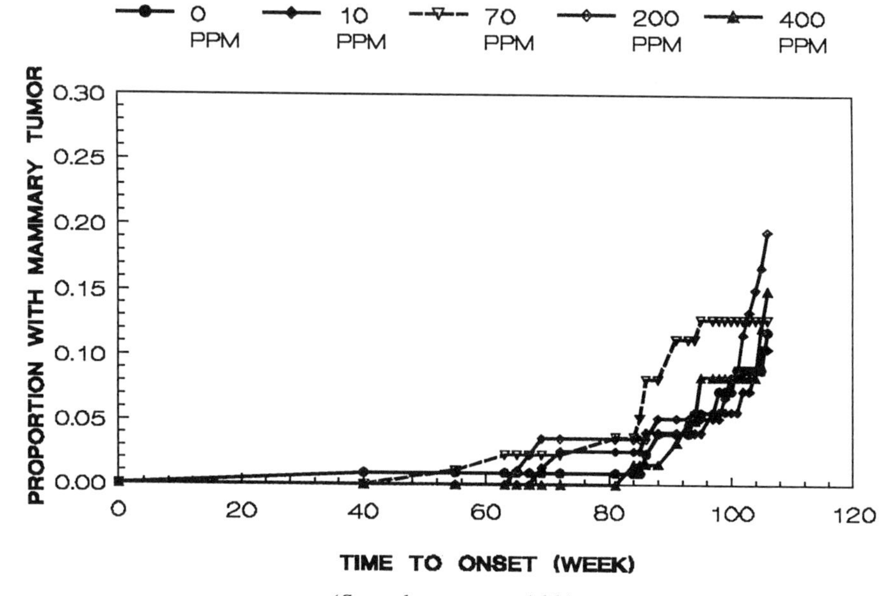

(*See* color on page 466.)

Pinter et al study (*9*). There was no significant effect in the mammary tumor incidences in the F344 females of the present study, either individually (trend p=0.0575 for the oncogenicity study and 0.1408 for the hormone study, both one-sided) or combined (trend p=0.2719, one-sided). In addition, the results show that there was no mammary tumorigenic effect of atrazine on treated F344 males in the present study (trend p=0.3197, one-sided).

The total hematopoietic neoplasm incidences in the present work does not show any significant increase in any group or a trend in either sex (trend p=0.4428 in the males and 0.4714 in the females, both one-sided).

Discussion

The studies reported here for male and female atrazine-treated F344 rats were performed to compare results from other 104-week studies (*4,8,9*) conducted in SD rats of both sexes, as well as those reported for a time-to-death study in F344 rats (*9*). Particularly, a comparison of the rates of mammary tumor induction between the F344 and SD females and between the F344 males from both studies was the primary focus. However, for the 104-week (no interval sacrifice) study in F344 males and females, all tissues were examined microscopically and other toxicological parameters, e.g. body weight gain, survival were also analyzed.

The dietary levels used in the present study (0, 10, 70, 200, and 400 ppm) achieved both NOEL and MTD in both sexes. That is, no treatment-related suppression of mean body weight or mean body weight gain was observed at feeding levels $\leq$ 70 ppm (NOEL) whereas feeding levels of 200 and 400 ppm caused significant decreases in those parameters at various intervals during the study in both males and females (Figures 2A and 2B). These results are consistent with the significant dose related body weight gain depression in both sexes as reported by Pinter et al (*9*).

There were no effects of atrazine on the survival of the male or female F344 rats fed any level of atrazine in this study. Pinter et al (*9*), in a non-GLP time-to-death study, in which F344 rats were fed 0, 375, or 750 ppm atrazine, showed a significant increase in survival of male rats in the 375 and 750-ppm groups compared to control. In that study, the last control male rat died at week 113 whereas the last high dose animal died at week 136. This difference between the two studies is particularly significant for analysis of any age-related tumors noted in treated male rats because there were no concurrent control group animals alive in the study for the 375 and 750-ppm animals after week 113. Therefore, no valid scientific comparison can be made to determine an oncogenic response for any age-related male tumor in the Pinter et al (*9*) study.

The female SD rat is well-recognized to have high spontaneous mammary tumor incidences (*19-22*). For example, among Color (Dye) studies (*23*), the spontaneous mammary tumor rates in untreated females were as high as 61% for benign and 42% for malignant mammary tumors. By contrast, female Fischer 344 rats do not have high spontaneous mammary tumor incidences (maximum 6-7% benign and 4.3% for malignant) (*20-22*).

The hormonal influence on mammary tumors in Sprague-Dawley rats is well documented in the literature (24-28). Mammary tumors in female Sprague-Dawley rats are primarily estrogen-dependent (*27-32*) and, in advanced age, these rats produce high levels of endogenous serum estrogen in response to alterations in estrous cycle activity resulting in the high background mammary tumor rate seen in this strain.

In contrast, Fischer 344 female rats develop high levels of endogenous serum progesterone as aging progresses, and this difference from the Sprague-Dawley rat may account for the low spontaneous rate of mammary tumors in the Fischer 344 strain (*33*). Results of selected serum steroid hormone determinations, vaginal cytology data, and estrous cycle evaluations to investigate strain differences are being compared and reported by Eldridge et al (*33)* and Simpkins et al (*34*) in this publication.

In the present study, there was no mammary tumorigenic effect on the female F344 rats exposed to any level of atrazine. By contrast, SD rats fed 400, but not 70 ppm, atrazine showed an earlier onset, but no significant increase mammary tumor bearing animals (*11,12*). The present results also contrast those of Mayhew and Wingard (*4*), where an earlier onset and/or increased incidence of mammary tumors at feeding levels of 70, 500, and 1000 ppm in the SD rat were reported. It should also be noted that the response at 70 ppm has not been duplicated in two subsequent two-year (*11,12*) and in a separate one-year (*35*) studies. Lastly, Pinter et al (*9*) did not observe an increased incidence in mammary tumors in female F344 rats fed 375 or 750 ppm atrazine.

Male rats from any feeding level in the present oncogenicity study did not show an increased incidence (or decreased latency) of any mammary neoplasia. These results contrast those reported by Pinter et al (*9*) who reported an increased incidence in mostly benign mammary gland tumors in male rats fed 750 ppm atrazine for about 136 weeks, a design significantly different from the standard bioassay designs. Although initially considered by the authors of that work to be a treatment-related effect, the finding is an artifact of the higher survival rate in that group (as noted above). Six of the eight mammary tumors in the males of that study occurred in the 750-ppm animals after the last male control animal had died. The other two mammary tumors in the 750 ppm treated male group occurred the same time as the one found in the control group. Since this tumor is age dependent in male rats (*36*), the increased survival in the high dose group masks any effect that may have been noted in control animals had they lived past 113 weeks. Pinter et al (*9*) used the interval-based prevalence method of Peto et al. (*37*) for their statistical evaluation of mammary tumors. In such methods one uses either pre-selected or isotonically (i.e. in the same direction) produced finite number of intervals instead of actual time points of death. When the survival patterns are not extremely disparate, such methods provide a surrogate of true survival adjusted methods such as logistic regression (*38*). However, in the study by Pinter et al (*9*), the control group males all died much earlier (week 113) than either the 375 ppm (week 126) or the 750-ppm group (week 136). As a consequence, the incidence data from those two treated groups get lumped in the same final interval as the control, thereby loosing the appropriate survival adjustment in the last phase of the study. Appropriate statistical analysis, as mentioned earlier, confirms that there was no significant trend (p=0.7971, one-sided exact) or high-dose increase (p=0.5457, one-sided exact) due to atrazine treatment in the Pinter et al work (*9*). These conclusions are further supported by results published by Solleveld, Haseman, and McConnell (*36*). These authors report that in 2320 untreated male rats which died before 100-116 weeks of age or which were killed at 110-116 weeks of age, 57 (2.5%) males with fibroadenoma or a carcinoma were found. In a separate life-span study (*36*) in which 529 male and 529 female F344 rats lived up to 140-150 weeks of age, it was shown that the age-specific prevalence rates of mammary gland tumors (fibroademonas or adenocarcinomas) rose markedly from week 111 onwards and resulted in 79 of 529 (14.9%) males showing a mammary gland tumor. Thus, the two mammary gland tumor prevalence rates in two cohorts of similar age differed by a factor of six in Solleveld, Haseman, and McConnell (*36*) and in the Pinter et al (9) as well.

For the significant leukemia/lymphoma finding in the females reported by Pinter et al (*9*), it is important to note that leukemia in female F344 rats is unique to that strain and should not be combined with lymphomas (*39,40*). That is, in the F344 rat, leukemia is a mononuclear cell type whereas that lymphoma is of either lymphocytic or histiocytic origin (*39,40*). The latter two are of different origins from leukemia as well as from each other and should not be combined. Indeed, when the two tumor types are separated, no significant trend or significant pairwise comparison exists. For the uterus, a significant trend occurs when adenocarcinoma or malignant tumors are analyzed by the Cochran-Armitage test for trend; however, no significant pairwise comparison is demonstrated. Further, it should be noted that there are several types of malignant uterine tumors of different origins that should not be combined (*39*). The work by Pinter et al (*9*) does not distinguish the origin of the uterine tumors, so that their statistical analysis is compromised at best. Lastly, the finding of leukemia, lymphoma, and uterine neoplasia are all age-related occurrences (*41*). Since survival data are not available from the female rats of their study, and because the specific origins of the tumors are not described, appropriate statistical analysis cannot be performed.

Conclusion

A strain and sex-specific mammary tumor response to high-dose atrazine (400 ppm, but not at 70 ppm) occurred in female SD rats, as evidenced by the earlier onset time for mammary tumors noted only in this strain. There was no oncogenic effect of any kind in either sex of the F344 rats in these studies which is consistent with the finding in male SD rats and mice. These results support the hypothesis that atrazine-induced mammary oncogenicity is specific to the SD female rat and is related to estrous cycle disruption and subsequent prolonged exposure to endogenous estrogen.

The authors thank Sandra L. Morseth and Cynthia Y. Liu for their help in preparation of this manuscript.

Literature Cited

1. Good, N.E. Plant Physiol. 1961, 36, pp 788-803.
2. Brusick, D.J. *Mutation Res.* **1994**, *317*, pp 133-144.
3. Stevens, J.Y.; Sumner, D.D. In *Herbicides;* Hayes, W.J., Laws, E.R.,Eds.; Handbook of Pesticide Toxicology; Academic Press: NY, 1991; Vol. 3, pp. 1317-1408.
4. Mayhew, D. A.; Wingard, B.L. Two-year feeding/oncogenicity study in rats administered atrazine. American Biogenics Corporation Report (Unpublished), 1986.
5. Innes, J.R.M.; Ulland, B.M.; Valerio, M.G.; Petrucelli, L.; Fishbein, L.; Hart,E.R.; Pallota, A.J.; Bates, R.R.; Falk, H.L.; Mitchell, I.; Peters, J. *J. Natl. Cancer Inst.* **1969**, *42, pp* 1104-1114.
6. Sumner, D.D., and Charles, J.M. (1981). Carcinogenicity study with atrazine technical in albino mice. Industrial Biotest Corporation Report (Unpublished), 1981.
7. Hazelette, J.R., and Green, J.D., Oncogenicity study in mice: Atrazine Technical. Report to CIBA-GEIGY Corporation (unpublished), 1987.
8. Thakur, A.K., Oncogenicity study in Sprague-Dawley rats with atrazine technical. Hazleton Washington report HWA 483-275 (Unpublished), 1992a.
9. Pinter, A.; Torok, G.; Borzsonyi, M.; Surjan, A.; Csik, M.; Kelecsenyi, Z.; Kocsis, Z. *Neoplasma.* **1990**, *37, pp* 533-544.
10. Thakur, A.K., Oncogenicity study in Fischer-344 rats with atrazine technical. Hazleton Washington Report HWA 483-277 (Unpublished), 1992b.
11. Thakur, A.K., Determination of hormone levels in Fischer-344 rats treated with atrazine technical. Hazleton Washington Report HWA 483-279 (Unpublished), 1991.
12. *Statistical Principles in Experimental Design;* Winer, B.J.; McGraw-Hill: NY, 1971, pp 149-220.
13. Dunnett, C.W.; *J. Am. Stat. Assoc.* **1955**, *50,* pp 1096-1121.
14. Dunnett, C.W.; *Biometrics.* **1964**, *20*, pp 482-491.
15. BMDP Statistical Software, Biomedical Data Processing, University of California Press, Berkeley, California, 1990.
16. Thakur, A.K.; Wetzel, L.T.; Stevens, J.T. *Proc. Biopharm. Section Am. Stat. Assoc.* **1990**, pp. 275-281.
17. Gart, J.J.; Krewski, D.; Lee, P.E.; Tarone, R.E.; Wahrendorf, J. *The Design and Analysis of Long-term Animal Experiments.* International Agency for Research on Cancer, Lyon, France: 1986, p. 86.
18. LogXact-Turbo, CyTel Software Corp., Cambridge, Massachussetts, 1993.
19. Lang, P.L.; *Spontaneous neoplastic lesions in the CDF (F-344)/CrlBR rat.* 1990, Charles River Laboratories.
20. McMartin, D.N.; Sahota, P.; Gunson, D.E.; Hsu, H.H.; Spaet, R.H. *Toxicol. Pathology.*
21. Haseman, J.F.; Huff, J.; Boorman, G.A. *Toxicol Pathol.* **1984**, *12*, pp 126-135.

22. Corning Hazleton Washington, Historical Control Database for Neoplastic Lesions in SD and F344 Rats from Two-year Chronic/Oncogenicity Studies (Unpublished), 1993.
23. Haseman, J.K.; Winbush, J.S.; O'Donnell, M.W. Use of Dual Control Groups to Estimate False Positive Rates in Laboratory Animal Carcinogenicity Studies. *Fundamental and Applied Toxicology.* **1986**, Vol. *7*, pg. 573-584.
24. Eldridge, J.C.; Wetzel, L.T.; Tisdel, M.O.; Leumpert, L.G. III. 1992a, EPA MRID# 42547106.
25. Eldridge, J.C.; Wetzel, L.T.; Tisdel, M.O.; Leumpert, L.G. III. 1992b, EPA MRID# 42547107.
26. Welsch, C.W. *Cancer Res.* **1985**, *43*, pp 3415-3443.
27. Thompson, H.J.; Ronan, A. *Carcinogenesis.* **1987**, *57*, pp 2003-2006.
28. Russo, J.; Gusterson, B.A.; Rogers, A.E.; Russo, I.H.; Wellings, S.R.; Van Swieten, M.J. *Lab Invest.* **1990**, *62*, pp 244-278.
29. Cutts, J.H.; Noble, R.L. *Cancer Res.* **1964**, *24*, pp 1116-1122.
30. Geschickter, C.F.; Byrnes, E.W. *Arch. Pathol.* **1942**,*33* pp 334-356,.
31. Shafie, S.M. *Science.* **1980**, *209*, pp 701-702.
32. Wise, P.M. *J. Steroid Biochem.* **1987**, *27,* pp 713-719.
33. Eldridge, J.C.; McConnell, R.F.; Wetzel L.T.; Tisdel, M.O. (This Volume).
34. Simpkins, J.W.; Eldridge, J.C.; Wetzel, L.T. (This Volume).
35. Peterson, J.C. and Turner, J.C., Ciba-Geigy Corporation Report (Unpublished), 1995.
36. Solleveld, H.A.; Haseman, J.K.; McConnell, E.E.; *J. Nat. Cancer Inst.* **1984**, *72*, pp 929-940.
37. Peto, R.; Pike, M.C.; Day, N.E.; Gray, R.G.; Lee, P.N.; Parrish, S.; Peto, J.; Richards, S.; Wahrendorf, J. In *Guidelines for simple, sensitive significance tests for carcinogenic effects in long-term animal experiments.* Long-term and short-term Screening Assays for Carcinogens: A Critical Appraisal. (IARC Monographs Supplement 2), Lyon, IARC: Lyon, France, 1980, pp 311-426.
38. Dinse, G.E.; Lagakos, S.W.; J. *Roy. Stat. Soc.* Series C (Appl. Stat.), **1983**, *32*, pp 236-248.
39. Thurman, J.D.; Bucci, T.J.; Hart, R.W.; Turturro, A. *Toxicol. Pathology*, **1994**, *22*, pp 1-9.
40. *Pathology of the Fischer Rat;* Boorman, G.A.; Eustis, S.C.; Elwell, M.R.; Montgomery, Jr., C.A.; MacKenzie, W.F., Eds.; Reference and Atlas; Academic Press: NY, 1990.
41. Stromburg, P.C. In *Atlas of Tumor Pathology of Fischer Rat*; Sherman, F.S.; Hildegard, M.; Gerd, G.R., Eds.; CRC Press: Boca Raton, FL, 1990, pp 505-526

Chapter 31

Role of Strain-Specific Reproductive Patterns in the Appearance of Mammary Tumors in Atrazine-Treated Rats

James W. Simpkins[1], J. Charles Eldridge[2], and Lawrence T. Wetzel[3]

[1]Department of Pharmacodynamics, University of Florida, Gainesville, FL 32610
[2]Department of Physiology and Pharmacology, Bowman Gray School of Medicine of Wake Forest University, Winston-Salem, NC 27157
[3]Department of Toxicology, Novartis Crop Protection, Inc., P.O. Box 18300 Greensboro, NC 27419

Atrazine has been a major agricultural herbicide in the U.S. for more than 25 years. It is used for the control of broadleaf and grass weeds in corn and sorghum crops. Because of its common use, the toxicity of atrazine has been the subject of many studies. Atrazine is not toxic with acute administration, with an oral and dermal LD_{50} of greater than 3,000 mg/kg. In tests of mutagenicity, atrazine have been negative in more than 50 tests *(1)*. Atrazine is not a teratogen or a reproductive toxin, and lacks carcinogenic activity in male and female mice and Fischer 344 (F-344) rats, as well as in male Sprague-Dawley (SD) rats. Five tests of the tumorogenicity of atrazine in SD rats have been conducted since the 1960s. Two of these tests, which assessed atrazine at doses up to 500 ppm, produced negative results, while 3 other studies have shown an earlier time of onset and/or an increased incidence of mammary tumors *(2-4)*. With the exception of one study *(4)*, the earlier onset of mammary tumors occurred at doses $\geq$ a maximum tolerated dose *(2,3)*. A no-observed-effect-level (NOEL) for tumorogenicity was established in all studies.

Atrazine is not a mutagen *(1)*, a direct acting carcinogen and it has no intrinsic estrogenic activity *(5,6)*.

The increased incidence and/or earlier age of appearance of mammary tumors in female SD, but not Fischer 344 rats warrants an evaluation of the strain-specificity of this response. The results discussed here present strong evidence that the specificity of the tumor-enhancing effects of atrazine in the female SD rat are the result of a treatment-related earlier appearance of persistent estrus in that strain.

Rodent Strains and Tumor Incidence

The background incidence of mammary tumors varies greatly among rodent strains *(7,8)*. The background incidence of spontaneous mammary tumors in 2 year old female SD rats ranges as high as 70 to 80% *(7,8)* which is in marked contrast to the very low incidence of spontaneous mammary tumors in F-344 rats *(2,3)*. Clearly, the SD female rat is very susceptible to factors which promote occurrence of mammary tumors, while the F-344 rat is comparatively resistant.

Endocrine Factors that Enhance the Growth of Mammary Tumors

It has been clearly established that the ovarian steroid hormone, estrogen, and the anterior pituitary peptide hormone, prolactin which plays a role in the normal development and physiology of the breast, are also involved in the rate of mammary tumor growth *(7)*. Chronic elevation of estrogens *(9-11)* or prolactin *(12,13)* increases the incidence and decrease the onset time of mammary tumors in either SD or F-344 rats. As such, treatment with agents that increase chronic secretion of estrogens and/or prolactin would be expected to increase the occurrence of mammary tumors in these strains.

Comparison of the SD and F-344 Rats as Surrogate Models for Human Assessment of Mammary Tumors

In women, the passage from normal menstrual cycles into reproductive senescence results from exhaustion of ovarian follicles, and is accompanied by a precipitous decline in ovarian steroid hormones, particularly, estradiol *(14)*. At the menopause, and for a considerable time thereafter, the hypothalmic-pituitary axis still has the capacity to regulate anterior pituitary hormone secretion. With the decline in estrogens, serum LH and FSH secretion increases markedly and can be suppressed by hormone replacement therapy *(14)*. In addition, regimens of estrogen treatment that induce an LH surge are also able to induce LH surges in postmenopausal women *(15)*. Therefore, the mechanism for transduction of the estrogen signal to an LH surge appears to be intact in postmenopausal women.

By contrast, in the aging SD and related strains of rats, the ovary retains a substantial number of follicles *(16,17)*. Reproductive senescence in the SD rat (Figure 1) appears to result from a breakdown in the capacity of the hypothalmus to convert the estrogen signal from the ovary into an LH surge sufficient to induce ovulation *(16-19)*. The breakdown is evidenced by a gradual transition from normal 4 to 5 day estrous cycles to extended periods of estrus with continuous endogenous estrogen secretion, and finally to a state of persistent estrus *(20,21)*. This persistent estrus state, which can last the remainder of the animal's life *(20,21)*, has a profile of moderate, continuous elevation and secretion of serum estradiol and low levels of serum progesterone *(21)*. Serum prolactin levels are elevated as a result of the increase in estradiol, which acts on the anterior pituitary to stimulate prolactin synthesis and secretion *(22)*.

Given this markedly different endocrine environment during reproductive decline in SD rats and humans, the SD rat seems to be a poor surrogate model for reproductive senescence in the human female.

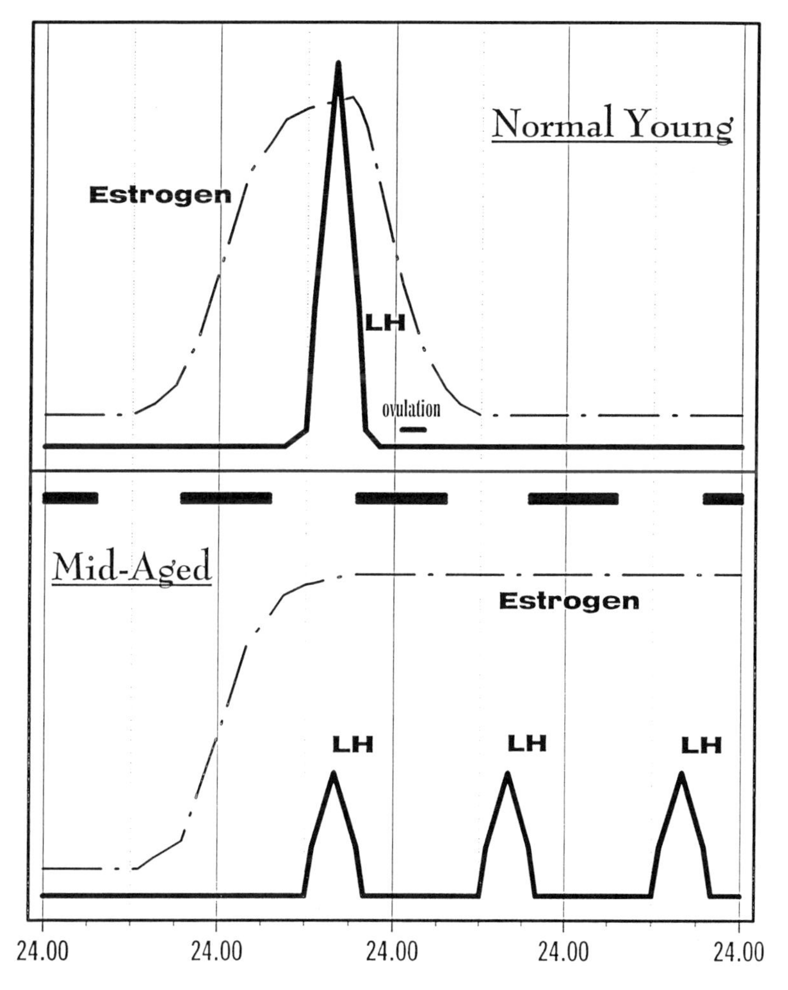

Figure 1. Schematic Representation of Preovulatory LH Surges in the Young Adult (Upper Panel) and Mid-Aged (Lower Panel) Female Sprague-Dawley (SD) Rat. In young SD rats, a rising estrogen secretion from developing ovarian follicles primes a surge of LH after noon of the day preceding ovulation. Estrogen secretion then declines as the follicular source disintegrates with ovulation. In middle age, declining hypothalamic function prohibits rising follicular estrogen from inducing a sufficient LH signal, so ovulation fails to occur and the ovarian follicles persist. Instead of normally cyclic elevations of estrogen, mid-aged SD rats maintain estrogen secretion at continuously elevated levels. In this case, vaginal cytology displays repeated days of heavy cornification.

The F-344 rat exhibits a late life reproductive senescence that runs a very different course than that observed in the SD rat. Through about 1.5 years of age, the majority of F-344 rats maintain normal 4 to 5 day estrous cycles *(23)*. By 2 years of age, the rats have entered a senescent reproductive pattern of normal estrous cycles interspersed with periods of extended maintenance of the corpus luteum and the resultant hypersecretion of ovarian progesterone *(23,21)*. This condition is appropriately called repeated pseudopregnancy. Anterior pituitary weights remain normal through 2 years of age, serum prolactin is slightly elevated and serum LH concentrations are slightly reduced *(23)*. More importantly, the hypothalamic-pituitary axis maintains the capacity to mediate the estrogen-induced hypersecretion of LH and normal ovulation that is common in aged F-344 rats *(23-25)*. The only known neuroendocrine defect in the F-344 rat is the inability to reduce episodic prolactin surges *(26,23)*, which maintains the corpus luteum *(27)*.

While not completely similar to the human in its pattern of reproductive senescence, the F-344 does share with the human female the following features; both have a late life reproductive senescence; both experience low estrogen levels during late life, and both maintain the ability to control LH secretion during reproductive senescence. As such, the F-344 rat more closely models the human female than does the SD rat.

Evaluation of the Mode of Action of Atrazine in the Strain-Specificity of the Mammary Tumorigenic Effects in the SD Rat

It has been proposed that the strain- and sex specificity of the earlier appearance of tumors with 2 years of high-dose atrazine feeding in the SD rat was a result of the superimposition of an atrazine effect on the early appearance of persistent estrus in the SD rat *(28,29,2,3)*. It is important to recognize that atrazine feeding results in an earlier appearance of a spontaneous reproductive senescence event in the SD rat, i.e., mammary tumors. Exposure to atrazine does not result in the development of a new mammary pathology. If this hypothesis is correct, then two predictions should follow. First, atrazine feeding at the MTD, a dose associated with an earlier appearance of mammary tumors in the SD female rat, should cause an earlier appearance of persistent estrus in the SD, but not in the F-344 rat. The chapter by Eldridge et al in this same volume documents that atrazine-induced early persistent estrus does occur. Second, atrazine treatment should induce an earlier appearance of a neuroendocrine deficit that would lead to the appearance of persistent estrus in SD female rats, i.e., a decrease or attenuation, of the proestrous LH surge. Indeed, a preliminary study by Cooper and colleagues (1996) (*30*) had suggested just this possibility. The evidence for the latter effect of atrazine treatment is the subject of the remainder of this chapter.

Effects of Atrazine Treatment on the Estrogen-Induced LH and Prolactin Surges: Acute Study

As an initial evaluation of the effects of atrazine treatment on the LH surge, female rats were ovariectomized and simultaneously implanted with an estradiol-containing sustained-release capsule (4 mg/ml sesame seed oil). This mode of estradiol administration produced levels seen during normal preovulatory surges of LH and has also been shown to produce daily surges of LH in young rats *(18,19)*.

Atrazine was then administered by gavage daily at a dose of 300 mg/kg body weight for 3 days. On the third day of treatment, animals were sacrificed by decapitation at 11:00, 13:00, 15:00, 18:00 and 22:00 hours, a time during which the LH surge was expected.

The vehicle-treated animals showed the expected estrogen-induced LH surge, with LH levels increasing at 15:00 h, peaking at 18:00 h and diminishing at 22:00 h (Figure 2). By contrast, the atrazine-treated rats failed to show an increase in serum LH at any interval through 18:00 h and exhibited only a slight increase in LH at 22:00 h.

Assessment of prolactin concentration in these same animals revealed that the estrogen-induced surge of this hormone was also blunted or perhaps delayed, as reflected by diminished prolactin levels at the times of the peak prolactin concentrations in control animals at 15:00 h and 18:00 h (Figure 3).

These results indicate that atrazine treatment at a level > the MTD was able to disrupt a neuroendocrine mechanism that transduces the estrogen signal to an LH response. As such, the hypothesis that atrazine blunts the preovulatory LH surge was supported. Additionally, in as much as the prolactin surge was also affected, it appears that treatment altered a very fundamental mechanism that converts the estrogen signal into neuroendocrine responses.

Dose-Dependent Effects of Atrazine on the Estrogen-Induced LH Surge: 4-Week Study

To determine the dose-dependent effects of atrazine on the estrogen-induced LH surge, an experiment similar in design to the acute study was conducted. Sprague-Dawley rats were treated by gavage, to 0,2.5,5.0,40, or 200 mg/kg/day doses of atrazine for 30 days. Three days prior to sampling, animals were ovariectomized and immediately implanted with a capsule containing estradiol as described above previously. On the 30th day of atrazine treatment, animals were sacrificed at 6 intervals, from 13:00 to 23:00.

As expected, the vehicle-treated control showed a "preovulatory-like" LH surge that peaked at 16:00 to 18:00 and declined thereafter (Figure 4). Responses to 2.5, 5 and 40 mg/kg were quite similar to vehicle-dosed animals. In contrast to these low doses of atrazine, the 200 mg/kg dose of atrazine caused a reduction of LH at the peak time of 16:00 (Figure 4).

Because the time of peak LH secretion in response to estrogen treatment is variable, another set of animals, dosed (0, 2.5, 5.0, 40 or 200 mg/kg/day) for 30 days, was sampled sequentially on the third day after ovariectomy and estrogen implantation. LH data were normalized to the time of the peak response and, when expressed on this basis, the effect of high dose atrazine treatment becomes even more clear. The control, 2.5 and 5 mg/kg doses of atrazine showed similar increases in LH secretion to peak levels and a decline over the next 4 h (Figure 5).

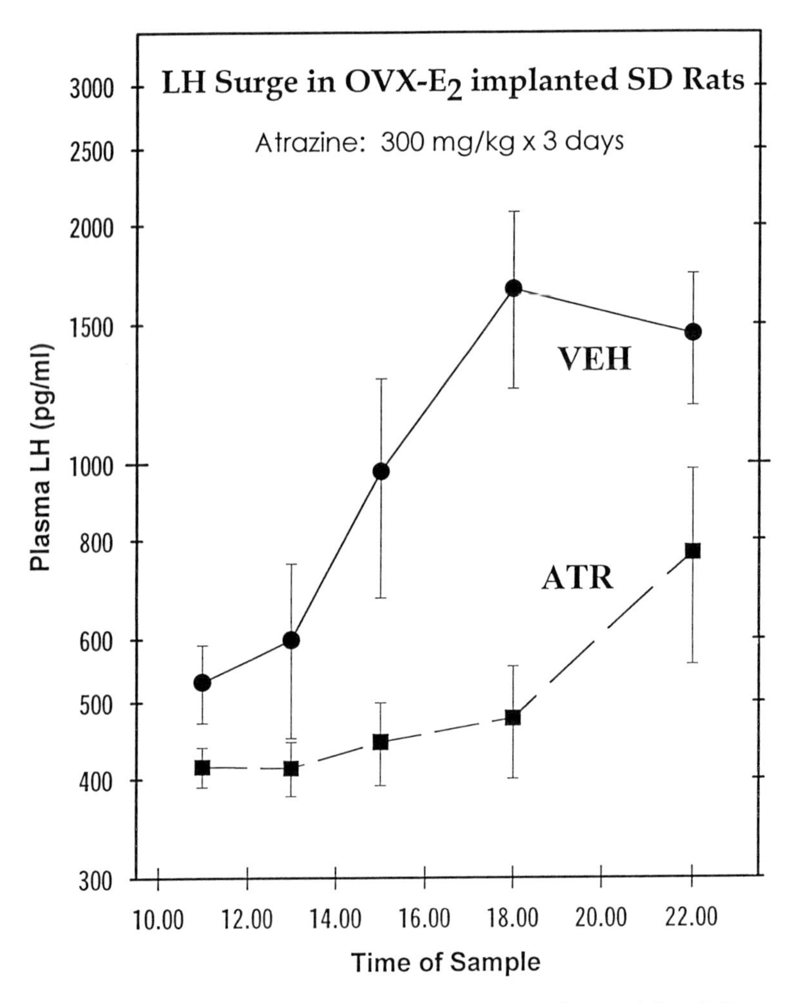

Figure 2. Suppression of the Estrogen-Induced LH Surge After 3 Days of Atrazine Treatment. Ovariectomized young adult SD rats were implanted with an estradiol-containing silastic capsule and were administered atrazine (ATR) by gavage, 300 mg/kg/day for 3 days, or gavage vehicle (VEH). Blood samples were collected after decapitation at the indicated times and plasmas were analyzed for LH. Points represent means + S.E.M. of 10 animals per interval. Group mean values in the ATR-treated animals were significantly different from VEH means at each time interval except 13.00 ($p<0.05$, Student's t).

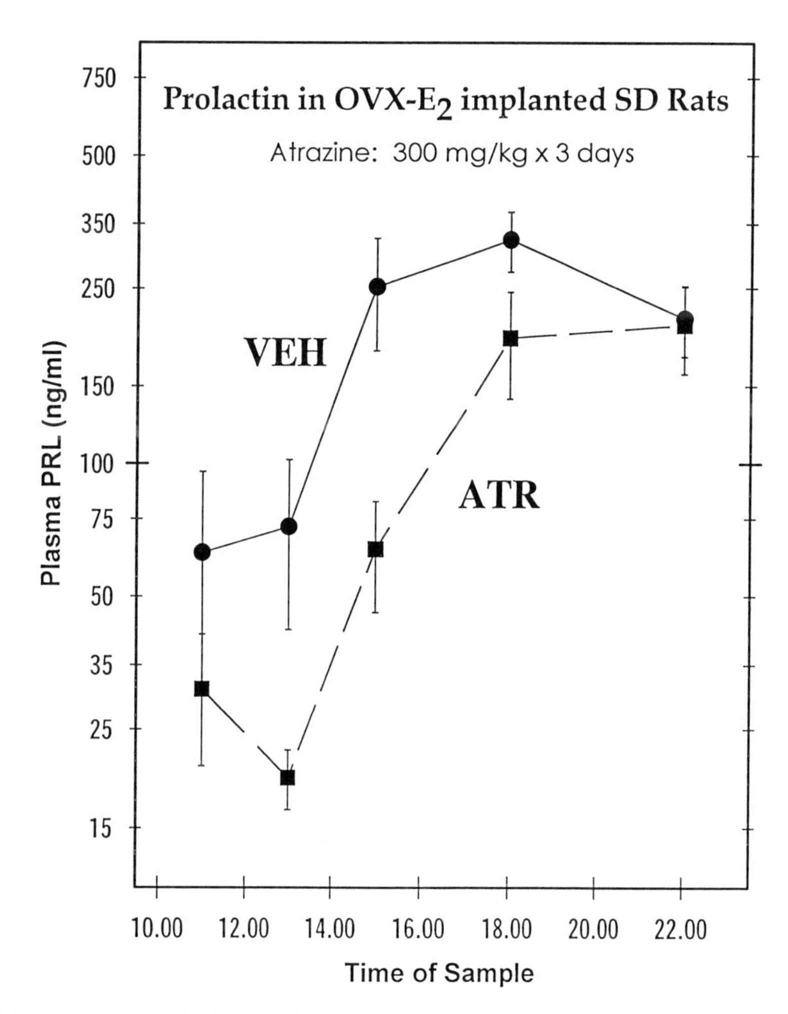

Figure 3. Altered Estrogen-Induced Prolactin (PRL) Surge After 3 Days of Atrazine Treatment. Plasma samples assayed for LH (Figure 2) were also assayed for PRL. Group mean values in the ATR-treated animals were significantly different from VEH means at 13:00, 15:00, and 18:00 ($p<0.05$, Student's t).

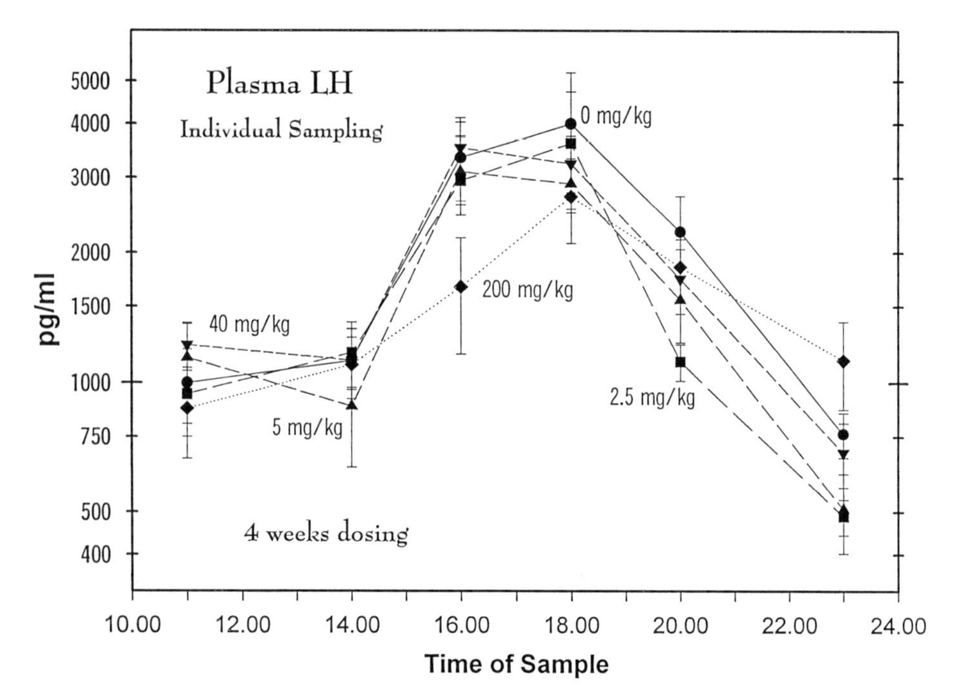

Figure 4. Suppression of the Estrogen-Induced LH Surge After 4 Weeks of Atrazine Treatment. Young adult SD female rats were administered atrazine by gavage, once daily at the doses indicated, for 28 days. All animals were then ovariectomized and given a silastic estradiol-containing implant. After 3 more days of atrazine dosing, animals were sacrificed by decapitation at the indicated times, and plasmas were assayed for LH. Each point represents the mean + S.E.M. of 10 (11:00 and 23:00) or 15 (other times) animals. The group mean values at 16:00 for the 200 mg/kg group was significantly less than control ($p<0.05$).

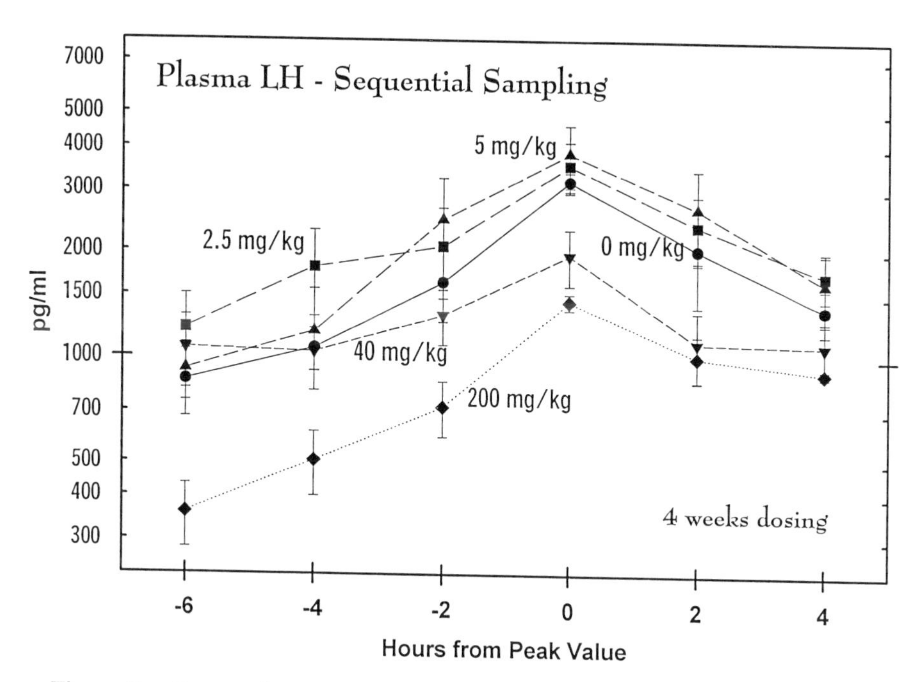

Figure 5. Suppression of the Estrogen-Induced LH Surge in Sequentially-Sampled Animals After 4 Weeks of Atrazine Treatment. Animals were part of the same treatment groups as those illustrated in Figure 4, but 10 of each dose were repeatedly sampled at the same time intervals as the other animals. Sequential LH plots were normalized so that each animal's peak value was placed at time 0. Points represent the mean + S.E.M. of 6-10 samples. By 2-way ANOVA, there was a significant overall effect of dosing at 40 and 200 mg/kg, compared to vehicle controls ($p<0.05$).

By contrast, the 40 and 200 mg/kg groups showed very little increase in LH for the 4 hours preceding the peak, and then a much more modest peak LH concentration (Figure 5).

This study demonstrated that atrazine treatment for 30 days at doses ≥ the MTD markedly blunted the estrogen-induced LH surge. Because aging SD rats spontaneously display a reduction of LH surges *(16-19)*, this effect of atrazine is consistent with the hypothesis that the triazine may selectively reduce the age at which persistent estrus occurs in the SD rat.

Dose-Dependent Effects of Atrazine on the Estrogen-Induced Prolactin Surge: 4-Week Study

Sera from blood samples collected from individually sacrificed animals (LH results presented in Figure 4) were also analyzed for prolactin content. As indicated by the results in Figure 3, estrogen-implanted ovariectomized rats normally display a surge of prolactin at approximately the same time as the LH surge.

The prolactin surge in animals treated with atrazine at doses of 2.5, 5, and 40 mg/kg for 30 days did not differ, in general, from that seen in control animals (Figure 6). Prolactin concentrations peaked at 14:00 to 16:00 and declined progressively thereafter. By contrast, the 200 mg/kg atrazine treatment group exhibited a delayed prolactin surge, with a progressive increase in the concentrations of the hormone through 18:00 and a decline thereafter (Figure 6). Serum prolactin levels were not measured in the serially sampled animals because of stress effects expected with repeated bleeding (*31*).

Effects of Atrazine Dosing on Circulating Estrogen Levels Resulting from Pellet Implantation

The final plasma sampled from the serially bled animals, and all samples from the individually sacrificed animals were assayed for estradiol levels to verify the success of the implant and to determine if atrazine treatment affected circulating estradiol levels.

Regardless of the dose of atrazine or the sampling time, plasma estradiol concentrations were not different (Figure 7). These data indicate that the implants were successful and that atrazine does not substantially affect the concentration of circulating estradiol. Therefore, treatment-induced changes in estradiol metabolism are not a likely explanation for the atrazine effects observed.

Discussion

Reproductive senescence patterns in the female SD rat are characterized by the mid-life appearance of persistent estrus and an associated continued exposure to plasma estradiol and prolactin *(16-21)*. This combination of persistent exposure to estradiol and prolactin, which are known to promote the growth of mammary tumors *(7,8)*, undoubtedly plays a major role in causing the substantial, spontaneous background incidence of mammary tumors in the SD rat *(2,3)*.

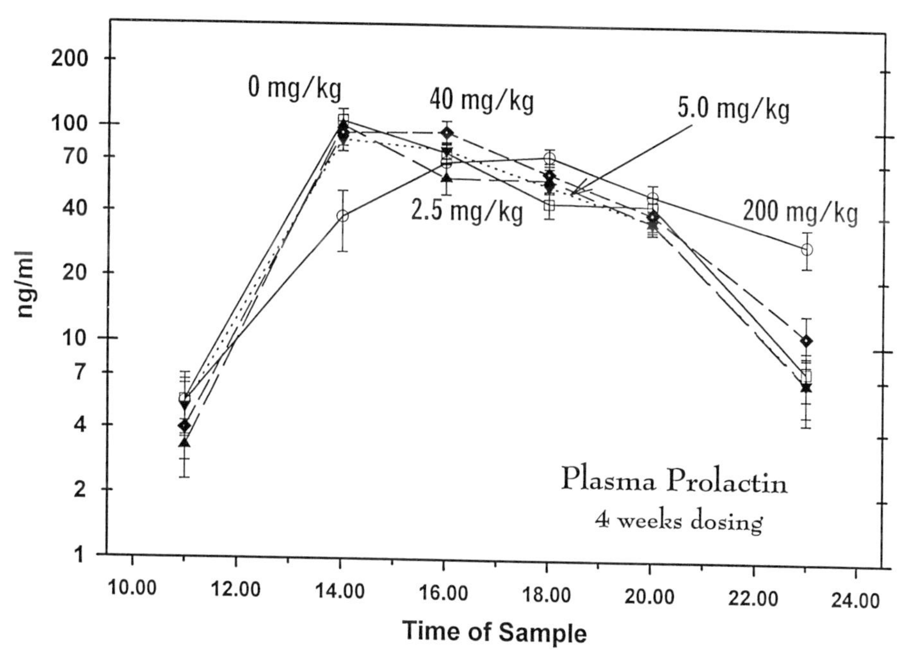

Figure 6. Altered Estrogen-Induced PRL Surge After 4 Weeks Atrazine Treatment. Plasma samples assayed for LH (Figure 4) were also assayed for PRL. Group mean values at 200 mg/kg were significantly lower at 14:00 and higher at 23:00 ($p<0.05$, 1-way ANOVA).

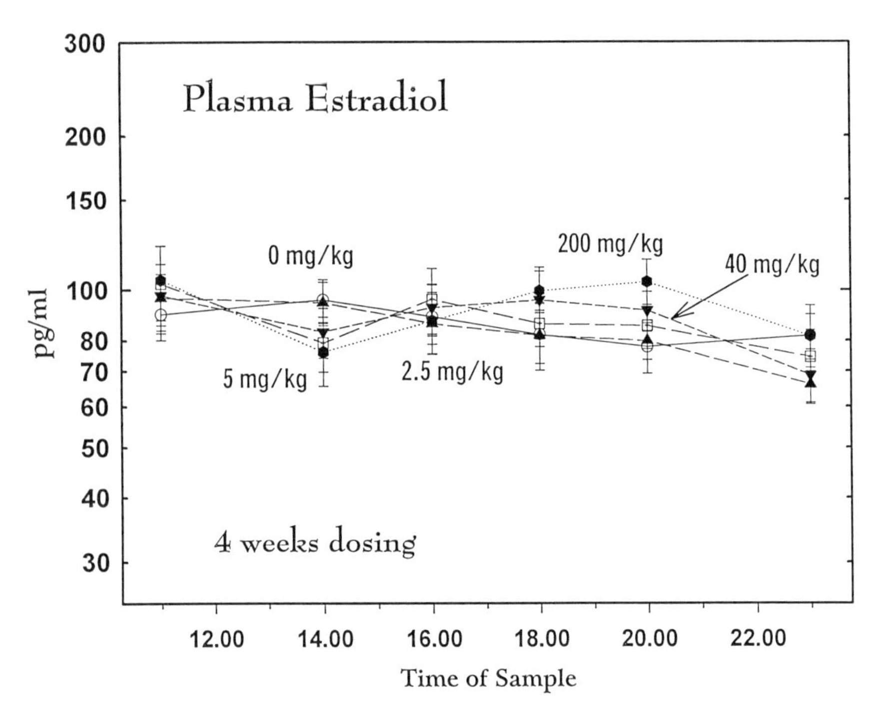

Figure 7. Plasma Estradiol Levels in SD Rats Treated with Atrazine for 4 Weeks. Plasma samples assayed for LH (Figure 4) and PRL (Figure 6) were also assayed for estradiol. There was no significant group mean variance at any time point, suggesting that atrazine dosing had no effect on estrogen levels produced by the silastic implants.

Two years of feeding with atrazine at or near the MTD has been shown to increase further the incidence and/or cause an earlier onset of mammary tumors in the SD female rat, but not in the SD male, or in either sex of the F-344 rat or in several strains of mice *(2-4)*. The explanation for the sex- and strain specificity of the mammary tumor response appears to be related to the pattern of reproductive senescence in female SD rats and the effects of atrazine exposure, at or near the MTD dose, on the time of occurrence of persistent estrus.

A report from Cooper, et.al *(30),* had suggested that very high doses of atrazine could acutely attenuate the LH surge. In the present studies, animals exposed to atrazine acutely or for 30 days also showed a blunting of the estrogen-induced LH surge. This effect consisted of a reduction in the amplitude of the LH surge with little or no affect on its timing. Such a reduction of the endogenous "preovulatory" LH surge by life-long feeding of atrazine would be expected to result in failure of the intact rats to ovulate, and thus to cause an earlier onset of the normally occurring persistent estrus state in SD female rats. This would in turn produce a longer duration of exposure to endogenous estrogen and prolactin and an expected effect on the onset of mammary tumors in the SD rat *(2-4).* Appearance of persistent estrus at an earlier age has been described in atrazine-treated SD rats *(28),* has an increase in the number of days that SD animals are exposed to estrogens *(28).*

The mode of action of atrazine on the estrogen-induced LH and prolactin surges is at present unknown. It is clear that atrazine does not exhibit estrogen agonist activity in *in vivo (5)* and *in vitro (6)* tests. Although a slight estrogen antagonist activity has been observed *(32),* the present results, are not likely attributable to a direct interaction of atrazine with hypothalamic or pituitary estrogen receptors. This is consistent with the observation that atrazine does not exert similar tumor promoting effects in other female rats or mice, as would be expected if atrazine directly interacted with the estrogen receptor. Rather, atrazine appears to interact with a component of the transduction mechanism that converts the estrogen signal into a neuroendocrine response. By so doing, the triazine reduces the estrogen-induced LH surge to levels that are insufficient to induce ovulation at an earlier age than untreated animals. Such an effect results in the persistence of the ovarian follicles and the persistent secretion of estradiol, a normal characteristic of the mid-life reproductive senescence in the SD rat, and this creates an earlier endocrine environment conducive to support mammary tumor growth in the SD female rat.

Literature Cited

1. Brusick, D. J. An assessment of the genetic toxicity of atrazine: relevance to human health and environmental effects. Mutation Research. **1994**, *317*, pp. 133-144.
2. Wetzel, T. L.; Luempert III, L. G.; Breckenridge, C. B.; Tisdel, M. O.; Stevens, J. T.; Thakur, A. K.; Extrom, P. J.; Eldridge, J. C. *Chronic effects of atrazine on estrus and mammary tumor formation in female Sprague-Dawley and Fischer 344 rats. J. Tox. Environ. Health.* **1994**, *43,* pp. 169-182.

3. Thakur, A. K. *Strain-dependent responses to long-term atrazine feeding in female Sprague-Dawley and Fischer 344 rats: Statistical analysis of mammary and pituitary tumors, body weight gain and survival.* Hazleton Report HWA. 1993, 483-275, 483-277, 483-278, 483-279.
4. Mayhew, D. *Two-year feeding/oncogenicity study in rats administered atrazine. American Biogenics Corporation Report.* **1986**.
5. Tennant, M. K.; Hill, D. S.; Eldridge, J. C.; Wetzel, L. T.; Breckenridge, C. B.; Stevens, J. T. *Chloro-s-triazine antagonism of estrogen action: limited interaction with estrogen receptor binding. J. Tox. Environ. Health.* **1994a**, *43,* pp. 197-211.
6. Safe, S.; Chen, L.; Liu, H.; Zacharewski, T. *Failure of atrazine and simazine to induce estrogenic responses in MCF-7 human breast cancer cells.* Report to Ciba-Geigy Corporation, 1995.
7. Welsch, C. W., In *Rodent models to examine in vivo hormonal regulation of mammary gland tumorigenesis;* Cellular and Molecular Biology of Mammary Cancer; Plenum Press: NY, 1987, pp 163-179.
8. Cutts, J. H.; Noble, R. L. *Estrone-induced mammary tumors in the rat. I. Induction and behavior of tumors. Cancer Res.* **1964**, *24*, pp. 1116-1123.
9. Welsch, C. W. *In Hormone and murine mammary tumorigenesis: an historical view;* Leung, B. S., Ed.; Hormonal Regulation of Mammary Tumors; Eden Press, Inc.: Montreal, 1982, Vol 1; pp 1-29.
10. Noble, R. L.; Cutts, J. H. *Mammary tumors of the rat: a review. Cancer Res.* **1959**, *19,* pp. 1125-1139.
11. Cutts, J. H. In *Estrogen-induced breast cancer in the rat.* Proceedings of the Canadian Cancer Research Conf., Honey Harbour; Pergamon Press, Inc.: Ontario, NY, 1965, pp. 50-68.
12. Welsch, C. W.; Adams, C.; Lambecht, L. K.; Hassett, C. C.; Brooks, C. L. *17 ß-Oestradiol and enovid mammary tumorigensis in C3H/HeJ female mice: contraction by concurrent 2-bromo-(α-ergocryptine. Br J. Cancer.* **1977**, *35*, pp. 322-328.
13. Welsch, C. W.; Jenkins, T. W.; Meites J. *Increased incidence of mammary tumors in the female rat grafted with multiple pituitaries. Cancer Res.* **1970**, *30*, pp. 024-1029.
14. Nicosia, S. V. *Ovarian changes during the climacteric, in Aging, Reproduction and the Climacteric;* Mastroianni, Jr., L., Paulsen, C.A., Eds.; Plenum Press: NY, 1986, pp 179-199.
15. Tanaka, Y.; Katayama, K. *Failure of positive feedback on the hypothalamus-pituitary-system in aged-women and its recovery in estrogen treatment.* Acta Obstet. Gynecol. Japan. 1982, *34,* pp. 1907-1915.
16. Meites, J.; Huang, H. H.; Simpkins, J. W. In *Recent studies on neuroendocrine control of reproductive senescence in the rats;* Schneider, E. L., Ed.; The Aging Reproductive System; Raven Press: NY, 1977, pp. 213-235.
17. Simpkins, J. W. In *Changes in hypothalamic hypophysiotropic hormones and neurotransmitters during aging;* Meites, J., Ed.; Neuroendocrinology of Aging; Plenum Press: NY, 1983, pp. 41-59.

18. Wise, P. M. *Norepinephrine and doparnine activity in microdissected brain areas of the middle-aged and young rat on proestrus. Biology Reproduction.* 1982, *27,* pp. 562-574.
19. Wise, P. M. *Estradiol-induced daily luteinizing hormone and prolactin surges in young and middle-aged rats: correlations with age-related changes in pituitary responsiveness and catecholarnine turnover rates in microdissected brain areas. Endocrinology.* **1984**, *115*, pp. 801-809.
20. Huang, H. H.; Steger, R. W.; Bruni, J. F.; Meites, J. *Patterns of sex steroid and gonadotropin secretion in aging female rats. Endocrinology.* **1978**, *103,* pp. 1855-1859.
21. Lu, K. H.; Hopper, B. R.; Vargo, T. M.; Yen, S. S. C. *Chronological changes in sex steroid, gonadotropin and prolactin secretion in aging female rats displaying different reproductive states. Biology of Reproduction.* **1979**, *21*, pp. 193-203.
22. Maurer, R. A. In *Regulation of prolactin gene expression;* Conn, P. M., Ed.; Cellular Regulation of Secretion and Release; Academic Press: 1982, pp. 167.
23. Estes, K. S.; Simpkins, J. W. *Age-related alternatives in catecholarnine activity within microdissected brain regions of ovariectomized Fischer 344 rats. J.Neuroscience Research.* **1984a**, *11*, pp. 405-417.
24. Lu, J. K. H.; Darnassa, D. A.; Gilman, D. P.; Judd, H. L.; Sawyer, C. H. *Differential patterns of gonadotropin responses to ovarian steroids and to LH-releasing hormone between constant-estrous and pseudopregnant states in aging rats. Biology Reproduction.* **1980**, *23*, pp. 345-351.
25. Estes, K. S.; Simpkins, J. W.; Kalra, S. P. *Normal LHRH neuronal function and hyperprolactinemia in old pseudopregnant Fischer 344 rats. Neurobiology of Aging.* **1982**, *3,* pp. 247-252.
26. Estes, K. S.; Simpkins, J. W. *Resumption of pulsatile luteinizing hormone release after a-adrenergic stimulation in aging constant estrous rats. Endocrinology.* **1982,** *111*, pp. 1776-1784.
27. Beach, J. E.; Tyrey, L.; Everett, J. W. *Serum prolactin and LH in early phases of delayed and direct pseudopregnancy in the rat. Endocrinology.* **1975**, *96*, pp. 243-1246.
28. Eldridge, J. C.; Wetzel, L. T.; Tisdel, M. O.; Luempert III, L. G. Determination of hormone levels in Sprague-Dawley rats treated with atrazine technical. HMA Study No. 483-278, 1993a.
29. Eldridge, J. C.; Wetzel, L. T.; Tisdel, M. O.; Luempert III, L. G. Determination of hormone levels in Fischer 344 rats treated with atrazine technical. HMA Study No. 483-279, 1993b.
30. Cooper, R. L.; Stoker, T. E.; Goldman, J. M.; Parrish, M. B.; Tyrey, L. *Effect of atrazine on ovarian function in the rat. Reproductive Toxicology.* **1996,** *10*, pp. 257-264.
31. Ajika, K.; Kalra, S. P.; Fawcet, C. P.; Krulick, L.; McCann, S. M. *The effects of stress and nembutal on plasma levels of gonadotropins and prolactin in ovariectomized rats. Endocrinology.* **1972,** *90*, pp. 707-715.
32. Tennant, M. K.; Hill, D. S.; Eldridge, J. C.; Wetzel, L. T.; Breckenridge, C. B.; Stevens, J. T. *Possible antiestrogenic properties of chloro-s-triazines in rat uterus. J. Tox. Environ. Health.* **1994b,** *43,* pp. 183-196.

Chapter 32

Appearance of Mammary Tumors in Atrazine-Treated Female Rats: Probable Mode of Action Involving Strain-Related Control of Ovulation and Estrous Cycling

J. Charles Eldridge[1], Robert F. McConnell[2], Lawrence T. Wetzel[3], and Merrill O. Tisdel[3]

[1]**Department of Physiology and Pharmacology, Bowman Gray School of Medicine of Wake Forest University, Winston-Salem, NC 27157**
[2]**Consulting Pathology Services, Flemington, NJ 08822**
[3]**Department of Toxicology, Novartis Crop Protection Inc., P.O. Box 18300 Greensboro, NC 27419**

Atrazine has been a principal broadleaf herbicide for over 25 years, used primarily for weed control in corn and sorghum. Atrazine is a chlorinated member of the s-substituted triazines that selectively inhibit electron transport systems in photosynthesis (*1*). Because of its specificity and sensitivity in plants (K_i in chloroplasts = 1.4×10^{-7} M) (*2*), atrazine poses a low toxic potential for humans, livestock and animal wildlife. The acute oral LD_{50} has been reported as ≈ 3000 mg/kg (1.39×10^{-2} mol/kg) in rodents, and a maximum tolerated dose (MTD) in chronically treated rats as approximately 25 mg/kg (1.16×10^{-4} mol/kg) (*3*). Atrazine is not a mutagen, teratogen or reproductive toxin (*4*).

Triazine herbicides have been tested in a wide variety of toxicological investigations since their introduction in 1958. A more extensive discussion of those results (*5*) is presented elsewhere in this volume, but in summary the compounds are generally free of toxic effects in test animals at chronic exposure levels > 0.5 mg/kg/day. However, long-term oral treatment, at doses approaching or exceeding the MTD, has been associated with a reduced onset time and/or increased final incidence, of mammary tumors in Sprague-Dawley (SD) female rats. Earlier onset or increased incidence have been observed in four of five studies of SD females, but a number of other lifetime exposure studies, using Fischer-344 (F344) female rats, male SD and F344 rats, or male and female CD-1 mice, have produced no increased appearance of mammary tumors (*3*). The negative results in other strains and species indicate that atrazine is not a direct-acting, genotoxic carcinogen. A weight-of-evidence evaluation of more than 50 direct tests of genotoxicity concluded that triazines are not genotoxic or mutagenic (*6,7*).

Mammary Tumors in Aging Sprague-Dawley Rats

A key observation in evaluating the relationship between mammary tumors and atrazine treatment was that the only responsive model, namely the SD female rat, has a high background incidence of these tumor types. For example, an average 61% mammary tumor incidence rate was recently reported *for the vehicle controls* in 12 long-term studies of various triazine herbicides (*8*). By contrast, a rat model with a normally low spontaneous incidence of mammary tumors (F344) had no response to atrazine exposure. Furthermore, careful morphologic examination found that the lesions in atrazine-treated animals were not histologically different from those in the controls (*3*). There were simply more animals with the same tumor type at a given age (Figure 1). It therefore seemed highly suggestive that the atrazine-associated tumor response, seen in only one sex of one strain of one species, was linked to the natural tendency of that strain to develop the tumors, and this prompted questions concerning the origin of spontaneous (i.e., those normally appearing in untreated animals) mammary tumors in the SD strain.

It has been clearly established that mammary tumors in rodents are promoted by hormones. The presence of estrogens and/or prolactin stimulates mammary tumor growth, while inhibition or removal of either hormone results in tumor regression (*9-11*). It is also well known that SD female rats undergo normal age-related estrous cycling changes resulting in persistent secretion of and continuous exposure to elevated estrogen and prolactin. Control of estrous cycling begins to decline at a relatively young age, with some deficits appearing as early as 8-10 months of age (*12,13*). Stimulation of pituitary gonadotropin release by gonadotropin-releasing hormone (GnRH) becomes compromised due to deficient function of certain noradrenergic neurons in the hypothalamic region of the brain (*14,15*). As the neurons become increasingly unable to respond to rising estrogen titers secreted from developing ovarian follicles (*16,17*), GnRH stimulation of follicle stimulating hormone (FSH) and luteinizing (LH) is subdued. Gonadotropin-stimulated ovulation is then delayed or absent, while the ovarian follicles remain functional for extended periods of time, secreting estrogen. Vaginal cytology reflects this endocrine *milieu* by displaying a highly keratinized epithelium called persistent vaginal cornification (PVC). Episodes of PVC are common in 9-15-month old SD female rats (*3,13,15*), but not F344 rats (*18*). Furthermore, other estrogen-responsive tissues, including uterus, mammary gland and anterior pituitary, become excessively stimulated. Tonic prolactin secretion is elevated which, together with persistent estrogen secretion, would promote mammary tumor growth. This also suggests why other animal models have a low spontaneous tumor incidence; aging F344 females exhibit lower tonic estrogen and prolactin levels (*18*), as do all male rodents.

Understanding these strain-related differences in reproductive aging leads to a prediction that a rodent strain whose control of ovarian function declines as described above for SD rats will demonstrate, with advancing age, a complete endocrine *milieu* supporting growth of mammary tumors, while another strain or animal model that does not change in such a manner will demonstrate a different

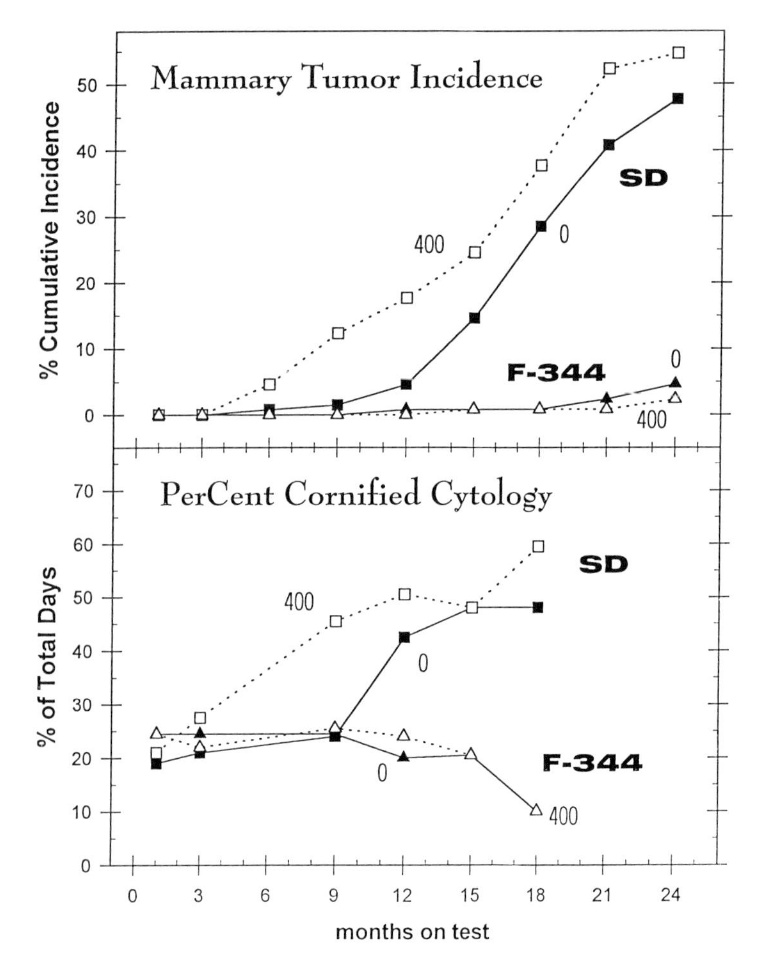

Figure 1 (Upper Panel). Cumulative Incidence of Mammary Tumors. Plots show percentage of total animals (130 per dose group) displaying palpable masses that were also confirmed as mammary tumors at necropsy. Female SD or F344 rats were fed a diet supplemented with atrazine: 0 ppm (filled symbols, solid lines) or 400 ppm (open symbols, dotted lines).

(Lower Panel). Percent of Total Days Sampled Displaying a Vaginal Cytology Pattern of Predominantly Dense, Cornified Cells. Symbols represent group means of percentage of days that a cornified vaginal smear was observed during 15-21-days just prior to sacrifice. Female SD and F344 rats were continuously fed a diet supplemented with atrazine at 0 ppm (filled symbols, solid lines) or 400 ppm (open symbols, dotted lines). Groups of 10/dose were sacrificed at the indicated intervals.

endocrine *milieu* which will not support mammary tumor growth with aging. The prediction yielded a correlated hypothesis for testing: high-dose, chronic atrazine exposure in a susceptible strain (SD) should enhance the demonstrable age-related changes at a faster rate but in the same direction as those which occur normally. A non-susceptible strain (F344) should show neither evidence of the tumor-supporting endocrine *milieu* nor enhancement or advancement of tumor formation above the normally occurring background rate with atrazine treatment.

Further Analysis of Strain-Related Responses to Atrazine Dosing

As described above, lifetime atrazine feeding studies of SD and F344 rats have been completed. Atrazine dosing at 400 ppm in the diet had advanced the rate of mammary tumor development in the susceptible SD strain (i.e., a high spontaneous background) but not in the non-susceptible F344 strain (i.e., a low spontaneous background) (Figure 1, upper panel). During this study, groups of each dose and strain were sacrificed and necropsied after 1, 3, 9, 12, 15, 18 and 24 months. Vaginal cytology was monitored daily during the final 2-3 weeks for those animals selected for sacrifice.

Substantial but predictable strain differences developing over time were noted in the appearance of cornified epithelium in vaginal cytology (Figure 1, lower panel). SD rats are prone to estrous cycle deterioration in mid-life (*12,13*), and in this study the percentage of days with a highly cornified vaginal cytology, indicative of persistent estrus, increased between 9 and 12 months in controls, but earlier in rats fed 400 ppm atrazine. This vaginal cytology change, among control as well as atrazine-treated animals, was suggestive of an endocrine *milieu* of persistent estrogen secretion. This biologic index of estrogen activity has been frequently observed in aging SD female rats as a consequence of estrogen secretion from ovarian follicles that resulted from a failure to ovulate in a timely fashion (*12,13*).

By contrast, the degree of vaginal cornification in F344 females had progressively declined in control rats and in females from the 400 ppm feeding level (Figure 1, lower panel). This suggested that atrazine contains no intrinsic estrogenic activity, because high dosing did not increase cornification over control levels, and additionally that the control of estrous cycling in the F344 strain is not so easily disrupted by xenobiotic exposure as it is in SD rats.

These dose- and strain-related changes to treatment were further explored by conducting a more detailed examination of potentially affected tissues. A total of 34 determinants were assessed in coded (origin of dose group unknown) histologic sections of ovary, uterus, vagina and mammary gland of each test animal, and each determinant was assigned a subjective score of 0 to 4. An analysis of 3 parameters in ovary and mammary tissue is presented here: mammary acinar-lobular development, mammary galactocele development, and ovarian follicular development.

Tissue Changes With Normal Aging and Combined With Atrazine Treatment

Sections of mammary tissue were assessed for acinar-lobular development, representing fundamental growth of the secretory apparatus of the gland, and results are illustrated in Figure 2 (upper panel). Because mammary gland maturation and function are completely dependent upon hormone exposure, rats have essentially no acinar-lobular development when very young, and a cumulative response to estrogen and prolactin becomes gradually evident over time in nulliparous animals (*11*). Among controls (0 ppm), normal acinar-lobular development progressed much more rapidly in SD than in F344 rats, beginning as early as 3 months on test (5 months of age). When SD rats were treated with 400 ppm atrazine, acinar-lobular development was stimulated earlier and to a greater degree than vehicle controls, and was essentially maximal by 9 months on test. Acinar-lobular development in F344 rats was much lesser than in SD rats, and was not at all affected by atrazine dosing (Figure 2, upper panel). This lack of treatment-related response in F344 underscored a low potential for direct atrazine action on mammary structure, and suggested rather that the observed responses in SD rats were more likely from endogenous sources (i.e., modulation of the animal's own endocrine secretions).

Galactoceles are milk-containing cystic structures not normally seen in mammary tissue of young nulliparous animals (see early 0 ppm control results, Figure 2, lower panel). These pathologic, non-neoplastic structures develop after prolonged, excessive mammary gland stimulation by estrogen and prolactin. Examination of mammary tissues from control SD rats showed a progressive appearance of galactoceles at 9-18 months, which coincided with other published studies of the timing of disrupted estrous cycling and persistent estrogen and prolactin secretion (*12,13*). Galactocele development was strikingly enhanced in 400-ppm atrazine-treated animals at all ages from 9 months onwards, and this finding suggests that the duration of estrogen and prolactin exposure was much greater in the treated animals. Enhanced galactocele development occurred later and to a lesser degree in the control F344 rats. In contrast to the 400-ppm-treated SD rats, there was no earlier onset of this finding in the 400-ppm group F344 females (Figure 2, lower panel). This result in F344 rats also confirmed the absence of direct estrogenic or prolactin-like activity of atrazine.

Ovarian tissues were also examined for signs of age- or treatment-related changes. Because persistent estrogen secretion in aging SD rats could arise from developed follicles that fail to ovulate, an evaluation was made of the two principal follicular types: Graafian (mature and presumably functional) follicles and atretic (regressed) follicles. The results of the combined two measures (Figure 3) showed that both SD and F344 control rats had similar levels of ovarian follicular development through 9 months on test. Follicular development among SD rats treated with 400 ppm atrazine began to increase more rapidly through 9 months on test and remained there through 18 months (Figure 3). Once again, this parameter of follicular development was unaffected by identical atrazine treatment in F344 rats.

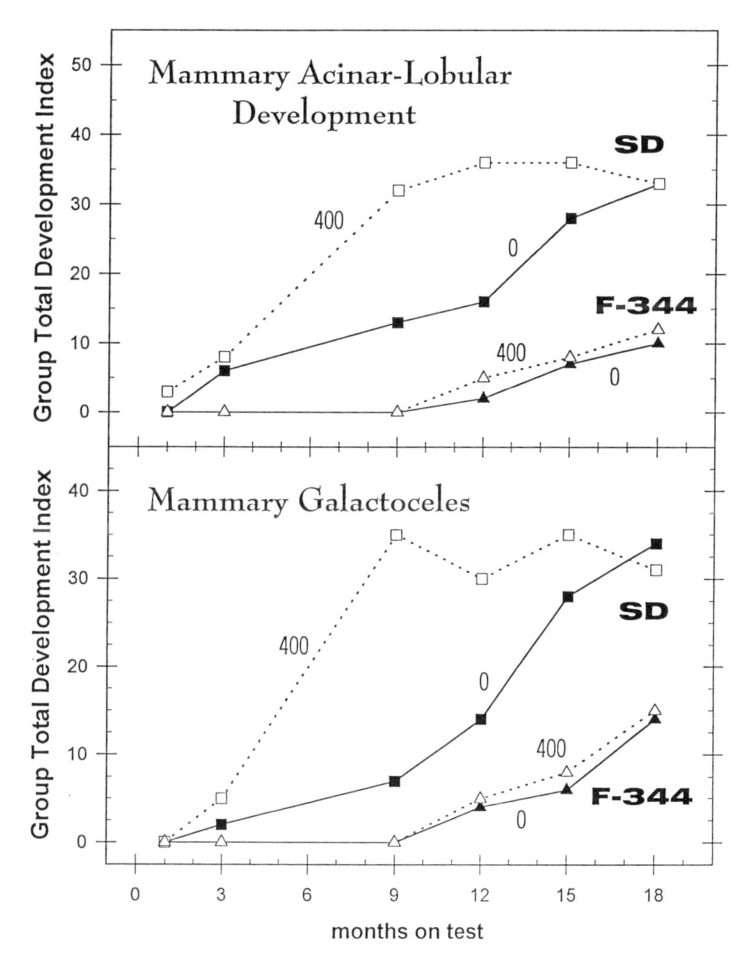

Figure 2 (Upper Panel). Development of Acinar-Lobular Structures in Mammary Glands. Symbols represent group total, where each animal was scored 0-4 for development of acinar-lobular structures in histologic sections of mammary tissues. Female SD and F344 rats were continuously fed a diet supplemented with atrazine at 0 ppm (filled symbols, solid lines) or 400 ppm (open symbols, dotted lines). Groups of 10/dose were sacrificed at the indicated intervals.

(Lower Panel). Galactocele Development in Mammary Glands. Symbols represent group total, where each animal was scored 0-4 for degree of galactocele development in histologic sections of mammary tissues. Female SD and F344 rats were continuously fed a diet supplemented with atrazine at 0 ppm (filled symbols, solid lines) or 400 ppm (open symbols, dotted lines). Groups of 10/dose were sacrificed at the indicated intervals.

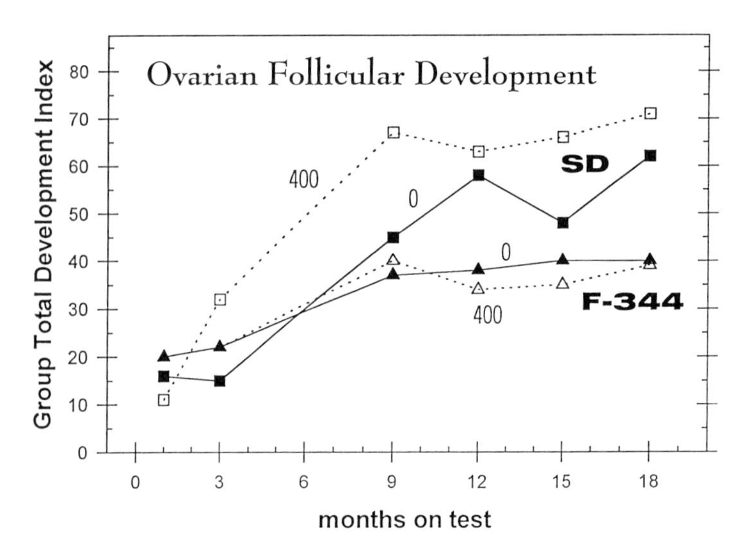

Figure 3 - Development and Density of Graafian and Atretic Follicles in Ovarian Tissue. Each animal was scored 0-4 to describe the size and density of mature Graafian follicles, and 0-4 to describe the density of atretic follicles, in histologic sections of ovary. The symbols represent the sum of each group's total for the two follicular types. Female SD and F344 rats were continuously fed a diet supplemented with atrazine at 0 ppm (filled symbols, solid lines) or 400 ppm (open symbols, dotted lines). Groups of 10/dose were sacrificed at the indicated intervals.

Discussion: Mode of Action of Long-Term, High-Dose Atrazine Feeding

Lifetime exposure of female SD rats to high levels of atrazine resulted in an earlier appearance of mammary tumor growth (Figure 1). However, because atrazine is neither estrogenic (*19,20*), nor a genotoxic, direct-acting carcinogen, and because the atrazine-associated tumor responses appeared only in female SD rats, a strain with a high, normally occurring incidence of mammary tumors, it became important to look for evidence that the test model's own endocrine system and hormonal *milieu* may have been perturbed by atrazine treatment.

Because mammary tumors in rodents are dependent upon support from estrogen and/or prolactin, the tissue samples were evaluated for histologic evidence that persistent secretion of these hormones might have occurred earlier in atrazine-treated animals. Results shown in Figure 1 (lower panel) and in Figure 2 clearly indicate that, with high-dose atrazine exposure, female SD rats were being exposed to significantly more endogenous estrogen and prolactin during the mid-portion (and probably earlier portions as well) of their lives.

Since the nature and degree of the treatment-related changes, including mammary tumor responses, were similar to those in control animals, only appearing sooner, a hypothesis was proposed that the type *and chronology* of reproductive senescence in SD rats is directly correlated with the potential for mammary tumor growth. That is, a direct linkage is proposed between the occurrence of endocrine changes in rats and mammary tumor growth, and that it can be modulated to occur at an earlier or later phase of life. If a strain, such as SD, undergoes age-related changes that result in persistent, earlier secretion of estrogens and prolactin, then a high incidence of mammary tumors is predictable during late life. If a strain, such as F344, does not change in a manner that stimulates estrogen and prolactin production, the potential for mammary tumor development is likely to be less. Long-term treatment with atrazine provided tests of this hypothesis by demonstrating that, as the endocrine *milieu* of one strain (SD) was shifted toward earlier persistent estrogen and prolactin secretion, mammary tumor development became concomitantly shifted. Furthermore, another rat strain (F344), whose endocrine environment failed to respond as the SD strain, was likewise unresponsive regarding mammary tumor development.

These experiments showed that atrazine dosing had no direct endocrine consequence or action; rather, that it modulated the test animals' own hormonal secretions toward their natural tendencies of age-related change. In the case of SD female rats, the tendency to begin a sequence of endocrine events ultimately leading to mammary tumor development was advanced to an earlier time as a result of dosing.

Tissues from SD (and F344) rats treated with 70 ppm atrazine were also examined in these studies. While not presented in this document, results showed changes consistently similar to those from the control animals. It is therefore proposed in SD rats, that, a threshold dose of chronic atrazine exposure above 70 ppm was necessary to perturb the endocrine system sufficiently to alter mammary tumor incidence. The incidence of mammary tumor development has already been observed not to increase at the 70 ppm exposure level (*3,5*).

Mammary tumor growth in senescent rodents is inexorably tied to the host endocrine environment, and to perturbations of that environment. A clue to understanding mammary tumor development comes from the fact that the altered endocrine environment of the aging SD female rat results from persistent estrogen and prolactin secretion. The altered environment originates from insufficient hypothalamic-pituitary support for timely ovulation. If neuroendocrine control in this model spontaneously declines in mid-age, it becomes logical to propose that high-dose atrazine treatment could perturb this control further and to force the decline at an earlier age. Evidence in support of this proposal is reported in a companion paper (*21*). Timely ovulation in rats requires complex neuroendocrine activity culminating in a massive surge of LH. Integrity of this LH surge declines in aging SD female rats, resulting in maintenance of ovarian follicles that fail to ovulate and that continue to secrete estrogen (*12,13*). Short-term treatment with atrazine at very high doses has been shown to suppress the LH surge as well (*21,22*). Thus, atrazine treatment-related alterations of preovulatory gonadotropin secretion could easily set in place a chain of events leading to the observed mammary tumor responses in SD rats.

Literature Cited

1. Good, N.E. *Inhibitors of the Hill reaction. Plant Physiol.* **1961**, *36*, pp. 788-803.
2. Tischer, W.; Strotmann, H. *Relationship between inhibitor binding by chloroplasts and inhibition of photosynthetic electron transport. Biochem Biophys Acta.* 1977, *460,* pp. 113-125.
3. Wetzel, L.T.; Luempert, L.G. III; Breckenridge, C.B.; Tisdel, M.O.; Stevens, J.T.; Thakur, A.K.; Extrom, P.J.; Eldridge, J.C. *Chronic effects of atrazine on estrus and mammary tumor formation in female Sprague-Dawley and Fischer-344 rats. J Toxicol Environ Health.* **1994**, *43*, pp. 169-182.
4. Stevens J.T.; Sumner D.D. In: *Handbook of Pesticide Toxicology. Volume 3: Classes of Pesticides;* Hayes, J., Ed.; Academic Press: New York, 1991, pp. 1317-1391.
5. Hauswirth, J.W.; Wetzel, L.T. Toxicity characteristics of the 2-chlorotriazines atrazine and simazine. This volume. 1997.
6. Franekic, J.; Hulina, G.; Kniewald, J.; Alacevic, M. *Atrazine and the genotoxicity of its metabolites. Environ Mol Mutagen.* **1989**, *14*, pp. 62.
7. Brusick, D.J. *An assessment of the genetic toxicity of atrazine: relevance to human health and environmental effects. Mutation Res.* **1994,** *317*, pp. 133-144.
8. Stevens, J.T.; Breckenridge, C.B.; Wetzel, L.T.; Gillis, J.H.; Luempert, L.G. III; Eldridge, J.C. *Hypothesis for mammary tumorigenesis in Sprague-Dawley rats exposed to certain triazine herbicides. J Toxicol Environ Health* **1994**, *43,* pp. 155-168.

9. Welsch, C.W. *Host factors affecting the growth of carcinogen-induced mammary carcinomas. A review and tribute to Charles Brenton Huggins. Cancer Res.* **1985**, *43*, pp. 3415-3443.
10. Thompson, H.J.; Ronan, A. *Effect of D, L-α-difluoromethylornithine and endocrine manipulation on the induction of mammary carcinogenesis by l-methyl, l-nitrosourea. Carcinogenesis.* **1987**, *57,* pp. 2003-2006.
11. Russo, J.; Gusterson, B.A.; Rogers, A.E.; Russo, J.; Wellings, S.R.; van Zweiten, M.J. *Comparative study of human and rat tumorigenesis. Lab Invest.* **1990**, *62,* pp. 244-278.
12. Simpkins, J.W. In: *Neuroendocrinology of Aging;* Changes in hypothalamic-hypophysiotropic hormones and neurotransmitters during aging; Meites, J., Ed.; Plenum Press: New York, 1983, pp. 41-59.
13. Wise, P.M.; Scarbrough, K.; Larson G.H.; Lloyd, J.M.; Weiland, N.G.; Chiu, S. *Neuroendo-crine influences on aging of the female reproductive system. Front Neuroendocr.* **1991**, *12*, pp. 323-356.
14. Estes, K.S.; Simpkins, J.W. *Resumption of pulsatile luteinizing hormone release after α-adrenergic stimulation in aging constant estrous rats. Endocrinology.* **1982a**, *111*, pp. 1776-1784.
15. Estes, K.S.; Simpkins, J.W. *Age-associated alterations in dopamine and norepinephrine activity within microdissected brain regions of ovariectomized Long-Evans rats. Brain Res.* **1984,** *298*, pp. 209-218.
16. Wise, P.M.. *Norepinephrine and dopamine activity in microdissected brain areas of the middle-aged and young rat on proestrus. Biol Reprod.* 1982, *27*, pp. 562-574.
17. Wise, P.M. *Estradiol-induced daily luteinizing hormone and prolactin surges in young and middle-aged rats: correlations with age-related changes in pituitary responsiveness and catecholamine turnover rates in microdissected brain areas. Endocrinology.* **1984**, *115*, pp. 801-809.
18. Estes, K.S.; Simpkins, J.W.; Kalra, S.P. *Normal LHRH neuronal function and hyperprolactin-emia in old pseudopregnant Fischer-344 rats. Neurobiol Aging.* **1982b**, *3*, pp. 247-252.
19. Tennant, M.K.; Hill, D.S.; Eldridge, J.C.; Wetzel, L.T.; Breckenridge, C.B.; Stevens, J.T. *Anti-estrogenic properties of chloro-s-triazines in rat uterus. J. Toxicol. Environ. Health.* **1994**, *43*, pp. 183-196.
20. Connor, K.; Howell, J.; Chen, I.; Liu, H.; Berhane, K.; Sciarretta, C.; Safe, S.; Zacharewski, T. *Failure of chloro-s-triazine-derived compounds to induce estrogen receptor-mediated responses in vivo and in vitro. Fund Appl Toxicol.* **1996**, *30*, pp. 93-101.
21. Simpkins, J.W.; Eldridge, J.C.; Wetzel, L.T. Role of strain-specific reproductive patterns in the appearance of mammary tumors in atrazine-treated female rats. This volume. 1997.
22. Cooper, R.L.; Stoker, T.E.; Goldman, J.M.; Parrish, M.B.; Tyrey, L. *Effect of atrazine on ovarian function in the rat. Reprod Toxicol.* **1996**, *10*, pp. 257-264.

Chapter 33

Failure of Chloro-*s*-triazine-Derived Compounds to Induce Estrogenic Responses In Vivo and In Vitro

K. Connor[1], J. Howell[1], S. Safe[1], I. Chen[1], H. Liu[1], K. Berhane[2], C. Sciarretta[2], and T. Zacharewski[2]

[1]Department of Veterinary Physiology and Pharmacology, Texas A & M University, College Station, TX 77843-4466
[2]Department of Pharmacology and Toxicology, Unviersity of Western Ontario, London, Ontario, N6A 5C1 Canada

The estrogenic activity of simazine and atrazine was investigated in the immature female Sprague-Dawley rat uterus, the estrogen-responsive MCF-7 human breast cancer cell line and PL3 yeast strain. Atrazine and simazine (50 - 300 mg/kg x 3) did not induce rat uterine wet weight, cytosolic progesterone receptor (PR) levels, or uterine peroxidase activity. In rats cotreated with 17β-estradiol (E2) plus atrazine or simazine, there was some inhibition of E2-induced uterine PR binding and peroxidase activity. In MCF-7 cells, simazine and atrazine did not affect E2-induced cell proliferation, nuclear PR levels or luciferase activity in cells transiently transfected with the Gal4-estrogen receptor chimera and a Gal4-regulated luciferase reporter gene, and no antiestrogenic activity was observed in cotreated cells, while growth was observed on similar media supplemented with 1 nM E2. These results indicate that atrazine and simazine are not estrogenic.

It has been reported that chloro-*s*-triazine herbicides affect hormonal homeostasis in rodents and several studies have investigated possible mechanisms associated with this activity (*1-6*). Female Sprague-Dawley rats maintained on dietary levels (0 - 400 ppm) of atrazine for 24 months developed a high incidence of mammary tumors in both control and treated animals; however, the onset time for mammary tumor formation was decreased in the latter treatment group (*1*). In contrast, no effects on mammary tumor formation were observed in female Fischer 344 rats maintained on the same dietary levels of atrazine and there was a significant decrease in pituitary tumors in the high dose atrazine group (440 ppm) (20/128) compared to spontaneous pituitary tumor formation in untreated (control) animals (33/130). Results of other studies showed that atrazine lengthened the estrous cycle in Sprague-Dawley rats (*3*). Tennant and coworkers (*2,3*) also reported a weak interaction between several chloro-*s*-triazines and the estrogen receptor (ER) using a competitive binding assay in which rat uterine cytosol was preincubated with the chloro-*s*-triazines prior to addition of [^{3}H]17β-estradiol (E2). ER binding was not observed in direct competition assays in which both hormone and triazines were added simultaneously. Subsequent studies in the female Sprague-Dawley rat uterus suggested that the chloro-*s*-triazines may be antiestrogenic. For example, atrazine, simazine and diaminochloro-*s*-triazine decreased E2-induced uterine wet weight increase and progesterone receptor (PR) levels in ovariectomized female Sprague-Dawley rats and E2-induced [^{3}H]thymidine uptake (uterine) in immature rats. The antiestrogenic activities of this compound could be the results of direct competition with the ER or through interactions between the ER and other endocrine response pathways.

Research in our laboratories has focused on the inhibition of E2-induced responses by 2,3,7,8-tetrachlorodibenzo-*p*-dioxin (TCDD) and related compounds which bind the aryl hydrocarbon (Ah) receptor (*7,8*). Ah receptor agonists inhibit a diverse spectrum of E2-induced responses in the rodent uterus and mammary and in human breast cancer cell lines. This study initially investigated the potential Ah receptor agonist activity of simazine and triazine as an explanation for their reported antiestrogenic activities. Using the female rat uterus, MCF-7 human breast cancer cells and an estrogen responsive yeast assay, the estrogenic and antiestrogenic activity of simazine and triazine were also investigated. The results of these studies show the chloro-*s*-triazines are neither ER or Ah receptor agonists; however, some dose-independent antiestrogenic activities were observed in the rat uterus (*9*).

Materials and Methods

***In Vivo* Studies.** Nineteen-day-old Sprague-Dawley rats were obtained from Harlan-Sprague-Dawley (Houston, TX) and allowed *ad libitum* access to food and water. Different concentrations of the triazines were dissolved in 1 ml DMSO and suspended in an appropriate volume of water and 5% hydroxypropyl cellulose. Twenty-one day-old rats (4 per treatment group) received 0.5 ml of a triazine suspension at various doses (po) or vehicle control for 3 days. E2 (in corn oil) was administered by ip injection (10 μg/kg) on days 21 - 23 and control rats received only corn oil. Animals were euthanized by carbon dioxide asphyxiation 20 hrs. after the last treatment and the uteri were quickly removed, cleansed of connective tissue, weighed, blotted, and placed in ice-cold buffer. The uteri were then bisected such that each half contained an entire uterine horn. Uterine tissue was subsequently utilized to investigate PR levels and uterine peroxidase activity using pooled samples for each treatment group as previously described (*9-12*). The results are expressed as means ± SD, and statistical differences were determined by Duncan's new multiple range test (ANOVA).

***In Vitro* Studies with MCF-7 Cells.** For the cell proliferation assay, MCF-7 cells were seeded at 50,000 cells/well in six-well plates with 2 ml of DME/F12 (without phenol red) supplemented with 2.2 g/liter $NaHCO_3$, 10 mg/liter apotransferrin, 200 mg/ml bovine serum albumin, and 5% DCC-fetal bovine serum (FBS). The cells were allowed to stabilize and attach for 24 hrs. The chemicals E2, simazine, atrazine, or their combination in DMSO were added to the media such that the final DMSO concentration was <0.1% for all experiments. After 11 days, the cell numbers were counted as previously described (*9*). PR binding assays were determined using MCF-7 cells treated with various chemicals for 3 days. Nuclear extracts were used to determine PR binding to the progesterone response element (PRE) as previously described (*13*). Transient transfection assays utilized MCF-7 cells which were transfected with 5 µg pCH110 (β-galactosidase expression vector, Pharmacia), 5 µg 17 m5-G-Luc (17-mer-regulated luciferase reporter gene provided by Dr. P. Chambon, IGBMC, Illkirch, France), 1 µg Gal4-HEGO (Gal4-estrogen receptor chimera), and 4 µg pBS (carrier DNA, Stratagene) per dish (*14*). All transient transfections were performed using the calcium phosphate coprecipitation technique as described.

Studies on different dose groups were performed in duplicate and two samples were taken from each extract. The results are reported as a % relative to the maximum induction observed using 1 nM E2 and are means ± SD for at least four determinations.

Results and Discussion

Initial studies showed that simazine and atrazine did not competitively bind to the Ah receptor and therefore the reported antiestrogenic activity was not associated with interactions between the Ah receptor and ER signalling pathways. The results in Figure 1 summarize the effects of simazine, atrazine, simazine plus E2 and atrazine plus E2 on uterine wet weight, peroxidase activity and cytosolic PR binding in immature female Sprague-Dawley rats. At doses as high as 300 mg/kg, atrazine and simazine did not significantly decrease uterine wet weight; however, both compounds decreased basal uterine peroxidase activity and cytosolic PR binding. These results are consistent with previous reports (3) which showed that atrazine alone decreased PR and [^{3}H]thymidine uptake but did not affect uterine wet weight. The potential antiestrogenic activity of atrazine and simazine was also investigated and the results (Figure 1) demonstrated some inhibition of E2-induced responses; however, the antiestrogenic effects were not dose-dependent for either compound. Previous studies using ovariectomized young adult female Sprague-Dawley rats (*3*) reported that atrazine and simazine significantly decreased E2-induced uterine wet weight and uterine PR levels indicating that the antiestrogenic effects of these chloro-*s*-triazines were also different in the 2 rat uterine models. The results of these *in vivo* studies suggest that atrazine and

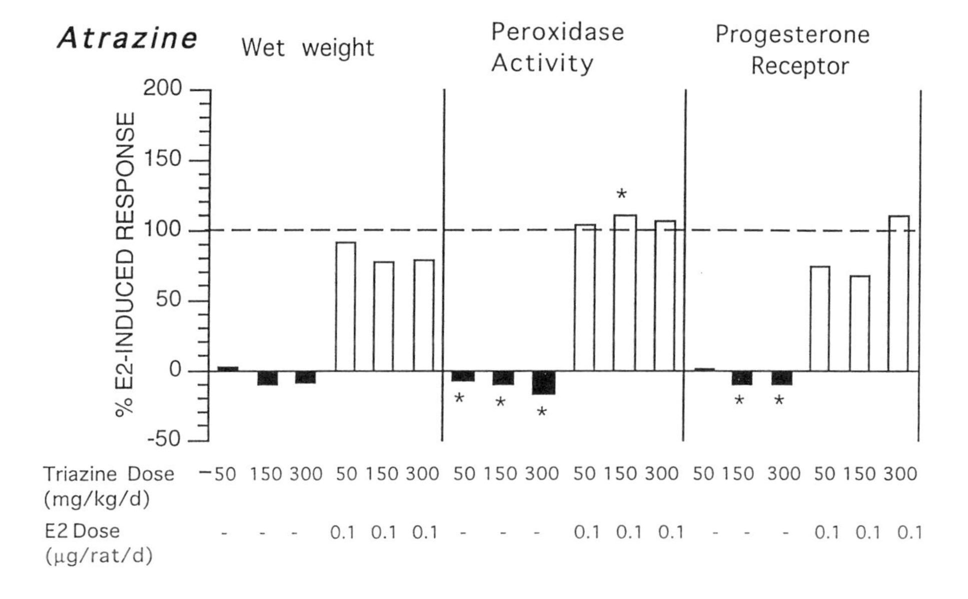

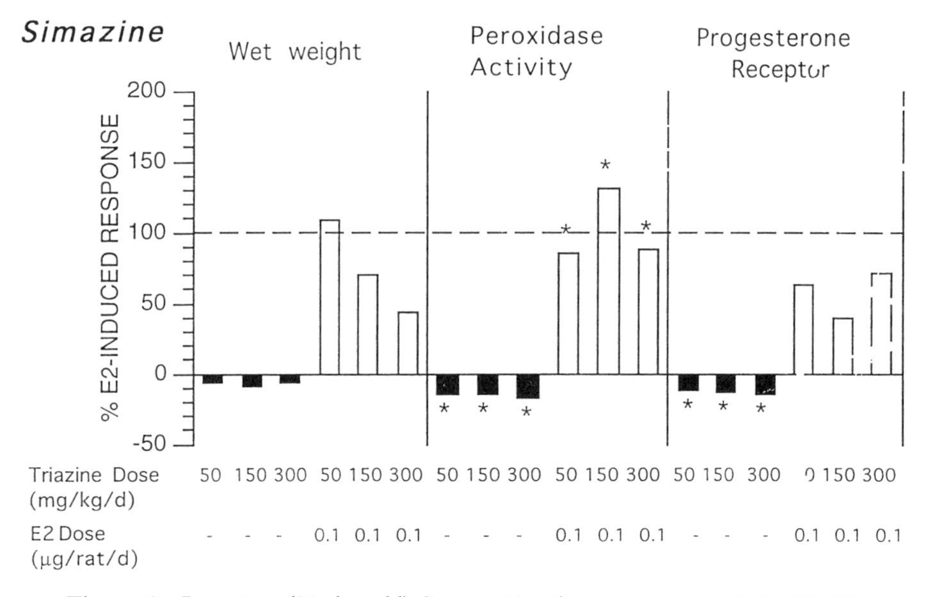

Figure 1. Immature (21 day-old) Sprague-Dawley rats were treated with 50, 150 or 300 mg/kg/d atrazine or simazine alone (solid bars) or in combination with E2 (open bars) for 3 days. Uterine wet weight, peroxidase activity and cytosolic PR levels were measured as described in Materials and Methods and are shown as % of E2-induced response. Dashed line indicates response caused by E2 treatment alone (100%) and solid line indicates levels in corn oil treated controls (0%). * indicates a significant treatment-related effect compared with corn oil controls (on solid bars) or compared with E2-treated groups (on open bars) ($p < 0.05$).

simazine do not exhibit estrogenic activity in the rat uterine model; however, the antiestrogenic activity of these compounds via direct competition for the ER cannot be excluded.

MCF-7 human breast cancer cells have been extensively used as an *in vitro* system for investigating the effects of both estrogens and antiestrogens on cell growth and gene expression. Previous studies in our laboratories have shown that E2 induced proliferation of MCF-7 cells and increased nuclear PR levels as determined by ligand binding or formation of a PR-PRE retarded band in gel electrophoretic mobility shift assays (*13*). The results summarized in Figure 2 show that simazine and atrazine do not induce proliferation of MCF-7 cells whereas 1 nM E2 induces a two-fold increase in MCF-7 cell growth. In cotreatment studies, the chloro-*s*-triazine did not inhibit E2-induced cell proliferation. The PR is also induced by E2 and this response is inhibited by classical antiestrogens which compete for ER binding sites, and by TCDD and related compounds (*13*). MCF-7 cells were treated with 10 nM E2, 1 μM simazine or atrazine, and E2 plus the chloro-*s*-triazines, and nuclear extracts were isolated. Nuclear PR levels were determined by incubating nuclear extracts with [^{32}P]PRE followed by a gel electrophoretic mobility shift assay which measured formation of a specifically-bound retarded [^{32}P]PRE-PR complex (using R5020 as the ligand). Analysis of the retarded [^{32}P]PRE-PR complexes show that E2 caused a 3.5-fold increase in PR-PRE band intensities whereas atrazine and simazine did not induce formation of this band. In cells cotreated with E2 plus atrazine or simazine, neither compound affected the E2-induced responses, thus confirming that atrazine and simazine did not elicit estrogenic or antiestrogenic responses in this cell line (*9*).

Several highly sensitive *in vitro* bioassays have been developed for investigating the estrogenic and antiestrogenic activities of various compounds and mixtures (*14-19*). In this study, a Gal4-HEGO chimeric receptor containing the ligand-binding domain of the human ER and the DNA binding domain of the yeast Gal4 nuclear transcription factor was utilized in a transient transfection assay in MCF-7 cells. The cells were cotransfected with a Gal4-regulated promoter (17m5-G-Luc) which contains 5 tandem 17-mer response elements upstream of the rabbit β-globin promoter linked to the firefly luciferase reported gene. Induction of luciferase activity in these cells is dependent on ligand-induced activation of the chimeric receptor by an estrogenic compound or mixture. The results in Figure 3 show that E2 caused a concentration-dependent increased of luciferase activity in MCF-7 cells transiently transfected with Gal4-HEGO and 17m5-G-Luc constructs. In contrast, using this same assay system, neither simazine or atrazine significantly induced luciferase activity. In cells cotreated with E2 plus simazine or atrazine at concentrations as high as 10 μM, there was no significant effect on E2-induced luciferase activity. Thus, simazine and atrazine do not activate this estrogen-responsive system suggesting that these chloro-2-triazines do not bind to the ER.

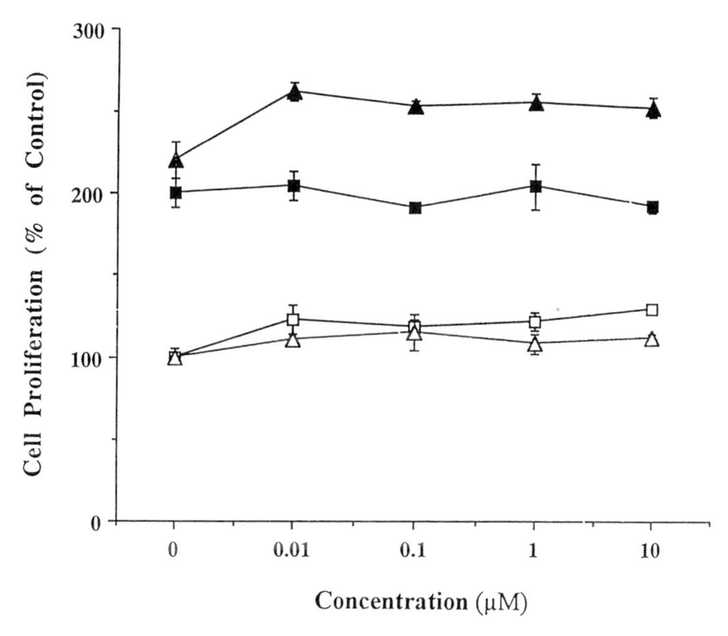

Figure 2. Effects of E2, atrazine (□) or simazine (△) alone or E2 plus simazine (▲) or atrazine (■) cotreatments on proliferation of MCF-7 cells. Cells were grown as described under Materials and Methods. Atrazine and simazine alone did not significantly affect basal or E2-induced MCF-7 cell growth.

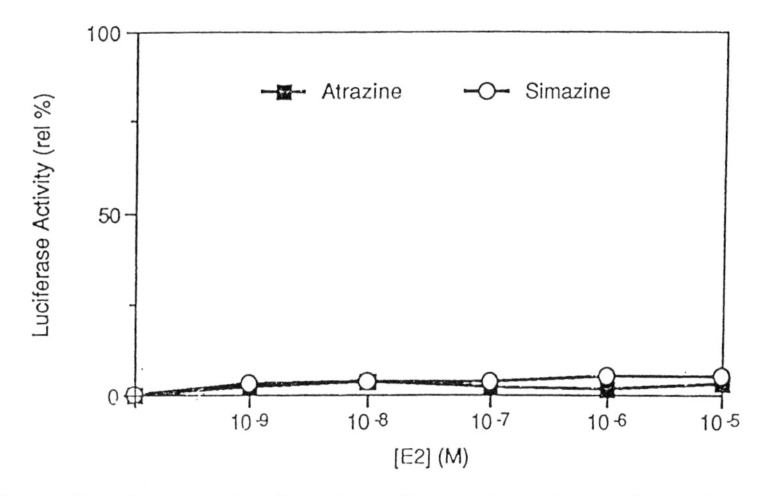

Figure 3. Concentration-dependent effects of atrazine and simazine on luciferase activity in MCF-7 cells transiently transfected with the Gal4-HEGO chimeric receptor and the 17m5-G-Luc reporter gene. The cells were treated with different concentrations of atrazine or simazine and luciferase activity was determined as described under Materials and Methods. Atrazine or simazine alone did not significantly affect basal or E2-induced luciferase activity. The results are reported relative to the maximal response (100%) observed with 1 nM E2.

Recent studies have also demonstrated that in an estrogen-responsive recombinant yeast strain, neither atrazine or simazine exhibited estrogenic or antiestrogenic activity (*9*). Thus, the results of these studies indicate that simazine and atrazine do not exhibit ER agonist activities for at least 7 different estrogen-regulated responses in both *in vivo* and *in vitro* assays (*9*). In MCF-7 cells cotreated with E2 plus the chloro-*s*-triazines, no significant inhibition of the E2-induced responses were observed whereas simazine and atrazine inhibited some E2-induced responses in the rat uterus. These data suggest that atrazine and simazine do not elicit their effects directly through the ER. However, this does not exclude a possible interaction between the chloro-*s*-triazines and ER-mediated responses through other pathways.

Acknowledgments

This work was supported by funds from the Canadian Network of Toxicology Centres, the University of Western Ontario [Academic Development Fund and Vice-President (Research) Special Competition], the Texas Agricultural Experiment Station, and the Ciba Geigy Corporation. T.Z. is supported by a PMAC-HRF/MRC Research Award in Medicine. The authors are grateful to Drs. P. Chambon and L. Murphy for providing various constructs.

Literature Cited

1. Wetzel, L.T.; Luempert, L.G.; Breckenridge, C.B.; Tisdel, M.O.; Stevens, J.T.; Thakur, A.J.; Extrom, P.C.; Eldridge, J.C. *J. Toxicol. Environ. Health* **1994,** *43,* pp. 169-182.
2. Tennant, M.K.; Hill, D.S.; Eldridge, J.C.; Wetzel, L.T.; Breckenridge, C.B.; Stevens, J.T. *J. Toxicol. Environ. Health* **1994,** *43,* pp. 197-211.
3. Tennant, M.K.; Hill, D.S.; Eldridge, J.C.; Wetzel, L.T.; Breckenridge, C.B.; Stevens, J.T. *J. Toxicol. Environ. Health* **1994,** *43,* pp. 183-196.
4. Stevens, J.T.; Breckenridge, C.B.; Wetzel, L.T.; Gillis, J.H.; Luempert III, L.G.; Eldridge, J.C. *J. Toxicol. Environ. Health* **1994,** *43,* pp. 139-153.
5. Eldridge, J.C.; Fleenor-Heyser, D.G.; Extrom, P.C.; Wetzel, L.T.; Breckenridge, C.B.; Gillis, J.H.; Luempert, L.G.; Stevens, J.T. *J. Toxicol. Environ. Health* **1994,** *43,* pp. 155-167.
6. Kniewald, J.; Mildner, P.; Kniewald, Z. *J. Steroid Biochem.* **1980,** *11,* pp. 833-838.
7. Safe, S.; Harris, M.; Biegel, L.; Zacharewski, T. In: *Biological Basis for Risk Assessment of Dioxins and Related Compounds;* Cold Spring Harbor Laboratory: Banbury Conference Proceedings #35, 1991, pp. 367-377.
8. Safe, S. *Pharmacol. Therap.* **1995,** *67,* pp. 247-281.
9. Connor, K.; Howell, J.; Chen, I.; Liu, H.; Berhane, K.; Sciarretta, C.; Safe, S.; Zacharewski, T. *Fund. Appl. Toxicol.* **1996,** *30,* pp. 93-101.

10. Dickerson, R.; Safe, S. *Toxicol. Appl. Pharmacol.* **1992,** *113,* pp. 55-63.
11. Astroff, B.; Safe, S. *Biochem. Pharmacol.* **1990,** *39,* pp. 485-488.
12. Lyttle, C.R.; DeSombre, E.R. *Proc. Natl. Acad. Sci. USA* **1977,** *74,* pp. 3162-3166.
13. Harper, N.; Wang, X.; Liu, H.; Safe, S. *Mol. Cell. Endocrinol.* **1994,** *104,* pp. 47-55.
14. Zacharewski, T.; Berhane, K.; Gillesby, B.; Burnison, K. *Environ. Sci. Tech.* **1995,** *29,* pp. 2140-2146.
15. Pierrat, B.; Heery, D.; Lemoine, Y.; Losson, R. *Gene* **1992,** *119,* pp. 237-245.
16. Gagne, D.; Balaguer, P.; Demirpence, E.; Trousse, F.; Nicolas, J.C.; Pons, M. *J. Biolum. Chem.* **1994,** *9,* pp. 201-209.
17. Klein, K.O.; Baron, J.; Colli, M.J.; McDonnell, D.P.; Cutler, G.B., Jr. *J. Clin. Invest.* **1994,** *94,* pp. 2475-2480.
18. Soto, A.M.; Sonnenschein, C.; Chung, K.L.; Fernandez, M.F.; Olea, N.; Serrano, F.O. *Environ. Health Perspect.* **1995,** *103,* pp. 113-122.
19. Jobling, S.; Reynolds, T.; White, R.; Parker, M.G.; Sumpter, J.P. *Environ. Health Perspect.* **1995,** *103,* pp. 582-587.

Chapter 34

A Pharmacodynamic Model of Atrazine Effects on Estrous Cycle Characteristics in the Sprague-Dawley Rat

Melvin E. Andersen[1], Harvey J. Clewell III[2], and Hugh A. Barton[1]

[1]ICF Kaiser Engineers, Inc., P.O. Box 14348, Research Triangle Park, NC 27709
[2]ICF Kaiser Engineers, Inc., 602 East Georgia Avenue, Ruston, LA 71270

Cessation of ovulation in Sprague-Dawley (S-D) rats begins by about 12 months of age, leading eventually to an anovulatory state, characterized first by persistent estrus (PE) and prolonged exposure to endogenous estradiol (E2). This first phase of reproductive senescence is due to the inability of the hypothalamus (HYPO) to support an effective gonandotropin releasing hormone (GnRH)-mediated, ovulation-inducing, luteinizing hormone (LH) surge, rather than ovarian exhaustion. We developed a pharmacodynamic estrous cycle (PD-EC) model for the S-D rat that focuses primarily on interactions between LH and E2. E2 has positive and negative feedback effects on LH release from the pituitary and also produces hypothalamic toxicity. E2 also mediates transcription in the HYPO leading to synaptic remodeling. Our model assumes that failure of the LH surge and ovulation ensue when cumulative-E2 toxicity leads to insufficient HYPO-E2 receptor reserve to accomplish remodeling in the intercycle period. The model was calibrated by examining data on altered cycle characteristics and PE induced by atrazine. The most intriguing model-derived insight was the prediction that both weak functional agonists and weak functional antagonists could lead to early onset PE. This model may be useful in evaluating toxic endpoints caused by various endocrine modulators in the S-D rat and for determining whether threshold doses are likely to be associated with those responses.

The ability of chemicals to cause toxicity by altering physiological processes regulated by various hormones has gained great visibility in the scientific and popular media. In contrast to previous high visibility environmental issues which have focused on toxic endpoints (e.g. cancer), specific chemicals (e.g. saccharin), or sites of contamination (e.g. hazardous waste sites) - the attention this time is focused on a general mode of action, often referred to as endocrine disruption. This simple description for a general mode of action leading to toxicity, encompasses a wide range of toxic endpoints and molecular mechanisms of action. These toxicities may arise via a range of molecular mechanisms including: 1) alterations in enzymes of hormonal biosynthesis or clearance, 2) alterations of the nervous system regulation of endocrine function, and 3) mimicking or inhibiting the activity of endogenous hormones in regulating cellular processes or altering gene expression. Furthermore, all of these effects occur in a physiological environment of sophisticated feedback control systems designed to maintain dynamic homeostasis.

A simplified description of part of this feedback control system for steroid sex hormone is illustrated in Figure 1. The gonads of both females and males produce steroid hormones [e.g. testosterone (T) in males, and E2 in females] that control the development of sperm or ova by the gonads. But, the steroid hormones are also released into the blood stream where they are distributed and can have effects on accessory sex organs and the brain. The brain, particularly the hypothalamus (HYPO), and the pituitary release peptide hormones, (e.g., LH and follicle stimulating hormone (FSH)) into the blood which regulate processes in the gonads. These interactive feedback processes are frequently referred to as the hypothalamic-pituitary-gonadal axis. Not shown in this illustration, but also present are neurological regulatory pathways, such as those important for maintenance of daily or circadian rhythms.

A pharmacodynamic model of atrazine effects on estrous cycling in S-D rats described here was designed to explore the hypothesis that atrazine interacts with the aging process in female S-D rats resulting in an earlier onset of persistent estrus. The persistent estrus state may be associated with the high incidence of spontaneous mammary tumors in this strain. Acceleration of persistent estrus in atrazine-treated S-D rats could result in increased incidence or earlier onset occurrence of mammary tumors.

Background on Atrazine

Atrazine (2-chloro-4-ethylamino-6-isopropylamino-*s*-triazine) is non-mutagenic in most test systems (*1*), but causes an increase in the prevalence and a decrease in latency of mammary tumors in female Sprague-Dawley rats (*2*). This strain of rat has a high incidence of spontaneous mammary tumors, frequently exceeding 50%. Atrazine did not increase mammary tumor incidence in Fischer 344 (F-344) rats which have a background incidence of less than 10%, in either sex of three strains of mice, or in male Sprague-Dawley rats (*2*). In general, all

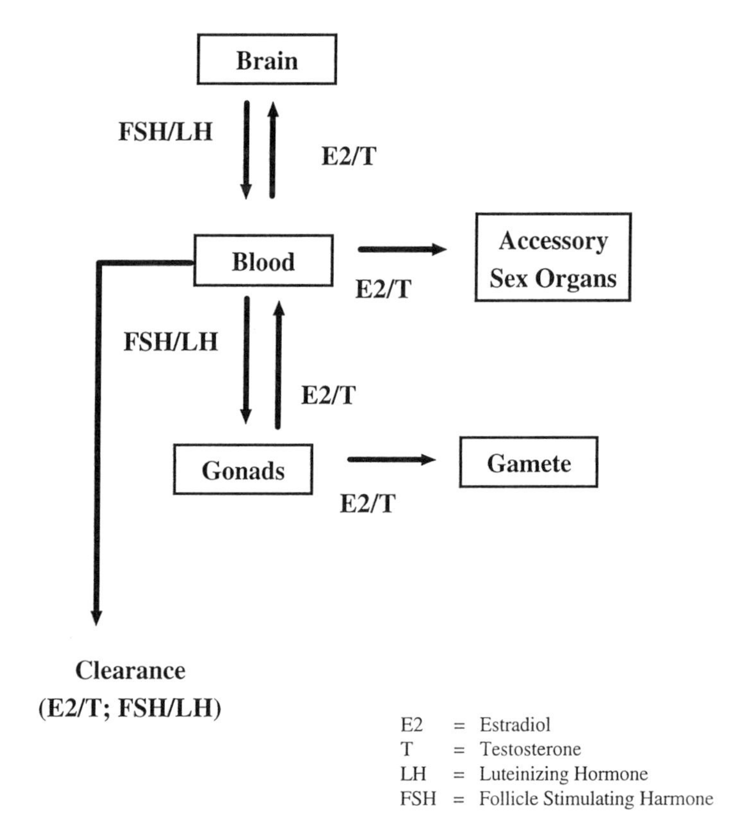

Figure 1: A generalized scheme of critical tissues and interactions between hypothalamic factors and gonadal hormones provides a framework for evaluating the role of chemicals as endocrine active compounds.

the 2-chloro-*s*-triazines increased mammary tumors in female Sprague-Dawley rats, while 2-thiomethyl and 2-methoxy-*s*-triazines did not increase the incidence of these tumors (*2*).

Despite the lack of genotoxicity, the high background of these tumors in the female Sprague-Dawley rats, and the absence of these tumors in male Sprague-Dawley rats or in several strains of mice, the US EPA has previously proposed regulating atrazine using a linearized multistage model for the tumor data coupled with standard default risk assessment procedures (*3*). This regulatory posture appears questionable for non-DNA reactive compounds, such as the triazine herbicides. With these latter chemicals, the appropriate risk assessment strategies should depend on knowledge of the mode of action in the test animal species, the manner in which the test compound alters the physiological system to enhance tumor incidence, and any differences in mechanisms of action and in pharmacokinetics between the test animal population and humans. The US EPA is currently evaluating new information on the mode of action of 2-chloro-*s*-triazines to determine the appropriate risk assessment methodology.

Atrazine is not a classic estrogenic agonist or antagonist, and does not likely bind to the ligand binding site of the estrogen receptor. *In vitro* atrazine did not displace E2 from the estrogen receptor in conventional ligand displacement incubations (*4*). In MCF-7 cells, atrazine neither induced cell proliferation nor inhibited E2-induced proliferation when both compounds were present simultaneously (*5*); similar results were found with simazine (2-chloro-4,6-bisethylamino-*s*-triazine). In addition, neither compound altered estrogen receptor mediated gene expression in MCF-7 or yeast cells modified with artificial constructs (*5*). However, atrazine, simazine, and the di-dealkylated metabolite, diamino-2-chloro-*s*-triazine, reduced E2 binding following preincubation of uterine cytosol with very high concentrations (*6*). In ovariectomized rats dosed *in vivo* with 50 or 300 mg atrazine/day for 2 days and then sacrificed, E2-binding to uterine receptors was reduced in a dose-dependent manner (*6*). Immature Sprague-Dawley rats treated with atrazine showed reduced uterine wet weight, and reductions in cytosolic progesterone receptors and uterine peroxidase activity (*6*), all of which are inducible by E2 (*5*).

However, recent studies have shown that atrazine may be having an effect on the hypothalamic/pituitary axis (*7*). Atrazine suppressed the estrogen-induced surge of LH in ovariectomized rats, but these animals could release LH in response to exogenous GnRH indicating their pituitary function was not impaired (*8*). Levels of two neurotransmitters, dopamine and norepinephrine, were not altered so the exact mechanism for atrazine's effects in the brain remains undetermined (*8*).

Estrous Cycle

The estrous cycle in rats lasts approximately 4-5 days, with 4 days being most common in Sprague-Dawley rats. There are 4 phases in the cycle, each lasting a day or less: diestrus I, diestrus II, proestrus, and estrus (*9*). Diestrus I, immediately following estrus and ovulation, is the period where there is growth and maturation of new corpora lutea, which produce the steroid hormone, progesterone. Diestrus II is the relatively quiescent stage characterized by slowly increasing E2 concentrations as the next group of follicles begin to mature. Beginning in late diestrus II, there is rapid growth of the follicles and a rapid increase in blood E2. On the proestrous day, the persistence of high E2 concentrations stimulates a surge of progesterone from the ovary and surges of LH, FSH, and prolactin from the pituitary. The LH surge, closely tied to the diurnal cycle, occurs in the afternoon of proestrus and promotes rapid growth and rupture of the ovarian follicles, leading to ovulation during the morning of estrus. The cyclical morphological changes in reproductive tissues are themselves dominated by the alterations in plasma E2. For instance, vaginal lining changes markedly during the cycle. Persistent vaginal estrus is the condition where vaginal tissues, in response to continuous exposure to E2, maintain a thick, stratified epithelium, normally conspicuous only in estrus.

Sprague-Dawley (*2*) and other strains of rats, including the Long-Evans (*10*), exhibit regular cyclicity until about 9 months of age, when they change to an irregular cycle, and then to acyclicity, characterized by persistent, anovulatory estrus. Persistent estrus is characterized by a reduction in the LH surge, inability to initiate ovulation, and continual high plasma levels of E2 and prolactin. These strains of rats have high background incidences of mammary tumors, which appear to be associated with these prolonged E2 and prolactin exposures during persistent estrus (*2*).

Reproductive Aging in Rats

Reproductive aging in humans occurs due to exhaustion of viable ova, but in S-D rats it results primarily from neuroendocrine failure (*11*). Induction of persistent estrus is one form of neuroendocrine mediated reproductive failure in rats. This failure is believed mediated by the cumulative toxicity of E2 on brain cells regulating the release of GnRH (*12*). Release of GnRH in turn is one of the signals controlling the release of LH which leads to ovulation.

The arcuate nucleus in the hypothalamus is believed to play a critical role in E2 positive feedback and in the LH surge (*13*). During normal cycles, this brain region undergoes phased synaptic remodeling (*14*). Persistent estrus in aging S-D rats appears to result from cumulative damage to the arcuate nucleus by E2. Impairment of the entercycle, synaptic remodeling processes results in acyclic behavior (*14*).

A pharmacodynamic PD-EC estrus cycle model, which has been described in detail elsewhere (*15*), was used as a framework to integrate the scientific data and relevant hypotheses for hormonal regulation of estrus cycling and reproductive neuroendocrine aging in rats. Modification of the model to incorporate a proposed interaction with atrazine is described below, along with potential uses for the model.

Methods

Pharmacodynamic Model for Estrous Cycle and Reproductive Aging: The computer code for the model was written in Advanced Continuous Simulation Language (ACSL, Mitchell & Gauthier Associates, Concord, MA) and run on a 486-66 MHz computer.

The model structure (Figures 2 and 3) describes the interactions between E2 and LH in the hypothalamic-pituitary-gonadal axis. Estradiol is produced in response to the growth of the ovaries which was described using an exponential Gompertz-function (Equations 1 and 2).

$$size = size0 + sizemax * exp(-a*exp(-b*t^n)) \tag{1}$$

$$dAE2/dt = k0*size-ke2*AE2 \tag{2}$$

The growth function parameter values (*a, b, n*) were determined by fitting E2 blood concentration data (Figure 4).

Circulating E2 concentrations (*AE2/vb*) affect the production of LH in a combined brain-pituitary "endocrine" compartment. The model currently describes average LH release without explicitly modeling GnRH release. Negative feedback by E2 on tonic LH release is described in the mass balance equation for LH in blood by a term describing the inhibition of LH release rate in terms of E2 concentrations [*Kilh/(CE2nM + Kilh)*] (Equation 3).

$$dALH/dt = k1lh*Kilh/(CE2nM + Kilh) - kelh*ALH + bolus\ input \tag{3}$$

The maximal LH release rate was estimated with data from ovariectomized rats (i.e. in the absence of E2). Clearance from blood of LH and E2 are described as first order (e.g., *kelh*ALH*).

Positive feedback by E2 on LH release described the cyclic LH surge that results in ovulation (Figure 5). As E2 concentrations in blood increase, there is increased binding to the E2 receptor (*R2*); this interaction was modeled to have an affinity of 1 nM (Equations 4 and 5). The ligand-receptor complex binds to DNA resulting in increased expression of a postulated estrus cycle-related protein (ECRP). As E2 exposure continues, levels of ECRP increase until a critical level has been reached resulting in an LH surge (Figure 4).

$$d(ECRP)/dt = k0ECRP*CE2R2/(CE2R2 + K2nM) - kECRP*ECRP \tag{4}$$

$$CE2R2 = Cr2*CE2nM/(KB2nM + CE2nM) \tag{5}$$

The surge was modeled as instantaneous release of LH into an intermediate holding compartment from which a first order rate constant describes its release into the blood (*bolus input*) (Figure 2). When a critical level of LH is achieved in the blood, ovulation occurs which results (in the model) in resetting the Gompertz function to zero and initiation of a new cycle

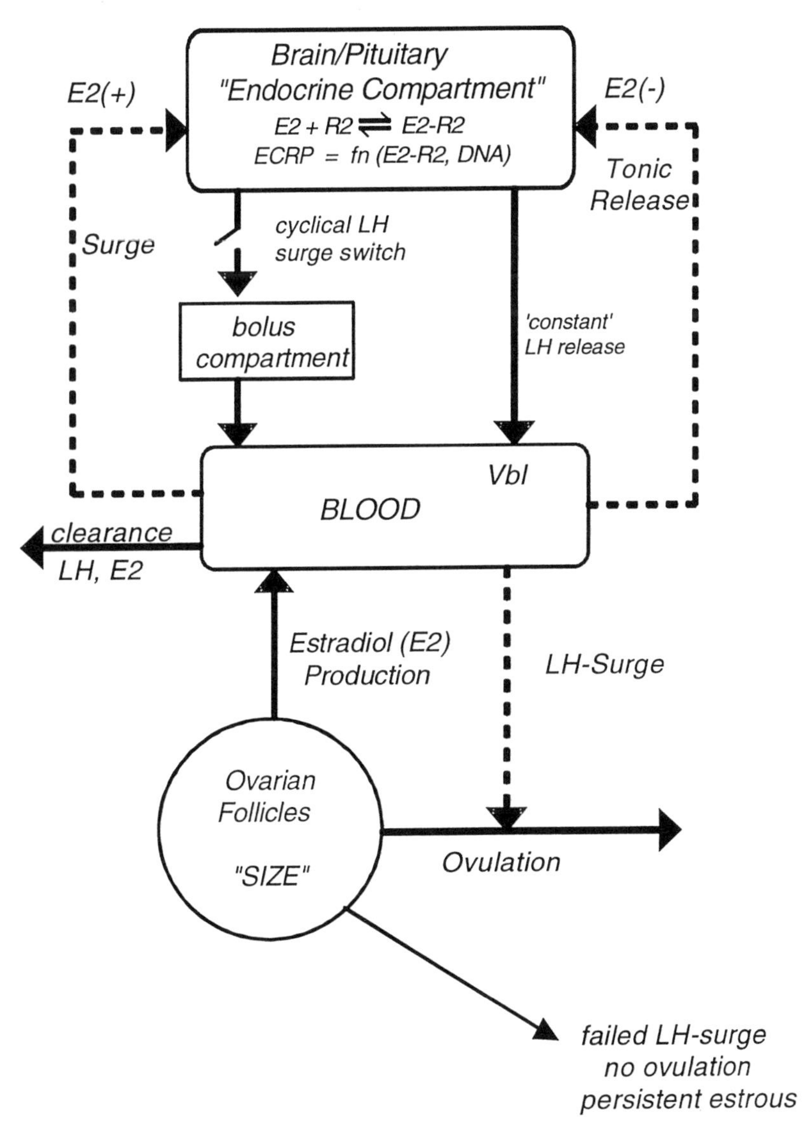

Figure 2: A General Schematic of the Simplified Estrus Cycle Model for the Sprague-Dawley Rat.

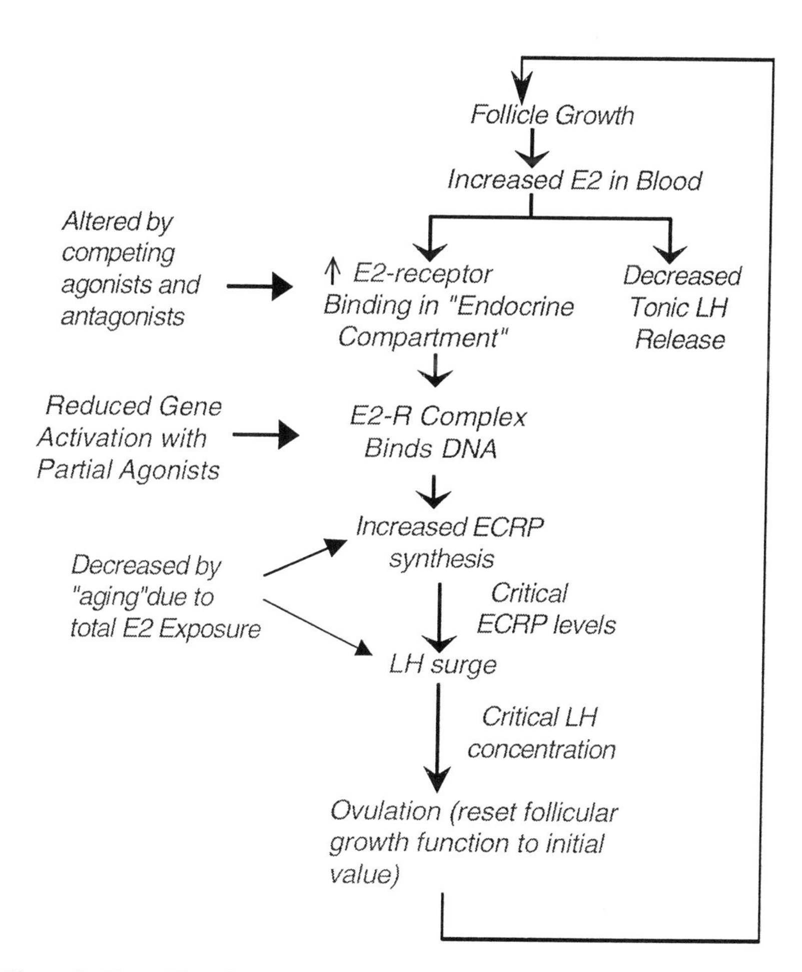

Figure 3: Event Flow in The Estrus Cycle Model: The points in the model that are involved in aging or interactions with xenobiotics are identified in this illustration of the flow of events in the model.

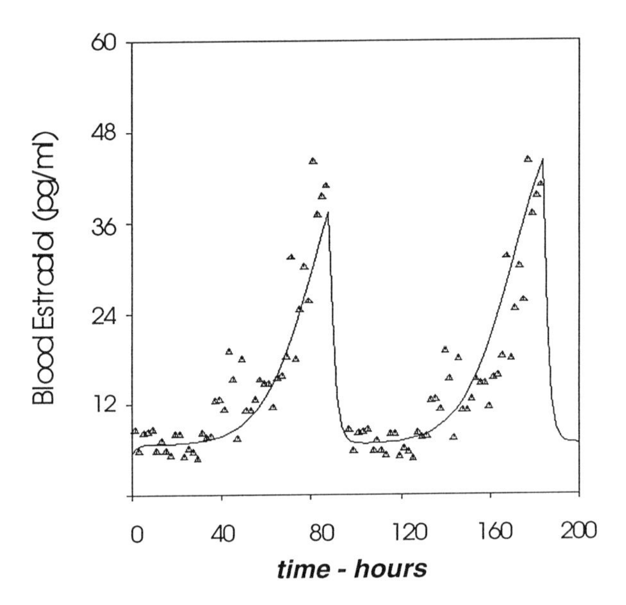

Figure 4: Estradiol Concentrations During the Estrus Cycle in the Rat. Data were extracted from Smith (1975) (*17*). The data in the paper covered a single 4-day cycle in female Sprague-Dawley rats. The single cycle values were duplicated for the period from 96-192 hours to generate representative behavior over 2-cycles.

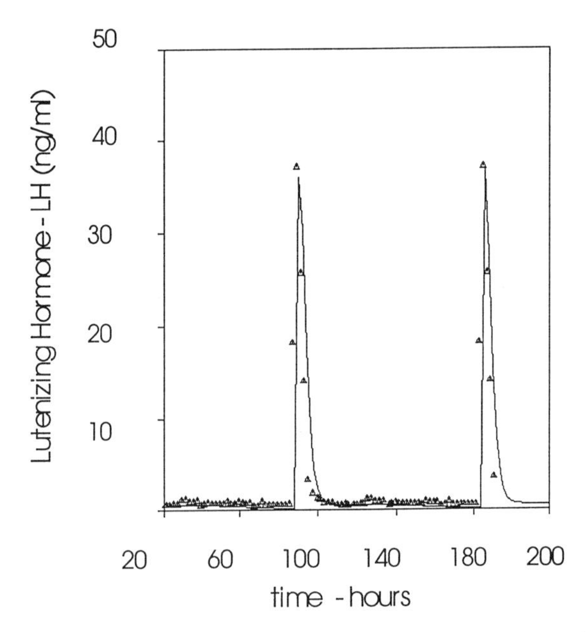

Figure 5: Luteinizing Hormone Concentrations During the Estrus Cycle in the Rat. Data from Smith (*17*); smooth curve generated by the PD-EC model. The peak in blood LH coincides with the point at which the ECRP in the model reaches a critical value and initiates the LH surge.

of follicular growth. This event is equivalent to ovulation in the animal. The model currently does not describe a luteal phase, but in rats this phase is extremely abbreviated in the absence of copulation or artificial stimulation of the cervix. Therefore, this model is appropriate for the typical chronic toxicity studies in which rats are not mated. Finally, to account for circadian control of neuroendocrine regulation, the LH surge must occur between 4:00 and 6:00 pm to produce ovulation the following morning. If the critical level of LH is reached after 6:00 pm, ovulation is delayed until the following day.

In the PD-EC model of reproductive senescence (*15*), the following assumptions are made: 1) neuroendocrine aging is assumed to occur due to the loss of intracycle resynthesis capacity for the hypothetical E2-regulated ECRP or related factors (Figures 2 and 3), 2) the root of neuroendocrine aging is cumulative E2 exposure reducing the capacity of the brain to synthesize ECRP or related factors, 3) the loss is related to the diminution of the maximal ECRP synthesis term by cumulative exposure to E2, and 4) the instantaneous rate of synthesis of ECRP depends on the maximal rate of synthesis and the concentration of E2-receptor complexes.

Data for Analysis: As female Sprague-Dawley rats age, they go from regular 4-day cycles, to variable length cycles, and finally to ovulatory failure with persistent estrous. Wetzel (*16*) estimated the percent of time in estrus at various ages for animals fed diets with atrazine at 0, 70, and 400 ppm. Cycle data from each dietary concentration (Table I) were fit to an empirical function to derive a continuous quantitative relationship between proportion of the cycle

Table I. Effect of Atrazine Administration to Sprague-Dawley Rats on the Percent Days of the Cycle in Estrus

	Percent Days in Estrus at Various Atrazine Feeding Levels		
[a]Time	*0 ppm*	*70 ppm*	*400 ppm*
1 mo.	19.0 ± 3.9 (10)	23.5 ± 5.0 (10)	22.0 ± 2.8 (10)
3 mo.	24.8 ± 7.7 (10)	25.2 ± 4.9 (10)	27.8 ± 7.5 (10)
9 mo.	24.2 ± 7.6 (10)	34.3 ± 9.0 (10)	44.8 ± 11.5 (10)
12 mo.	42.9 ± 10.1 (10)	47.2 ± 13.7 (10)	53.3 ± 11.2 (10)
15 mo.	44.4 ± 12.2 (10)	42.7 ± 12.6 (10)	49.6 ± 12.2 (10)
18 mo.	44.9 ± 5.7 (10)	57.2 ± 12.5 (10)	55.9 ± 20.7 (10)
24 mo.	47.8 ± 18.9 (5)	50.0 ± 27.3 (4)	24.0 ± 0.0 (2)

[a]Data reproduced from Wetzel (*16*) in reference section with permission.

in estrus and the duration of dosing with atrazine. The functional relationship used was:

$$estrous\ days = normal + maxincrease*(t-1)n/(mpn + (t-1)n) \qquad (6)$$

where *normal* (= 22%) is the baseline proportion of days in estrus, *maxincrease* is the maximum increase in the proportion of days in estrus (= 29%), and t is age in months. The time variable is adjusted by 1 month to account for the age at sexual maturity (about 1 month) when cycling first begins. A common shape parameter, *n*, was estimated for all three dose groups, but *mp*, the age at the midpoint of the transition, was estimated separately for each group. The estimated values of *mp* were 10.5, 8.5, and 6.5 for the 0, 70, and 400 ppm groups, respectively. It is assumed that the measured maximum percent days in estrus (estimated by vaginal cytology) coincides with the anovulatory, constant estrus condition in these rats.

Atrazine disruption of cstrous cycling in the S-D rat is assumed to be related to its ability to suppress or inactivate one or more aspects of estrogen promotion of factors in the hypothalamus critical for normal LH surges.

Model Alterations to Incorporate Atrazine. A non-competitive functional inhibition by atrazine is assumed to be related to its ability to interact with and remove some of the E2-receptors from the pool of active receptors involved in intracycle remodeling. Equation 7 describes the rate of change of concentration of the E2 receptor over time.

$$d(cr2t)/dt = k0r2 - ker2*cr2t - katraz*catraz*cr2t \qquad (7)$$

This equation has terms for receptor synthesis, *k0r2*, for basal receptor degradation, *ker2*, and for the postulated inactivation of the receptor due to its reaction with atrazine, *katraz*. The degradation rate depends on the atrazine concentration, *catraz*, which is expressed simply as ppm in feed, and the E2-receptor concentration. The values for the atrazine related model constants appear in Table II; all other constants are the same as originally described by Andersen (*15*).

Table II. Biologically Based Dose-Response Model Parameters

	E2- Receptor Atrazine Reaction Parameters:	
katraz	(second order reaction rate constant - /ppm/hr)	0.0000225
catraz	(atrazine concentration in feed - ppm)	0 - 2000
k0r2	(synthesis rate for receptor - nm/hr)	0.1
ker2	(degradation constant - /hr)	0.1

Results and Discussion

Modeling atrazine

Using a dietary input of 400 ppm and the time to loss of regular cycling equal to 6.5 months, *katraz* was estimated to be 2.25 x 10^{-5}/(ppm atrazine/hr). Given that estimate, time of loss of cyclicity can be simulated from the model for any other level of atrazine exposure. Figure 6 shows the change in cycling expected at 400 ppm atrazine with the specified model parameters and compares that with the behavior for the control rat simulation in the absence of added atrazine. Failure of cycling occurs when E2 exposures diminish the ability of the central nervous system to resynthesize sufficient factors in the intercycle period to reach a critical value. If the critical level is not achieved, there is a failure of the LH surge mechanism.

By looking at various output values from the model, the mechanism of the earlier onset of reproductive senescence by atrazine in the S-D rat can be understood. Figure 7 shows a plot of the maximum achievable resynthesis rate of these hypothalamic factors at various times after the initiation of treatment with 0, 200, 400, 800, or 2000 ppm atrazine. (Figure 7 does not present data for levels that have been shown not to affect the estrous cycle.) Atrazine treatment could enhance the decline of hypothalamic capacity to resynthesize critical neuroendocrine factors. Eventually the intercycle resynthesis rate is too low to allow an LH surge and the rat becomes acyclic. In very young cycling rats, the simulations presented here indicate that close to 2000 ppm atrazine in the diet would be required to block ovulation. Since ovulation is quantal in nature, i.e., either ovulation occurs or it doesn't, lower doses would not block either the LH surge or ovulation. Thus, for any given age there is a threshold dose below which atrazine would not have any effect on the estrous cycle.

Dose Surrogates for Mammary Tumorigenicity

The proper dose surrogate must be related to the mode of action of atrazine. Thus, it is important to recognize that atrazine appears to promote tumor formation by increasing the duration of exposure to endogenous estrogen. The appropriate dose surrogates, then, should be related to E2 exposures or, more correctly, E2/prolactin exposures in the constant estrus phase where the cyclic compensatory progesterone exposures do not act to counterbalance E2 and prolactin. These hormone concentrations are not known with precision for the entire lifetime of the rat. Neither is the exact functional relationship between these exposures and tumor outcome clearly understood.

One dose surrogate that is available from the model is days in constant estrus. To determine the number of constant estrous days, based on the model, we estimate the days in constant estrus from the calculated failure-to-cycle (Figure 6) to an age at which the rats are presumed to pass on into complete ovarian failure without E2 production. The days in constant estrous dose surrogate, calculates the additional time (above background) from predicted failure of cycling until 18-months (Figure 8).

The danger of unopposed estrogen, i.e., estrogen exposure in the absence of progesterone exposure, is significant in S-D rats. However, the mechanism by which atrazine leads to increased E2 exposure appears to be idiosyncratic to these rodents. First, atrazine reactivity appears to disturb the level of some critical component in the hypothalamus associated with cycling. (Our description focused on the possibility that atrazine reacts to reduce receptor

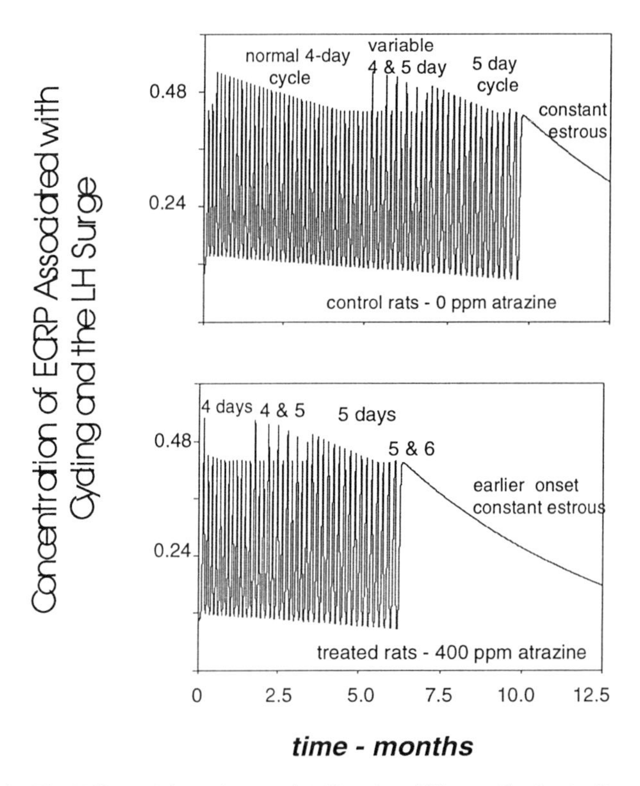

Figure 6: The Effect of Atrazine on the Simulated Estrus Cycles in Rats from the PD-EC Model. The plots show the cyclical nature of the hypothetical estrus cycle-related protein. Panel A: Model simulations of estrus cycle aging in the SD rat resulting in loss of cycling at about 9.5 months. Panel B: As in A, except *katraz* = 0.0000225 and *catraz* = 400. Persistent estrus occurs when the rats are about 6.5 months old with these parameter values.

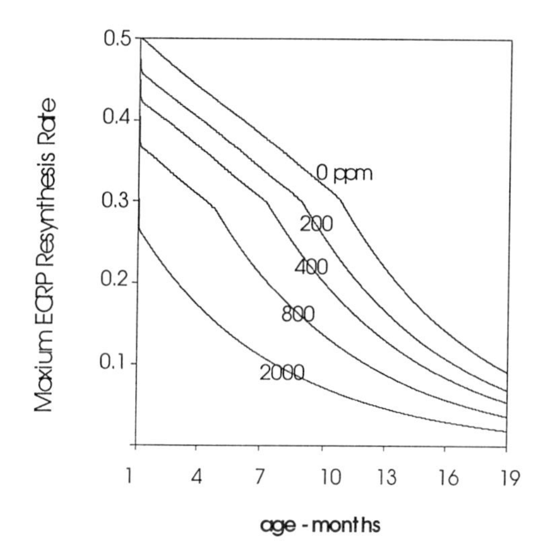

Figure 7: Maximum resynthesis rate of ECRP during atrazine exposures. The differences in these curves primarily reflect the predicted reduction in receptor concentration by atrazine exposures with the residual slope related to the E2-dependent alterations aging. Together these processes produce a more rapid onset of persistent estrus in the presence of atrazine.

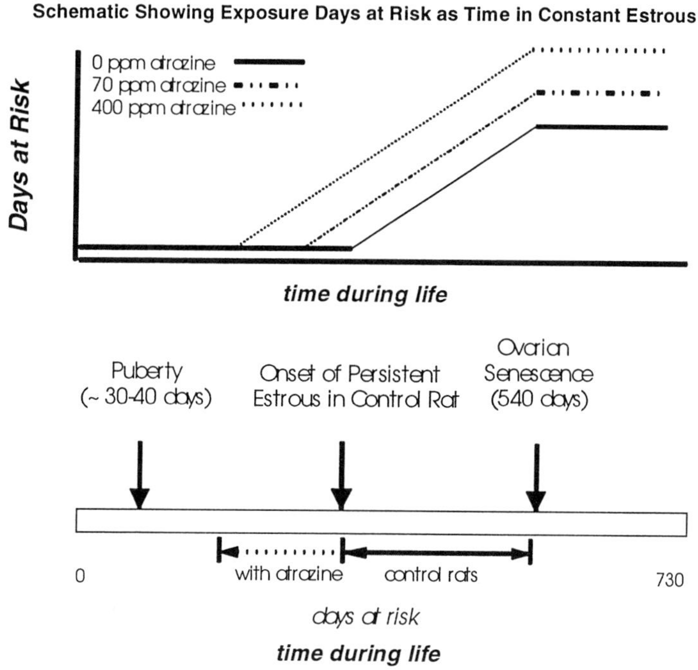

Schematic Showing Exposure Days at Risk as Time in Constant Estrous

Days at Risk

0 ppm atrazine
70 ppm atrazine
400 ppm atrazine

time during life

Puberty (~ 30-40 days)

Onset of Persistent Estrous in Control Rat

Ovarian Senescence (540 days)

0

with atrazine

control rats

730

days at risk

time during life

Figure 8: Schematic of the increase in the dose surrogate for days at risk after conversion from normal-to-acyclic behavior in the estrus PD-EC model. The addition of atrazine to the diet accelerated early onset persistent estrus and increases the days at risk.

concentration in target tissues although other possibilities for target sites may exist in the hormonal transcriptional activation systems.) This disturbance acts synergistically with normal E2-related cumulative toxicity to speed up neuroendocrine senescence in these rats (Figures 6 and 7). In the S-D rat, the senescence leads to ovulatory failure and constant estrus, involving tissue exposure to both E2 and prolactin. The resulting promotional activity occurs because the reproductive failure is at the neuroendocrine level while intact ovaries continue to produce E2 without the cyclic LH surge needed to convert the follicular structures to progesterone-secreting corpora lutea. In women, reproductive senescence is not due to a neuroendocrine failure; ovulation ceases due to ovulatory exhaustion which results in decreased estrogen exposure after menopause.

Our formulation of a PD-EC model indicated that levels of atrazine that might impair resynthesis of ECRP or a related factor would alter normal estrous cycling. This interference could arise from weak functional antagonists, as modeled here, or from partial agonists that interact with, but do not fully activate the signalling pathways required for the production of the ECRP (*15*).

Neuroendocrine failure in the SD rat could be considered a critical effect of chronic atrazine exposure upon which a risk assessment could be based. This mode-of-action supports a non-linear low-dose extrapolation for atrazine implying that a methodology such as margin of exposure approach using the benchmark dose would be appropriate. As additional research on atrazine's mechanism of action becomes available, the model could be further modified to reflect these findings.

Acknowledgments

We gratefully acknowledge support for this work by Ciba-Geigy Corporation.

Literature Cited

(1) Brusick D. *An assessment of the genetic toxicity of atrazine: Relevance to human health and environmental effects. Mutat Res.* **1994,** *317,* pp. 133-144.

(2) Stevens, J. T.; Breckenridge, C. B.; Wetzel, L. T.; Gillis, J. H.; Luempert, L. G., III; Eldridge, J. C. *Hypothesis for mammary tumorigenesis in Sprague-Dawley rats exposed to certain triazine herbicides. J Toxicol Environ Health.* **1994**, *43,* pp. 139-153.

(3) EPA. (1994). The Triazine Herbicides Atrazine, Simazine and Cyanizine. Position Document 1. Initiation of Special Review. *Fed Regist*, Nov 9, 1994.

(4) Tennant, M. K.; Hill, D. S.; Eldridge, J. C.; Wetzel, L. T.; Breckenridge, C. B.; Stevens, J. T. *Chloro-s-triazine antagonism of estrogen action: limited interaction with estrogen receptor binding. J.Toxicol Environ Health.* **1994**, *43*, pp. 197-211.

(5) Connor, K.; Howell, J.; Chen, I.; Liu, H.; Berhane, K.; Sciarretta, C.; Safe, S.; Zacharewski. T. *Failure of chloro-s-triazine-derived compounds to induce estrogen receptor-mediated response in vivo and in vitro. Fundam Appl Toxicol.* **1996**, *30*, pp. 93-101.

(6) Tennant, M. K.; Hill, D. S.; Eldridge, J. C.; Wetzel, L. T.; Breckenridge, C. B.; Stevens, J. T. *Possible antiestrogenic properties of chloro-s-triazines in rat uterus. J Toxicol Environ Health.* **1994b**, *43*, pp. 183-196.

(7) Cooper, R. L.; Stoker, T. E.; Goldman, J. M.; Hein, J.; Tyrey, L. *Atrazine disrupts hypothalamic control of pituitary-ovarian function. Fundam Appl Toxicol.* **1996**, *30*, pp. 339.

(8) Cooper, R. L.; Parrish, M. B.; McElroy, W. K.; Rehnberg, G. L.; Hein, J. F.; Goldman, J. M.; Stoker, T. E.; Tyrey, L. *Effect of atrazine on the hormonal control of the ovary. Abstract. Society of Toxicology. Baltimore*, **1995**.

(9) McConnell, D. R. In *Safety Requirements for Contraceptive Steroids;* Michal, F., Ed.; General observations on the effects of sex steroids in rodents with emphasis on long-term oral contraceptive studies; Cambridge University Press: Cambridge, **1989**, pp. 211-229.

(10) Lu, K. H.; Hopper, B. R.; Vargo, T. M.; Yen, S. S. C. *Chronological changes in sex steroid, gonadotropin and prolactin secretion in aging rats displaying different reproductive states. Biol Reprod.* **1979**, *21*, pp. 193-203.

(11) Finch, C. E.; Felicio, L. S.; Mobbs, C. V.; Nelson, J. F. *Ovarian and steroidal influences on neuroendocrine aging processes in female rodents. Endocrinology Rev 5.* **1984**, pp. 467-497.

(12) Desjardins, G. C.; Beaudet, A.; Meaney, M. J.; Brawer, J. R. *Estrogen-induced hypothalamic beta-endorphin neuron loss: a possible model of hypothalamic aging. Exper Gerontol.* **1995**, *30*, pp. 253-267.

(13) Brawer, J. R.; Beaudet, A.; Desjardins, G. C.; Schipper, H. M. *Pathologic effect of estradiol on the hypothalamus. Bio Reprod.* **1993**, *49*, pp. 647-652.

(14) Keefe, D.; Garcia-Segura, L. M.; Naftolin, F. *New insights into estrogen action on the brain. Neurobiol Aging.* **1994**, *15*, pp. 495-497.

(15) Andersen, M. E.; Clewell, H. J., III; Gearhart, J.; Allen, B. C.; Barton, H. A. (1997). A pharmacodynamic model of the rat estrous cycle in relation to endocrine disruptors. *J. Toxicol. Environ. Health*, in press.

(16) Wetzel, L. T.; Luempert, L. G., III; Breckenridge, C. B.; Tisdel, M. O.; Stevens, J. T.; Thakur, A. K.; Extrom, P. J.; Eldridge, J. C. *Chronic effects of atrazine on estrus and mammary tumor formation in female Sprague-Dawley and Fischer 344 rats J Toxicol Environ Health.* **1994**, *43*, pp. 169-182.

(17) Smith, M. S.; Freeman, M. E.; Neill, J. D. *The control of progesterone secretion during the estrous cycle and early pseudopregnancy in the rat: prolactin, gonadotropin and steroid levels associated with rescue of the corpus luteum of pseudopregnancy. Endocrinology.* **1975**, *96*, pp. 219-226.

Chapter 35

Probabilistic Risk Assessment for Atrazine and Simazine

Robert L. Sielken, Jr., Robert S. Bretzlaff, and Ciriaco Valdez-Flores

Sielken, Inc., 3833 Texas Avenue, Suite 230, Bryan, TX 77802

Probability distributions have been used for the triazine herbicides, atrazine and simazine to more accurately characterize the uncertainty and variability associated with the individual components of exposure (e.g., drinking water concentrations, food residue concentrations, and pesticide handling exposures associated with different user subpopulations, crops, product formulations, techniques of mixing/loading and application, and types of protective clothing). Furthermore, distributional techniques have made it possible to more realistically combine exposures from multiple years, subpopulations, exposure pathways, and chemicals.

Using new studies on potential human exposure through diet, drinking water and pesticide handling, safety margins and margins of exposure have been calculated for atrazine and simazine. Margins of safety calculated for potential exposure through diet and drinking water ranged from 25,000 to 1,000,000 depending on the exposure pathway. Margin of safety per pesticide handler ranged from 1,400 to 33,000. Results indicated that neither occupation exposure, dietary exposure nor environmental exposure to atrazine and simazine alone or to atrazine and simazine combined are likely to produce adverse health consequences.

Human exposure and risk assessments often rely on the use of default assumptions. The limitations associated with the use of default assumptions and the appropriateness of those assumptions when detailed exposure or dose-response data are available, have stimulated the development of new quantitative risk assessment methods (*1*) which use probabilistic techniques and distributional characterizations of dose-response, exposure, and risk. These methods include estimates of uncertainty and the variability of the risks among individuals in the population.

Several new studies have been conducted to characterize the potential human exposure to atrazine and simazine through diet (*2,3,4*), drinking water (*5*) and as a result of pesticide handling (6-*8*). In this report, probabilistic techniques are used to describe the distributional characteristics of potential human exposure from these sources (*2-8*). These analyses have been conducted separately for each exposure pathway (i.e., diet, water or pesticide handling) for atrazine or simazine. The distributional characteristics of aggregate exposure to either atrazine or simazine (diet, water, and pesticide handling combined) were expressed as lifetime average daily doses (LADDs). Aggregate exposure to atrazine and simazine combined have also been determined as proposed by new Federal Regulation (*9*). Ciba has also conducted extensive research to characterize the dose-response relationships for long-term toxicity of atrazine and simazine in the Sprague-Dawley rat (*10*).

Methods

Toxicity Endpoints. In this analysis, two toxicity endpoints are used to evaluate potential human exposure to atrazine and simazine. The first is based on the reference dose (RfDs) for chronic toxicity using a 100X safety factor established by EPA for atrazine [0.035 mg/kg/day] (*11*) and simazine [0.005 mg/kg/day] (*12*). For occupational exposure, a safety factor of 10 was used and the reference doses for atrazine and simazine are then 0.35 and 0.05 mg/kg/day, respectively.

EPA has recently considered an alternative approach to risk characterization based on a benchmark dose (ED_{10}) method outlined in their new proposed cancer guidelines (*13*). Therefore, the second exposure characterization in this report is based on a benchmark dose for the mammary tumor responses observed in the carcinogenicity studies conducted on atrazine (*15-17*) and simazine (*18*) in the Sprague-Dawley rat. The ED_{10} is defined as that dose which caused a 10% increase in the incidence of mammary tumors above the control group incidence in the Sprague-Dawley rat; the lower 95% confidence limit of a distribution of ED_{10}s is used in the current evaluation. The lower 95th percent confidence limit of this distribution of ED_{10}s is a conservative, worst case estimate of the dose that causes a 10% increase in tumor incidence in the carcinogenicity studies on atrazine (*14-17*) and simazine (*18*).

The ED_{10}-based benchmark doses for atrazine and simazine were 1.4 and 2.6 mg/kg/day, respectively. The benchmark doses for atrazine and simazine for non-occupational exposure were therefore 0.014 and 0.026

mg/kg/day respectively, when a 100 fold safety factor was used (i.e., $ED_{10}/100$). The corresponding values for atrazine and simazine for occupational exposure were 0.14 and 0.26 mg/kg/day, respectively, based on a 10 fold safety factor.

Exposure Characterization

Exposure is characterized for the water, diet, and pesticide handling exposure pathways separately and then combined for atrazine and simazine individually and then collectively. The analysis is based on data provided by Ciba is to EPA between March 23, 1995 and October 31, 1996.

Exposure is characterized by distributions of individual doses for specified populations. The distributions describe the probability that an individual selected at random from the population will receive different exposures via each of the three exposure pathways and via the combined pathways. The distributions incorporate the variability in exposure from individual to individual and the uncertainty associated with the characterization of the exposure pathway. The distributions indicate the dose level that is most likely to occur, the range from the lowest to the highest exposures expected in the population, and the relative likelihood of the different exposures in that range. Rather than focusing on an average exposure, the distribution describes the relative frequency of each exposure value. Instead of focusing on upper or lower bounds, the distributional characterization of the dose from exposure focuses on the best estimate of the probability of each possible exposure level.

Distributions of the dose from exposure are provided for national, regional, and state populations, different sources of drinking water (ground water, surface water, and blends of ground and surface water), and several subpopulations of pesticide handlers that reflect the different uses, product formulations, and tasks.

Margin of Exposure

The margin of exposure is expressed in this chapter as follows:

$$\text{Margin of Exposure} = ED_{10}/(\text{Dose from Exposure}).$$

Water. Distributional analysis of exposure indicates that for atrazine at least 95% of the estimated lifetime average daily doses (LADDs) from drinking water ingestion have a margin of exposure of at least 50,000 in the 18 major use states combined for atrazine. For simazine, the corresponding margin of exposure is at least 200,000 in the 18 major use states combined.

These analyses incorporate the distributions of chemical concentration in community water supplies.

Sensitivity analyses evaluated the quantitative impacts of using distributional data for the following parameters:

- drinking water consumption and body weight distributions
- age-dependent drinking water consumption and body weight distributions
- separate concentration distributions for ground water, surface water, and blends of ground and surface water
- seasonal variability
- year-to-year variability
- exposure duration distributions.

Diet. Distributional analysis of exposure to atrazine in the U.S.A. and its four regions (northwest, north central, southern, or western) indicates that at least 95% of the estimated LADDs from dietary consumption have a margin of exposure of at least 300,000 for atrazine in each of the four regions and 330,000 for atrazine in the U.S.A. For simazine, the corresponding margin of exposure is at least 1,750,000 for each of the four regions and at least 2,000,000 for the U.S.A.

These analyses incorporate the residue concentration distributions for the dietary components of dairy cows and poultry supplying meat, milk, and eggs. (Residue concentrations in human foods are all nondetects.) These analyses also incorporate distributional characterizations of the relative amounts of dietary components consumed by dairy cows.

Sensitivity analyses evaluated the quantitative impacts of using distributional data for the following parameters:

- human dietary consumption distributions
- age-dependent human dietary consumption distributions

Worker. The distributions of exposure indicate that at least 95% of the estimated lifetime average daily doses (LADDs) associated with pesticide handling have a margin of exposure of at least between 500 and 11,000 for atrazine and between 10,000 and 20,000 for simazine, depending on the product use (e.g., corn, sod, etc.) (Table I).

These analyses incorporate the following components:

- distributions of absorbed dose per pound of active ingredient applied for each body part, product formulation, method of mixing/loading, and method of application
- relative numbers of workers in each subpopulation based on pesticide handling activity or activities and methods of mixing/loading and application for each product use and each product formulation
- distributions of the pounds of active ingredient applied for each product use and each product formulation
- distributions of adult body weight.

Sensitivity analyses evaluated the quantitative impact of the following parameter:

- year-to-year variability in a body part exposure value instead of assuming the same body part exposure value for every year.

Table I: Non-Occupational Exposure Assessment: Atrazine and Simazine at the 95th Percentile (Percent of Reference Dose, Percent of Benchmark Dose, Margin of Safety, Margin of Exposure)

	Reference Dose (NOEL/Safety Factor)		**Benchmark Dose (ED_{10}/Safety Factor)**	
Exposure Pathway	*LADD as % Reference Dose*	*MOS = NOEL/Exposure*	*LADD as % Benchmark Dose*	*MOE = ED_{10}/Exposure*
		Atrazine		
Diet	0.01%	1,000,000	0.03%	333,333
Water	0.08%	125,000	0.2%	50,000
Diet + Water	0.09%	111,111	0.2%	50,000
		Simazine		
Diet	0.02%	500,000	0.005%	2,000,000
Water	0.3%	33,333	0.05%	200,000
Diet + Water	0.3%	33,333	0.06%	166,667
		Atrazine + Simazine		
Diet	0.04%	250,000	0.04%	250,000
Water	0.4%	25,000	0.3%	33,333
Diet + Water	0.4%	25,000	0.3%	33,333

[1]The margin of safety when atrazine and simazine are combined is based on the atrazine RfD. The margin of exposure is calculated as follows:

$$Margin\ of\ Exposure = 1 / [(Atrazine\ Dose / Atrazine\ ED_{10}) + (Simazine\ Dose / Simazine\ ED_{10})]$$

For each pesticide use, the whole population of pesticide handlers and each of the relevant subpopulations (including potentially sensitive subpopulations) are evaluated. For example, for crops, the following subpopulations are explicitly evaluated:

(1) growers
(2) growers who do mixing/loading
(3) growers who do applications
(4) growers who do both mixing/loading and application
(5) commercial pesticide handlers
(6) commercial pesticide handlers - ground application
(7) commercial ground mixer/loaders
(8) commercial ground applicators
(9) commercial pesticide handlers - aerial application
(10) commercial aerial mixer/loaders
(11) commercial aerial applicators (pilots).

These subpopulations are also further subdivided by

(a) product formulation (flowable formulation or "FF" and water dispersible granules or "WDG"),
(b) type of mixing/loading operation, and type of application.

Probabilistic techniques allow the exposure characterizations for individual subpopulations to be properly aggregated into a population characterization which reflects the relative subpopulation sizes and the different exposure distributions in each subpopulation.

Results and Discussion

The results for atrazine and simazine are summarized in Table I for diet, water, or a combination of diet and water. The data show that 95% of the time the LADD contribution from water is less than 0.08% of the RfD or 0.2% of the benchmark dose for atrazine and less than 0.03% of the RfD or 0.03% of the benchmark dose for simazine. Similarly, 95% of the time the LADDs contribution from diet accounted for less than 0.04% of the RfD or the benchmark dose for atrazine and simazine. Even when diet and water exposure pathways are combined for atrazine, for simazine or for the aggregate exposure to atrazine and simazine, exposures are always less than 0.5% of either the respective RfDs or benchmark doses. Margins of safety or margins or exposure range from 25,000 to 2,000,000 depending on the exposure pathway(s).

Stated another way, when aggregate exposure is considered, 95% of human exposure is 25,000 fold less than that dose which had no effect in chronic toxicity studies and 33,333 fold less than the benchmark dose. Very small incremental exposure accrue by combining diet and water or by combining atrazine and simazine. Sensitivity analysis of food intake characteristics and hence dietary exposure of selected subpopulations; such as infants and children did not produce different results because both the reference dose and the benchmark dose assume chronic and/or lifetime exposure. Even larger margins of safety may be expected for infants and children if toxicity endpoints derived from subchronic studies are used.

The percent of the reference dose or the percent of the benchmark dose for pesticide handlers is somewhat larger than for dietary and water sources of exposure. However, 95% of the distribution of LADDs was less than 1% of the RfD for atrazine or simazine using flowable formulations (Table II). Results using granular formulations are similar or less than those obtained with flowable formulations. When dose is expressed as a percent of the benchmark dose, the LADDs for atrazine account for only 0.1 to 1.9% of benchmark dose. For simazine, the LADDs account for 0.1% or less of the benchmark dose. Depending on the exposure source, margins of safety or margins of exposure for occupational exposure range from 500 to 33,000.

There is a very small incremental effect of exposure via diet or water by combining atrazine and simazine. Since pesticide handlers tend to have exposures that are greater than those typically seen via water or diet, adding the latter two exposure routes to pesticide handlers has a minimal incremental effect.

Overall, the results indicate that neither occupational nor exposure to atrazine or simazine via diet or drinking water is likely to cause adverse health effects in the United States population because exposure to atrazine or simazine, either separately or combined, represents only a small fraction of the reference and benchmark doses. Correspondingly, large margins of safety exist between human exposure and those doses that have either no effect or minimal effect in chronic toxicity studies on atrazine and simazine.

Literature Cited

1. National Research Council. (1994). Science and Judgement in Risk Assessment. National Academy Press, Washington, DC.
2. Bray, L. (1996). Revised Dietary Exposure Assessment for Atrazine, Ciba-Geigy, ABR-96105. Submission by Ciba Crop Protection, Ciba-Geigy Corporation.
3. Bray, L. (1996). Updated Dietary Exposure Assessment for Atrazine, Ciba-Geigy, ABR-96009. Submission by Ciba Crop Protection, Ciba-Geigy Corporation.
4. Bray, L. (1996). Updated Dietary Exposure Assessment for Simazine, Ciba-Geigy, ABR-96093. Submission by Ciba Crop Protection, Ciba-Geigy Corporation.
5. Clarkson, J. R. (1996). Human Exposure to Atrazine and Simazine via Ground and Surface Drinking Water: Update 1. Montgomery Watson, Wayzata, Minnesota. (Submission by Ciba Crop Protection, Ciba-Geigy Corporation: Atrazine/Simazine: Atrazine/Simazine Response to the United States Environmental Protection Agency's Position Document 1: Initiation of Special Review: November 23, 1994; Public Docket OPP-30000-60)

Table II: Occupational Exposure Assessment: Atrazine and Simazine at the 95th Percentile (Percent of Reference Dose, Percent of Benchmark Dose, Margin of Safety, Margin of Exposure)

	Reference Dose (NOEL/Safety Factor)		Benchmark Dose (ED_{10}/Safety Factor)	
Exposure Pathway	*LADD as % Reference Dose*	*MOS = NOEL/Exposure*	*LADD as % Benchmark Dose*	*MOE = ED_{10}/Exposure*
Atrazine (Flowable Formulation)				
Lawn Care	0.08%	12,500	0.2%	5,000
Sorghum	0.05%	20,000	0.1%	10,000
Corn	0.05%	20,000	0.1%	10,000
Sod	0.1%	10,000	0.3%	3,333
Vegetation Mgmt.	0.1%	10,000	0.3%	3,333
Hawaiian Sugar	0.4%	2,500	1.0%	1,000
N. American Sugar	0.7%	1,429	1.9%	526
Simazine (Flowable Formulation)				
Corn	0.3%	3,333	0.06%	16,667
Sod	0.7%	1,429	0.1%	10,000
Atrazine (Water Dispersible Granule Formulation)				
Lawn Care	0.08%	12,500	0.2%	5,000
Sorghum	0.04%	25,000	0.1%	10,000
Corn	0.03%	33,333	0.09%	11,111
Sod	0.08%	12,500	0.2%	5,000
Vegetation Mgmt.	0.1%	10,000	0.3%	3,333
Hawaiian Sugar	0.3%	3,333	0.8%	1,250
N. American Sugar	0.6%	1,667	1.5%	667
Simazine (Water Dispersible Granule Formulation)				
Corn	0.3%	3,333	0.05%	20,000
Sod	0.6%	1,667	0.1%	10,000

6. Selman, F. (1996). A Revised Assessment of Worker Exposure For Atrazine in Response to the U.S. Environmental Protection Agency Issuance of the "Triazine Herbicides Position Document 1 Initiation of Special Review." Ciba-Geigy, ABR-96071. Submission by Ciba Crop Protection, Ciba-Geigy Corporation.
7. Selman, F. (1996). A Revised Assessment of Worker Exposure For Simazine in Response to the U.S. Environmental Protection Agency Issuance of "The Triazine Herbicides Position Document 1 Initiation of Special Review." Ciba-Geigy, ABR-96072. Submission by Ciba Crop Protection, Ciba-Geigy Corporation.
8. Selman, F. (1996). Supplemental Data and Evaluation of Exposure to Lawn Care Operators Using Atrazine in the Southern United States, Supplement to ABR-95038 - Assessment of Worker Exposure for Atrazine in Response to the U.S. Environmental Protection Agency Issuance of the "Triazine Herbicides Position Document - Initiation of Special Review." Ciba-Geigy, ABR-96069, Submission by Ciba Crop Protection, Ciba-Geigy Corporation.
9. Food Quality Protection Act, H.R. 1627 and S. 1166 enacted 8/1996.
10. Breckenridge, C., (1996). Summary of Additional Comments on the Response to the Special Review Position Document 1 for Pesticide Products Containing Atrazine and Simazine. Supplement 1. Submission by Ciba Crop Protection, Ciba-Geigy Corporation.
11. Atrazine: Integrated Risk Information Systems (IRIS), Cincinnati, OH, U.S. Environmental Protection Agency, 1989.
12. Rinde, E., Peer Review of Simazine, Health Effects Division, U.S. Environmental Protection Agency, 1989.
13. Proposed Guideline for Carcinogen Risk Assessment, U.S. Environmental Protection Agency, April, 1996.
14 Twenty-Four Month Combined Chronic Oral Toxicity and Oncogenicity Study in Rats Utilizing Atrazine Technical. Project No. 410-1102. American Biogenics Corporation, April 29, 1986.
MRID No. 00158930.
15. Thakur, A. K. (1991). Determination of Hormone Levels in Sprague-Dawley Rats Treated with Atrazine. Technical Hazleton Washington Report 483-278. EPA MRID No. 42085001.
16. Thakur, A. K. (1992). Oncogenicity Study in Sprague Dawley Rats Treated with Atrazine. Technical Hazleton Washington Report 483-275. EPA MRID No. 4214610.
17. Breckenridge, C. (1996). Summary of Additional Comments on the Response to the Special Review Position Document 1 for Pesticide Products Containing Atrazine and Simazine. Supplement 11. Submission by Ciba Crop Protection, Ciba-Geigy Corporation.
18. McCormick, G. C. (1988). Simazine Technical: Combined Chronic Toxicity/Oncogenicity Study in Rats. Ciba-Geigy Corp. unpublished final report.

Glossary

Common Name	CAS Name	Structure
Atrazine	6-chloro-*N*-ethyl-*N*'-(1-methylethyl-1,3,5-triazine-2,4-diamine	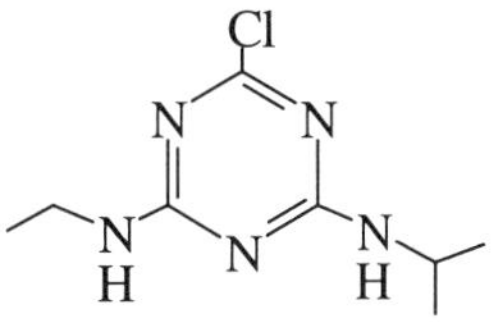 Cl, N, N, N, H, N, N, H
Ametryn	*N*-ethyl-*N*'-(1-methylethyl)-6-(methylthio)-1,3,5-triazine-2,4-diamine	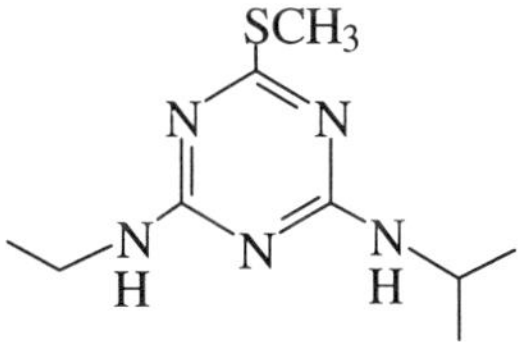 SCH_3, N, N, N, H, N, N, H
Simazine	6-chloro-*N*,*N*'-diethyl-1,3,5-triazine-2,4-diamine	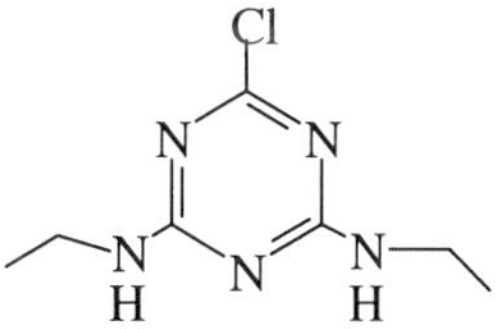 Cl, N, N, N, H, N, N, H
Cyanazine	2-[[4-chloro-6-(ethylamino)-1,3,5-triazin-2-yl]amino]-2-methylpropionitrile	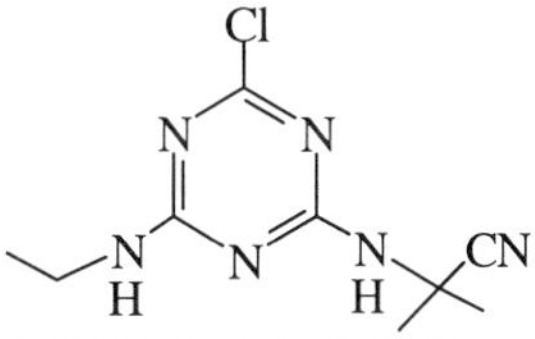 Cl, N, N, N, H, N, N, H, CN
Desethylatrazine	6-chloro-*N*-(1-methylethyl)-1,3,5-triazine-2,4-diamine	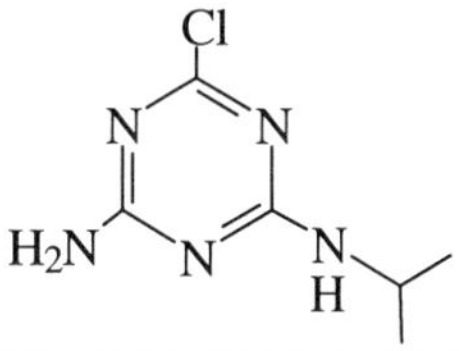 Cl, N, N, H_2N, N, N, H
Desethylametryn	6-(methylthio)-*N*-(1-methylethyl)-1,3,5-triazine-2,4-diamine	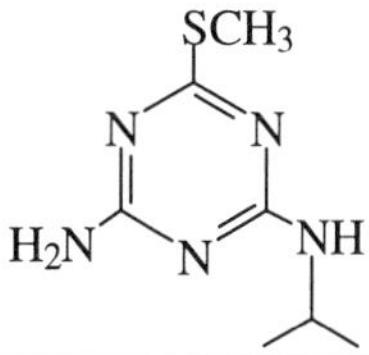 SCH_3, N, N, H_2N, N, NH
Desisopropylatrazine *Des-2-methylpropionitrile cyanazine *desethylsimazine	6-chloro-*N*-ethyl-1,3,5-triazine-2,4-diamine	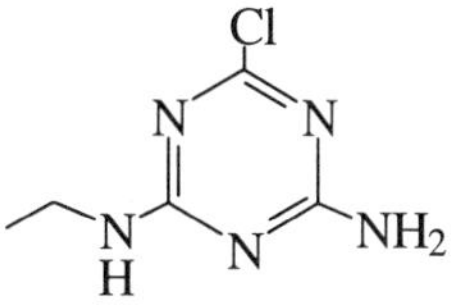 Cl, N, N, N, H, N, NH_2

*Common metabolite

Desisopropylametryn	6-(methylthio)-*N*-ethyl-1,3,5-triazine-2,4-diamine	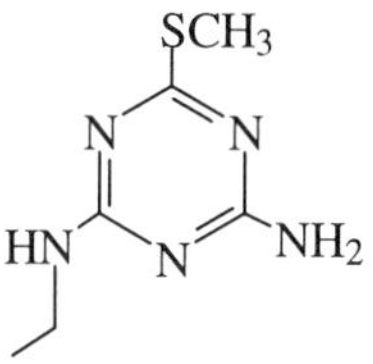
*Diaminochloro-s-triazine	6-chloro-1,3,5-triazine-2,4-diamine	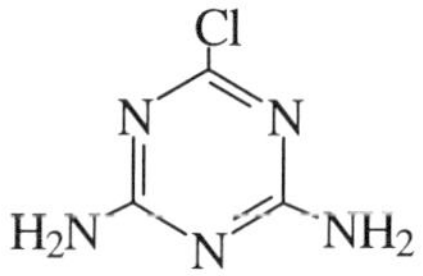
Diaminoametryn	6-(methylthio)-1,3,5-triazine-2,4-diamine	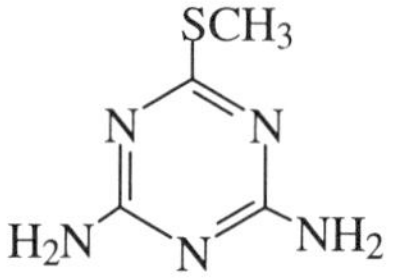
Atrazine mercapturate	*N*-acetyl-S-[4-(ethylamino)-6-[(1-methylethyl)amino]-1,3,5-triazin-2-yl, L-cysteine	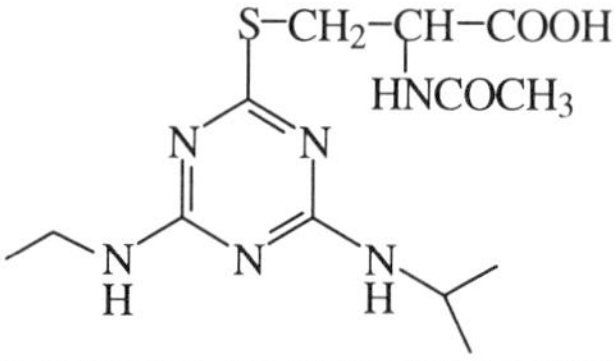
Desethylatrazine mercapaturate	*N*-acetyl-S-[4-amino-6-[(1-methylethyl)amino]-1,3,5-triazin-2-yl, L-cysteine	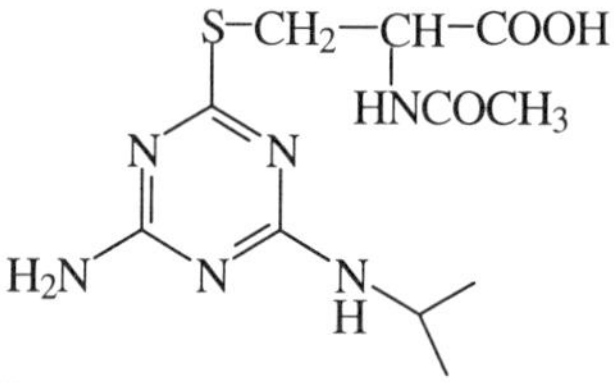
Desisopropylatrazine mercapturate *Desethylsimazine mercapturate *Des-2-methylpropionitrile cyanazine mercapturate	*N*-acetyl-S-[4-amino-6-ethylamino] -1,3,5-triazin-2-yl, L-cysteine	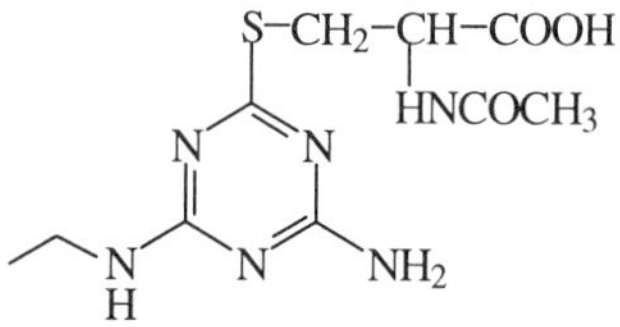
Diaminochloro triazine mercapturate	*N*-acetyl-S-(4,6-diamino-1,3,5-triazin-2-yl),L-cysteine	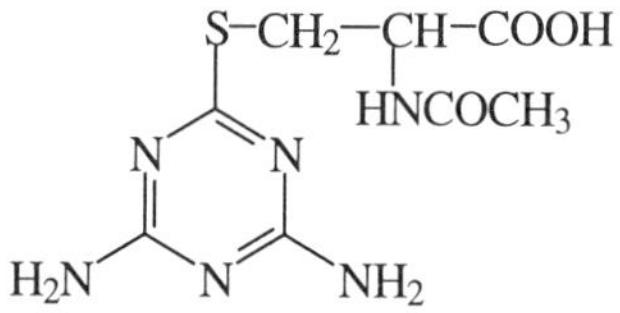

*Common metabolite

Ammeline	4,6-diamino-1,3,5-triazin-2(1*H*)-one	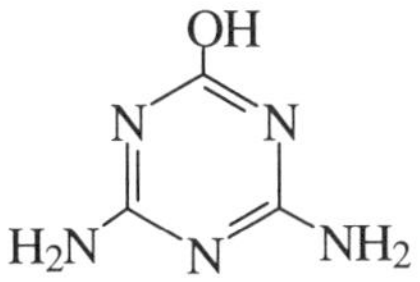
Ammelide	6-amino-1,3,5-triazine-2,4(1*H*,3*H*)-dione	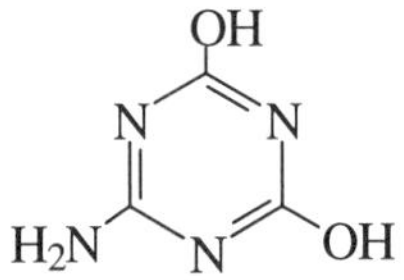
Desisopropylhydroxy atrazine *Desethylhydroxy simazine *Des-2-methylpropionitrile hydroxy cyanazine	4-amino-6-(ethylamino)-1,3,5-triazin-2(1*H*)-one	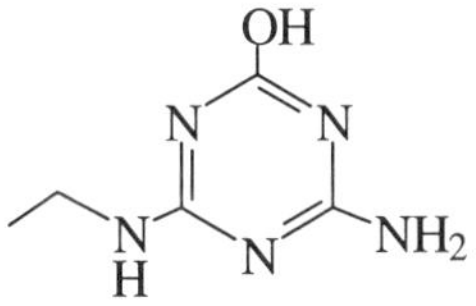
Desethylhyroxy-atrazine *Desethylehydroxy ametryn	4-amino-6-[(1-methylethyl)amino]-1,3,5-triazine-2(1*H*)-one	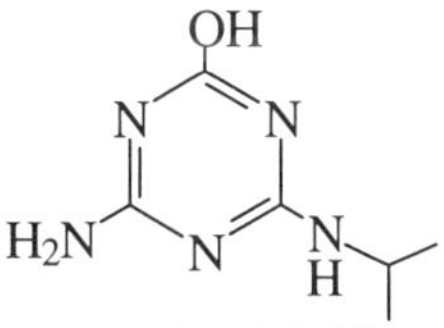
Hydroxyatrazine *Hydroxyametryn	4-(ethylamino)-6-[(1-methylethyl)amino]-1,3,5-triazine-2(1*H*)-one	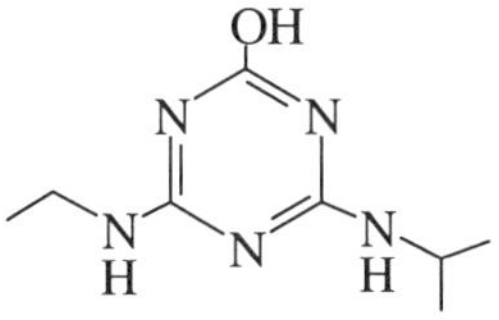
Cyanuric Acid	1,3,5-triazine-2,4,6(1*H*,3*H*,5*H*)-trione	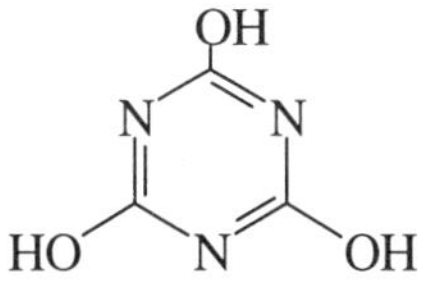
Hydroxysimazine	4,6-bis(ethylamino)-1,3,5-triazine-2(1*H*)-one	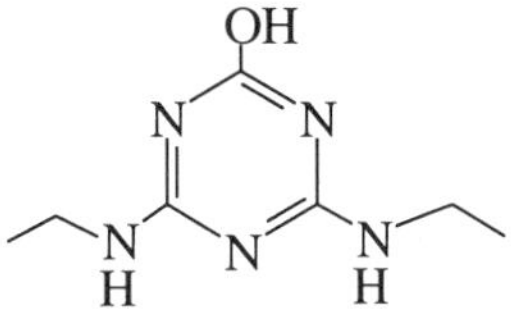

*Common metabolite

	N-(1-methylethyl)-1,3,5-triazine-2,4,6-triamine	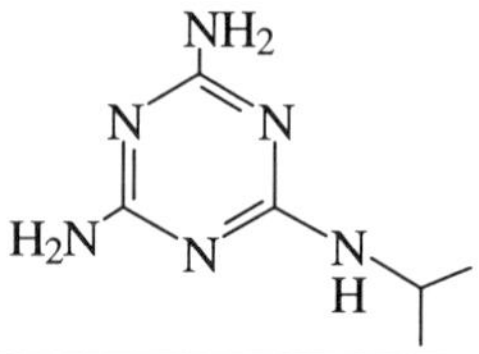
Aminoatrazine	*N*-ethyl-*N*'-(1-methylethyl)-1,3,5-triazine-2,4,6-triamine	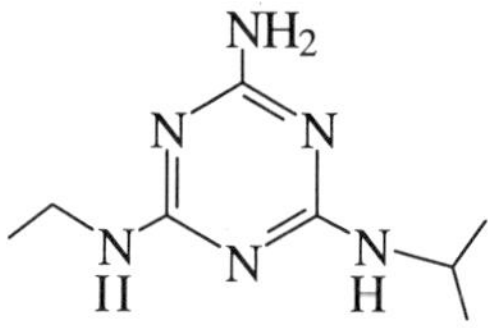
	N-ethyl-1,3,5-triazine-2,4,6-triamnie	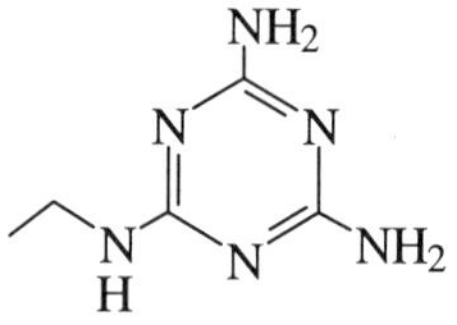
Melamine	1,3,5-triazine-2,4,6-triamine	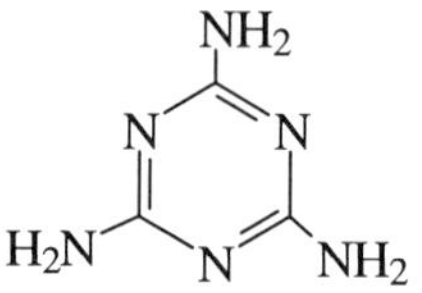

*Common metabolite

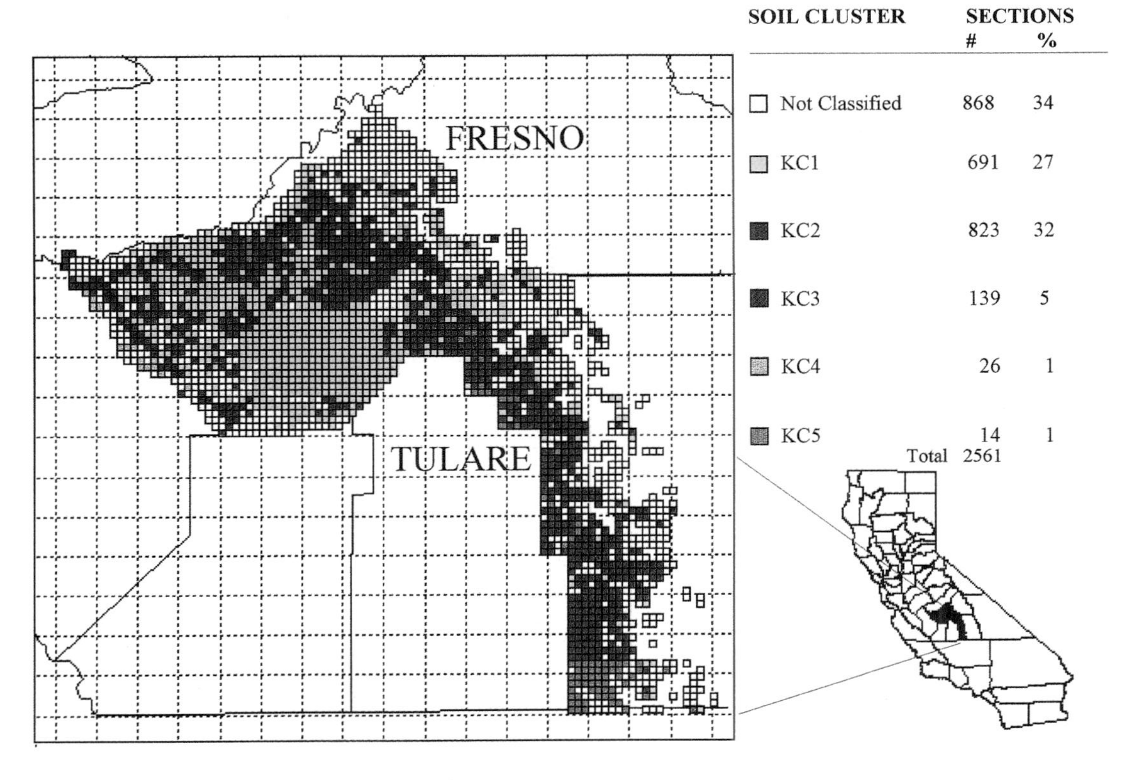

Figure 1. Use of PCA classification to identify sections of land in Fresno and Tulare counties into vulnerable soil clusters.
(Adapted with permission from reference 1. Copyright 1994 Kluwer Academic.)

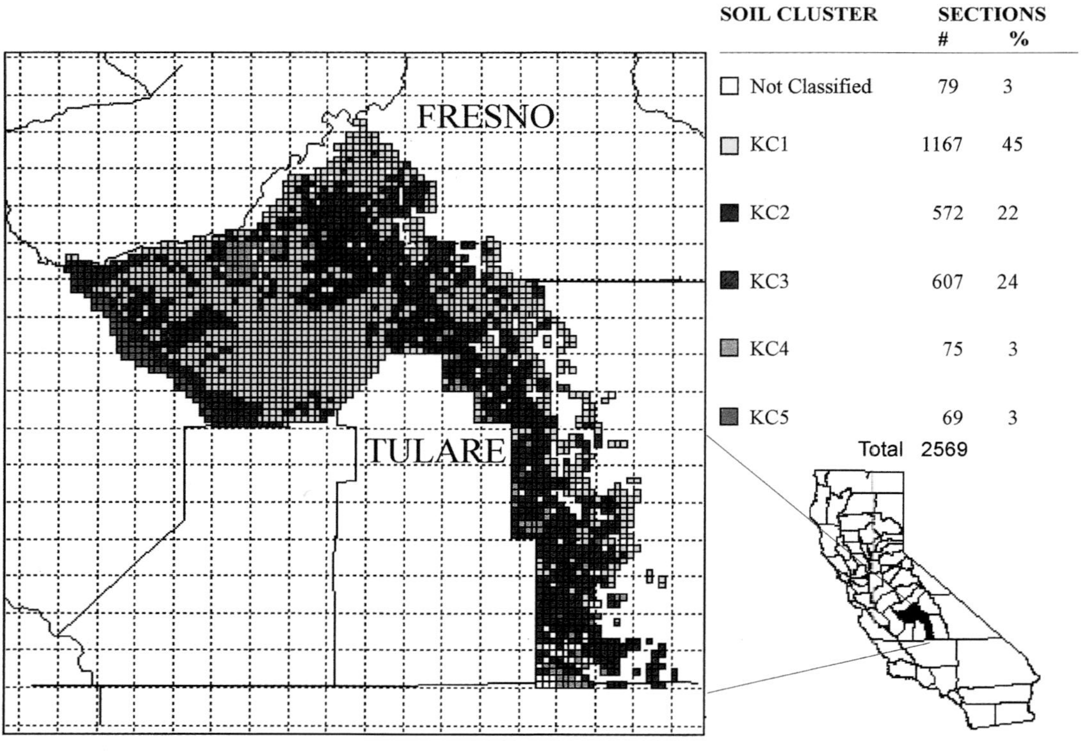

Figure 2. Use of CVA classification to identify sections of land in Fresno and Tulare counties into vulnerable soil clusters.
(Adapted with permission from reference 2. Copyright 1997 Kluwer Academic.)

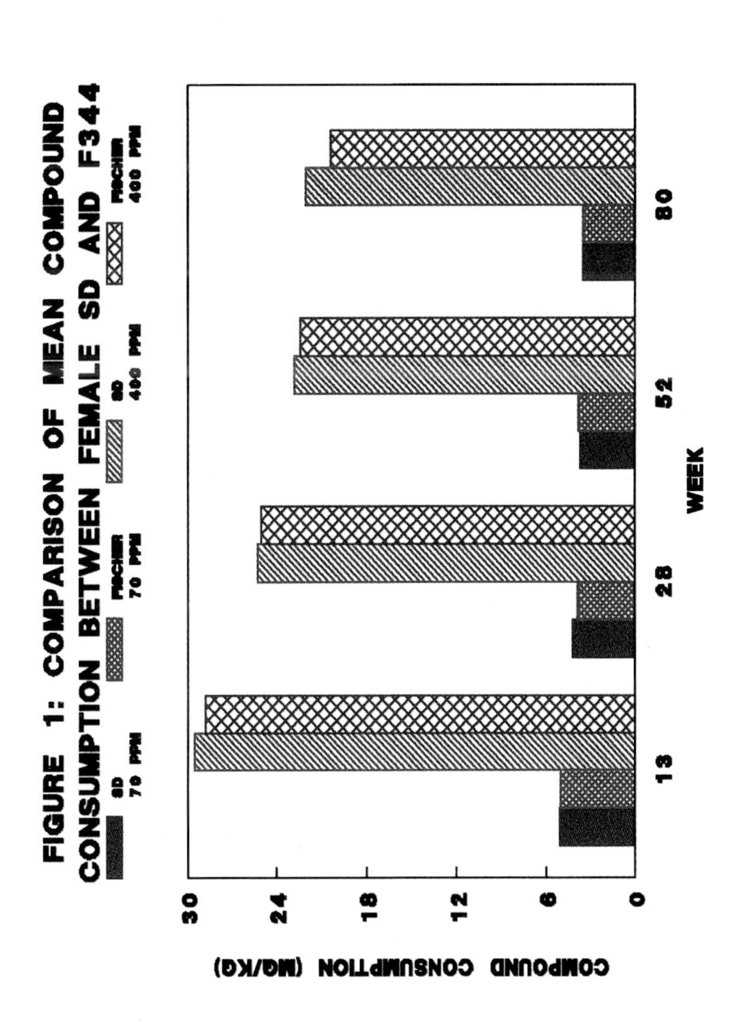

FIGURE 1: COMPARISON OF MEAN COMPOUND CONSUMPTION BETWEEN FEMALE SD AND F344

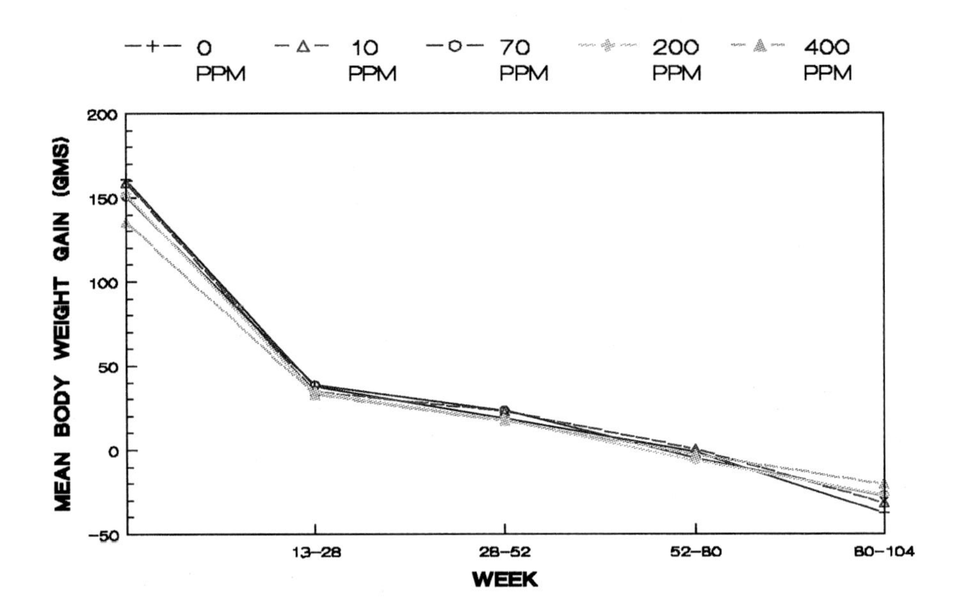
FIGURE 2A: F344 BODY WEIGHT GAIN: MALE
0 PPM
10 PPM
70 PPM
200 PPM
400 PPM
MEAN BODY WEIGHT GAIN (GMS)
200
150
100
50
0
-50
13-28
28-52
52-80
80-104
WEEK

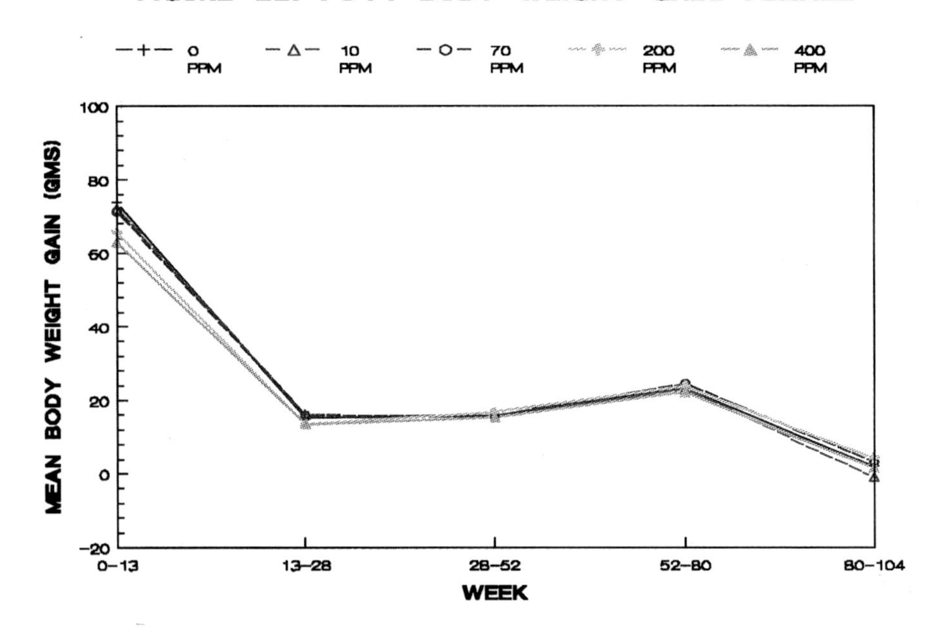
FIGURE 2B: F344 BODY WEIGHT GAIN: FEMALE
0 PPM
10 PPM
70 PPM
200 PPM
400 PPM
MEAN BODY WEIGHT GAIN (GMS)
100
80
60
40
20
0
-20
0-13
13-28
28-52
52-80
80-104
WEEK

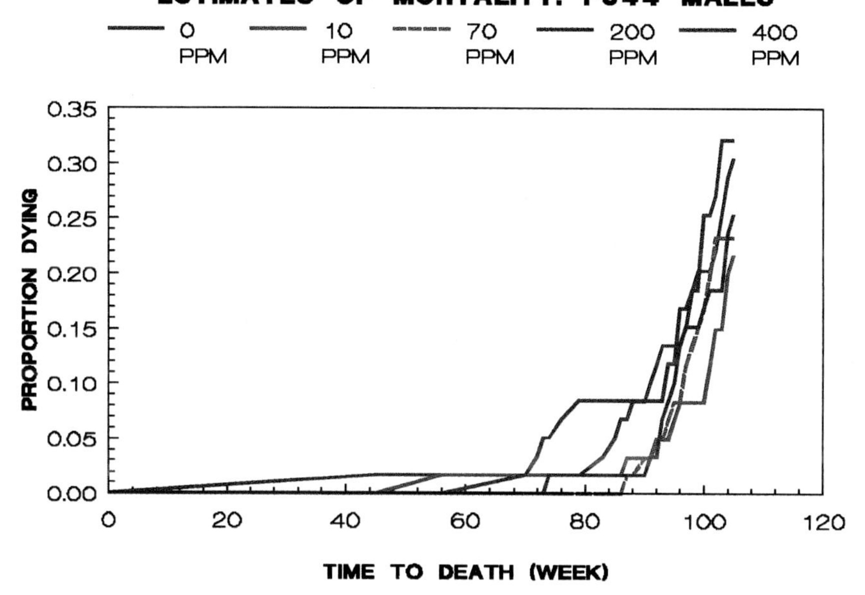
FIGURE 3A: KAPLAN-MEIER PRODUCT LIMIT ESTIMATES OF MORTALITY: F344 MALES
0 PPM
10 PPM
70 PPM
200 PPM
400 PPM
PROPORTION DYING
0.35
0.30
0.25
0.20
0.15
0.10
0.05
0.00
0
20
40
60
80
100
120
TIME TO DEATH (WEEK)

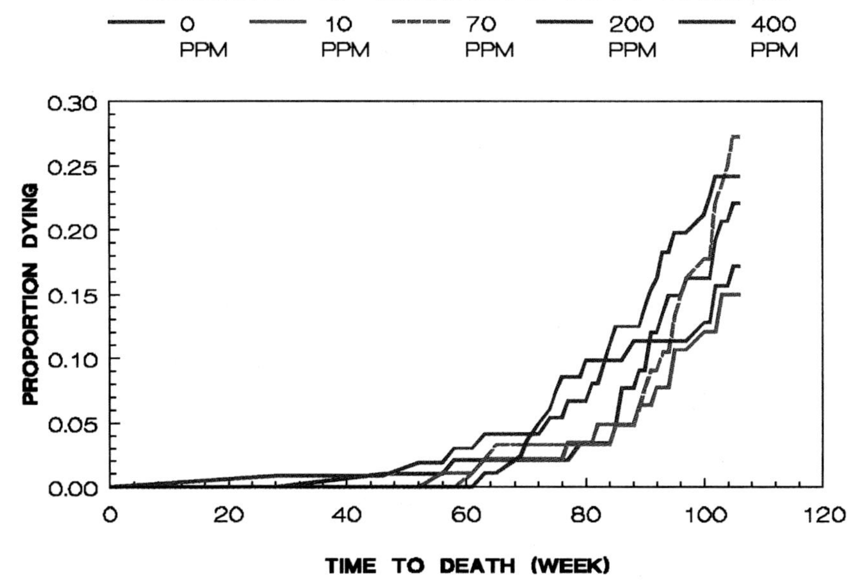
FIGURE 3B: KAPLAN-MEIER PRODUCT LIMIT ESTIMATES OF MORTALITY: F344 FEMALES
0 PPM
10 PPM
70 PPM
200 PPM
400 PPM
PROPORTION DYING
0.30
0.25
0.20
0.15
0.10
0.05
0.00
0
20
40
60
80
100
120
TIME TO DEATH (WEEK)

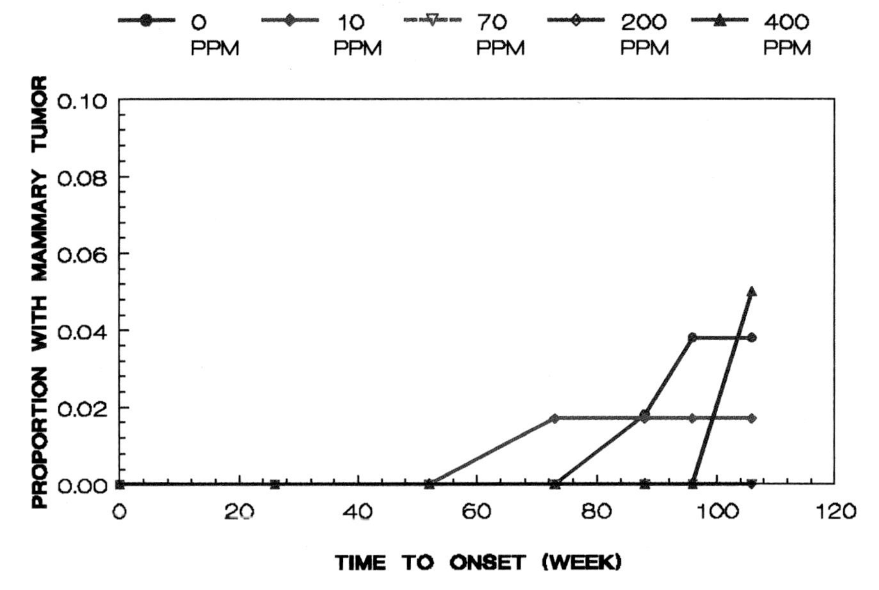
FIGURE 4A: KAPLAN-MEIER PRODUCT LIMIT ESTIMATES OF MAMMARY TUMORS : F344 MALES
0 PPM
10 PPM
70 PPM
200 PPM
400 PPM
PROPORTION WITH MAMMARY TUMOR
0.10
0.08
0.06
0.04
0.02
0.00
0
20
40
60
80
100
120
TIME TO ONSET (WEEK)

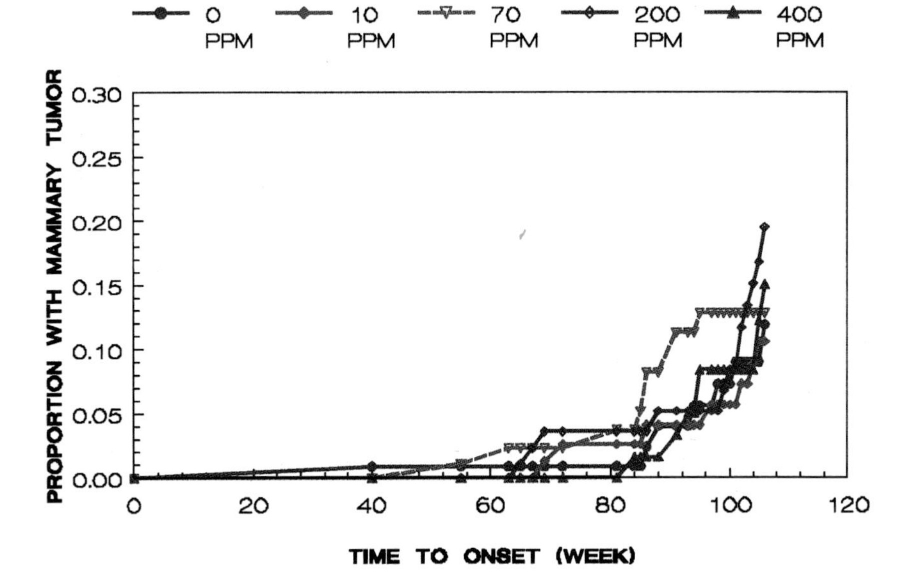
FIGURE 4B: KAPLAN-MEIER PRODUCT LIMIT ESTIMATES OF MAMMARY TUMOR: F344 FEMALES
0 PPM
10 PPM
70 PPM
200 PPM
400 PPM
PROPORTION WITH MAMMARY TUMOR
0.30
0.25
0.20
0.15
0.10
0.05
0.00
0
20
40
60
80
100
120
TIME TO ONSET (WEEK)

INDEXES

Author Index

Subject Index

B

C

D

E

L

M

N

O

S

U

V

W